Animal Models of Diabetes

Frontiers in Research

Second Edition

Animal Models of Diabetes

Frontiers in Research

Second Edition

Edited by
Eleazar Shafrir

CRC Press is an imprint of the
Taylor & Francis Group, an **informa** business

CRC Press
Taylor & Francis Group
6000 Broken Sound Parkway NW, Suite 300
Boca Raton, FL 33487-2742

First issued in paperback 2019

ISBN-13: 978-0-8493-9534-5 (hbk)
ISBN-13: 978-0-367-38925-3 (pbk)

Library of Congress Cataloging-in-Publication Data

Animal models of diabetes : frontiers in research / editor, Eleazar Shafrir. -- 2nd ed.
p. ; cm.
Includes bibliographical references and index.
ISBN-13: 978-0-8493-9534-5 (hardcover : alk. paper)
ISBN-10: 0-8493-9534-8 (hardcover : alk. paper)
1. Diabetes--Animal models. I. Shafrir, Eleazar.
[DNLM: 1. Diabetes Mellitus. 2. Disease Models, Animal. WK 810 A598 2007]

RC660.A66 2007
616.4'62--dc22 2006035038

Visit the Taylor & Francis Web site at
http://www.taylorandfrancis.com

and the CRC Press Web site at
http://www.crcpress.com

Table of Contents

The Editor

Eleazar Shafrir graduated from the Hebrew University in Jerusalem, Israel, where he earned a Ph.D. in biochemistry and an M.Sc. in medical sciences. He was professor and head of the Department of Biochemistry (1970–1992) of the Hadassah University Hospital. Professor Shafrir founded the Diabetes Research Center at the medical school in 1982 and headed it until his retirement in 1994. He continues his research activity as a senior investigator and consultant at the Diabetes Center of the Hadassah University Hospital. Professor Shafrir, with his colleagues, has convened eight workshops on "Lessons from Animal Diabetes" in 1982–1997 in Europe, Israel, and Japan. He was awarded the Claude Bernard Medal and Lectureship by the European Association for the Study of Diabetes (EASD) in 1991, entitled "Understanding of diabetes through studies of its etiopathology in animals." He was conferred honorary membership in EASD in 1997. He has published numerous research papers and reviews on animal diabetes and its implications for human diabetes, and has been particularly interested in the development of diabetes in domesticated desert rodents, spiny mice, and *Psammomys* gerbils.

Preface

The present edition of *Animal Models of Diabetes*, first issued in 2001, is motivated by the wide-ranging expansion and intensity of diabetes research comprising numerous diabetic animals. This is a welcome response to the rise in global awareness of the increasing incidence of diabetes. The gain in experience and new information on existing and new animals with diabetes required an update and a review of the accrued data to promote new avenues of investigation.

The new observations in this volume are contributed by prominent experts in the field. The accumulation of scattered data and their lucid presentation will promote understanding of the etiopathology of diabetes and offers a new grasp of insulin action, its negative feedback leading to insulin resistance, and its detrimental outcomes. There is also new knowledge on specific complications of diabetes offering an incentive to test advanced modalities to prevent and inhibit their occurrence.

This edition contains new and updated, well-referenced reviews on rodent diabetes, pigs in diabetes research, nutritionally induced diabetes in the gerbil *Psammomys obesus*, Cohen rats, and animals with induced obesity. It also offers interesting observations on retinopathy in spontaneous diabetes resembling human lesions.

There is no need to underscore the enormous contribution of diabetes research in animals to the understanding of various aspects of human diabetes. Although none of the animals fully represents the human syndrome, the various mechanisms and complications in animals are worth testing because human diabetes also shows many variations. This knowledge will help in the elucidation of general pathogenesis of this disease and clarify the processes of development of diabetic lesions and complications.

Eleazar Shafrir

Contributors

Sofianos Andrikopoulos
University of Melbourne, Department of Medicine (AH/NH)
Heidelberg Repatriation Hospital
Heidelberg Heights, Victoria, Australia

D. Bailbe
Laboratoire Physiopathologie de la Nutrition
Université Paris
Paris, France

Elizabeth P. Blankenhorn
Department of Microbiology and Immunology
Drexel University College of Medicine
Philadelphia, Pennsylvania

Yanyun Chen
Lilly Research Laboratories, a division of Eli Lilly and Co.
Indianapolis, Indiana

Streamson Chua, Jr.
Albert Einstein College of Medicine
New York, New York

Tamer Coskun
Lilly Research Laboratories, a division of Eli Lilly and Co.
Indianapolis, Indiana

C. Cuzin-Tourrel
Laboratoire Physiopathologie de la Nutrition
Université Paris
Paris, France

M. Dolz
Laboratoire Physiopathologie de la Nutrition
Université Paris
Paris, France

Paul Ernsberger
Department of Nutrition
Case Western Reserve University School of Medicine
Cleveland, Ohio

Barbara C. Fam
University of Melbourne, Department of Medicine (AH/NH)
Heidelberg Repatriation Hospital
Heidelberg Heights, Victoria, Australia

M. H. Giroix
Laboratoire Physiopathologie de la Nutrition
Université Paris
Paris, France

Dale L. Greiner
Department of Medicine
University of Massachusetts Medical School
Worcester, Massachusetts

Barbara C. Hansen
Obesity, Diabetes and Aging Research Center
Department of Internal Medicine and Pediatrics
College of Medicine, University of South Florida
Tampa, Florida

Mark Heiman
Lilly Research Laboratories, a division of Eli Lilly and Co.
Indianapolis, Indiana

Lieselotte Herberg
Diabetes Research Institute
Heinrich–Heine University
Düsseldorf, Germany

Hiroshi Ikegami
Kinki University School of Medicine
Osaka, Japan

Akihiro Kakehashi
Department of Ophthalmology
Omiya Medical Center
Jichi Medical School
Saitama, Japan

Rony Kalman
Diabetes Center
Department of Medicine
Hadassah University Hospital
Jerusalem, Israel

Kazuya Kawano
Tokushima Research Institute
Otsuka Pharmaceutical Co. Ltd
Tokushima, Japan

Sandra E. Kelly
Alberta Institute for Human Nutrition
University of Alberta
Edmonton, Alberta, Canada

M. Kergoat
Merck-Santé
Centre de Recherche
Chilly-Mazarin, France

Richard J. Koletsky
Department of Nutrition
Case Western Reserve University School of Medicine
Cleveland, Ohio

Marianne O. Larsen
Incretin Pharmacology
Diabetes Research Unit
Novo Nordisk
Maløv, Denmark

Edward H. Leiter
The Jackson Laboratory
Bar Harbor, Maine

Susumu Makino
KAC Co., Ltd.
Shiga, Japan

Taku Masuyama
Toxicology Research Laboratories
Central Pharmaceutical Institute
Japan Tobacco Inc.
Kanagawa, Japan

John P. Mordes
Department of Medicine
University of Massachusetts Medical School
Worcester, Massachusetts

J. Movassat
Laboratoire Physiopathologie de la Nutrition
Université Paris
Paris, France

Esther Orlanski
The Diabetes Center
Department of Internal Medicine
Hadassah-Hebrew University Medical Center
Jerusalem, Israel

Claes-Göran Östenson
Department of Molecular Medicine and Surgery
Karolinska Institute
Karolinska University Hospital
Stockholm, Sweden

Richard G. Peterson
Indiana University School of Medicine and Research and Development, PreClinOmics, Inc.
Indianapolis, Indiana

B. Portha
Laboratoire Physiopathologie de la Nutrition
Université Paris
Paris, France

Philippe Poussier
Departments of Medicine and Immunology
University of Toronto Faculty of Medicine
Toronto, Ontario, Canada

Spencer D. Proctor
Alberta Institute for Human Nutrition
University of Alberta
Edmonton, Alberta, Canada

Bidda Rolin
Incretin Pharmacology
Diabetes Research Unit
Novo Nordisk
Maløv, Denmark

Aldo A. Rossini
Department of Medicine
University of Massachusetts Medical School
Worcester, Massachusetts

James C. Russell
Alberta Institute for Human Nutrition
University of Alberta
Edmonton, Alberta, Canada

P. Serradas
Laboratoire Physiopathologie de la Nutrition
Université Paris
Paris, France

Eleazar Shafrir
Diabetes Center
Department of Medicine
Hadassah University Hospital
Jerusalem, Israel

Masami Shinohara
CLEA Japan, Inc.
Planning and Development Section
Tokyo, Japan

Dana Sindelar
Lilly Research Laboratories, a division of Eli Lilly and Co.
Indianapolis, Indiana

Shigehisa Taketomi
Hamari Chemicals Ltd.
Osaka, Japan

Hidenori Taniguchi
Higashiosaka City General Hospital
Osaka, Japan

Xenia T. Tigno
Obesity, Diabetes and Aging Research Center
Department of Internal Medicine and Pediatrics
College of Medicine, University of South Florida
St. Petersburg, Florida

Rodney A. Velliquette
Department of Nutritional Sciences
Penn State University
University Park, Pennsylvania

Sarah Weksler-Zangen
The Diabetes Center
Department of Internal Medicine
Hadassah-Hebrew University Medical Center
Jerusalem, Israel

David H. Zangen
Pediatric Endocrine Unit
Department of Pediatrics
Hadassah-Hebrew University Medical Center
Jerusalem, Israel

Ehud Ziv
Diabetes Center
Department of Medicine
Hadassah University Hospital
Jerusalem, Israel

1 Rat Models of Type 1 Diabetes: Genetics, Environment, and Autoimmunity

John P. Mordes, Philippe Poussier, Aldo A. Rossini, Elizabeth P. Blankenhorn, and Dale L. Greiner

CONTENTS

ABSTRACT

The study of rat models continues to be a productive component of efforts to understand and prevent human type 1 diabetes (T1D); complements to the standard BB rat have been developed, and all have contributed insights into immunopathogenesis. In addition, genetic analysis of the rat has accelerated, and rats are proving useful to study the interactions of genetics and environment that may be critical for disease expression in humans. Finally, data generated in the rat have correctly predicted the outcome of several human diabetes prevention trials.

ABBREVIATIONS

AG aminoguanidine
BB rat BioBreeding rat
BB/Wor rat BB Worcester rat
BBDP rat diabetes-prone BB rat
BBDR diabetes-resistant BB rat
BCG bacille Calmette–Guerin
Cblb Casitas B-lineage lymphoma b
CFA complete Freund's adjuvant
DC dendritic cell
EMC-D encephalomyocarditis virus
GAD glutamic acid decarboxylase
Gimap5 GTPase of the immune associated protein 5 family
IFN interferon
iNOS inducible nitric oxide synthase
IGF-2 insulin-like growth factor 2
KDP rat Komeda diabetes-prone rat
KRV Kilham rat virus
LETL rat Long–Evans Tokushima lean rat
LCMV lymphocytic choriomeningitis virus
LKLF lung Krüppel-like factor
MHC major histocompatibility complex
NAD nicotinamide adenine dinucleotide
NK natural killer

NKT	natural killer T
NO	nitric oxide
NOD mouse	nonobese diabetic mouse
PP	pancreatic polypeptide
PFU	plaque-forming units
Poly I:C	polyinosinic:polycytidylic acid
RCMV	rat cytomegalovirus
SNP	single nucleotide polymorphism
STZ	streptozotocin
TCR	T cell receptor
TLR	toll-like receptor
Treg	regulatory T cell
TNF-α	tumor necrosis factor-α
STZ	streptozotocin
TNF	tumor necrosis factor
VAF	viral antibody free
WF rat	Wistar Furth rat

INTRODUCTION

Type 1 diabetes* (T1D) comprises ~10% of all diabetic syndromes. Onset is typically during childhood, and prevalence is increasing (Green et al. 2001). The disease begins with inflammatory infiltration of the islets of Langerhans (insulitis) and culminates in selective destruction of insulin-producing beta cells (Atkinson and Eisenbarth 2001). Considered to be autoimmune in origin, T1D is strongly associated with the major histocompatibility complex (MHC), is T cell-dependent, and is ameliorated by immunosuppression. It often occurs with other autoimmune diseases that affect the gut and thyroid. Despite intensive study, however, T1D remains refractory to prevention (Allen et al. 1999; Lampeter et al. 1998; Skyler et al. 2002) by methods other than unacceptably toxic immunosuppression (Parving et al. 1999).

Studying T1D in humans is difficult because the population is outbred and randomly exposed to chemical and microbiological agents. To circumvent these constraints, investigators continue to advocate the study of animal models that can more readily be tested, biopsied, and autopsied (Leiter and Von Herrath 2004). The most widely used model is the nonobese diabetic (NOD) mouse (Anderson and Bluestone 2005). Although NOD and various NOD congenic mice continue to provide extensive valuable information, there is growing recognition of the advisability of studying alternative yet complementary systems (Greiner et al. 2001). Inbred rats provide one such system.

The BioBreeding (BB) rat (Mordes et al. 2001) is the most extensively studied rat model of T1D, but a predisposition to autoimmune beta cell destruction is common to several rat strains. These are listed in table 1.1. The table highlights a signature feature of rat models of T1D. Rat autoimmune diabetes may occur "spontaneously" in viral-antibody-free housing, as is the case with the NOD mouse.

* More strictly speaking, type 1A diabetes (American Diabetes Association 2006).

TABLE 1.1
Rat Models of T1D

Type	Strain	MHC	Ref.
Spontaneous	BBDP and BBdp	$RT1^{u/u/u}$	(Mordes et al. 2001)
	BBDR.*lyp/lyp*	$RT1^{u/u/u}$	(Bieg et al. 1997)
	Long–Evans Tokushima lean (LETL)	$RT1^{u/u/u}$	(Kawano et al. 1991)
	Komeda diabetes prone (KDP)	$RT1^{u/u/u}$	(Yokoi 2005)
	LEW.1AR1/Ztm-*iddm*	$RT1^{u/u/a}$	(Jörns et al. 2005)
Induced	BBDR	$RT1^{u/u/u}$	(Mordes et al. 2001)
	LEW.1AR1	$RT1^{u/u/a}$	(Wedekind et al. 2005)
	PVG	$RT1^{c/c/c}$	(Penhale et al. 1990)
	PVG.$RT1^{u}$	$RT1^{u/u/u}$	(Ellerman and Like 2000)
	PVG.R8	$RT1^{a/u/u}$	(Ellerman and Like 2000)
	WAG	$RT1^{u/u/u}$	(Ellerman and Like 2000)
	WF.*iddm4*^{d} congenic	$RT1^{u/u/u}$	(Mordes et al. 2002)
Both	LEW.1WR1	$RT1^{a/u/u}$	(Mordes et al. 2005)
Transgenic	None	—	—

Notes: Models of autoimmune diabetes in the rat. All are characterized by the presence of pancreatic insulitis and beta cell destruction. The major histocompatibility locus (MHC) in the rat is designated *RT1* and the haplotypes are designated *RT1 A*, *B/D*, and *C*. *RT1 A* and *RT1 C* are class I loci; the class II loci are in linkage disequilibrium and designated *B/D*. With the exception of the PVG model induced by the combination of irradiation and thymectomy, all rat models of autoimmune diabetes express the *RT1 B/D*^{u} class II allele. Methods of induction include regulatory T cell depletion, toll-like receptor (TLR) ligation, and certain viral infections (see text).

Alternatively, diabetes may appear only in response to immunological perturbation, which is generally not the case in nontransgenic mouse models (table 1.1).

SPONTANEOUS AUTOIMMUNE DIABETES IN THE RAT

Salient characteristics of some of the principal animal models of spontaneous type 1-like diabetes in the rat are listed in table 1.2, together with corresponding characteristics of the human disease.

THE DIABETES-PRONE BB RAT

"BB rats" are derived from a Canadian colony of outbred Wistar rats in which spontaneous hyperglycemia and ketoacidosis occurred in the 1970s (Mordes et al. 2001). Affected animals were the founders for two colonies later used to establish all other BB rat colonies. The colony in Worcester, Massachusetts, has been inbred and the spontaneously diabetic animals are formally designated "BBDP/Wor"

TABLE 1.2
Comparative Clinical and Genetic Features of Spontaneous Autoimmune Diabetes

	Human	NOD Mouse	BBDP/Wor	LEW.1AR1/Ztm-*iddm*	LEW.1WR1	KDP
Age at onset	Adolescence	Up to 6 months	7–14 weeks	6–12 weeks	8–12 weeks	8–16 weeks
Prevalence	1:600	>90% in females	>90%	60%	2.5%	~70%
Ketosis	Severe	Mild	Severe	Very severe	Severe	Mild to severe
Insulin deficiency	Severe	Mild to severe	Severe	Severe	Severe	Severe
Associated autoimmune diseases	Thyroiditis, celiac disease, vitiligo, PA, polyendocrine syndromes	Sialadenitis, thyroiditis	Thyroiditis	Thyroid, adrenal, salivary glands free of infiltration	Collagen-induced arthritis	Thyroid and kidney infiltrates present
Autoantibodies	Insulin, GAD, ICA, ICSA, BSA, CPH, EC, IA-2, IAA	Insulin, GAD, ICA	ICA present; anti-GAD Abs and IAA controversial	ICA and antibodies to GAD, IA-2 not found	Unknown	Unknown
MHC genes	HLA-DQ and DR	Unique I-A^{g7}; absent I-E	RT1u/u/u	RT1u/u/a	RT1a/u/u	RT1u/u/u
Non-MHC genes	*CTLA4*, *PTPN22*. Perhaps ≥16 other loci	β2 Microglobulin; possibly *CTLA4*; ≥27 loci	>15 Loci. *Gimap5* mutation causes lymphopenia	At least two loci	No data	Cblb
Gender effect	M = F	F > M	M = F	M = F	M = F	M = F
Response to general immunosuppression	Cyclosporin prolongs endogenous insulin production if given at onset	Cyclosporin, tacrolimus prevent diabetes	Cyclosporin, tacrolimus, thymectomy, ALS, radiation prevent diabetes	Unknown	Unknown	Unknown
Response to environmental perturbation	Diet and viral infection implicated in pathogenesis, but no definitive proof	More than 150 interventions prevent disease	LCMV prevents disease. Certain diets reduce diabetes frequency	Increased frequency in clean environment	Increased by poly I:C and viral infection	Unknown

Notes: Characteristics of spontaneous autoimmune diabetes mellitus in humans, NOD mice, and three rat model systems. The BBDR.*lyp/lyp* congenic rat (Bieg et al. 1997) is not listed but shares many of the features of the BBDP rat. ALS: antilymphocyte serum; BSA: bovine serum albumin; CPH: carboxypeptidase H; EC: endothelial cell surface; GAD: glutamic acid decarboxylase; HLA: human leukocyte antigen; IAA: insulin autoantibodies; ICA: islet cell autoantibodies; ICSA: islet cell surface antibodies; LCMV: lymphocytic choriomeningitis virus; PA: pernicious anemia.

(see www.biomere.com). A second remains in Ottawa, Canada; the "BBdp" rats from this colony are outbred. Several of the tertiary BB rat colonies have given rise to immunologically and genetically distinct BB rat substrains (Prins et al. 1991). These include BB/OK, BB/Pfd, BB/S, and BB.SHR rats (Klöting et al. 1998; Kovács et al. 2001; Lucke et al. 2003; Mathieu et al. 1994; Schröder et al. 2002). Rats of Worcester origin sold in Europe (by Taconic Europe, formerly Møllegaard) are designated BB/Mol or BB/Wor//MolTac (Mordes et al. 2001). Results of studies that use BB rats of different origins may not be directly comparable. Recently, a number of BB congenic rats have been developed (Bieg et al. 1997; Klöting et al. 2001; Mordes et al. 2002). Here the term "BB rat" is used when referring to findings that appear to apply generically to all rats derived from the original founders; more specific designations are used when referring to findings that may be substrain specific.

Clinical and Immunological Pathology

Histopathology

In inbred BBDP/Wor rats and outbred BBdp rats of both sexes, the cardinal pathological finding is pancreatic insulitis, or inflammatory infiltration of the islets of Langerhans. Serial studies have shown that insulitis starts 2 to 3 weeks before overt diabetes. It is followed by selective destruction of beta cells and diabetes in rats between 50 and 90 days of age (Guberski 1994).

After the onset of hyperglycemia, residual end-stage islets with few or no inflammatory cells are observed. These are small and comprise predominantly non-beta cells. Islet alpha, delta, and pancreatic polypeptide (PP) cell numbers and morphology appear to be preserved. There are no published data on islet ghrelin cells in the rat. Unless treated with insulin, hyperglycemic BB rats quickly develop fatal diabetic ketoacidosis. A variable fraction of BBDP/Wor rats also develop spontaneous lymphocytic thyroiditis, but these rats do not progress spontaneously to clinical hypothyroidism (Colzani et al. 1999; Rajatanavin et al. 1991; Simons et al. 1998).

Immunopathology of Insulitis

Subpopulations of lymphoid cells present at different stages of insulitis have been characterized (Crisá et al. 1992). Macrophages are among the earliest, possibly the first, of the cellular elements observed (Hanenberg et al. 1989). Alternatively, the earliest infiltrating elements may include dendritic cells, which may have enhanced antigen-presenting capability in the BB rat (Tafuri et al. 1993). $CD4^+$ and $CD8^+$ T cells, natural killer (NK) cells, and, to a lesser extent, B cells subsequently infiltrate the islets. Recent data suggest that eosinophils may also play a role in the insulitis lesion in BB rats, possibly due to overexpression of the chemokine eotaxin in these animals (Hessner, Wang, Meyer, et al. 2004). Expression of inducible nitric oxide synthase (iNOS) is absent in young BBDP rats and increases with age (Pieper et al. 2004). In addition, data from the diabetes-prone BB/S strain suggest that lower islet antioxidant enzyme activities could be a factor in the development of disease (Sigfrid et al. 2004).

The natural course of insulitis in the spontaneously diabetic BB rat is different from that of the NOD mouse; in the rat, there is little or no persistent infiltration adjacent to the islet ("peri-insulitis") before progression to frank insulitis and overt diabetes. Insulitis in BB rats is morphologically similar to that observed in human T1D and features a predominance of Th1-type lymphocytes (Kolb et al. 1996; Zipris et al. 1996).

Pathogenesis

The inciting event that leads to beta cell autoreactivity in the spontaneously diabetic BB rat is not known, but probably involves the presentation of autoantigen by *RT1^u* molecules. It is not implausible that, for unknown reasons, BB rat islets initially trigger an innate inflammatory reaction that subsequently leads to a destructive, beta cell-specific T cell response. Such a sequence of pancreatic events would be reminiscent of those recently described in the skin of mice prone to psoriasis (Zenz et al. 2005). Alternatively, immunogenic presentation of autoantigen in association with *RT1^u* molecules by dendritic cells in pancreatic lymph nodes may be the initiating event, as has been suggested for the NOD mouse (Höglund et al. 1999). The identity of the primary autoantigen or autoantigens is unknown (Mordes et al. 2001). Insulin, glutamic acid decarboxylase (GAD), and perhaps sulfatide (Blomqvist et al. 2003; Buschard et al. 2005) are among the candidate autoantigens.

Substantial data document that the immunopathogenic process in the BBDP/Wor rat originates at the level of bone marrow progenitor cells (Naji et al. 1983; Nakano et al. 1988). Reciprocal bone marrow and thymus transplants have been used to reveal the presence of a thymic microenvironmental defect of bone marrow origin (Georgiou et al. 1988; Georgiou and Bellgrau 1989), leading to the hypothesis that intrathymic antigen-presenting cells are abnormal in BBDP/Wor rats. This abnormality is associated with a limited peripheral T cell receptor (TCR) β chain variable region repertoire associated with diabetogenicity (Gold and Bellgrau 1991). Defective expression of insulin-like growth factor 2 (IGF-2) in the thymus has also been suggested to play a role in BBDP rat pathogenesis (Geenen and Brilot 2003; Kecha-Kamoun et al. 2001). Additional abnormalities in populations of NK cells, NKT cells, and intraepithelial lymphocytes (reviewed in Mordes et al. 2001) have also been observed, but their contribution to pathogenesis is unclear. Increased neonatal beta cell apoptosis has also been suggested to play a role in diabetes susceptibility in the BB rat and NOD mouse (Trudeau et al. 2000).

Many observations document a role for T cells in the pathogenesis of spontaneous BB rat diabetes (Mordes et al. 2001). Mitogen-activated spleen cells from acutely diabetic donors accelerate disease in young diabetes-prone BB rats and transfer it to MHC-compatible naïve recipients. Neonatal thymectomy, injections of antilymphocyte serum, and depletion of $CD8^+$ T cells (but not NK cells) all prevent the disease, as do many standard immunosuppressive drugs and immunomodulatory modalities (Mordes et al. 2001). Particular attention has been focused on the role of the inflammatory cytokine IL-1β in the pathogenesis of beta cell destruction (Reimers et al. 1996). Although it is clearly a potential beta cell toxin, the exact mechanism by which it acts is complex and appears to involve a series of cumulative alterations in protein expression (Sparre et al. 2004). T cell hyporesponsiveness, defined as an increased activation threshold for upregulation of activation markers following T cell receptor cross-linking, is reportedly a characteristic not only of BB rats but also of many other rodent strains susceptible to autoimmunity (Lang and Bellgrau 2002).

Lymphopenia and Treg Function

The most obvious and problematic immunopathology in all spontaneously diabetic BB rats is profound T cell lymphopenia (Elder and Maclaren 1983; Ramanathan and Poussier 2001; Yale et al. 1985), which is due to a frameshift mutation in the *Gimap5* (GTPase of the immune associated protein 5) gene (Hornum et al. 2002; MacMurray et al. 2002). The number of $CD4^+$ T cells is severely reduced (Jackson et al. 1981) and the $CD8^+$ T cell subset is nearly completely absent (Jackson et al. 1983; Poussier et al. 1982; Woda et al. 1986). The phenotypic T cell abnormalities in BBDP/Wor rats are due to a severely reduced life span of peripheral T cells. The majority of cultured BBDP/Wor T cells undergo apoptosis within 24 hours (Ramanathan et al. 1998). Recent thymic emigrants *in vivo* exhibit a preapoptotic $\alpha\beta TCR^{low}B220^+CD4^-CD8^-$ phenotype and undergo apoptosis in the liver (Iwakoshi et al. 1998). The presence of lymphopenia does not in itself confer susceptibility to autoimmunity in the rat (Colle et al. 1992; Joseph et al. 1993; Moralejo et al. 2003), but spontaneous diabetes in BBDP/Wor rats requires that they be lymphopenic (Awata et al. 1995). This pathology is not characteristic of either NOD mice or humans with T1D and has compromised the acceptance of diabetes-prone BB rats as a model for the human disease.

Soon after the recognition of lymphopenia, it was discovered that transfusion of putative regulatory T cells (Tregs) can prevent diabetes in the BBDP rat (Burstein et al. 1989; Rossini et al. 1983). The phenotype of the protective cells was initially defined as $CD4^+ART2^+$ (Mordes et al. 1987b). ART2 (formerly designated RT6) is a rat maturational T cell alloantigen with nicotinamide adenine dinucleotide (NAD) glycohydrolase activity that appears to identify cells with immunoregulatory properties (Bortell et al. 1999). It exists in both cell-surface and soluble forms (Bortell et al. 2001). BBDP rats are severely deficient in peripheral $ART2^+$ T cells (Greiner et al. 1986). Early reconstitution of lymphopenic BBDP/Wor rats with normal histocompatible T cells protects the recipients from diabetes. This protection occurs despite the absence of physical or functional depletion of islet cell-specific T cell precursors among donor cells (Ramanathan and Poussier 1999).

In humans and mice, populations of Treg cells express the Foxp3 transcription factor and have a $CD4^+CD25^+$ cell surface phenotype. The regulatory potential of cells expressing this phenotype has been demonstrated in the rat using a model in which diabetes in young BBDP rats is accelerated by the transfer of purified diabetogenic T cells (Lundsgaard et al. 2005). Small numbers of $CD4^+CD25^+Foxp3^+$ cells prevented the accelerated diabetes when transferred either prior to or at the same time as diabetogenic T cells. $CD4^+CD25^+$ cells with low Foxp3 expression were much less effective.

As a corollary, it was hypothesized that the *Gimap5* mutation in the BBDP rat leads to impaired differentiation of these regulatory populations. Interestingly, BBdp rats have normal numbers of functional $CD8^-CD4^+CD25^+Foxp3^+$ thymocytes that fail to undergo appropriate expansion and survive; Foxp3 transcription is reduced in mature $CD8^-CD4^+CD25^+$ T cells, suggesting that these cells comprise mostly recently activated T cells. Hence, it appears that the BBdp rat *Gimap5* mutation alters the survival and function of regulatory $CD4^+CD25^+$ T cells post-thymically, thereby leading to expansion of diabetogenic T cells within the peripheral $CD4^+CD25^+$ subset (Poussier et al. 2005). There appears to be no association between *Gimap5* polymorphisms and human T1D (Payne et al. 2004).

In addition to effects on Treg function, the *Gimap5* mutation also appears to affect dendritic cells (DCs) in the BBDP rat. In comparison with nonlymphopenic controls, including BBDR rats, BBDP rat DCs rather paradoxically exhibit less MHC class II expression (Sommandas et al. 2005b). They also lack normal capability to terminally differentiate into mature T cell stimulatory DC, but this characteristic is not linked exclusively to the *Gimap5* mutation (Sommandas et al. 2005a).

Diet and Gut Pathology

Diet, particularly its protein content, can influence the frequency of diabetes in the BBdp rat (Scott et al. 1997). Disease penetrance can be substantially reduced by the substitution of *l*-amino acids, soy protein, or hydrolyzed casein for intact protein in the diet (Li et al. 1995). In the case of the *l*-amino acid-based diet, the protective effect was abrogated by milk protein supplementation (Elliott and Martin 1984). Hydrolyzed casein diets have been associated with some degree of islet neogenesis in the BBdp rat (Wang et al. 2000). The effects of such modification are not uniform, however. In one study, BBDP rats were not protected from diabetes when fed an amino acid-based diet. (Simonson et al. 2002)

Antibodies directed against wheat globulin are also detectable in BBdp rats (MacFarlane et al. 2003). This finding is of interest for several reasons: (1) dietary wheat protein has been associated with a high rate of diabetes in BBdp rats; (2) there is a high prevalence of celiac disease among persons with T1D (MacFarlane et al. 2003); and (3) early cereal consumption has been identified as a potential environmental precipitant of human T1D (Norris et al. 2003).

More generally, it is hypothesized that defects in gut immunity may play a role in the development of T1D generally and in the diabetes-prone BB rat in particular (Scott and Marliss 1991). BBDP/Wor rats are deficient in gut NK cell number and function in the period preceding diabetes onset (Todd et al. 2004), and BBdp rats have an impairment in the development of intestinal TCRγδ T cells (Ramanathan, Marandi, et al. 2002). There is also histological evidence of enteropathy and mucosal inflammation in the BBdp rat (Graham et al. 2004; Hardin et al. 2002; Najjar et al. 2001). In addition, alterations in intestinal disaccharidase activity, changes in intestinal peroxidase activity, glucagon-like peptide 1 anomalies, and perturbation of intestinal permeability and mucin content occur in BBdp rats (Malaisse et al. 2004; Meddings et al. 1999). In addition, early increased intestinal permeability has been described in BBDP rats, and it is thought that exposure to antigens traversing the gut could trigger autoimmunity (Neu et al. 2005). Consistent with this view, it has been shown that in BBdp rats, protective hydrolyzed casein diets, which reduce the penetrance of spontaneous diabetes, may prevent the increase in gut permeability (Courtois et al. 2005).

Humoral Immunity

There has been little recent study of autoantibodies in the BB rat. BBDP rats do circulate autoantibodies reactive against lymphocytes, gastric parietal cells, smooth muscle, and thyroid colloid, but islet cell cytoplasmic antibodies (ICA) (Like et al. 1982) and antiadrenal antibodies are probably not present. Antilymphocyte antibodies can be present before diabetes onset and their presence predicts the development of spontaneous diabetes in Ottawa BBdp rats (Bertrand et al. 1994). Antiendothelial cell autoantibodies that appear to be capable of inducing endothelial leakiness are also detectable (Doukas et al. 1996).

Autoantibodies important for predicting onset of human T1D may or may not have counterparts in BB rats. Anti-GAD antibodies are reportedly absent in BB/d (Davenport et al. 1995) and other BB rats (Mackay et al. 1996), but may be present in the BB/OK strain (Ziegler et al. 1994). Another human autoantibody, IA 2, is either absent (DeSilva et al. 1996) or present at low titer (Myers et al. 1998) in BB rats. Autoantibodies against heat shock protein 65 (hsp-65) (Mackay et al. 1996) are absent. Insulin autoantibodies (IAA) have been reported in BB/W/D rats (Wilkin et al. 1986), but not in BBDP/Wor rats (Markholst et al. 1990).

Genetics

As indicated in table 1.1, autoimmune diabetes in rats appears to require at least one gene associated with the rat MHC. The primary components of the rat MHC, designated *RT1*, include two class I loci, *A* and *C*, and two class II loci in linkage disequilibrium, designated *B*/*D* (Gunther and Walter 2001). In the BB rat, the *RT1 B*/*D* region is designated *Iddm1*, and expression of diabetes requires the presence of at least one class II *RT1 B*/*D*^{u} allele (Colle 1990; Fuks et al. 1990).

As shown in table 1.1, autoimmune diabetes has been observed in several non-BB strains, and the presence of a class II *RT1 B*/*D*^{u} haplotype is characteristic of nearly all of them (Ellerman and Like 2000). The class II allele in the BBdp rat is not a unique isoform; *u* alleles from normal rat strains also confer susceptibility when moved to the BBdp strain (Colle et al. 1986). The enrichment for DQβ chain alleles with uncharged amino acids at position 57—characteristic of Caucasian humans (Nepom et al. 1996) and of NOD mice (Corper et al. 2000)—is not found in the BB rat. The region around position 57 of the class II β-chains of normal LEW and BUF rats is identical to that found in BBDP/Wor rats (Chao et al. 1989).

As noted above, the lymphopenia of the BBDP/Wor rat is due to a recessive mutation in a gene now designated *Gimap5* (Hornum et al. 2002; MacMurray et al. 2002). Transgenic complementation using a congenic F344.lyp rat established that the mutation is responsible for lymphopenia and established its functional importance (Michalkiewicz et al. 2004) Absence of the protein encoded by *Gimap5* in T cells causes mitochondrial dysfunction, increased mitochondrial levels of stress-inducible chaperonins, and T cell-specific spontaneous apoptosis (Pandarpurkar et al. 2003). A similar lymphopenia phenotype in the mouse has been attributed to a mutation in the LKLF gene, but LKLF appears not to be defective in the BBdp rat (Diessenbacher et al. 2003).

In analyses of (BBDP/Wor × WF) × WF rats, which do not become spontaneously diabetic but do become diabetic in response to perturbation (v.i.), a locus on chromosome 4 (*Iddm4*) was found to have significant linkage to diabetes (Martin, Blankenhorn, et al. 1999; Martin, Maxson, et al. 1999). This locus has been confirmed, and the data suggest that it is a major, but incompletely penetrant, non-MHC determinant of autoimmune diabetes in the BBDP/Wor rat (Blankenhorn et al. 2005; Hornum et al. 2004; Mordes et al. 2002). Recently, the curators of the Rat Genome Database have redesignated *Iddm4* as *Iddm14* (http://rgd.mcw.edu).

The existence of rat diabetes resistance genes has been inferred from crosses between diabetes-prone BB and resistant non-BB rats (Colle et al. 1992; Hornum et al. 2001; Jacob et al. 1992). Using the BB/OK rat and crosses to several strains including DA and SHR, Klöting et al. have also reported several other disease-modifying loci (Klöting et al. 1998, 2001; Klöting and Kovacs 1998a, 1998b; Van den Brandt et al. 1999). Congenic BB.SHR rats carrying a segment of SHR chromosome 6 show reduced diabetes frequency. Candidate genes in this interval include Yy1, a multifunctional transcription factor, and Pref-1, an epidermal growth factor-like repeat protein (Klöting and Klöting 2004). Congenic BB.SHR rats have also been used to identify loci relevant to obesity and metabolic syndrome (Klöting et al. 2004, 2005).

BB Rats and Translational Modeling Fidelity

Prevention

Studies in diabetes-prone BB rats (Laupacis et al. 1983; Like et al. 1983) and NOD mice (Mori et al. 1986) revealed that generalized immunosuppression with cyclosporine can prevent autoimmune diabetes. In the rodent trials, primary prevention was almost uniformly successful, therapy could be brief, and no toxicities were reported. In human clinical trials, cyclosporine was documented clearly to ameliorate T1D and preserve insulin secretory capability when given soon after onset (Mahon et al. 1993); however, disease recurred when therapy was discontinued and long-term results were disappointing (De Filippo et al. 1996). Therapy was sometimes complicated by early (Feutren and Mihatsch 1992; Rodier et al. 1991) and late (Parving et al. 1999) drug-induced nephrotoxicity.

In studies designed to induce immunological tolerance, it was found that modest doses of parenteral (Atkinson et al. 1990) or oral (Zhang et al. 1991) insulin prevent diabetes in NOD mice. Parenteral insulin also prevents autoimmunity in diabetes-prone BB rats (Gotfredsen et al. 1985; Visser, Klatter, et al. 2003), but to be fully effective the intervention may require long-term therapy and relatively substantial doses of insulin, leading to beta cell involution. Oral tolerance with insulin in diabetes-prone BB rats with or without adjuvant is ineffective (Mordes et al. 1996b) and may exacerbate the disease (Bellmann et al. 1998). There has been one report of reduced incidence of diabetes in the BBDP rat following immunization with insulin B chain (Song et al. 1999), but this has not been confirmed. Results in humans suggest that neither parenteral (Skyler et al. 2002) nor oral (Chaillous et al. 2000; Skyler et al. 2005) insulin prevents or delays diabetes onset. A variant strategy for preservation of beta cells in the acutely diabetic BB rat combines "functional rest" achieved with intensive treatment with insulin plus a potassium channel opener such as diazoxide (Rasmussen et al. 2000). Another unusual towering regimen that reduces the penetrance of diabetes in the BBDP rat is oral administration of a peptide of human heat shock protein 60 (DiaPep277) plus a hydrolyzed casein diet (Brugman et al. 2004). Whether these regimens might prove useful in humans is not known.

Nicotinamide prevents diabetes in NOD mice (Elliott et al. 1993) but not in diabetes-prone BB/Mol rats (Hermitte et al. 1989) or in humans (Lampeter et al. 1998; Philips and Scheen 2002). GAD and bovine serum albumin are candidate autoantigens in humans that are effective in the prevention of NOD mouse diabetes (Atkinson and Leiter 1999), but are ineffective in the BB/Wor//Mol rat (Petersen et al. 1997). A number of bacterial vaccines impair the development of inflammatory responses and prevent rodent diabetes. These include complete Freund's adjuvant (CFA) (Qin et al. 1993) and bacille Calmette–Guerin (BCG) (Yagi et al. 1991) in the NOD mouse and CFA (Sadelain et al. 1990) and OK-432 (Satoh et al. 1988) in BB rats. BCG in humans is ineffective (Allen et al. 1999); no data are available for the other reagents.

Prolongation of breast feeding appears to delay the onset and reduce the frequency of diabetes in BBDP rats (Visser, Groen, et al. 2003), but the importance of limited-duration breast feeding as a risk factor in human diabetes is unclear (Kimpimäki et al. 2001; Visalli et al. 2003). Certain defined diets and diets low in essential fatty acids reduce the prevalence of disease in BBDP/Wor rats, but do not prevent it (Mordes et al. 2001). Proteins present in cow's milk are hypothesized to play a role in the pathogenesis of T1D in humans, but it is not yet known whether elimination of these proteins from the diets of children at risk will reduce the frequency of diabetes. Although dietary modification can reduce spontaneous diabetes expression in BBDP/Wor rats, the agent of protection is not elimination of cow's milk protein (Malkani et al. 1997). The addition of BSA or intact milk protein does not abrogate the effectiveness of a protective diet. Interestingly, feeding neonatal BBdp rats lipopolysaccharide, a ligand of toll-like receptor 4 (TLR4), increases insulitis severity and the cumulative frequency of diabetes (Scott et al. 2002).

A preventive strategy unique to the spontaneously diabetic BB rat model is transfusion of Tregs to overcome the effects of lymphopenia before the initiation of insulitis (Burstein et al. 1989). Retrovirus-transduced lymphocytes designed to secrete high levels of IL-4 have also been shown to ameliorate BB rat diabetes, suggesting that this animal system could serve as a platform for gene therapy studies (Zipris and Karnieli 2002). There is, however, no documented efficacy of any form of transfusion therapy in human T1D (Mordes et al. 1986) and, interestingly, it has been recently shown that T cell reconstitution after the initiation of insulitis actually can precipitate onset of diabetes through the recruitment of donor T cells to the autoimmune process (Ramanathan and Poussier 1999). Treg infusion is under active consideration as a therapeutic intervention in patients at risk for autoimmune disease (Earle et al. 2005).

One of the most promising treatments for autoimmunity is costimulation blockade (Carreno and Collins 2002), and studies in the BBDP rat support that hope. Two anti-CD28 monoclonal antibodies (mAb) with different functional activities can completely prevent diabetes in BBDP rats. Blockade of the CD40–CD154 pathway was also effective, but less so. These data suggest that the effectiveness of costimulation blockade in the treatment of T1D is dependent on the costimulatory pathway targeted (Beaudette-Zlatanova et al. 2006).

BBDP/Wor rat diabetes can also be prevented by intrathymic transplantation of islets (Posselt et al. 1992) and low doses of the TLR3 ligand polyinosinic:polycytidylic acid (poly I:C) (Sobel, Goyal, et al. 1998), but high-dose poly I:C accelerates diabetes (Sobel et al. 1994). Spontaneous diabetes in BB rats is also prevented or retarded by pentoxifylline (Visser et al. 2002), gliotoxin (Larsen et al. 2000; Liu et al. 2000), tumor necrosis factor-α (TNF-α) (Satoh et al. 1990), lymphotoxin (Takahashi et al. 1993), interferon-α (IFN-α) (Sobel, Creswel, et al. 1998), and IFN-γ (Nicoletti et al. 1998; Sobel and Newsome 1997); paradoxically, anti-IFN-γ is also effective (Nicoletti et al. 1997). Treatment with IL-2 (Burstein et al. 1987) or with an antioxidant (Iovino et al. 1999) is not effective. Cytokine treatment can also induce Fas-expression and iNOS, causing generation of nitric oxide (NO), which is toxic for beta cells. The iNOS inhibitor aminoguanidine (AG) delays diabetes onset, but does not reduce diabetes incidence (Kuttler et al. 2003). More promising are studies of other agents that limit NO production; the dithiocarbamate analogue NOX-200 reportedly prevents diabetes in BBDP rats as monotherapy and in combinations with immunosuppression (Pieper et al. 2004). The relevance of these preventive strategies to human diabetes is not yet clear.

Cure

Cure of BB rat diabetes has been achieved with pancreas and islet transplantation together with immunosuppression (Dugoni, Jr. and Bartlett 1990), antiadhesion molecule antibodies (Tori et al. 1999), or costimulatory blockade (Kover et al. 2000). The last finding is of particular interest in light of the resistance of the NOD mouse to tolerance-based transplantation (Markees et al. 1999) and the fact that resistance in the NOD is genetically distinct from its susceptibility to autoimmune diabetes (Pearson, Markees, Serreze, et al. 2003; Pearson, Markees, Wicker, et al. 2003). Arguably, the BB rat could be the preferred small-animal model for the study of islet transplantation tolerance induction. Islet regeneration in the course of beta cell destruction has also been studied in the BBdp rat (Wang et al. 2005), as has hepatic gene therapy using adenoviral delivery of a glucose-regulated insulin transgene in BBDP/Wor rats (Olson et al. 2003).

Spontaneous diabetes in BBDP/Wor rats occurs in more than half of animals housed in "conventional" vivaria (Guberski 1994), whereas diabetes is uncommon in NOD mice exposed to most microbiological agents (Serreze and Leiter 2001). Like the NOD mouse, however, diabetes does occur earlier and with a higher frequency in viral antibody-free (VAF) vivaria than in less clean environments. Deliberate infection with lymphocytic choriomeningitis virus (LCMV) prevents diabetes—perhaps by deleting effector T cells (Dyrberg et al. 1988), but infection with Kilham rat virus (KRV) and other viruses does not affect disease frequency or age at onset (Guberski et al. 1991). In contrast, rat cytomegalovirus (RCMV) accelerates onset of diabetes without infecting pancreatic islets (Hillebrands et al. 2003; van der Werf et al. 2003).

BB Rats in the Study of Diabetic Complications

BB rats continue to be a preferred platform for the study of diabetic complications, including retinopathy (Ellis et al. 2002), nephropathy (Carlson et al. 2004; Inkinen et al. 2005; Jörns et al. 2004), bone abnormalities (Beam et al. 2002; Follak et al.

2005), erectile dysfunction (Podlasek et al. 2003), cognitive impairment (Li et al. 2002), vasculopathy (Kwan 1999), and cardiovascular disease (Broderick and Hutchison 2004; Villanueva et al. 2003). In the area of neuropathy, comparison of the BBDP rat model of T1D with the obese BBZDR model of type 2 diabetes (Tirabassi et al. 2004) has revealed potentially important differences in the pathophysiology of this complication in the two disease states. These differences could reflect the absence of C-peptide secretion in the models of T1D (Kamiya et al. 2005; Schmidt et al. 2004; Stevens et al. 2004). Studies of diabetic autonomic neuropathy in the BB rat have revealed defective glucagon responses to hypoglycemia, a major complication of longstanding diabetes (Mundinger et al. 2003).

Spontaneously Diabetic LETL and KDP Rats

The first model of spontaneous autoimmune diabetes in a nonlymphopenic rat was the Long–Evans Tokushima lean (or LETL) rat (Kawano et al. 1991), which initially developed diabetes at a rate of 15–20%. Subsequently, two substrains were established (Komeda et al. 1998). In the nonlymphopenic Komeda diabetes-prone (KDP) substrain, the frequency of diabetes is ~70%, and all rats have mild to severe insulitis by 120–220 days of age. They also exhibit lymphocytic infiltration of thyroid and kidney. The Komeda nondiabetic (KND) substrain is disease free. Genetic analysis indicated that diabetes in the KDP rat required the presence of the *RT1 B/D*^{u} class II MHC and at least one additional recessive gene. The diabetogenic allele at that genetic locus was identified as a nonsense mutation in *Cblb* (Casitas B-lineage lymphoma b), a member of the *Cbl/Sli* family of ubiquitin-protein ligases (Yokoi et al. 2002).

Transgenic complementation significantly reduced diabetes frequency in rescued animals, and it was hypothesized that in the KDP rat the MHC determines the tissue specificity of the disease and *Cblb* facilitates the activation of autoreactive T cells (Yokoi 2005). *Cblb* expression is regulated by CD28 and CTLA-4 and is critical for establishing the threshold for T cell activation (Li et al. 2004). *Cblb* is one of few non-MHC genes definitively linked to autoimmune diabetes in any strain or species. Data on the potential role of *Cblb* in human diabetes are limited. In a Danish study of 480 families, a single nucleotide polymorphism (SNP) in exon 12 of the *Cblb* gene was found to be associated to T1D (Bergholdt et al. 2005), but another analysis found no evidence of genetic association (Payne et al. 2005). *Cblb* has not been linked to diabetes in the BB rat.

Congenic LEW Rats with Spontaneous Diabetes

LEW.1AR1/Ztm-*iddm* Rats

Another new rat model of T1D arose through a spontaneous mutation in the MHC congenic LEW.1AR1 strain (Lenzen et al. 2001). The MHC haplotype (*RT1 A*^{a}*B/D*^{u}*C*^{u}) of these animals includes a class II *u* allele. These are designated LEW.1AR1/Ztm-*iddm* rats. The cumulative frequency of diabetes was initially ~20% by two months of age, but the penetrance of the disease has increased to 60% or more (Jörns et al. 2005). Both sexes are affected equally. Diabetic animals are ketonuric and the disease is frequently fatal even when treated with insulin. The animals are not lymphopenic and circulate normal numbers of ART2$^+$ T cells. Islet pathology is analogous to that reported in the BB rat. Islets of affected animals are initially infiltrated with macrophages followed by CD8$^+$ T cells (Jörns et al. 2004). IL-1β and TNF-α are expressed by the infiltrating cells, as are iNOS and procaspase 3. It appears that beta cells die via apoptosis, and classical end-stage islets are characteristic of chronically diabetic animals. T cells from LEW.1AR1/Ztm-*iddm* rats can adoptively transfer diabetes to athymic recipients (Wedekind et al. 2005). There are no reports to date on diabetes prevention trials in the LEW.1AR1/Ztm-*iddm* rat.

Genetic analysis of the LEW.1AR1/Ztm-*iddm* rat is at an early stage, but the disease appears to be inherited as an autosomal recessive trait (Weiss et al. 2005). Analysis of two backcross populations revealed three susceptibility loci. Two map to chromosome 1 and the third to chromosome 20, within the MHC. Homologous regions in the human genome contain the *IDDM1*, *5*, *8*, and *17* loci. The parental LEW.1AR1 strain has remained free of spontaneous diabetes.

Unlike BBDP rats and NOD mice, LEW.1AR1/Ztm-*iddm* show no evidence of autoimmunity directed against the adrenal gland, thyroid, or salivary glands. Gut, heart, liver, and lung are also normal, but after six months of diabetes, pathological changes compatible with a diabetic nephropathy are observed (Jörns et al. 2004).

LEW.1WR1 Rats

Spontaneous diabetes has also appeared in a different colony of MHC congenic LEW rats that are designated LEW.1WR1. The MHC haplotype ($RT1.A^uB/D^uC^a$) of these animals again includes a class II *u* allele, but the class I haplotype is different from that of the LEW.1AR1/Ztm-*iddm* rat (table 1.1). In the colony of LEW.1WR1 rats maintained at BRM, Inc. (Worcester, Massachusetts), spontaneous diabetes was absent from acquisition until 1999, but now occurs with a cumulative frequency of ~2% at a median age of 59 days. The disease is characterized by hyperglycemia, glucosuria, ketonuria, and polyuria. Both sexes are affected and islets of acutely diabetic rats are devoid of beta cells, whereas alpha and delta cell populations are spared. The peripheral lymphoid phenotype is normal, including the fraction of ART2$^+$ Tregs. The LEW.1WR1 rat is also susceptible to collagen-induced arthritis but free of spontaneous thyroiditis. Genetic analysis of the LEW.1WR1 has only recently started. Comparative analysis of the LEW.1WR1 and LEW.1AR1/Ztm-*iddm* strains promises to be informative, particularly in regard to the class I MHC. As will be discussed later, the penetrance of the diabetes phenotype in the LEW.1WR1 rat can be greatly increased by immunological perturbation.

INDUCED AUTOIMMUNE DIABETES IN THE RAT

Why Study Induced Diabetes?

Many lines of evidence, including a concordance rate in monozygotic twins that averages only ~50% (Kyvik et al. 1995; Redondo et al. 2001), suggest that T1D is caused by nongenetic environmental factors operating in a genetically susceptible host to initiate a destructive immune process (Åkerblom et al. 2002; Hawa et al. 2002). Many human T1D susceptibility genes may operate only in the context of environmental perturbation, amplifying the immune response and the rate of disease progression (Hawa et al. 2002). Disease is thought likely to be due to interaction with the environment of alleles at many loci scattered throughout the genome (Todd 1999).

Candidate perturbants include toxins, diet, vaccination, and infection. There are few convincing data to support toxins (Helgason and Jonasson 1981) and strong data against an effect of vaccination (Hviid et al. 2004). Dietary analyses are under way but thus far not conclusive (Couper et al. 1999; Norris et al. 2003).

Strong epidemiologic evidence suggests that viral infection is perhaps the most important of the candidate perturbants (Pietropaolo and Trucco 1996; Yoon and Jun 2004), particularly in populations in which the incidence of diabetes is increasing (Laron 2002). Nonetheless, definitive studies linking diabetes pathogenesis to infection are lacking, as are documented mechanisms. Efforts to obtain unequivocal identification of causative infectious agents have not been successful; indeed, evidence can be adduced to suggest that infectious agents can initiate, enhance, or even abrogate autoimmunity (Bach 2002) (the "hygiene hypothesis"). A possible explanation recently outlined by Filippi and von Herrath (2005) says that the link between infections and autoimmunity is multifaceted. Viral footprints might be hard to detect systemically or in the target organ once autoimmunity has been initiated, and several infections might have to act in concert to precipitate clinical autoimmunity. In addition, viral infections alone might not be able to induce disease in the absence of other inflammatory factors, a concept termed the "fertile field hypothesis" (Von Herrath et al. 2003).

Against the background of a complex human dataset and the need to surmount the complexities inherent in studying an outbred population exposed at random to environmental agents, there has been increasingly animated interest in appropriate and tractable inbred animal model systems that develop (or fail to develop) autoimmune diabetes in response to a range of environmental interactions.

Induced Autoimmune Diabetes

Nontransgenic mouse models of induced T1D are few and most are not widely used. Immunization with a peptide fragment of a heat shock protein induces a T1D-like syndrome in normal mice (Elias et al. 1995), as can certain orally ingested autoantigenic peptides (Blanas et al. 1996). NOD mice model interactions with the environment largely in support of the hygiene hypothesis because the great majority of perturbants, including viral infection, reduce the frequency of diabetes and often prevent it entirely (Atkinson and Leiter 1999). A toxin-inducible model is the mouse treated with multiple small doses of streptozotocin (STZ) (Mordes et al. 2000). Because this induction method can lead to diabetes in NOD-*scid* mice that have no functional lymphocytes, its relevance as a model of T1D is doubtful (Gerling et al. 1994).

In contrast, susceptibility to induced autoimmune diabetes in the rat appears to be relatively common. As shown in table 1.1, with one exception, the susceptible rat strains all share the class II *u* MHC allele. Activated spleen cells from $RT1^c$ PVG rats depleted of regulatory T cells (Treg cells) are capable of the adoptive transfer of diabetes (McKeever et al. 1990). Diabetes also occurs in PVG rats after thymectomy plus sublethal irradiation (Penhale et al. 1990; Stumbles and Penhale 1993), whereas other rat strains, including the *RT1[u]* WAG, are resistant. The induction system in PVG rats is exceptional with respect to MHC and to the intensity of the induction process, but it continues to be used. Thymectomy plus irradiation-induced diabetes in a cohort of nonlymphopenic (WF × BBDP)F2 animals has been used in genetic analyses. These revealed linkage between regions on chromosomes 1, 3, 4, 6, 9, and 16 and expression of autoimmune diabetes (Ramanathan, Bihoreau, et al. 2002). Other systems, in wider use, employ viral infection, TLR ligation, Treg depletion, and combinations of these perturbants.

Immunological Perturbation and Diabetes Induction in the BBDR/Wor Rat

BBDR/Wor rats comprise a distinct inbred strain of BB rats (Guberski 1994; Mordes et al. 2001). They were derived from BBDP/Wor forebears at the fifth generation of inbreeding by selection for the absence of disease. They share the *RT1[u]* MHC haplotype of the spontaneously diabetic BB rat and are not lymphopenic because the breeding process at that time selected against the recessive *Gimap5* mutation. They circulate normal numbers of $CD4^+$, $CD8^+$, and $ART2^+$ T cells and never become spontaneously diabetic in VAF vivaria (Butler et al. 1991). They are also susceptible to collagen-induced arthritis (Cremer et al. 2004; Watson et al. 1990; Ye et al. 2004).

Treg Depletion

As noted earlier, spontaneous diabetes in lymphopenic BBDP/Wor rats can be prevented by transfusion with $CD4^+ART2^+$ Treg cells. Conversely, BBDR/Wor rats housed in conventional, non-VAF conditions become diabetic if treated with a depleting anti-$ART2^+$ antibody (Greiner et al. 1987). Cyclophosphamide (Like et al. 1985) and low-dose irradiation (Handler et al. 1989) can also precipitate diabetes in this animal under these conditions, but whether the mechanism involves Treg populations is unknown.

In VAF conditions, BBDR/Wor rats rarely become diabetic when treated with a depleting anti-$ART2^+$ antibody alone. In this case, coadministration of an innate immune system activator like poly I:C, a synthetic double-stranded polyribonucleotide ligand of TLR3, is required (Guberski et al. 1991; Thomas et al. 1991).

Induction of diabetes in BBDR rats by immunological perturbation has permitted kinetic analyses of islet pathology (Jiang et al. 1990). These revealed a prodromal period of 10 days during which no morphologic abnormalities of the pancreas were detected. This was followed by a second phase of early insulitis in which a few islets were infiltrated by macrophages and T cells. The lesions rapidly progressed, and by day 18 insulitis was generalized and intense.

TLR Ligation

It has been known for some time that poly I:C alone or in synergy with Treg depletion can induce diabetes in the BBDR/Wor rat. Poly I:C binds to TLR3 (Alexopoulou et al. 2001) and acts much as a "viral-mimetic" agent (Doukas et al. 1994). Treatment of BBDR/Wor rats with low doses of poly I:C (5 μg/g) induces diabetes in ~20% of animals (Thomas et al. 1991). Higher doses of poly I:C (10 μg/g) induce diabetes in nearly all BBDR/Wor rats (Sobel et al. 1992). Parenthetically, high-dose poly I:C accelerates diabetes in the BBDP/Wor rat (Ewel et al. 1992; Sobel et al. 1995), an effect associated with increased levels of IFN-α (Sobel et al. 1994); at lower doses, it is protective (Sobel, Goyal, et al. 1998).

Viral Infection

In contrast to viral infection in NOD mice, infection of the BBDR/Wor rat with KRV induces diabetes. Naturally occurring infection in a closed colony of inbred animals affected ~1% of one generation before its eradication by caesarian rederivation of the colony (Guberski et al. 1991). Intentional intraperitoneal infection (10^7 PFU of KRV-UMass) induces diabetes in ~30% of BBDR/Wor rats (Guberski et al. 1991). Infection is associated with the development of pancreatic insulitis but not with infection of islet cells or with the development of exocrine pancreatitis (Brown et al. 1993). The latter characteristic distinguishes this system from that of encephalomyocarditis virus (EMC-D) in mice, which appears not to have a primary autoimmune component to pathogenesis but to be largely dependent on destruction of beta cells by the replication of the virus within them (Yoon and Jun 2004). The ability to induce autoimmune diabetes in BBDR/Wor rats is virus specific; infection with the closely homologous parvovirus H-1, which is ~98% identical to KRV at the level of DNA, fails to induce diabetes despite the induction of robust cellular and humoral immune responses (Zipris et al. 2003). KRV infection increases serum IL-12 p40 in treated animals, and it increases IL-12 p40, IP-10, and IFN-γ mRNA transcript levels, particularly in the pancreatic lymph nodes (Zipris et al. 2005).

The mechanism by which KRV induces diabetes is under study. One possible mechanism is alteration of the immunoregulatory environment. The virus does infect T and B lymphocytes (McKisic et al. 1995), although it does not cause the severe T lymphopenia characteristic of the spontaneously diabetic BBDP rat. Infected lymphocytes are phenotypically normal but have diminished proliferative and cytolytic responses (McKisic et al. 1995). KRV could induce diabetes in the BBDR rat by changing the immunoregulatory environment of these animals, specifically by reducing the frequency of $CD4^+CD25^+$ Treg cells (Zipris et al. 2003).

Yoon and colleagues have tested the alternative hypothesis that diabetes might be the result of molecular mimicry. This hypothesis was disconfirmed in studies in which BBDR/Wor rats were injected with viral vectors encoding KRV proteins (Chung et al. 2000). No diabetes occurred despite the generation of cellular and humoral immune responses to those proteins. This group also investigated the role of macrophage-derived factors after infection (Mendez et al. 2004). They observed that the expression of iNOS increased early in the course of infection and that inhibition of iNOS by aminoguanidine prevented KRV-induced diabetes.

Immunological Perturbants Enhance Diabetogenicity of Viral Infection

Noninfectious environmental perturbants can synergize with KRV to induce diabetes. Depletion of ART2$^+$ Treg cells in the BBDR/Wor rat synergizes with KRV infection and leads to a high frequency of autoimmune diabetes (Ellerman et al. 1996). The combination of KRV infection and a brief course of poly I:C (one incapable of inducing diabetes by itself) induces diabetes in 100% of BBDR/Wor rats (Ellerman et al. 1996; Zipris et al. 2003). More recent studies indicate that other TLR ligands can also synergize with KRV infection; these included heat-killed *Escherichia coli* and *Staphylococcus aureus*, which are natural TLR agonists. TLR ligation is known to lead to a pleiotrophic immune response and antigen presenting cell activation that could clearly "disequilibrate" an immune system genetically predisposed to autoimmunity. Consistent with this view, it is known that macrophage depletion renders BBDR/Wor rats resistant to the induction of diabetes in response to KRV plus poly I:C (Chung et al. 1997).

Genetic Analysis

The susceptibility of BBDR/Wor rats to induced autoimmune diabetes has proven useful in the identification of disease susceptibility loci. In analyses of (BBDR/Wor × WF) × WF backcross rats, none became spontaneously diabetic whereas ~25% did so after treatment with Treg depletion and poly I:C. Genome-wide scan and an identity-by-descent analysis of these animals revealed that they share the genetically dominant diabetes susceptibility locus designated *Iddm4* that is also critical for diabetes in BBDP/Wor rats (Martin, Maxson, et al. 1999; Mordes et al. 2002). The BB/Wor-origin allele at *Iddm4* has 79% sensitivity and 80% specificity in prediction of diabetes in rats segregating for this locus. Comparably treated MHC-identical WF rats (*Iddm4^w*) resist diabetes induction. A WF rat congenic for the BB/Wor-derived allele at *Iddm4* (WF.*Iddm4^d*) has been developed for further analysis of this locus and identification of the gene (Martin, Moxson, et al. 1999; Mordes et al. 2002). Interestingly, WF.*iddm4^d* congenic rats that become diabetic in response to Treg depletion plus poly I:C do not become diabetic after KRV infection plus poly I:C (Blankenhorn et al. 2005). A genome-wide scan of (BBDR/Wor × WF)F2 rats revealed that diabetes is induced by KRV in BBDR/Wor rats only if the BB-origin alleles of *Iddm4* and at least one additional susceptibility gene, designated *Iddm20*, are present.

BBDR Rats and Translational Modeling Fidelity

BBDR/Wor rats induced to become diabetic develop hyperglycemia under well controlled circumstances and within a defined time frame, usually two to four weeks. Accordingly, they provide an opportunity to assess preventive intervention strategies. Few such studies have been reported to date, but it is known that intrathymic islet transplantation (Battan et al. 1994) and hydrolyzed casein diets (Malkani et al. 1997) do not prevent induction of diabetes in animals treated subsequently with Treg depletion and poly I:C. Intervention based on the administration of parenteral insulin is effective in similarly treated BBDR/Wor rats, but only at high doses of insulin sufficient to cause beta cell involution (Gottlieb et al. 1991). Rats treated in this way that remain nondiabetic nonetheless exhibit thyroiditis and insulitis and harbor autoreactive cells that can transfer diabetes to naïve recipients. Diets low in essential fatty acids do prevent disease in Treg-depleted BBDR/Wor rats (Lefkowith et al. 1990).

The effects of immunomodulation and costimulation blockade have also been studied in the BBDR rat treated with Treg depletion plus poly I:C. The immunomodulatory drug FTY720 reportedly prevents diabetes reliably when administered before or early in the course of induction (Popovic et al. 2004). Anti-CD154 mAb delays onset of diabetes, whereas CTLA4-Ig, anti-CD134L mAb, or anti-CD28 mAb have little or no effect (Beaudette-Zlatanova et al. 2006). These results differ from those obtained with the same reagents in the BBDP rat (Beaudette-Zlatanova et al. 2006), and they suggest that the effectiveness of costimulation blockade in the treatment of autoimmune diabetes is dependent on the costimulatory pathway targeted and the mechanism of induction, stage, intensity, and duration of the pathogenic process.

The observation that treatment with aminoguanidine, an inhibitor of iNOS, can prevent diabetes subsequent to KRV infection (Mendez et al. 2004) raises the intriguing possibility that some cases of autoimmune diabetes may be preventable by vaccination or antiviral therapy. With respect to suitability as a model, KRV-infected BB rats do suffer from one obvious limitation: KRV is a rat-specific parvovirus. The only significant human parvoviral pathogen is B19. There are no epidemiological associations between B19 infection and T1D (Chen and Howard 2004), but, intriguingly, a single case report has recently appeared (Munakata et al. 2005).

Immunological Perturbation and Diabetes Induction in Other Rat Strains

TLR Ligation

Susceptibility to induced autoimmune diabetes appears to be common in rat strains that express a class II *u* MHC haplotype. Table 1.1 summarizes some of the characteristics of these systems. Although the doses and duration of treatment were variable among the studies, the importance of the class II MHC haplotype and the diversity of susceptible strains are apparent.

It is particularly interesting to note that in the case of the BBDP and LEW.1AR1/Ztm-*iddm* rats, both of which develop spontaneous autoimmune diabetes with high penetrance, treatment with poly I:C does not have a significant effect on disease frequency or kinetics (Lenzen et al. 2001). In contrast, treatment with poly I:C induces diabetes in up to 20% of the nondiabetic parental LEW.1AR1 strain (Wedekind et al. 2005) and up to 80% (Wedekind et al. 2005) or 100% (Mordes et al. 2005) of the LEW.1WR1 strain. Depending on dosing, it also induces diabetes in 20–68% of PVGR8 rats (Ellerman and Like 2000), which share the same MHC haplotype as the LEW.1AR1 rat. Studies in the LEW.1WR1 rat suggest that the effect exhibits some specificity; the TLR4 ligand LPS does not affect disease penetrance (Mordes et al. 2005). The mechanisms underlying these observations are unknown, but clearly a diverse array of background genes in these strains must modulate the basic susceptibility conferred by the *RT1 B/D*^{u} MHC, and a different set of genes determines susceptibility to an environmentally induced immune response that converts susceptibility to clinically apparent disease.

Viral Infection

KRV induces insulitis and autoimmune diabetes in LEW.1WR1 rats (Ellerman et al. 1996). Preliminary data suggest that LEW.1WR1 rats, like diabetes-resistant BBDR/Wor rats, also resist diabetes induction when treated with H-1 or vaccinia (Mordes et al. 2003). Surprisingly, however, unlike BBDR/Wor rats, LEW.1WR1 rats developed autoimmune diabetes after infection with rat cytomegalovirus (RCMV). No treated rats had exocrine pancreatitis. The underlying mechanisms and susceptibility genes that account for this differential autoimmune response to infection in BBDR versus LEW.1WR1 rats are unknown. The combination of KRV infection and a brief course of poly I:C (one incapable of inducing diabetes by itself) induces diabetes in 100% of BBDR/Wor and LEW.1WR1 rats and a fraction of PVG.*RT1^u* rats (Ellerman et al. 1996; Zipris et al. 2003).

Comparative study of the BBDR/Wor and LEW.1WR1 systems should be of particular interest in light of recent data suggesting a possible role for CMV in human T1D (Hiemstra et al. 2001). In addition, preliminary results suggest that maternal immunization protects weanling LEW.1WR1 rats from cytomegalovirus-induced diabetes (Tirabassi et al. 2005). Were even a small number of cases of human T1D to be preventable by immunization, the effect would be important (England and Roberts 1981).

A WORKING HYPOTHESIS OF AUTOIMMUNE DIABETES IN THE RAT

The enlarging repertoire of rat strain resources for the study of autoimmune diabetes allows for the expansion and strengthening of a working hypothesis of autoimmune diabetes that was originally derived from studies in the BB/Wor rat (Mordes et al. 1996a). As depicted schematically in figure 1.1, this hypothesis holds that diabetes in the rat results from an imbalance between (1) beta cell-cytotoxic effector cells and (2) regulatory cells that normally prevent disease. The hypothesis predicts the existence of at least two and perhaps three defects in all susceptible rat strains.

The first defect is genetic, involves the class II *u* allele of the MHC, and leads to the generation of autoreactive effector cells, not only in the BB rat but also in a several other rat strains (table 1.1). The second defect leads to amplification of the autoreactive population and/or to a regulatory cell deficiency. The second defect can be genetic or acquired. In the diabetes-prone BB rat, this defect is congenital lymphopenia, which can be overcome by the transfusion of $CD4^+ART2^+$ Treg cells. In the case of the KDP rat (Yokoi et al. 1997) a loss-of-function mutation in the *Cblb* gene appears to predispose to abnormal T cell activation (Yokoi et al. 2002). In terms of our hypothesis, the KDP rat combines an MHC-dependent genetic predisposition with a "disequilibrating" genetic defect leading to autoreactive T cell activation. In the LEW.1AR1/Ztm-*iddm* rat, a secondary locus may be located on chromosome 1.

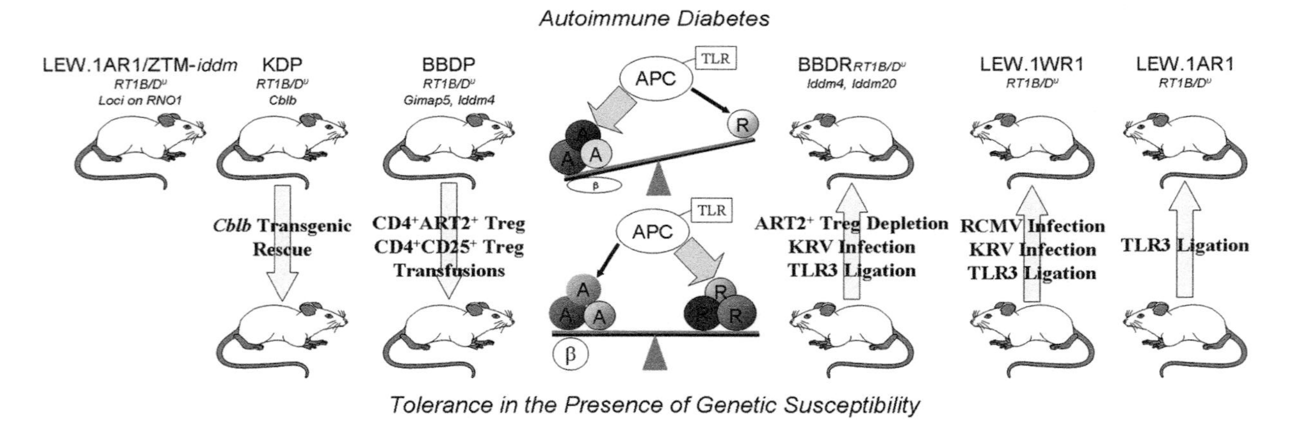

FIGURE 1.1 The balance hypothesis of autoimmune diabetes based on studies of the rat. (Mordes et al. 1996a; Mordes et al. 1987a.) The upper half of the central teeter-totter depicts an imbalance between autoreactive (A) and regulatory (R) cell populations leading to destruction of pancreatic beta cells. The lower half of the figure depicts a balance between these forces. All of the rat strains shown share genetic susceptibility to autoimmunity based on the presence of the *RT1 B/D^u* MHC class II allele. Other genes and loci associated with the individual strains are indicated beneath the MHC. BBDP and BBDR rats share the *Iddm4* locus. The *Iddm20* locus is specific for virus-induced diabetes in the BBDR rat. In the case of the BBDP rat on the left, the presence of a mutation in the *Gimap5* gene causes lymphopenia and an imbalance between autoreactive and regulatory cell populations, leading to diabetes. Restorative transfusions of MHC-compatible lymphocytes restore immunological balance and prevent diabetes. In the case of the BBDR rat on the right, autoreactive cells are present but do not lead to the expression of diabetes unless the balance is perturbed. Perturbants that lead to diabetes in some but not all strains include regulatory cell depletion with anti-ART2.1 monoclonal antibody, infection with Kilham rat virus, infection with rat cytomegalovirus (RCMV), and injection of the toll-like receptor 3 (TLR3) ligand poly I:C. Some of these inducing agents can act synergistically (e.g., poly I:C and KRV). Agents capable of preventing diabetes in the LEW.1AR1/Ztm-*iddm* rat have not been reported.

Alternatively, the second defect can be a genetically determined response to an environmental signal. The array of relevant environmental perturbants and the number of rat strains that they affect are growing. These include KRV infection in BBDR/Wor and LEW.1WR1 rats, RCMV infection in LEW.1WR1 rats, and TLR3 ligation in an array of rats (table 1.3), including the nondiabetic LEW.1AR1 rat. Synergistic combinations of these agents have only begun to be studied, but clearly may involve the interplay of viral and bacterial pathogens.

The third "defect" in T1D may also be genetic or environmental. It would account for the fact that common MHC alleles coupled with perturbants or responses that cause systemic responses (e.g., TLR activation, lack of Tregs, infection with a virus that targets subsets of immune cells) results nonetheless in a very restricted inflammatory response localized to islets. This unique integration that characterizes the genetic response to a random environmental event (or even the absence of that event) has not been carefully addressed. The data suggest that beta cells may somehow be in part "responsible" for their own targeting. Whether they overexpress chemotactic factors at a certain stage of their development or activate surrounding DCs are only two of many potentially interesting unanswered questions. Something must, however, make beta cells more palatable for immune cells than a muscle, skin, or even neighboring alpha cells.

WHAT DOES AUTOIMMUNE DIABETES IN THE RAT TEACH US?

The wisdom of continuing to study animal models of T1D has become a matter of debate (Leiter and Von Herrath 2004; Roep and Atkinson 2004). On the one hand, decades of animal studies have failed to reveal the cause or cure of T1D. It has been disappointing that genes like *Cblb* and *Gimap5* are not clearly associated with human diabetes and that therapies effective in rodent models have not met expectations in human trials. On the other, the inexorably rising incidence (Green et al. 2001) of the disease during decades of human research suggests it may be imprudent to abandon any avenue of inquiry (Lam-Tse et al. 2002; Rossini 2004).

We espouse the latter viewpoint. When the full repertoire of rodent model datasets—rat and mouse—is analyzed, useful insights can be gained. There have been correct predictions of human trials' successful and unsuccessful outcomes for T1D prevention in the animal literature (Greiner et al. 2001; Mordes et al. 2004). There are syntenic relationships among diabetes susceptibility loci (Hessner, Wang, and Ghosh 2004), and a hyporesponsive T cell functional phenotype is reported to be shared by multiple strains of autoimmune rats and mice (Lang and Bellgrau 2002).

It is particularly intriguing to speculate, based on examination of the enlarging rat and mouse datasets, on the role of infection in T1D. Those animals with genetic susceptibility to autoimmune diabetes within the MHC and an arguably "dysfunctional" immune system include the NOD mouse (Anderson and Bluestone 2005), the lymphopenic BBDP rat, the LEW.1AR1/Ztm-*iddm* rat, and the *Cblb*-deficient KDP rat. Each has seen the penetrance of disease increase as they were bred into clean environments, identifying a class of animals to which the hygiene hypothesis (Bach 2002; Tulic et al. 2004) may apply. Conversely, animals with genetic susceptibility

TABLE 1.3
Clinical and Genetic Features of Some Induced Autoimmune Diabetes Syndromes

	BBDR Rat (Mordes et al. 2001)	LEW.1WR1 Rat (Mordes et al. 2005)	LEW.1AR1 Rat (Wedekind et al. 2005)	PVG/c Rat (Penhale et al. 1990)	YOS, WF, PVG.RT1^{u}, WF.Iddm4^{d}, PVGR8 (Ellerman and Like, 2000; Mordes et al. 2002)
Methods of induction	ART2^{+} Treg depletion TLR ligation (poly I:C) Kilham rat virus Combination treatments	ART2^{+} Treg depletion TLR ligation (poly I:C) Virus: KRV; RCMV Combination treatments	Poly I:C	Thymectomy plus irradiation	Poly I:C; poly I:C plus Treg depletion; poly I:C plus KRV; thymectomy plus irradiation
Latency to onset	2–4 weeks	2–4 weeks	~7 weeks	3–8 weeks	2–4 weeks
Ketosis	Severe	Severe	Not stated	Severe	Severe
Associated autoimmune diseases	Thyroiditis; collagen-induced arthritis	Collagen-induced arthritis	Not known	Thyroiditis	Not known
Autoantibodies	Endothelial cell	Not known	Not known	Not known	Not known
MHC genes	RT1$^{u/u/u}$	RT1$^{u/u/a}$	RT1$^{a/u/u}$	RT1$^{c/c/c}$	RT1$^{u/u/u}$
Non-MHC genes	*Iddm4* and others	Not known	Not known	Not known	—
Gender effect	M = F	M = F	Not stated	M > F	M = F

Notes: Comparative features of diabetic syndromes with autoimmune features in several rat model systems. KRV = Kilham rat virus; RCMV = rat cytomegalovirus; TLR = toll-like receptor; Treg = regulatory T cell; poly I:C = polyinosinic:polycytidylic acid.

to autoimmune diabetes within the MHC and an arguably "functional" immune system include the BBDR, LEW.1WR1, and LEW.1AR1 strains. Each has a low to zero rate of spontaneous diabetes in a clean environment, but each can be perturbed by infection or innate immune activation to express autoimmune diabetes. These observations may represent a separate class of animals to which the "fertile field" hypothesis (Filippi and Von Herrath 2005) may apply.

To conclude, we believe that, in their aggregate, the available data gathered from the study of newer rat models of autoimmune diabetes suggest a promising "environmental genetics" approach to modeling T1D. This approach may allow for coordinated analysis of the influence of genetic makeup on the balance hypothesis, coordinating data on the initiation of autoimmune processes with the response of the individual to environmental agents that perturb the immune system in a diverse array of rat strains. This kind of analysis may help unravel the puzzling possible role of diet and viral infection in T1D and provide answers to questions ranging from why identical twins and triplets can be discordant for the disease to why well-reasoned interventions may fail in human trials.

HUSBANDRY

BBDP, BBDR, and LEW.1WR1 rats can be obtained commercially in the United States from BRM, Inc. Worcester, MA (www.biomere.com) and in Europe from Taconic Europe (www.m-b.dk). BBDR rats can be obtained by contacting the Animal Resources Division of Health Canada at 613-946-5372. LEW.1AR1 rats are available through the Institute of Laboratory Animal Science and Central Animal Facility, Hannover Medical School, Carl-Neuberg-Str. 1 30625 Hannover, Germany. Detailed protocols for the husbandry of the BB rat and protocols for diabetes induction have been published (Whalen et al. 1996). Among the protocols described is transfusion for the prevention of spontaneous diabetes in BBDP breeding stock. Husbandry recommendations for other strains are available from individual vendors.

ACKNOWLEDGMENTS

Supported in part by grant DK49106 (DLG, JPM, EPB), DK36024 (DLG), DK25306 (AAR, JPM), DK066447 (JPM) and an institutional Diabetes and Endocrinology Research Center (DERC) grant P30-DK32520 from the National Institutes of Health. Also supported in part by grants (to PP) from Genome Canada and the Canadian Institutes of Health Research (MOP 77713).

REFERENCES

Åkerblom HK, Vaarala O, Hyöty H, et al. (2002) Environmental factors in the etiology of T1D. *Am J Med Genet* 115:18–29.

Alexopoulou L, Holt AC, Medzhitov R, et al. (2001) Recognition of double-stranded RNA and activation of NF-kappaB by Toll-like receptor 3. *Nature* 413:732–738.

Allen HF, Klingensmith GJ, Jensen P, et al. (1999) Effect of bacillus Calmette–Guerin vaccination on new-onset T1D—A randomized clinical study. *Diabetes Care* 22:1703–1707.

American Diabetes Association. (2006) Diagnosis and classification of diabetes mellitus. *Diabetes Care* 29 (Suppl. 1):S43–S48.

Anderson MS, Bluestone JA. (2005) The NOD mouse: A model of immune dysregulation. *Annu Rev Immunol* 23:447–485.

Atkinson M, Leiter EH. (1999) The NOD mouse model of insulin dependent diabetes: As good as it gets? *Nat Med* 5:601–604.

Atkinson MA, Eisenbarth GS. (2001) T1D: New perspectives on disease pathogenesis and treatment. *Lancet* 358:221–229.

Atkinson MA, Maclaren NK, Luchetta R. (1990) Insulitis and diabetes in NOD mice reduced by prophylactic insulin therapy. *Diabetes* 39:933–937.

Awata T, Guberski DL, Like AA. (1995) Genetics of the BB rat: Association of autoimmune disorders (diabetes, insulitis, and thyroiditis) with lymphopenia and major histocompatibility complex class II. *Endocrinology* 136:5731–5735.

Bach JF. (2002) Mechanisms of disease: The effect of infections on susceptibility to autoimmune and allergic diseases. *N Engl J Med* 347:911–920.

Battan R, Mordes JP, Abreau S, et al. (1994) Evidence that intrathymic islet transplantation does not prevent diabetes or subsequent islet graft destruction in RT6-depleted diabetes resistant BioBreeding/Worcester rats. *Transplantation* 57:731–736.

Beam HA, Parsons JR, Lin SS. (2002) The effects of blood glucose control upon fracture healing in the BB Wistar rat with diabetes mellitus. *J Orthop Res* 20:1210–1216.

Beaudette-Zlatanova BC, Whalen B, Zipris D, et al. (2006) Costimulation and autoimmune diabetes in BB rats. *Am J Transplant* 6:894–902.

Bellmann K, Kolb H, Rastegar S, et al. (1998) Potential risk of oral insulin with adjuvant for the prevention of Type I diabetes: a protocol effective in NOD mice may exacerbate disease in BB rats. *Diabetologia* 41:844–847.

Bergholdt R, Taxvig C, Eising S, et al. (2005) *CBLB* variants in T1D and their genetic interaction with CTLA4. *J Leukocyte Biol* 77:579–585.

Bertrand S, Vigeant C, Yale J-F. (1994) Predictive value of lymphocyte antibodies for the appearance of diabetes in BB rats. *Diabetes* 43:137–142.

Bieg S, Möller C, Olsson T, et al. (1997) The lymphopenia (*lyp*) gene controls the intrathymic cytokine ratio in congenic BioBreeding rats. *Diabetologia* 40:786–792.

Blanas E, Carbone FR, Allison J, et al. (1996) Induction of autoimmune diabetes by oral administration of autoantigen. *Science* 274:1707–1709.

Blankenhorn EP, Rodemich L, Martin-Fernandez C, et al. (2005) The rat diabetes susceptibility locus *Iddm4* and at least one additional gene are required for autoimmune diabetes induced by viral infection. *Diabetes* 54:1233–1237.

Blomqvist M, Kaas A, Mansson JE, et al. (2003) Developmental expression of the type I diabetes related antigen sulfatide and sulfated lactosylceramide in mammalian pancreas. *J Cell Biochem* 89:301–310.

Bortell R, Kanaitsuka T, Stevens LA, et al. (1999) The RT6 (Art2) family of ADP-ribosyltransferases in rat and mouse. *Mol Cell Biochem* 193:61–68.

Bortell R, Waite DJ, Whalen BJ, et al. (2001) Levels of Art2^{+} cells but not soluble Art2 protein correlate with expression of autoimmune diabetes in the BB rat. *Autoimmunity* 33:199–211.

Broderick TL, Hutchison AK. (2004) Cardiac dysfunction in the euglycemic diabetic-prone BB Wor rat. *Metabolism* 53:1391–1394.

Brown DW, Welsh RM, Like AA. (1993) Infection of peripancreatic lymph nodes but not islets precedes Kilham rat virus-induced diabetes in BB/Wor rats. *J Virol* 67:5873–5878.

Brugman S, Klatter FA, Visser J, et al. (2004) Neonatal oral administration of DiaPep277, combined with hydrolyzed casein diet, protects against T1D in BB-DP rats. An experimental study. *Diabetologia* 47:1331–1333.

Burstein D, Handler ES, Schindler J, et al. (1987) Effect of interleukin-2 on diabetes in the BB/Wor rat. *Diabetes Res* 5:163–167.

Burstein D, Mordes JP, Greiner DL, et al. (1989) Prevention of diabetes in the BB/Wor rat by a single transfusion of spleen cells: Parameters that affect the degree of protection. *Diabetes* 38:24–30.

Buschard K, Blomqvist M, Osterbye T, et al. (2005) Involvement of sulfatide in beta cells and type 1 and type 2 diabetes. *Diabetologia* 48:1957–1962.

Butler L, Guberski DL, Like AA. (1991) Changes in penetrance and onset of spontaneous diabetes in the BB/Wor rat. In *Lessons from animal diabetes III*, ed. Shafrir E. London: Smith–Gordon, 50–53.

Carlson EC, Vari RC, Audette JL, et al. (2004) Significant glomerular basement membrane thickening in hyperglycemic and normoglycemic diabetic-prone BB Wistar rats. *Anat Rec* 281A:1308–1318.

Carreno BM, Collins M. (2002) The B7 family of ligands and its receptors: New pathways for costimulation and inhibition of immune responses. *Annu Rev Immunol* 20:29–53.

Chaillous L, Lefèvre H, Thivolet C, et al. (2000) Oral insulin administration and residual b-cell function in recent-onset T1D: A multicentre randomized controlled trial. Diabetes Insulin Orale Group. *Lancet* 356:545–549.

Chao NJ, Timmerman L, McDevitt HO, et al. (1989) Molecular characterization of MHC class II antigens (beta 1 domain) in the BB diabetes-prone and -resistant rat. *Immunogenetics* 29:231–234.

Chen S, Howard O. (2004) Parvovirus B19 infection. *N Engl J Med* 350:598.

Chung YH, Jun HS, Kang Y, et al. (1997) Role of macrophages and macrophage-derived cytokines in the pathogenesis of Kilham rat virus induced autoimmune diabetes in diabetes-resistant biobreeding rats. *J Immunol* 159:466–471.

Chung YH, Jun HS, Son M, et al. (2000) Cellular and molecular mechanism for Kilham rat virus-induced autoimmune diabetes in DR-BB rats. *J Immunol* 165:2866–2876.

Colle E. (1990) Genetic susceptibility to the development of spontaneous insulin-dependent diabetes mellitus in the rat. *Clin Immunol Immunopathol* 57:1–9.

Colle E, Fuks A, Poussier P, et al. (1992) Polygenic nature of spontaneous diabetes in the rat: Permissive MHC haplotype and presence of the lymphopenic trait of the BB rat are not sufficient to produce susceptibility. *Diabetes* 41:1617–1623.

Colle E, Guttmann RD, Fuks A, et al. (1986) Genetics of the spontaneous diabetic syndrome. Interaction of MHC and non-MHC-associated factors. *Mol Biol Med* 3:13–23.

Colzani RM, Alex S, Dunn AD, et al. (1999) The oral administration of human thyroglobulin does not affect the incidence of lymphocytic thyroiditis in the BioBreeding Worcester rat. *Thyroid* 9:831–835.

Corper AL, Stratmann T, Apostolopoulos V, et al. (2000) A structural framework for deciphering the link between I-A^{g7} and autoimmune diabetes. *Science* 288:505–511.

Couper JJ, Steele C, Beresford S, et al. (1999) Lack of association between duration of breast-feeding or introduction of cow's milk and development of islet autoimmunity. *Diabetes* 48:2145–2149.

Courtois P, Nsimba G, Jijakli H, et al. (2005) Gut permeability and intestinal mucins, invertase, and peroxidase in control and diabetes-prone BB rats fed either a protective or a diabetogenic diet. *Dig Dis Sci* 50:266–275.

Cremer MA, Ye XJ, Myers LK, et al. (2004) T cell immunity to type II collagen in the biobreeding rat: The identification and characterization of *RT1*^{u}-restricted T cell epitopes on α1(II). *J Immunol* 173:1795–1801.

Crisá L, Mordes JP, Rossini AA. (1992) Autoimmune diabetes mellitus in the BB rat. *Diabetes/Metab Rev* 8:9–37.

Davenport C, Lovell H, James RFL, et al. (1995) Brain-reactive autoantibodies in BB/d rats do not recognize glutamic acid decarboxylase. *Clin Exp Immunol* 101:127–135.

De Filippo G, Carel JC, Boitard C, et al. (1996) Long-term results of early cyclosporin therapy in juvenile IDDM. *Diabetes* 45:101–104.

DeSilva MG, Jun HS, Yoon JW, et al. (1996) Autoantibodies to IA-2 not detected in NOD mice or BB rats. *Diabetologia* 39:1237–1238.

Diessenbacher P, Bartels K, Koch-Nolte F, et al. (2003) T-cell survival regulator LKLF is not involved in inappropriate apoptosis of diabetes-prone BBDP rat T cells. *Ann NY Acad Sci* 1010:548–551.

Doukas J, Cutler AH, Mordes JP. (1994) Polyinosinic:polycytidylic acid is a potent activator of endothelial cells. *Am J Pathol* 145:137–147.

Doukas J, Majno G, Mordes JP. (1996) Antiendothelial cell autoantibodies in BB rats with spontaneous and induced IDDM. *Diabetes* 45:1209–1216.

Dugoni WE Jr., Bartlett ST. (1990) Evidence that cyclosporine prevents rejection and recurrent diabetes in pancreatic transplants in the BB rat. *Transplantation* 49:845–848.

Dyrberg T, Schwimmbeck PL, Oldstone MBA. (1988) Inhibition of diabetes in BB rats by virus infection. *J Clin Invest* 81:928–931.

Earle KE, Tang Q, Zhou X, et al. (2005) *In vitro* expanded human $CD4^{+}CD25^{+}$ regulatory T cells suppress effector T cell proliferation. *Clin Immunol* 115:3–9.

Elder ME, Maclaren NK. (1983) Identification of profound peripheral T lymphocyte immunodeficiencies in the spontaneously diabetic BB rat. *J Immunol* 130:1723–1731.

Elias D, Marcus H, Reshef T, et al. (1995) Induction of diabetes in standard mice by immunization with the p277 peptide of a 60-kDa heat shock protein. *Eur J Immunol* 25:2851–2857.

Ellerman KE, Like AA. (2000) Susceptibility to diabetes is widely distributed in normal class II^{u} haplotype rats. *Diabetologia* 43:890–898.

Ellerman KE, Richards CA, Guberski DL, et al. (1996) Kilham rat virus triggers T-cell-dependent autoimmune diabetes in multiple strains of rat. *Diabetes* 45:557–562.

Elliott RB, Martin JM. (1984) Dietary protein: A trigger of insulin dependent diabetes in the BB rat? *Diabetologia* 26:297–299.

Elliott RB, Pilcher CC, Stewart A, et al. (1993) The use of nicotinamide in the prevention of T1D. *Ann NY Acad Sci* 696:333–341.

Ellis EA, Guberski DL, Hutson B, et al. (2002) Time course of NADH oxidase, inducible nitric oxide synthase and peroxynitrite in diabetic retinopathy in the BBZ/WOR rat. *Nitric Oxide* 6:295–304.

England WL, Roberts SD. (1981) Immunization to prevent insulin-dependent diabetes mellitus? The economics of genetic screening and vaccination for diabetes. *Ann Int Med* 94:395–400.

Ewel CH, Sobel DO, Zeligs BJ, et al. (1992) Poly I:C accelerates development of diabetes mellitus in diabetes-prone BB rat. *Diabetes* 41:1016–1021.

Feutren G, Mihatsch MJ. (1992) Risk factors for cyclosporine-induced nephropathy in patients with autoimmune diseases. *N Engl J Med* 326:1654–1660.

Filippi C, Von Herrath M. (2005) How viral infections affect the autoimmune process leading to T1D. *Cell Immunol* 233:125–132.

Follak N, Klöting I, Merk H. (2005) Influence of diabetic metabolic state on fracture healing in spontaneously diabetic rats. *Diab/Metab Res Rev* 21:288–296.

Fuks A, Ono SJ, Colle E, et al. (1990) A single dose of the MHC-linked susceptibility determinant associated with the *RT1*^{u} haplotype is permissive of insulin-dependent diabetes mellitus in the BB rat. *Expl Clin Immunogenet* 7:162–169.

Geenen V, Brilot F. (2003) Role of the thymus in the development of tolerance and autoimmunity towards the neuroendocrine system. *Ann NY Acad Sci* 992:186–195.

Georgiou HM, Bellgrau D. (1989) Thymus transplantation and disease prevention in the diabetes-prone BioBreeding rat. *J Immunol* 142:3400-3405.

Georgiou HM, Lagarde AC, Bellgrau D. (1988) T cell dysfunction in the diabetes-prone BB rat. A role for thymic migrants that are not T cell precursors. *J Exp Med* 167:132–148.

Gerling IC, Friedman H, Greiner DL, et al. (1994) Multiple low-dose streptozocin-induced diabetes in NOD-*scid/scid* mice in the absence of functional lymphocytes. *Diabetes* 43:433–440.

Gold DP, Bellgrau D. (1991) Identification of a limited T-cell receptor β chain variable region repertoire associated with diabetes in the BB rat. *Proc Natl Acad Sci USA* 88:9888–9891.

Gotfredsen CF, Buschard K, Frandsen EK. (1985) Reduction of diabetes incidence of BB Wistar rats by early prophylactic insulin treatment of diabetes-prone animals. *Diabetologia* 28:933–935.

Gottlieb PA, Handler ES, Appel MC, et al. (1991) Insulin treatment prevents diabetes in RT6-depleted diabetes resistant BB/Wor rats. *Diabetologia* 34:296–300.

Graham S, Courtois P, Malaisse WJ, et al. (2004) Enteropathy precedes T1D in the BB rat. *Gut* 53:1437–1444.

Green A, Patterson CC, Eurodiab Tiger Study Group. (2001) Trends in the incidence of childhood-onset diabetes in Europe 1989–1998. *Diabetologia* 44:B3–B8.

Greiner DL, Handler ES, Nakano K, et al. (1986) Absence of the RT-6 T cell subset in diabetes-prone BB/W rats. *J Immunol* 136:148–151.

Greiner DL, Mordes JP, Handler ES, et al. (1987) Depletion of RT6.1^{+} T lymphocytes induces diabetes in resistant BioBreeding/Worcester (BB/W) rats. *J Exp Med* 166:461–475.

Greiner DL, Rossini AA, Mordes JP. (2001) Translating data from animal models into methods for preventing human autoimmune diabetes mellitus: *Caveat emptor* and *primum non nocere*. *Clin Immunol* 100:134–143.

Guberski DL. (1994) Diabetes-prone and diabetes-resistant BB rats: Animal models of spontaneous and virally induced diabetes mellitus, lymphocytic thyroiditis, and collagen-induced arthritis. *ILAR News* 35:29–37.

Guberski DL, Thomas VA, Shek WR, et al. (1991) Induction of T1D by Kilham's rat virus in diabetes resistant BB/Wor rats. *Science* 254:1010–1013.

Gunther E, Walter L. (2001) The major histocompatibility complex of the rat (*Rattus norvegicus*). *Immunogenetics* 53:520–542.

Handler ES, Mordes JP, McKeever U, et al. (1989) Effects of irradiation on diabetes in the BB/Wor rat. *Autoimmunity* 4:21–30.

Hanenberg H, Kolb-Bachofen V, Kantwerk-Funke G, et al. (1989) Macrophage infiltration precedes and is a prerequisite for lymphocytic insulitis in pancreatic islets of pre-diabetic BB rats. *Diabetologia* 32:126–134.

Hardin JA, Donegan L, Woodman RC, et al. (2002) Mucosal inflammation in a genetic model of spontaneous type I diabetes mellitus. *Can J Physiol Pharmacol* 80:1064–1070.

Hawa MI, Beyan H, Buckley LR, et al. (2002) Impact of genetic and non-genetic factors in T1D. *Am J Med Genet* 115:8–17.

Helgason T, Jonasson MR. (1981) Evidence for a food additive as a cause of ketosis-prone diabetes. *Lancet* 2:716–720.

Hermitte L, Vialettes B, Atlef N, et al. (1989) High-dose nicotinamide fails to prevent diabetes in BB rats. *Autoimmunity* 5:79–86.

Hessner M, Wang X, Ghosh S. (2004) Genetics of T1D. In *Diabetes mellitus. A fundamental and clinical text*, ed. LeRoith D, Taylor SI, Olefsky JM. New York: Lippincott, 483–498.

Hessner MJ, Wang XJ, Meyer L, et al. (2004) Involvement of eotaxin, eosinophils, and pancreatic predisposition in development of T1D mellitus in the BioBreeding rat. *J Immunol* 173:6993–7002.

Hiemstra HS, Schloot NC, Van Veelen PA, et al. (2001) Cytomegalovirus in autoimmunity: T cell cross-reactivity to viral antigen and autoantigen glutamic acid decarboxylase. *Proc Natl Acad Sci USA* 98:3988–3991.

Hillebrands JL, van der Werf N, Klatter FA, et al. (2003) Role of peritoneal macrophages in cytomegalovirus-induced acceleration of autoimmune diabetes in BB-rats. *Clin Develop Immunol* 10:133–139.

Höglund P, Mintern J, Waltzinger C, et al. (1999) Initiation of autoimmune diabetes by developmentally regulated presentation of islet cell antigens in the pancreatic lymph nodes. *J Exp Med* 189:331–339.

Hornum L, DeScipio C, Markholst H, et al. (2004) Comparative mapping of rat *Iddm4* to segments on HSA7 and MMU6. *Mamm Genome* 15:53–61.

Hornum L, Lundsgaard D, Markholst H. (2001) An F344 rat congenic for BB/DP rat-derived diabetes susceptibility loci *Iddm1* and *Iddm2*. *Mamm Genome* 12:867–868.

Hornum L, Rømer J, Markholst H. (2002) The diabetes-prone BB rat carries a frameshift mutation in *Ian4*, a positional candidate of *iddm1*. *Diabetes* 51:1972–1979.

Hviid A, Stellfeld M, Wohlfahrt J, et al. (2004) Childhood vaccination and T1D. *N Engl J Med* 350:1398–1404.

Inkinen K, Soots A, Krogerus L, et al. (2005) Cytomegalovirus enhance expression of growth factors during the development of chronic allograft nephropathy in rats. *Transplant Int* 18:743–749.

Iovino G, Kubow S, Marliss EB. (1999) Effect of α-phenyl-*N-tert*-butylnitrone on diabetes and lipid peroxidation in BB rats. *Can J Physiol Pharmacol* 77:166–174.

Iwakoshi NN, Goldschneider I, Tausche F, et al. (1998) High-frequency apoptosis of recent thymic emigrants in the liver of lymphopenic diabetes-prone BioBreeding rats. *J Immunol* 160:5838–5850.

Jackson R, Kadison P, Buse J, et al. (1983) Lymphocyte abnormalities in the BB rat. *Metabolism* 32 (Suppl. 1):83–86.

Jackson R, Rassi N, Crump T, et al. (1981) The BB diabetic rat: profound T-cell lymphocytopenia. *Diabetes* 30:887–889.

Jacob HJ, Pettersson A, Wilson D, et al. (1992) Genetic dissection of autoimmune type I diabetes in the BB rat. *Nat Genet* 2:56–60.

Jiang Z, Handler ES, Rossini AA, et al. (1990) Immunopathology of diabetes in the RT6-depleted diabetes-resistant BB/Wor rat. *Am J Pathol* 137:767–777.

Jörns A, Gunther A, Hedrich HJ, et al. (2005) Immune cell infiltration, cytokine expression, and beta-cell apoptosis during the development of T1D in the spontaneously diabetic LEW.1AR1/Ztm-*iddm* rat. *Diabetes* 54:2041–2052.

Jörns A, Kubat B, Tiedge M, et al. (2004) Pathology of the pancreas and other organs in the diabetic LEW.1AR1/Ztm-*iddm* rat, a new model of spontaneous insulin-dependent diabetes mellitus. *Virchows Arch Int J Pathol* 444:183–189.

Joseph S, Diamond AG, Smith W, et al. (1993) BB-DR/Edinburgh: A lymphopenic, nondiabetic subline of BB rats. *Immunology* 78:318–328.

Kamiya H, Murakawa Y, Zhang W, et al. (2005) Unmyelinated fiber sensory neuropathy differs in type 1 and type 2 diabetes. *Diabetes/Metab Res Rev* 21:448–458.

Kawano K, Hirashima T, Mori S, et al. (1991) New inbred strain of Long–Evans Tokushima lean rats with IDDM without lymphopenia. *Diabetes* 40:1375–1381.

Kecha-Kamoun O, Achour I, Martens H, et al. (2001) Thymic expression of insulin-related genes in an animal model of autoimmune T1D. *Diabetes/Metab Res Rev* 17:146–152.

Kimpimäki T, Erkkola M, Korhonen S, et al. (2001) Short-term exclusive breastfeeding predisposes young children with increased genetic risk of type I diabetes to progressive beta-cell autoimmunity. *Diabetologia* 44:63–69.

Klöting N, Klöting I. (2004) Congenic mapping of T1D-protective gene(s) in an interval of 4 Mb on rat chromosome 6q32. *Biochem Biophys Res Commun* 323:388–394.

Klöting I, Kovacs P. (1998a) Genes of the immune system cosegregate with the age at onset of diabetes in the BB/OK rat. *Biochem Biophys Res Commun* 242:460–463.

Klöting I, Kovacs P. (1998b) Phenotypic differences between diabetes-prone BB rat sublines cosegregate with loci on chromosomes X and 10. *Biochem Mol Biol Int* 45:865–870.

Klöting I, Schmidt S, Kovacs P. (1998) Mapping of novel genes predisposing or protecting diabetes development in the BB/OK rat. *Biochem Biophys Res Commun* 245:483–486.

Klöting I, Van den Brandt J, Kuttler B. (2001) Genes of SHR rats protect spontaneously diabetic BB/OK rats from diabetes: Lessons from congenic BB.SHR rat strains. *Biochem Biophys Res Commun* 283:399–405.

Klöting N, Wilke B, Klöting I. (2004) Phenotypic and genetic analyses of subcongenic BB.SHR rat lines shorten the region on chromosome 4 bearing gene(s) for underlying facets of metabolic syndrome. *Physiol Genomics* 18:325–330.

Klöting N, Wilke B, Klöting I. (2005) Alleles on rat chromosome 4 (D4Got41-Fabp1/Tacr1) regulate subphenotypes of obesity. *Obesity Res* 13:589–595.

Kolb H, Wörz-Pagenstert U, Kleemann R, et al. (1996) Cytokine gene expression in the BB rat pancreas: Natural course and impact of bacterial vaccines. *Diabetologia* 39:1448–1454.

Komeda K, Noda M, Terao K, et al. (1998) Establishment of two substrains, diabetes-prone and nondiabetic, from Long–Evans Tokushima lean (LETL) rats. *Endocrinology J* 45:737–744.

Kovács P, Van den Brandt J, Bonné ACM, et al. (2001) Congenic BB.SHR rat provides evidence for effects of a chromosome 4 segment (*D4Mit6-Npy* similar to 1 cm) on total serum and lipoprotein lipid concentration and composition after feeding a high-fat, high-cholesterol diet. *Metabolism* 50:458–462.

Kover KL, Geng ZH, Hess DM, et al. (2000) Anti-CD154 (CD40L) prevents recurrence of diabetes in islet isografts in the DR-BB rat. *Diabetes* 49:1666–1670.

Kuttler B, Steveling A, Klöting N, et al. (2003) Aminoguanidine downregulates expression of cytokine-induced Fas and inducible nitric oxide synthase but not cytokine-enhanced surface antigens of rat islet cells. *Biochem Pharmacol* 66:2437–2448.

Kwan CY. (1999) Membrane abnormalities of vascular smooth muscle of mesenteric arteries of spontaneous diabetic BB rats. *J Smooth Muscle Res* 35:77–86.

Kyvik KO, Green A, Beck-Nielsen H. (1995) Concordance rates of insulin dependent diabetes mellitus: A population-based study of young Danish twins. *Br Medical J* 311:913–917.

Lampeter EF, Klinghammer A, Scherbaum WA, et al. (1998) The Deutsche Nicotinamide Intervention Study—An attempt to prevent T1D. *Diabetes* 47:980–984.

Lam-Tse WK, Lernmark Å, Drexhage HA. (2002) Animal models of endocrine/organ-specific autoimmune diseases: Do they really help us to understand human autoimmunity? *Springer Semin Immunopathol* 24:297–321.

Lang J, Bellgrau D. (2002) A T-cell functional phenotype common among autoimmune-prone rodent strains. *Scand J Immunol* 55:546–559.

Laron Z. (2002) Interplay between heredity and environment in the recent explosion of type 1 childhood diabetes mellitus. *Am J Med Genet* 115:4–7.

Larsen B, Liu HG, Jackman S, et al. (2000) Effect of gliotoxin on development of diabetes mellitus in diabetes-prone BB/Wor rats. *Ann Clin Lab Sci* 30:99–106.

Laupacis A, Stiller CR, Gardell C, et al. (1983) Cyclosporin prevents diabetes in BB Wistar rats. *Lancet* 1:10–12.

Lefkowith J, Schreiner G, Cormier J, et al. (1990) Prevention of diabetes in the BB rat by essential fatty acid deficiency. Relationship between physiological and biochemical changes. *J Exp Med* 171:729–743.

Leiter EH, Von Herrath M. (2004) Animal models have little to teach us about T1D: 2. In opposition to this proposal. *Diabetologia* 47:1657–1660.

Lenzen S, Tiedge M, Elsner M, et al. (2001) The LEW.1AR1/Ztm-iddm rat: A new model of spontaneous insulin-dependent diabetes mellitus. *Diabetologia* 44:1189–1196.

Li DD, Gál I, Vermes C, et al. (2004) Cutting edge: Cbl-b: One of the key molecules tuning CD28- and CTLA-4-Mediated T cell costimulation. *J Immunol* 173:7135–7139.

Li XB, Scott FW, Park YH, et al. (1995) Low incidence of autoimmune type I diabetes in BB rats fed a hydrolyzed casein-based diet associated with early inhibition of non-macrophage-dependent hyperexpression of MHC class I molecules on beta cells. *Diabetologia* 38:1138–1147.

Li ZG, Zhang W, Grunberger G, et al. (2002) Hippocampal neuronal apoptosis in T1D. *Brain Res* 946:221–231.

Like AA, Anthony M, Guberski DL, et al. (1983) Spontaneous diabetes mellitus in the BB/W rat. Effects of glucocorticoids, cyclosporin-A, and antiserum to rat lymphocytes. *Diabetes* 32:326–330.

Like AA, Appel MC, Rossini AA. (1982) Autoantibodies in the BB/W rat. *Diabetes* 31:816–820.

Like AA, Weringer EJ, Holdash A, et al. (1985) Adoptive transfer of autoimmune diabetes in BioBreeding/Worcester (BB/W) inbred and hybrid rats. *J Immunol* 134:1583–1587.

Liu HG, Jackman S, Driscoll H, et al. (2000) Immunologic effects of gliotoxin in rats: Mechanisms for prevention of autoimmune diabetes. *Ann Clin Lab Sci* 30:366–378.

Lucke S, Klöting I, Pusch A, et al. (2003) Endocrine pancreas histology of congenic BB-rat strains with reduced diabetes incidence after genetic manipulation on chromosomes 4, 6 and X. *Autoimmunity* 36:143–149.

Lundsgaard D, Holm TL, Hornum L, et al. (2005) *In vivo* control of diabetogenic T-cells by regulatory $CD4^{+}CD25^{+}$ T-cells expressing *Foxp3*. *Diabetes* 54:1040–1047.

MacFarlane AJ, Burghardt KM, Kelly J, et al. (2003) A T1D-related protein from wheat (*Triticum aestivum*)-cDNA clone of a wheat storage globulin, Glb1, linked to islet damage. *J Biol Chem* 278:54–63.

Mackay IR, Bone A, Tuomi T, et al. (1996) Lack of autoimmune serological reactions in rodent models of insulin dependent diabetes mellitus. *J Autoimmunity* 9:705–711.

MacMurray AJ, Moralejo DH, Kwitek AE, et al. (2002) Lymphopenia in the BB rat model of T1D is due to a mutation in a novel immune-associated nucleotide (*Ian*)-related gene. *Genome Res* 12:1029–1039.

Mahon JL, Dupre J, Stiller CR. (1993) Lessons learned from use of cyclosporine for insulin-dependent diabetes mellitus: The case for immunotherapy for insulin-dependent diabetics having residual insulin secretion. *Ann NY Acad Sci* 696:351–363.

Malaisse WJ, Courtois P, Scott FW. (2004) Insulin-dependent diabetes and gut dysfunction: The BB rat model. *Horm Metab Res* 36:585–594.

Malkani S, Nompleggi D, Hansen JW, et al. (1997) Dietary cow's milk protein does not alter the frequency of diabetes in the BB rat. *Diabetes* 46:1133–1140.
Markees TG, Serreze DV, Phillips NE, et al. (1999) NOD mice have a generalized defect in their response to transplantation tolerance induction. *Diabetes* 48:967–974.
Markholst H, Klaff LJ, Klöppel G, et al. (1990) Lack of systematically found insulin autoantibodies in spontaneously diabetic BB rats. *Diabetes* 39:720–727.
Martin A-M, Blankenhorn EP, Maxson MN, et al. (1999) Nonmajor histocompatibility complex-linked diabetes susceptibility loci on chromosomes 4 and 13 in a backcross of the DP BB/Wor rat to the WF rat. *Diabetes* 48:50–58.
Martin A-M, Maxson MN, Leif J, et al. (1999) Diabetes-prone and diabetes-resistant BB rats share a common major diabetes susceptibility locus, *iddm4*: Additional evidence for a "universal autoimmunity locus" on rat chromosome 4. *Diabetes* 48:2138–2144.
Mathieu C, Kuttler B, Waer M, et al. (1994) Spontaneous reestablishment of self-tolerance in BB/PFD rats. *Transplantation* 58:349–354.
McKeever U, Mordes JP, Greiner DL, et al. (1990) Adoptive transfer of autoimmune diabetes and thyroiditis to athymic rats. *Proc Natl Acad Sci USA* 87:7718–7722.
McKisic MD, Paturzo FX, Gaertner DJ, et al. (1995) A nonlethal rat parvovirus infection suppresses rat T lymphocyte effector functions. *J Immunol* 155:3979–3986.
Meddings JB, Jarand J, Urbanski SJ, et al. (1999) Increased gastrointestinal permeability is an early lesion in the spontaneously diabetic BB rat. *Am J Physiol* 276:G951–G957.
Mendez II, Chung YH, Jun HS, et al. (2004) Immunoregulatory role of nitric oxide in Kilham rat virus-induced autoimmune diabetes in DR-BB rats. *J Immunol* 173:1327–1335.
Michalkiewicz M, Michalkiewicz T, Ettinger RA, et al. (2004) Transgenic rescue demonstrates involvement of the *Ian5* gene in Tcell development in the rat. *Physiol Genom* 19:228–232.
Moralejo DH, Park HA, Speros SJ, et al. (2003) Genetic dissection of lymphopenia from autoimmunity by introgression of mutated *Ian5* gene onto the F344 rat. *J Autoimmunity* 21:315–324.
Mordes J, Flanagan J, Leif J, Greiner D, Kislauskis E, Rossini A, Blankenhorn E, Hillebrands JL, Guberski D. (2003) Autoimmune diabetes in the LEW.1WR1 rat after infection with Kilham rat virus (KRV) or rat cytomegalovirus (RCMV). *Diabetes* 52 (Suppl 1): A528.
Mordes JP, Bortell R, Doukas J, et al. (1996a) The BB/Wor rat and the balance hypothesis of autoimmunity. *Diabetes/Metab Rev* 2:103–109.
Mordes JP, Bortell R, Groen H, et al. (2001) Autoimmune diabetes mellitus in the BB rat. In *Animal models of diabetes: A primer*, ed. Sima AAF, Shafrir E. London: Harwood Academic Publishers, 1–41.
Mordes JP, Brown R, Szymanski I, Lundstrom R, Rossini AA. (1986) White blood cell (WBC) transfusion therapy in monozygotic twins discordant for insulin-dependent diabetes mellitus (IDDM). *Clin Res* 34:550A.
Mordes JP, Desemone J, Rossini AA. (1987a) The BB rat. *Diabetes/Metab Rev* 3:725–750.
Mordes JP, Gallina DL, Handler ES, et al. (1987b) Transfusions enriched for W3/25^{+} helper/inducer T lymphocytes prevent spontaneous diabetes in the BB/W rat. *Diabetologia* 30:22–26.
Mordes JP, Greiner DL, Rossini AA. (2000) Animal models of autoimmune diabetes mellitus. In *Diabetes mellitus. A fundamental and clinical text*, ed. LeRoith D, Taylor SI, Olefsky JM. New York: Lippincott Williams and Wilkins. 430–441.
Mordes JP, Guberski DL, Leif JH, et al. (2005) LEW.1WR1 rats develop autoimmune diabetes spontaneously and in response to environmental perturbation. *Diabetes* 54:2727–2733.

Mordes JP, Leif J, Novak S, et al. (2002) The *iddm4* locus segregates with diabetes susceptibility in congenic WF.*iddm4* rats. *Diabetes* 51:3254–3262.

Mordes JP, Schirf B, Roipko D, et al. (1996b) Oral insulin does not prevent insulin-dependent diabetes mellitus in BB rats. *Ann NY Acad Sci* 778:418–421.

Mordes JP, Serreze DV, Greiner DL, Rossini AA. (2004) Animal models of autoimmune diabetes mellitus. In *Diabetes mellitus. A fundamental and clinical text*, ed. LeRoith D, Taylor SI, Olefsky JM. New York: Lippincott Williams and Wilkins, 591–610.

Mori Y, Suko M, Okudaira H, et al. (1986) Preventive effects of cyclosporin on diabetes in NOD mice. *Diabetologia* 29:244–247.

Munakata Y, Kodera T, Saito T, et al. (2005) Rheumatoid arthritis, T1D, and Graves' disease after acute parvovirus B19 infection. *Lancet* 366:780.

Mundinger TO, Mei Q, Figlewicz DP, et al. (2003) Impaired glucagon response to sympathetic nerve stimulation in the BB diabetic rat: effect of early sympathetic islet neuropathy. *Am J Physiol* 285:E1047–E1054.

Myers MA, Laks MR, Feeney SJ, et al. (1998) Antibodies to ICA512/IA-2 in rodent models of IDDM. *J Autoimmunity* 11:265–272.

Naji A, Silvers WK, Barker CF. (1983) Bone marrow transplantation in adult diabetes-prone rats. *Surg Forum* 34:374–376.

Najjar SM, Broyart JP, Hampp LT, et al. (2001) Intestinal aminooligopeptidase in diabetic BioBreed rat: Altered posttranslational processing and trafficking. *Am J Physiol* 280:G104–G112.

Nakano K, Mordes JP, Handler ES, et al. (1988) Role of host immune system in BB/Wor rat: Predisposition to diabetes resides in bone marrow. *Diabetes* 37:520–525.

Nepom BS, Nepom GT, Coleman M, et al. (1996) Critical contribution of β chain residue 57 in peptide binding ability of both HLA-DR and -DQ molecules. *Proc Natl Acad Sci USA* 93:7202–7206.

Neu J, Reverte CM, Mackey AD, et al. (2005) Changes in intestinal morphology and permeability in the biobreeding rat before the onset of T1D. *J Pediat Gastroenterol Nutr* 40:589–595.

Nicoletti F, Zaccone P, Di Marco R, et al. (1997) Prevention of spontaneous autoimmune diabetes in diabetes-prone BB rats by prophylactic treatment with antirat interferon-gamma antibody. *Endocrinology* 138:281–288.

Nicoletti F, Zaccone P, Di Marco R, et al. (1998) Paradoxical antidiabetogenic effect of gamma-interferon in DP-BB rats. *Diabetes* 47:32–38.

Norris JM, Barriga K, Klingensmith G, et al. (2003) Timing of initial cereal exposure in infancy and risk of islet autoimmunity. *JAMA* 290:1713–1720.

Olson DE, Paveglio SA, Huey PU, et al. (2003) Glucose-responsive hepatic insulin gene therapy of spontaneously diabetic BB/Wor rats. *Hum Gene Ther* 14:1401–1413.

Pandarpurkar M, Wilson-Fritch L, Corvera S, et al. (2003) *Ian4* is required for mitochondrial integrity and T cell survival. *Proc Natl Acad Sci USA* 100:10382–10387.

Parving HH, Tarnow L, Nielsen FS, et al. (1999) Cyclosporine nephrotoxicity in type 1 diabetic patients. A 7-year follow-up study. *Diabetes Care* 22:478–483.

Payne F, Smyth DJ, Pask R, et al. (2004) Haplotype tag single nucleotide polymorphism analysis of the human orthologues of the rat T1D genes *Ian4* (*Lyp/Iddm1*) and *Cblb*. *Diabetes* 53:505–509.

Payne F, Smyth DJ, Pask R, et al. (2005) No evidence for association of the TATA-box binding protein glutamine repeat sequence or the flanking chromosome 6q27 region with T1D. *Biochem Biophys Res Commun* 331:435–441.

Pearson T, Markees TG, Serreze DV, et al. (2003) Genetic disassociation of autoimmunity and resistance to costimulation blockade-induced transplantation tolerance in nonobese diabetic mice. *J Immunol* 171:185–195.

Pearson T, Markees TG, Wicker LS, et al. (2003) NOD congenic mice genetically protected from autoimmune diabetes remain resistant to transplantation tolerance induction. *Diabetes* 52:321–326.

Penhale WJ, Stumbles PA, Huxtable CR, et al. (1990) Induction of diabetes in PVG/c strain rats by manipulation of the immune system. *Autoimmunity* 7:169–179.

Petersen JS, MacKay P, Plesner A, et al. (1997) Treatment with GAD65 or BSA does not protect against diabetes in BB rats. *Autoimmunity* 25:129–138.

Philips JC, Scheen AJ. (2002) Infocongress. Study of the prevention of T1D with nicotinamide: Positive lessons of a negative clinical trial (ENDIT). *Rev Med Liege* 57:672–675.

Pieper GM, Henderson JD, Jr., Roza AM, et al. (2004) A dithiocarbamate analogue decreases intraislet cell infiltration and the incidence of diabetes mellitus in the genetic diabetes-prone BB rat. *Pancreas* 28:E16–E25.

Pietropaolo M, Trucco M. (1996) Viral elements in autoimmunity of type I diabetes. *Trends Endocrinol Metab* 7:139–144.

Podlasek CA, Zelner DJ, Harris JD, et al. (2003) Altered sonic hedgehog signaling is associated with morphological abnormalities in the penis of the BB/WOR diabetic rat. *Biol Reprod* 69:816–827.

Popovic J, Kover KL, Moore WV. (2004) The effect of immunomodulators on prevention of autoimmune diabetes is stage dependent: FTY720 prevents diabetes at three different stages in the diabetes-resistant BioBreeding rat. *Pediat Diabetes* 5:3–9.

Posselt AM, Barker CF, Friedman AL, et al. (1992) Prevention of autoimmune diabetes in the BB rat by intrathymic islet transplantation at birth. *Science* 256:1321–1324.

Poussier P, Nakhooda AF, Falk JF, et al. (1982) Lymphopenia and abnormal lymphocyte subsets in the "BB" rat: Relationship to the diabetic syndrome. *Endocrinology* 110:1825–1827.

Poussier P, Ning T, Murphy T, et al. (2005) Impaired post-thymic development of regulatory $CD4^+ 25^+$ T cells contributes to diabetes pathogenesis in BB rats. *J Immunol* 174:4081–4089.

Prins J-B, Herberg L, Den Bieman M, Van Zutphen BFM. (1991) Genetic characterization and interrelationship of inbred lines of diabetes-prone and not diabetes-prone BB rats. In *Lessons from animal diabetes*, ed. Shafrir E. London: Smith–Gordon, 19–24.

Qin H-Y, Sadelain MWJ, Hitchon C, et al. (1993) Complete Freund's adjuvant-induced T cells prevent the development and adoptive transfer of diabetes in nonobese diabetic mice. *J Immunol* 150:2072–2080.

Rajatanavin R, Appel MC, Reinhardt W, et al. (1991) Variable prevalence of lymphocytic thyroiditis among diabetes-prone sublines of BB/Wor rats. *Endocrinology* 128:153–157.

Ramanathan S, Bihoreau MT, Paterson AD, et al. (2002) Thymectomy and radiation-induced T1D in nonlymphopenic BB rats. *Diabetes* 51:2975–2981.

Ramanathan S, Marandi L, Poussier P. (2002) Evidence for the extrathymic origin of intestinal TCRgammaδ^+ T cells in normal rats and for an impairment of this differentiation pathway in BB rats. *J Immunol* 168:2182–2187.

Ramanathan S, Norwich K, Poussier P. (1998) Antigen activation rescues recent thymic emigrants from programmed cell death in the BB rat. *J Immunol* 160:5757–5764.

Ramanathan S, Poussier P. (1999) T cell reconstitution of BB/W rats after the initiation of insulitis precipitates the onset of diabetes. *J Immunol* 162:5134–5142.

Ramanathan S, Poussier P. (2001) BB rat *lyp* mutation and T1D. *Immunol Rev* 184:161–171.

Rasmussen SB, Sorensen TS, Hansen JB, et al. (2000) Functional rest through intensive treatment with insulin and potassium channel openers preserves residual β-cell function and mass in acutely diabetic BB rats. *Horm Metab Res* 32:294–300.

Redondo MJ, Yu L, Hawa M, et al. (2001) Heterogeneity of type I diabetes: Analysis of monozygotic twins in Great Britain and the United States. *Diabetologia* 44:354–362.

Reimers JI, Andersen HU, Mauricio D, et al. (1996) Strain-dependent differences in sensitivity of rat β-cells to interleukin 1β *in vitro* and *in vivo*—Association with islet nitric oxide synthesis. *Diabetes* 45:771–778.

Rodier M, Ribstein J, Parer-Richard C, et al. (1991) Renal changes associated with cyclosporine in recent type I diabetes mellitus. *Hypertension* 18:334–340.

Roep BO, Atkinson M. (2004) Animal models have little to teach us about T1D: 1. In support of this proposal. *Diabetologia* 47:1650–1656.

Rossini AA. (2004) From beast to bedside: A commentary. *Diabetologia* 47:1647–1649.

Rossini AA, Mordes JP, Pelletier AM, et al. (1983) Transfusions of whole blood prevent spontaneous diabetes in the BB/W rat. *Science* 219:975–977.

Sadelain MW, Qin HY, Sumoski W, et al. (1990) Prevention of diabetes in the BB rat by early immunotherapy using Freund's adjuvant. *J Autoimmunity* 3:671–680.

Satoh J, Seino H, Shintani S, et al. (1990) Inhibition of T1D in BB rats with recombinant human tumor necrosis factor-α. *J Immunol* 145:1395–1399.

Satoh J, Shintani S, Oya K, et al. (1988) Treatment with streptococcal preparation (OK-432) suppresses anti-islet autoimmunity and prevents diabetes in BB rats. *Diabetes* 37:1188–1194.

Schmidt RE, Dorsey DA, Beaudet LN, et al. (2004) Experimental rat models of types 1 and 2 diabetes differ in sympathetic neuroaxonal dystrophy. *J Neuropath Exp Neurol* 63:450–460.

Schröder D, Ratke M, Bauer UC, et al. (2002) Prophylactic insulin treatment of DP BB/OK rats by application of a sustained release insulin implant. *Autoimmunity* 35:143–153.

Scott FW, Cloutier HE, Kleemann R, et al. (1997) Potential mechanisms by which certain foods promote or inhibit the development of spontaneous diabetes in BB rats. Dose, timing, early effect on islet area, switch in infiltrate from Th1 to Th2 cells. *Diabetes* 46:589–598.

Scott FW, Marliss EB. (1991) Conference summary: Diet as an environmental factor in development of insulin-dependent diabetes. *Can J Physiol Pharmacol* 69:311–319.

Scott FW, Rowsell P, Wang GS, et al. (2002) Oral exposure to diabetes-promoting food or immunomodulators in neonates alters gut cytokines and diabetes. *Diabetes* 51:73–78.

Serreze DV, Leiter EH. (2001) Genes and pathways underlying autoimmune diabetes in NOD mice. In *Molecular pathology of insulin-dependent diabetes mellitus*, ed. Von Herrath MG. Geneva: Karger Press, 3:1–67.

Sigfrid LA, Cunningham JM, Beeharry N, et al. (2004) Antioxidant enzyme activity and mRNA expression in the islets of Langerhans from the BB/S rat model of T1D and an insulin-producing cell line. *J Mol Med* 82:325–335.

Simons PJ, Delemarre FG, Jeucken PH, et al. (1998) Preautoimmune thyroid abnormalities in the biobreeding diabetes-prone (BB-DP) rat: A possible relation with the intrathyroid accumulation of dendritic cells and the initiation of the thyroid autoimmune response. *J Endocrinol* 157:43–51.

Simonson W, Ramanathan S, Bieg S, et al. (2002) Protein-free diets do not protect high-incidence diabetes-prone BioBreeding rats from diabetes. *Metabolism* 51:569–574.

Skyler JS, Brown D, Chase HP, et al. (2002) Effects of insulin in relatives of patients with T1D mellitus. *N Engl J Med* 346:1685–1691.

Skyler JS, Krischer JP, Wolfsdorf J, et al. (2005) Effects of oral insulin in relatives of patients with type 1 diabetes: The Diabetes Prevention Trial—Type 1. *Diab Care* 28:1068–1076.

Sobel DO, Azumi N, Creswell K, et al. (1995) The role of NK cell activity in the pathogenesis of poly I:C accelerated and spontaneous diabetes in the diabetes-prone BB rat. *J Autoimmunity* 8:843–857.

Sobel DO, Creswell K, Yoon JW, et al. (1998) Alpha interferon administration paradoxically inhibits the development of diabetes in BB rats. *Life Sci* 62:1293–1302.

Sobel DO, Ewel CH, Zeligs B, et al. (1994) Poly I:C induction of α-interferon in the diabetes-prone BB and normal Wistar rats: Dose–response relationships. *Diabetes* 43:518–522.

Sobel DO, Goyal D, Ahvazi B, et al. (1998) Low-dose poly I:C prevents diabetes in the diabetes prone BB rat. *J Autoimmunity* 11:343–352.

Sobel DO, Newsome J. (1997) Gamma interferon prevents diabetes in the BB rat. *Clin Diag Lab Immunol* 4:764–768.

Sobel DO, Newsome J, Ewel CH, et al. (1992) Poly I:C induces development of diabetes mellitus in BB rat. *Diabetes* 41:515–520.

Sommandas V, Rutledge EA, Van Yserloo B, et al. (2005a) Aberrancies in the differentiation and maturation of dendritic cells from bone-marrow precursors are linked to various genes on chromosome 4 and other chromosomes of the BB-DP rat. *J Autoimmunity* 25:1–12.

Sommandas V, Rutledge EA, Van Yserloo B, et al. (2005b) Defects in differentiation of bone-marrow derived dendritic cells of the BB rat are partly associated with IDDM2 (the *lyp* gene) and partly associated with other genes in the BB rat background. *J Autoimmunity* 25:46–56.

Song HY, Abad MM, Mahoney CP, et al. (1999) Human insulin B chain but not A chain decreases the rate of diabetes in BB rats. *Diabetes Res Clin Pract* 46:109–114.

Sparre T, Christensen UB, Gotfredsen CF, et al. (2004) Changes in expression of IL-1β influenced proteins in transplanted islets during development of diabetes in diabetes-prone BB rats. *Diabetologia* 47:892–908.

Stevens MJ, Zhang W, Li F, et al. (2004) C-peptide corrects endoneurial blood flow but not oxidative stress in type 1 BB/Wor rats. *Am J Physiol* 287:E497–E505.

Stumbles PA and Penhale WJ. (1993) IDDM in rats induced by thymectomy and irradiation. *Diabetes* 42:571–578.

Tafuri A, Bowers WE, Handler ES, et al. (1993) High stimulatory activity of dendritic cells from diabetes-prone BioBreeding/Worcester rats exposed to macrophage-derived factors. *J Clin Invest* 91:2040–2048.

Takahashi K, Satoh J, Seino H, et al. (1993) Prevention of type I diabetes with lymphotoxin in BB rats. *Clin Immunol Immunopathol* 69:318–323.

Thomas VA, Woda BA, Handler ES, et al. (1991) Altered expression of diabetes in BB/Wor rats by exposure to viral pathogens. *Diabetes* 40:255–258.

Tirabassi RS, Flanagan JF, Wu T, et al. (2004) The BBZDR/Wor rat model for investigating the complications of type 2 diabetes mellitus. *ILAR J* 45:292–302.

Tirabassi RS, Guberski DL, Leif JH, Winans DA, Mordes JP. (2005) Maternal immunization protects weanling LEW.1WR1 rats from cytomegalovirus-induced diabetes. *Diabetes* 54 (Suppl 1):A94.

Todd DJ, Forsberg EM, Greiner DL, et al. (2004) Deficiencies in gut NK cell number and function precede diabetes onset in BB rats. *J Immunol* 172:5356–5362.

Todd JA. (1999) From genome to aetiology in a multifactorial disease, T1D. *BioEssays* 21:164–174.

Tori M, Ito T, Yumiba T, et al. (1999) Significant role of intragraft lymphoid tissues in preventing insulin-dependent diabetes mellitus recurrence in whole pancreaticoduodenal transplantation. *Microsurgery* 19:338–343.

Trudeau JD, Dutz JP, Arany E, et al. (2000) Neonatal β-cell apoptosis—A trigger for autoimmune diabetes? *Diabetes* 49:1–7.

Tulic MK, Fiset PO, Manoukian JJ, et al. (2004) Role of toll-like receptor 4 in protection by bacterial lipopolysaccharide in the nasal mucosa of atopic children but not adults. *Lancet* 363:1689–1697.

Van den Brandt J, Kovács P, Klöting I. (1999) Congenic diabetes-prone BB.Sa and BB.Xs rats differ from their progenitor strain BB/OK in frequency and severity of insulin-dependent diabetes mellitus. *Biochem Biophys Res Commun* 263:843–847.

van der Werf N, Hillebrands JL, Klatter FA, et al. (2003) Cytomegalovirus infection modulates cellular immunity in an experimental model for autoimmune diabetes. *Clin Dev Immunol* 10:153–160.

Villanueva DS, Poirier P, Standley PR, et al. (2003) Prevention of ischemic heart failure by exercise in spontaneously diabetic BB Wor rats subjected to insulin withdrawal. *Metabolism* 52:791–797.

Visalli N, Sebastiani L, Adorisio E, et al. (2003) Environmental risk factors for T1D in Rome and province. *Arch Dis Child* 88:695–698.

Visser J, Groen H, Klatter F, et al. (2002) Timing of pentoxifylline treatment determines its protective effect on diabetes development in the BioBreeding rat. *Eur J Pharmacol* 445:133–140.

Visser J, Groen H, Klatter F, et al. (2003) The diabetes-prone BB rat model of IDDM shows duration of breastfeeding to influence T1D development later in life. *Diabetologia* 46:1711–1713.

Visser J, Klatter F, Vis L, et al. (2003) Long-term prophylactic insulin treatment can prevent spontaneous diabetes and thyroiditis development in the diabetes-prone BioBreeding rat, while short-term treatment is ineffective. *Eur J Endocrinol* 149:223–229.

Von Herrath MG, Fujinami RS, Whitton JL. (2003) Microorganisms and autoimmunity: Making the barren field fertile? *Nat Rev Microbiol* 1:151–157.

Wang GS, Gruber H, Smyth P, et al. (2000) Hydrolyzed casein diet protects BB rats from developing diabetes by promoting islet neogenesis. *J Autoimmunity* 15:407–416.

Wang GS, Rosenberg L, Scott FW. (2005) Tubular complexes as a source for islet neogenesis in the pancreas of diabetes-prone BB rats. *Lab Invest* 85:675–688.

Watson WC, Thompson JP, Terato K, et al. (1990) Human HLA-DRβ gene hypervariable region homology in the biobreeding BB rat: Selection of the diabetic-resistant subline as a rheumatoid arthritis research tool to characterize the immunopathologic response to human type II collagen. *J Exp Med* 172:1331–1339.

Wedekind D, Weiss H, Jörns A, et al. (2005) Effects of polyinosinic-polycytidylic acid and adoptive transfer of immune cells in the Lew.1AR1-iddm rat and in its coisogenic LEW.1AR1 background strain. *Autoimmunity* 38:265–275.

Weiss H, Bleich A, Hedrich HJ, et al. (2005) Genetic analysis of the LEW.1AR1-*iddm* rat: An animal model for spontaneous diabetes mellitus. *Mamm Genome* 16:432–441.

Whalen BJ, Mordes JP, Rossini AA. (1996) The BB rat as a model of human insulin-dependent diabetes mellitus. In *Animal models for autoimmune and inflammatory disease*, ed. Coligan JE, Kruisbeek AM, Margulies DH, Shevach EM, Strober W. New York: John Wiley & Sons, 15.3.1–15.3.15.

Wilkin T, Kiesel U, Diaz J-L, et al. (1986) Autoantibodies to insulin as serum markers for autoimmune insulitis. *Diabetes Res* 3:173–174.

Woda BA, Like AA, Padden C, et al. (1986) Deficiency of phenotypic cytotoxic-suppressor T lymphocytes in the BB/W rat. *J Immunol* 136:856–859.

Yagi H, Matsumoto M, Suzuki S, et al. (1991) Possible mechanism of the preventive effect of BCG against diabetes mellitus in NOD mouse. I. Generation of suppressor macrophages in spleen cells of BCG-vaccinated mice. *Cell Immunol* 138:130–141.

Yale J-F, Grose M, Marliss EB. (1985) Time course of the lymphopenia in BB rats. Relation to the onset of diabetes. *Diabetes* 34:955–959.

Ye XJ, Tang B, Ma Z, et al. (2004) The roles of interleukin-18 in collagen-induced arthritis in the BB rat. *Clin Exp Immunol* 136:440–447.

Yokoi N. (2005) Identification of a major gene responsible for T1D in the Komeda diabetes-prone rat. *Exp Anim* 54:111–115.

Yokoi N, Kanazawa M, Kitada K, et al. (1997) A non-MHC locus essential for autoimmune type I diabetes in the Komeda diabetes-prone rat. *J Clin Invest* 100:2015–2021.

Yokoi N, Komeda K, Wang HY, et al. (2002) *Cblb* is a major susceptibility gene for rat T1D mellitus. *Nat Genet* 31:391–394.

Yoon J-W, Jun H-S. (2004) Role of viruses in the pathogenesis of type 1 diabetes mellitus. In *Diabetes mellitus. A fundamental and clinical text*, ed. LeRoith D, Taylor SI, Olefsky JM. New York: Lippincott Williams & Wilkins, 575–590.

Zenz R, Eferl R, Kenner L, et al. (2005) Psoriasis-like skin disease and arthritis caused by inducible epidermal deletion of Jun proteins. *Nature* 437:369–375.

Zhang ZJ, Davidson L, Eisenbarth G, et al. (1991) Suppression of diabetes in nonobese diabetic mice by oral administration of porcine insulin. *Proc Natl Acad Sci USA* 88:10252–10256.

Ziegler M, Schlosser M, Hamann J, et al. (1994) Autoantibodies to glutamate decarboxylase detected in diabetes-prone BB/OK rats do not distinguish onset of diabetes. *Exp Clin Endocrinol* 102:98–103.

Zipris D, Greiner DL, Malkani S, et al. (1996) Cytokine gene expression in islets and thyroids of BB rats: Interferon gamma and IL-12 p40 mRNA increase with age in both diabetic and insulin-treated nondiabetic BB rats. *J Immunol* 156:1315–1321.

Zipris D, Hillebrands JL, Welsh RM, et al. (2003) Infections that induce autoimmune diabetes in BBDR rats modulate $CD4^{+}CD25^{+}$ T cell populations. *J Immunol* 170:3592–3602.

Zipris D, Karnieli E. (2002) A single treatment with IL-4 via retrovirally transduced lymphocytes partially protects against diabetes in BioBreeding (BB) rats. *JOP* 3:76–82.

Zipris D, Lien E, Xie JX, et al. (2005) TLR activation synergizes with Kilham rat virus infection to induce diabetes in BBDR rats. *J Immunol* 174:131–142.

2 The NOD Mouse and Its Related Strains

Hidenori Taniguchi, Susumu Makino, and Hiroshi Ikegami

CONTENTS

INTRODUCTION

Autoimmune type 1 diabetes (T1D) is caused by autoimmune destruction of insulin-producing beta-cells of the pancreas in genetically susceptible individuals. Understanding of the genetics and autoimmune mechanisms of the disease has been greatly facilitated by an animal model, the nonobese diabetic (NOD) mouse, which spontaneously develops T1D. The origins and characteristics of this strain and a closely related strain, the nonobese nondiabetic (NON) mouse, are summarized in this chapter.

ESTABLISHMENT OF NOD AND ITS RELATED STRAINS: ORIGIN OF THE NOD MOUSE

The NOD mouse was established as an inbred strain of mice with spontaneous development of autoimmune T1D by Dr. Susumu Makino in Shionogi Laboratory (Makino et al. 1980). Discovery of a mouse with T1D, which eventually led to the development of the NOD mouse strain, was rather paradoxical. The origin of the NOD traces back to a mouse with cataracts in Jcl:ICR mice in 1966; an inbred strain of mice with cataracts and small eyes, the CTS mouse, was established by brother–sister mating (Ohtori 1968). The designation of "CTS" was originally for "cataract and small eyes," but it was subsequently redesignated "cataract Shionogi." The presence of cataract in CTS mice is an autosomal dominant trait, whereas small eyes are an autosomal recessive trait. Thus, mice homozygous for the CTS allele at cataract locus (*Cs*) show cataracts and small eyes, whereas heterozygous mice show cataracts, but not small eyes. At the fourth generation of inbreeding of CTS, a control line with normal eyes was separated and an inbred NCT (noncataract) strain was established (figure 2.1) (Makino et al. 1988).

At the sixth generation, selective breeding for euglycemic and hyperglycemic mice was performed with an expectation of obtaining hyperglycemic mice, prompted by the fact that the cataract is one of the common complications in human diabetes. At this stage, it was not clear whether mating was strictly brother–sister mating or only selective breeding was performed for elevated or normal glucose levels. After selective breeding for hyperglycemia and euglycemia for about 10 generations, these two sublines were transferred from the toxicology department to the laboratory animal department because of limitation of space for animals in the former department.

Dr. Makino at the Laboratory Animal Department took over this project and strict brother–sister mating was started. Two sister strains, one with slightly higher fasting blood glucose levels (~150 mg/dL) and the other with normal fasting blood glucose levels (approximately 100 mg/dL) and both free of cataracts, were obtained after selective breeding for 13 generations (Makino et al. 1980). At the 20th generation in selective breeding, a mouse with polyuria, polydipsia, and weight loss was found

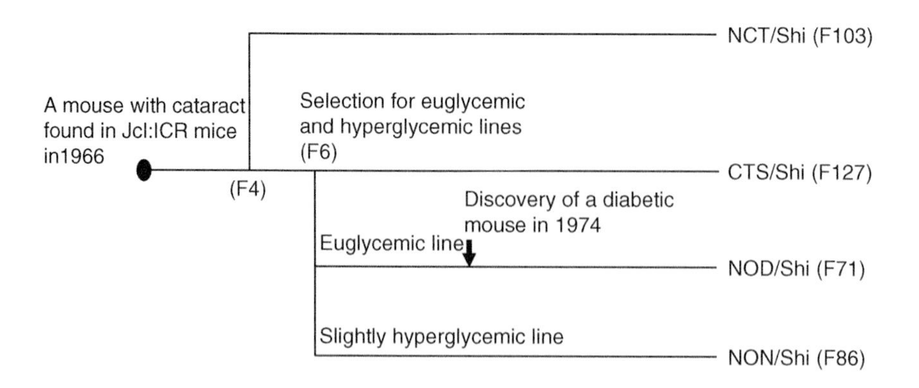

FIG. 2.1 Genealogy of NOD and its related strains.

in the line with normal, but not increased, fasting blood glucose levels. This is a rather paradoxical and unexpected observation in that, contrary to the initial expectation of obtaining diabetic mice in the hyperglycemic line, a diabetic mouse was detected in the normoglycemic line.

Inbreeding was continued with this mouse to establish an inbred strain with spontaneous development of diabetes, which culminated in a strain currently known as the NOD mouse. Conversely, a slightly hyperglycemic subline, subsequently designated the NON mouse, never developed overt diabetes. As expected from the history described earlier, glucose tolerance in NON mice is not completely normal, but slightly impaired as compared with other control laboratory strains. However, selection for hyperglycemia in this line was discontinued after discovery of a diabetic mouse subsequently established as NOD. NON mice, therefore, should not be regarded as a model for type 2 diabetes or glucose intolerance because the phenotype for glucose intolerance is not fixed in this strain.

MAINTENANCE OF THE NOD MOUSE

NOD mice are derived from Jcl:ICR mice, which are a closed colony derived from so-called "Swiss" mice. The name ICR denotes the Institute of Cancer Research in Philadelphia, Pennsylvania, where a closed colony of Swiss mice was maintained by random breeding. These mice were subsequently transferred to commercial breeders and are now commercially available as Jcl:ICR (CLEA, Japan) and CD-1:Crj (Charles River, Japan). Jcl:ICR (or CD-1) are known to breed well by producing large litters. It was therefore fortunate that NOD mice were derived from Jcl:ICR mice in terms of maintenance of the strain. In fact, NOD mice produced an average of 9.1 offspring per litter as compared with 6.6 offspring per litter in C57BL/6J mice in Shionogi Aburahi Laboratories (table 2.1). Therefore, it is relatively easy to maintain a colony.

TABLE 2.1
Reproductive Efficiency of NOD/Shi, NON/Shi, and C57BL/6J Strains

Strain		Birth Order					Reproductive Efficiency*
		I	II	III	IV	V	
C57BL/6J	No. of female mice	25	25	24	22	18	
	Average litter size	6.3	7	6.5	6.4	6.6	26.1
	Weaning rate (%)	82.6	91.9	87.5	89.6	98.6	
NOD/Shi	No. of female mice	43	22	5	1	0	
	Average litter size	8.4	10.4	11	2	—	8.3
	Weaning rate (%)	70.7	81.7	27	0	—	
NON/Shi	No. of female mice	37	31	28	15	7	
	Average litter size	6.5	8.4	8.2	5.9	6.1	16.7
	Weaning rate (%)	85.3	88.7	82.9	70	56.8	

* Total number of weanlings obtained from one female mouse.

TABLE 2.2
Commercial Breeders for NOD Mice

Strain Name	Breeder	Address
NOD/ShiJic	Clea Japan, Inc.	1-2-7, Higashiyama, Meguroku, Tokyo 153-8533, Japan Tel: +81 (0) 3 5704 7011 Fax: +81(0) 3 3791 2859 URL: www.clea-japan.com/index.html
NOD/Bom	Bomholtgard Breeding and Research Center Ltd	Bomholtvej 10, PO BOX 39, DK8680 Ry, Denmark Tel: +45 86 84 1211 Fax: +45 86 84 1699
NOD/RijHsd	Harlan UK Ltd	Shaw's Farm, Blackthorn, Bicester, Oxon OX25 1TP, England Tel: +44 (0) 1869 243 241 Fax: +44 (0) 1869 246 759 URL: www.harlaneurope.com/main.htm
NOD/OrlIco	IFFA CREDO	BP 0109-69592, l'Arbresle Cedex, France Tel: +33 74 01 6969 Fax: +33 74 01 6999
NOD/LtSanIbm	RCC Biotechnology and Breeding Division	BRL Biological Research Laboratories Ltd, Wolferstrasse 4, CH-4414 Fullinsdorf, Switzerland Tel: +41 61 906 4242 Fax: +41 61 901 2565 URL: www.rcc.ch/
NOD/MrkTac	Taconic	One Hudson City Center, Hudson, NY 12534, USA Tel: +1 518-697-3900 Fax: +1 518-697-3910 URL: www.taconic.com/

Most female NOD mice, however, raise at most two litters due to the development of diabetes. Diabetic females show very low reproductive capacity. Although most of the mothers before the onset of diabetes can nurse their pups well, those with overt diabetes cannot complete weaning of their pups. As a consequence, the total efficiency of breeding is much lower (about 30%) than that in usual laboratory strains, such as C57BL/6 mice (table 2.1). If female NOD mice in a colony continuously breed well (delivering more than three litters), then there is a possibility that the incidence of diabetes in the colony will go down due to selection of mice less prone to diabetes as mating pairs, as will be discussed in the next section.

The NOD mouse is at present commercially available from the breeders listed in table 2.2. At the time of writing, a NOD mouse of 4 to 6 weeks old is supplied by Clea Japan at a price of approximately 6,200 yen ($52). The cost of maintenance of nondiabetic NOD mice is similar to that of usual laboratory strains, except for the cost for test strips for screening of urinary glucose. For diabetic NOD mice, the cost is higher because frequent changes of cage bedding (almost every day) are required due to polyuria in diabetic mice. Each regular cage usually houses three to five mice; only one diabetic NOD mouse should be housed because the cage would become very dirty with multiple diabetic mice due to polyuria, and such an environment would lead to an unhealthy condition and predispose to infectious disease. Because of this and low reproductive efficiency, approximately three times more space is required to maintain a NOD colony than that for usual laboratory strains.

CHARACTERISTICS OF THE NOD MOUSE

The NOD mouse spontaneously develops T1D prior to the development of diabetes; infiltration of mononuclear cells into the pancreatic islets (insulitis) is observed. Insulitis is not observed before three weeks of age, but appears spontaneously at around four weeks of age. The frequency of insulitis reaches 70–90% by 3 weeks of age, and almost 100% of mice of both sexes develop insulitis by 20 weeks of age (Makino, Hayashi, et al. 1985). Mononuclear cells infiltrating the islets are mostly T cells (CD4+ and CD8+), but B cells, dendritic cells, and macrophages are also observed. Despite massive infiltration of mononuclear cells into the islets, beta-cells remain intact until 12–15 weeks of age, when destruction of beta-cells becomes aggressive and overt diabetes develops.

Although almost all NOD mice of both sexes develop insulitis, only some of them develop overt diabetes, suggesting that several factors modify the process from insulitis to beta-cell destruction and the development of overt diabetes. After the onset of overt diabetes, marked polyuria and polydipsia develop, and mice lose weight and die within one to two months unless treated with a daily injection of insulin (Makino et al. 1980), as in the case of human T1D.

ENVIRONMENTAL EFFECT ON INCIDENCE OF DIABETES

The cumulative incidence of T1D in the original colonies at Shionogi Aburahi Laboratories was approximately 80% in females and less than 20% in males at 30 weeks of age (Makino et al. 1980). The incidence in males subsequently increased and currently the cumulative incidence of T1D at 30 weeks of age is 90% in females and 50% in males. A sex-related difference in the frequency of diabetes has been observed in most NOD colonies. Orchiectomy in males has been reported to increase the frequency of diabetes, while oophorectomy in females has been reported to decrease the incidence of diabetes (Makino et al. 1981). Prevention of diabetes in female NOD mice by treatment with androgen has also been reported (Fox 1992), suggesting the modifying effect of sex steroids on the development of diabetes. The maternal environment has also been suggested to be important in the development of diabetes. Kagohashi and colleagues suggest that the maternal environment modulates the immune response, leading to the early onset of insulitis. Changes in immune response brought about by maternal factors during the perinatal period might further affect progression to overt diabetes (Kagohashi, Udagawa, Abiru, et al. 2005; Kagohashi, Udagawa, Moriyama, et al. 2005).

Although the NOD mouse is an inbred strain with more than 70 generations of brother–sister mating, considerable variation in the incidence of diabetes has been reported among different colonies (Leiter et al. 1990; Pozzilli et al. 1993). Diet (Coleman et al. 1990; Elliott et al. 1988; Giulietti et al. 2004; Zella and Deluca

2003; Zella et al. 2003), room temperature (Williams et al. 1990), and pathogens (Ohsugi et al. 1994) have been suggested as factors influencing the incidence of diabetes, but the extract reason is still unknown. The following are recommendations to maintain a constant incidence of diabetes in the NOD mouse:

The colony should be raised under specific pathogen-free (SPF) conditions because a lower incidence of diabetes has been reported in NOD mice in conventional conditions (Ohsugi, Kurosawa 1994) and in those infected with viruses (Oldstone 1988; Wilberz et al. 1991).

The same diet should be maintained because changes in the incidence of diabetes have been reported in NOD mice fed a different diet.

It was previously reported that 100% of female NOD mice maintained in germ-free conditions developed T1D (Suzuki et al. 1987). This observation, however, was based on a relatively small number of animals (N = 15 for germ-free conditions). In our experience with a larger number of animals at Shionogi Aburahi Laboratories, no such increase in the incidence of diabetes was observed under germ-free conditions sufficient for the maintenance of NOD mouse colonies (table 2.3).

Even in the same colony maintained under the conditions described previously, the incidence of diabetes readily goes down unless careful breeding is performed. Although the exact reason for the variation in the incidence of T1D in genetically homogenous NOD mice is unknown, mating should be performed between NOD mice—both or at least one of which had developed or subsequently developed T1D—in order to maintain a high incidence of diabetes. In practice, since the reproductive capacity of diabetic NOD female is low, the development of diabetes in mice used for mating should be followed up and only pups from parents or a parent that subsequently developed diabetes should be used for breeding of the next generation. Deviation of the phenotypes by random mating is observed not only in NOD mice, but also in other inbred animal models with polygenic, multifactorial inheritance.

It is therefore recommended that the incidence of diabetes in each NOD colony should be monitored and used as a reference incidence instead of referring to the incidence of diabetes in other colonies reported in the literature. This is particularly important in experiments where the incidence of diabetes is used as the outcome. In addition, the incidence and degree of insulitis should also be monitored and used as a phenotype. Even in colonies with a low incidence of diabetes, insulitis is similarly observed as in the original NOD colony.

TABLE 2.3
Cumulative Incidence of Diabetes at 30 Weeks of Age in NOD Mice under SPF and GF Conditions

Conditions	Female	Male
Germ free (GF)	70.2% (47)	26.8% (41)
Specific pathogen free (SPF)	81.4% (43)	13.3% (45)

GENETICS OF TYPE 1 DIABETES: THE NOD MOUSE AS A MODEL FOR MULTIFACTORIAL DISEASE

Inheritance of T1D in NOD mice is multifactorial, as in the case of human T1D, with genetic and environmental factors contributing to disease development. The contribution of environmental factors is evident from the fact that not all NOD mice develop T1D despite their identical genetic background. The variation in the incidence of diabetes among colonies or even within the same colony as described earlier provides other evidence for an environmental effect on the development of diabetes. The identity of such environmental factors is not fully characterized; however, environmental factors are thought to alter autoimmune responses to beta-cell antigens and thereby to modulate the development of diabetes, with some reports on viral infection (Oldstone 1988; Wetzel et al. 2006; Wilberz et al. 1991), diet (Coleman et al. 1990; Elliott et al. 1988; Giulietti et al. 2004; Zella and Deluca 2003; Zella et al. 2003), and room temperature (Williams et al. 1990). Identification and elucidation of the role of environmental factors would be beneficial for the prevention of T1D in humans (Todd 1991).

On the other hand, the contribution of multiple genes on different chromosomes to disease susceptibility has been directly demonstrated in NOD mice (Ghosh et al. 1993; Hattori et al. 1986; Prochazka et al. 1989; Todd et al. 1991; Wicker et al. 1987). This has greatly contributed to understanding of the genetics of T1D as well as that of multifactorial disease in humans in general. The first genome scan of NOD mice was performed about 15 years ago and demonstrated that multiple recessive loci, termed insulin-dependent diabetes or *Idd*) loci, were linked to diabetes resistance or susceptibility. To date, at least 20 loci have been mapped to the mouse genome and are thought to play a role in this complex, multigenic process (table 2.4) (Hall et al. 2003; Rogner et al. 2001).

CONGENIC STRAINS

The contribution of multiple genes to disease predisposition in the NOD mouse has been clearly demonstrated using congenic strategy. Congenic strains are strains differing at a locus or a chromosomal segment encompassing the locus, with all other background genes identical (figure 2.2). For example, a NOD mouse strain congenic for the major histocompatibility complex (MHC) of the C57BL/6 (B6) mouse means that the MHC region on chromosome 17 is from the B6 strain, but all other genetic background is identical to that of the NOD mouse.

In practice, such congenic strains can be produced as follows (conventional congenic strategy; figure 2.3a): NOD mice (recipient strain) are crossed with B6 mice (donor strain) to produce F1 mice (donated N1 generation in congenic strain), and the F1 mice are backcrossed to NOD mice to produce the first backcross generation (donated N2 generation). N2 mice with B6 alleles in the MHC are selected for backcrossing with NOD mice and N3 mice are produced. This process is repeated. Successive backcrossing with NOD mice gradually increases the proportion of NOD genetic background with the B6 MHC retained by selection. A sufficient number of backcrosses (to at least N8, usually N12) will eventually replace almost all background genes with the NOD genome, but only the MHC from the B6 strain is retained

TABLE 2.4
Susceptibility Loci for Type 1 Diabetes Mapped in the NOD Mouse

Locus	Chromosome	Insulitis	Diabetes	Susceptibility to Type 1 Diabetes*	Ref.
Idd1	17	+**	+	NOD (+); B6, B10, C3H, NON (–)	1–4
Idd2	9	–	+	NOD > B10, NON	2,5,6
Idd3	3	+	+	NOD > B6, B10, NON	5,8
Idd4	11	–	+	NOD > B10	5,7
Idd5	1	+	+	NOD > B10	9,10
Idd6	6	–	+	NOD > B10	5
		+	?	NOD > *Mus Spretus*	11
		–	+	NON > NOD	6
		–	+	PWK > NOD	12
Idd7	7	–	+	B10, NON > NOD	5,6
Idd8	14	–	+	B10 > NOD	5
Idd9	4	?	+	NOD > B6, B10. NON	5,6,13
Idd10	3	+	+	NOD > B6, B10	5,6,8,14,15
Idd11	4	?	+	NOD > B6+SJL	16
Idd12	14	?	+	NOD > B6+SJL	16
Idd13	2	–	+	NOD > NOR	17
Idd14	13	–	+	NOD > NON	6
Idd15	5	–	+	NOD > NON	6
Idd16	17	?	+	NOD > CTS	18
Idd17	3	?	+	NOD > B10	15
Idd18	3	?	+	NOD > B6	15
Idd19	6	+	+	C3H > NOD	19
Idd20	6	+	+	NOD > C3H	19
Idd21	18	?	+	NOD > ABH	20

*With the exception of *Idd1*, none of the *Idd* loci is absolutely required for susceptibility to diabetes or insulitis. Strength of susceptibility to diabetes conferred by NOD alleles is therefore expressed relative to that of strains with which NOD mice are outcrossed. In most cases, of course, NOD alleles are more susceptible to insulitis and diabetes than alleles of diabetes-resistant control strains. In some strain combinations, however, alleles from diabetes-resistant strains are more susceptible than NOD alleles (e.g. *Idd7* and *Idd8*). In the case of *Idd6*, alleles of NON and PWK strains are more susceptible, but alleles of B10 and Mus Spretus are more susceptible to diabetes than NOD alleles.

**(+): linkage with insulitis or diabetes; (–): lack of linkage with insulitis or diabetes; (?): data not available.

1. Hattori, M., et al., *Science.* 231(4739): 733–735. 1986.
2. Prochazka, M., et al., *Science* 237(4812): 286–289. 1987.
3. Wicker, L. S., et al., *J Exp Med* 165(6): 1639–1654. 1987.
4. Ikegami, H., et al., *Diabetologia* 31(4): 254–258. 1988.

5. Ghosh, S., et al., *Nat Genet* 4(4): 404–409. 1993.
6. Mcaleer, M. A., et al., *Diabetes* 44(10): 1186–1195. 1995.
7. Todd, J. A., et al., *Nature* 351(6327): 542–547. 1991.
8. Wicker, L. S., et al., *J Exp Med* 180(5): 1705–1713. 1994.
9. Garchon, H. J., et al., *Nature* 353(6341): 260–262. 1991.
10. Cornall, R. J., et al., *Nature* 353(6341): 262–265. 1991.
11. De Gouyon, B., et al., *Proc Natl Acad Sci USA* 90(5): 1877–1881. 1993.
12. Melanitou, E., et al., *Genome Res* 8(6): 608–620. 1998.
13. Rodrigues, N. R., et al., *Mamm Genome* 5(3): 167–170. 1994.
14. Prins, J. B., et al., *Science* 260(5108): 695–698. 1993.
15. Podolin, P. L., et al., *J Immunol* 159(4): 1835–1843. 1997.
16. Podolin, P. L., et al., *Mamm Genome* 9(4): 283–286. 1998.
17. Morahan, G., et al., *Proc Natl Acad Sci USA* 91(13): 5898–5902. 1994.
18. Serreze, D. V., et al., *J Exp Med* 180(4): 1553–1558. 1994.
19. Ikegami, H., et al., *J Clin Invest* 96(4): 1936–1942. 1995.
20. Rogner, U. C., et al., *Genomics* 74(2): 163–171. 2001.
21. Hall, R. J., et al., *Mamm Genome* 14(5): 335–339. 2003.

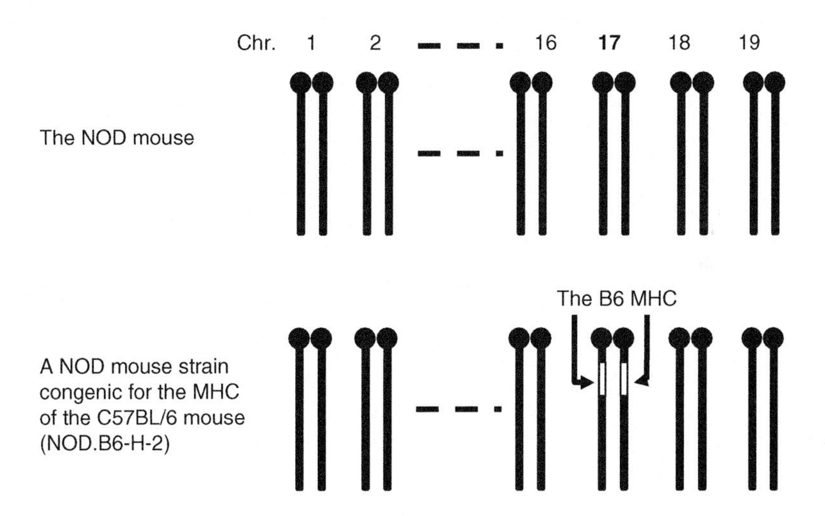

FIG. 2.2 Congenic strains.

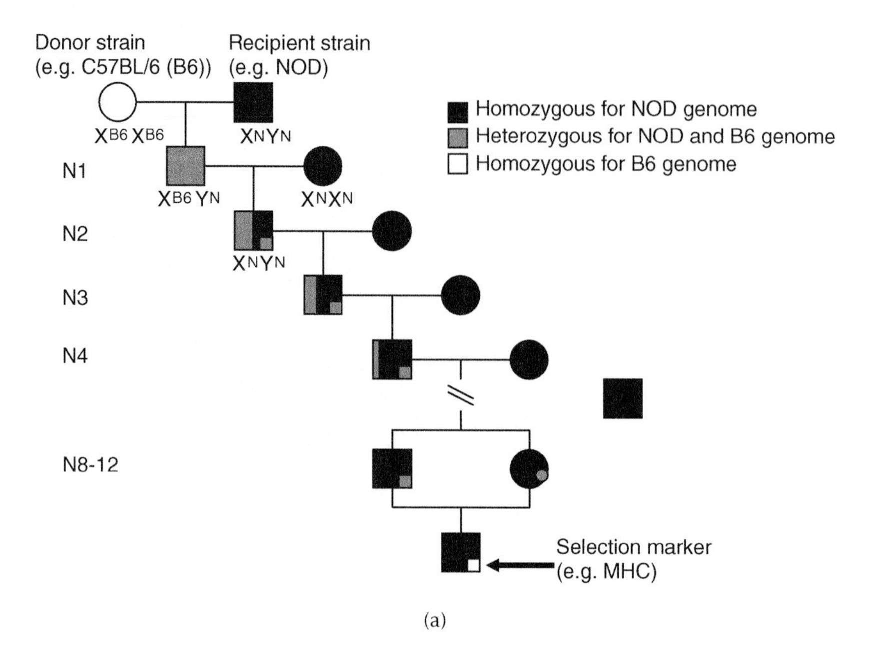

(a)

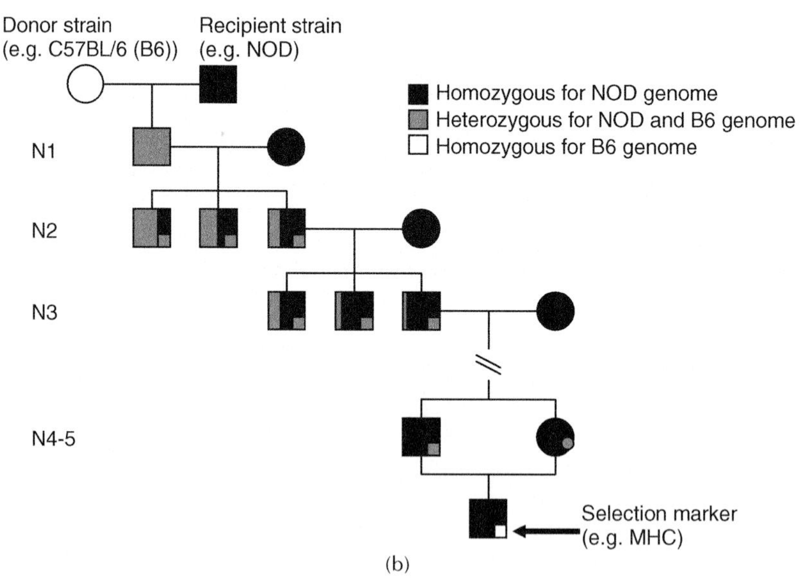

(b)

FIG. 2.3 (a) Conventional congenic protocol, (b) speed congenic protocol.

by selection. By this strategy, the effects of a single locus or a single chromosomal segment on disease predisposition can be assessed.

Theoretically, half of the donor genome will be replaced by the recipient genome in each generation; therefore, 99.2% ($1 - \{1/2\}^{8-1}$) of the genome will be homozygous for the NOD genome at N8 generation and 99.95% ($1 - \{1/2\}^{12-1}$) at N12 generation in NOD MHC congenic strains. Classically, N12 generation (or at least N8 generation) was proposed to be required for sufficient replacement of background genes in congenic strains. Although theoretical calculations indicate that more than 99% of the background genes are identical to the recipient strain after N8 generation, in the case of polygenic diseases it is possible that a certain locus critical for the disease susceptibility is located in <1% of the genome of donor origin. Furthermore, the percentage of the genome of donor origin can be larger than the theoretical value in some of the congenic mice, making the possibility of involvement of the undesired genome in congenic mice higher.

Thus, we should be careful in studying the effect of a locus on the susceptibility to a polygenic disease, such as T1D, by congenic strategy. It would therefore be desirable that certain loci critical for disease susceptibility, such as *Idd1* in the MHC of the NOD mouse, are confirmed to be of recipient origin (Ikegami et al. 1995). Alternatively, intercrossing of congenic mice heterozygous for a certain locus should be performed and the incidence of diabetes should be compared relative to the inheritance of alleles at the locus of interest.

Recent progress in genotyping with a microsatellite marker throughout the genome has made it possible to monitor background genes in each generation and to use mice with the least undesired genome for mating to generate the next generation. By this method, termed "speed congenic" strategy (figure 2.3b), it is now possible to establish congenic strains in a much shorter period compared with the classical congenic strategy. In our experience, background genes can completely be replaced with recipient strain by N5 generation. "Speed congenic" strategy is not only fast, but also accurate and reliable in that replacement of background genes with the recipient genome can clearly be monitored. This is particularly useful in polygenic diseases, such as T1D, in which susceptibility loci have been mapped to the genome, because markers linked to susceptibility loci can be used for monitoring background genes. In this way, the possibility of involvement of the undesired genome in congenic mice with polygenic diseases, as mentioned before, can be eliminated.

Congenic NOD strains have been established by classical or speed congenic strategies and have contributed greatly to understanding the genetics of not only T1D, but also polygenic, multifactorial diseases in general.

MHC GENES

The MHC gene is necessary, but not sufficient for disease development. As summarized in table 2.5, none of the NOD mouse strain congenic for the MHC from control strains was reported to develop T1D (Ikegami et al. 1993, 1996; Podolin et al. 1993; Prochazka et al. 1989; Wicker et al. 1992), indicating that the NOD MHC is essential for disease development. However, a control strain, B6, congenic for the NOD MHC, B6.NOD-H-2, does not develop T1D despite the presence of the diabetogenic MHC from the NOD mouse, indicating that the NOD MHC alone is not sufficient for disease development. A similar finding was also reported in B10.NOD-H-2 congenic mice (Wicker et al. 1994). This is a typical example of polygenic inheritance of a disease: Any one of the genes

TABLE 2.5
NOD Strains Congenic for MHC

Strains	Donor	MHC (H-2)	Recipient	Generations	Cumulative Incidence % (n, age)	Ref.
NOD.B10-H-2	C57BL6/10	b	NOD	N6F1	0% (0/21, 5–13 mo)	*J Exp Med*, 178:793, 1993
				N12F1	0% (0/6, 8–12 mo)	*J Exp Med*, 176:67, 1992
NOD.H-2^{i7}	B10.D2(R107)	i7	NOD	N6F1	0% (0/40, 5–13 mo)	*J Exp Med*, 178:793, 1993
NOD.H-2^{i5}	B10.A(5R)	i5	NOD	N6F1	0% (0/39, 5–13 mo)	*J Exp Med*, 178:793, 1993
NOD.H-2^{h4}	B10.A(4R)	h4	NOD	N6F1	0% (0/54, 5–13 mo)	*J Exp Med*, 178:793, 1993
NOD.H-2^{h2}	B10.A(2R)	h2	NOD	N6F1	0% (0/38, 5–13 mo)	*J Exp Med*, 178:793, 1993
NOD.H-2^{k}	B10.BR	k	NOD	N6F1	0% (0/31, 5–13 mo)	*J Exp Med*, 178:793, 1993
NOD.NON-H-2	NON	nbl	NOD	N6F1	0% (0/6, 41 weeks)	*Diabetes*, 38:1446, 1989
				N15F2	0% (0/28, 30 weeks)	Kino et al. (unpublished data)

that predispose to the disease is sufficient for disease susceptibility and combinations of multiple susceptibility genes are necessary for disease development (Mohan et al. 1999).

MULTIPLE COMPONENTS IN MHC-LINKED SUSCEPTIBILITY GENES

Although data on MHC congenic NOD strains indicate that the NOD MHC is diabetogenic and MHC from control strains confers resistance to diabetes, the data do not show identity of MHC-linked susceptibility gene(s) (*Idd1*). The strongest candidate gene for *Idd1* is class II MHC of the NOD mouse. The NOD MHC class II molecule (I-Ag^7) is unique and differs from any known I-A from control laboratory strains. It is also lack of I-E expression due to a deletion mutation in the promoter region of the gene encoding the α chain of the I-E molecule. The I-Ag^7 molecule shows a strong structural resemblance to a human MHC associated with T1D. The I-Ag^7 molecule was shown to possess a nonaspartic residue at position 57 and probably acts by selecting an autoreactive T cell repertoire and by presenting autoantigens to T cells.

Due to strong linkage disequilibrium within the MHC, however, the NOD MHC is transmitted to offspring as a set of genes, termed a haplotype, and therefore the independent effect of class II MHC from other nearby genes within the MHC cannot be assessed in usual breeding studies. One way to overcome this problem and to demonstrate the effect of class II MHC directly on susceptibility to T1D is to use intra-MHC recombinants. Although such recombinations cannot be easily obtained in usual breeding studies between two inbred strains due to the low frequency of recombination within the MHC, recombination events have occurred during many historical meioses among ancestral haplotypes contained in an original outbred colony of the NOD mouse, giving rise to intra-MHC recombinants.

By screening NOD-related strains, such intra-MHC recombinants were identified in a sister strain, the CTS mouse, and some mice in outbred Jcl:ICR mice, from which NOD mice were derived. In particular, the CTS mouse was found to have the same class II MHC as the NOD mouse, but different class I MHC (Ikegami et al. 1990). With this recombinant MHC, it became possible to dissect out the effect of class II MHC, previously reported NOD strains congenic for the MHC, which were completely resistant to diabetes. The NOD strain congenic for CTS MHC, NOD.CTS-H-2, developed diabetes, indicating that the CTS MHC contained an MHC-linked susceptibility gene (*Idd1*) for type 1 diabetes (Ikegami et al. 1995).

To our surprise, however, the incidence of diabetes in NOD.CTS-H-2 mice was much lower than that in the NOD parental strain (Ikegami et al. 1990), indicating that MHC-linked susceptibility consists of multiple components. Since class II MHC is identical in the CTS and the NOD mouse, the data indicate that class II MHC is diabetogenic, but that other genes adjacent to, but distinct from, class II MHC also contribute to disease susceptibility. A combination of the former (*Idd1*) and latter (*Idd16*) components is necessary for full expression of the MHC-linked susceptibility to T1D in the NOD mouse (figure 2.4). At present, it is reported that the class I *K* gene may be a responsible candidate gene for the *Idd16* by the congenic mapping and the nucleotide sequence analysis (Hattori et al. 2003; Inoue et al. 2004).

NON-MHC GENES: MULTIPLE SUSCEPTIBILITY GENES ON OTHER *Idd* LOCI

By genome scan in crosses of NOD with B10.H-2^{g7} mice, strong linkage of T1D was identified in the central part of chromosome 3 near the marker *D3Nds1* (figure 2.4) (Todd et al. 1991). Subsequent studies in congenic strains, however, revealed that localization was caused by the combined effect of two independent loci, *Idd3* and *Idd10*, flanking *D3Nds1* (figure. 2.4) (Wicker et al. 1994). In fact, NOD mice congenic for B10 or B6 alleles at *D3Nds1* were not protected against diabetes, indicating lack of a susceptibility gene in the *D3Nds1* region, where strong

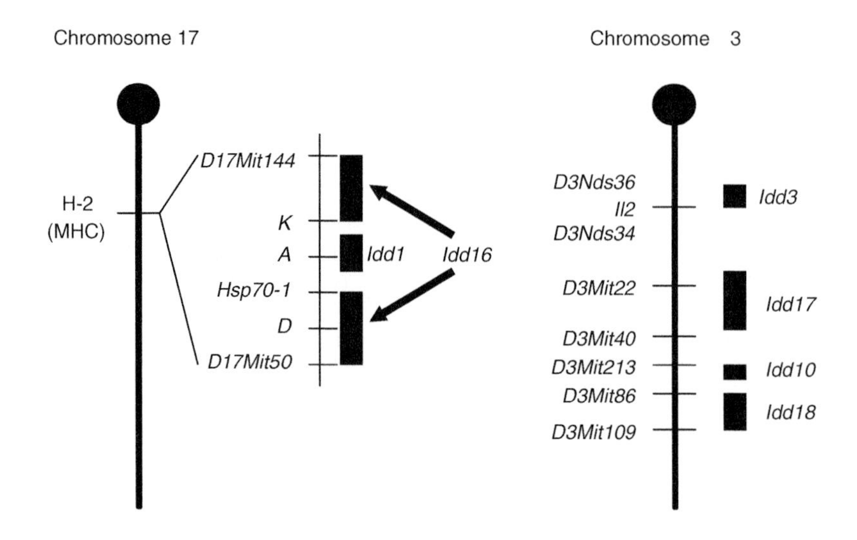

FIG. 2.4 Multiple susceptibility genes on chromosome 17 and chromosome 3. Congenic mapping revealed the existence of at least two (*Idd1* and *Idd16*), and probably more, susceptibility genes for Type 1 diabetes on chromosome 17. Similarly, at least four (*Idd3, 10, 17, 18*) susceptibility genes have been mapped on chromosome 3 by congenic mapping.

linkage was initially detected. Fine mapping in additional congenic strains has now revealed the contribution of at least four independent loci (*Idd3, 10, 17, 18*) to disease susceptibility, with no susceptibility gene in the initial location of peak linkage (figure 2.4) (Podolin et al. 1997, 1998).

Recently, in those *Idd* loci, *Il2* (interleukin 2) gene (*Idd3*) and *Cd101* (*Idd10*) gene (Penha-Goncalves et al. 2003) are thought to be candidates and under investigation by many laboratories. *Idd5* in chromosome 1 has been also revealed to include more than two independent loci, *Idd5.1* and *Idd5.2*, by fine congenic mapping like *Idd3*. Recently, *CTLA4* (cytotoxic T-lymphocyte-associated antigen 4) gene, an important negative regulator of T cell response, is suggested to be a candidate gene for *Idd 5.1*. Ueda et al. reported that susceptibility was associated with variation in *CTLA4* gene splicing with reduced production of a splice form. This splice variant of *CTLA4*, named ligand-independent CTLA4 (liCTLA4), lacks exon 2, including the binding domain to the costimulatory ligands B7-1 and B7-2 (Ueda et al. 2003). liCTLA4 is expressed as a protein in primary T cells and strongly inhibits T cell responses. Expression of liCTLA4, but not full length CTLA4 (flCTLA4), was higher in memory/regulatory T cells from diabetes-resistant NOD congenic mice compared to susceptible NOD mice. Transgenic expression of liCTLA4 in autoimmune prone Ctla4 -/- mice inhibited spontaneous T cell activation and prevented early lethality in the Ctla4 -/- mice. Thus, increased expression and negative signaling delivered by the liCTLA4 may play a critical role in regulating the development of T1D (Vijayakrishnan et al. 2004, 2005). The *Nramp1/Slc11a1* (solute carrier family 11 member 1) gene is thought to be a candidate gene for *Idd 5.2*.

Recently, it has been reported that NOD mice in which Slc11a1 is silenced by RNAi using lentiviral transgenesis showed reduced frequency of type 1 diabetes, mimicking the protective Idd5.2 region and demonstrating a role for Slc11a1 in modifying susceptibility to type 1 diabetes (Kissler et al. 2006). These data, together with identification of multiple susceptibility genes in the MHC as mentioned in the previous section, emphasize the power and importance of the congenic strategy in fine mapping and further characterization of susceptibility genes for polygenic diseases, such as T1D.

THE NON MOUSE: ORIGIN

As mentioned earlier, two sublines were separated at the sixth generation during the establishment of the CTS strain to produce euglycemic and hyperglycemic lines (figure 2.1). The NOD mouse was established from one diabetic female mouse found in the former line, an unexpected finding as mentioned previously. Contrary to the initial expectation, the latter line never developed overt diabetes and was subsequently designated as the NON mouse. Due to the selection for hyperglycemia in the initial part of inbreeding, glucose tolerance in the NON mouse is not completely normal, but slightly impaired. Selection for hyperglycemia, however, was discontinued when a diabetic mouse was found in the former line.

The NON strain, therefore, should not be regarded as a model for type 2 diabetes or glucose intolerance because the phenotype for glucose intolerance is not fixed in this strain and is not being maintained by selective breeding. This is in clear contrast to the NSY mouse (Ueda et al. 1995), which was established as an inbred animal model for type 2 diabetes by selection for glucose intolerance. NON mice are not available for commercial breeders, but they are maintained at Shionogi Aburahi Laboratories in Japan and Jackson Laboratory in the United States.

PHENOTYPES: GLUCOSE TOLERANCE

Although the phenotype for glucose tolerance has not been fixed in the NON mouse, glucose tolerance in this strain as a group is reported to be slightly impaired. Mild glucose intolerance as reported in the NON mouse, however, has also been reported in some inbred laboratory strains, such as C57BL/6 mice (Kaku et al. 1988). Glucose intolerance in NON mice is observed as early as 8–10 weeks of age. Unlike human type 2 diabetes, however, glucose intolerance in NON mice does not deteriorate with age, but rather improved in males and females. This is in clear contrast to the NSY mouse, an animal model of type 2 diabetes, in which age dependent deterioration of glucose tolerance is observed, as in the case of type 2 diabetes in humans (Ueda et al 1995). In addition, marked interindividual variation in glucose tolerance is reported in NON mice, probably because selective breeding for hyperglycemia was discontinued a long time ago and therefore glucose tolerance is more markedly impaired in male NON mice than in female mice.

To evaluate glucose tolerance, an intraperitoneal glucose tolerance test (ipGTT) is usually performed. Glucose (2 g/kg) is injected i.p. in overnight-fasted mice, and blood glucose levels are monitored at 0, 30, 60, (90), and 120 min after glucose administration (Ueda et al. 1995). The easiest way to monitor blood glucose is to use a glucose meter developed for self-monitoring of blood glucose in diabetic patients. At each time point for measuring blood glucose, a small cut is made in the tail with a razor blade, and a small amount of blood is aspirated with a test strip attached to the glucose monitoring device. Most sensors with a capillary aspiration system can measure blood glucose level with a small amount of blood (as little as 3–5 μL) and the results can be obtained in 10–30 sec. Diabetes is usually diagnosed when blood glucose level at 120 min after glucose load is equal to or above 200 mg/dL (11.1 m*M*).

RENAL LESIONS

PAS-positive deposits in glomerular capillaries have been reported in the NON mouse (Muraoka et al. 1992). The deposits were reported to be positive for IgM as well as PAS and to consist of fine, lipid-like droplets on electron microscopic observation. These characteristics are similar to those observed in lipoprotein glomerulopathy in humans (Saito et al. 1989). The frequency of the lesion is 37% in females and 31% in males at 16–24 weeks of age, and 78% in females and 53% in males at 36–38 weeks of age. Since this lesion is observed in NON mice with normal glucose tolerance, it is not related to diabetes or glucose intolerance and therefore is different from diabetic nephropathy.

GENETIC BACKGROUNDS

The recent identification of highly polymorphic microsatellite markers throughout the genome has made it possible to compare genetic identity and differences among different inbred strains (Dietrich et al. 1994). Despite the close relationship between the NOD and NON strains as sister strains, allelic variation in microsatellite markers between NOD and NON strains is as large as that between NOD and inbred laboratory strains, such as BALB/c and C3H/He (Dietrich et al. 1994). This may be partly due to the nonstrict process of the initial part of inbreeding as described previously. If strict brother–sister mating had been performed from the initiation of inbreeding of these strains, there would have been much smaller differences. These differences made it possible to use NON mice as an outcross partner with NOD mice for mapping of susceptibility genes for T1D (Ikegami et al. 1989; Prochazka et al. 1989; McAleer et al. 1995). A partner for insulitis and diabetes between NOD and NON mice was suggested by breeding studies (Makino, Muraoka, et al. 1985; McAleer et al. 1995) and examination of congenic strains (Ikegami et al. 1993; McAleer et al. 1995).

APPLICATION OF NOD AND RELATED STRAINS FOR BIOMEDICAL RESEARCH

Mice have been the most useful experimental tool for the majority of immunologists and study of their immune responses has yielded tremendous insight into the workings of the human immune system. Recently, it has been suggested, however, that from 65 million years of evolution there are significant differences in innate and adaptive immunity, including balance of leukocyte subsets, defensins, Toll receptors, inducible NO synthase, the NK inhibitory receptor families Ly49 and KIR, FcR, Ig subsets, the B cell and T cell signaling pathway components, cytokines and cytokine receptors, Th1/Th2 differentiation, costimulatory molecule expression and function, Ag-presenting function of endothelial cells, and chemokine and chemokine receptor expression. Such differences should be taken into account when using mice as preclinical models of human disease (Mestas et al. 2004; Roep et al. 2004).

There are, however, sufficient similarities between the pathogenesis of T1D in humans, mice, and rats to justify the efforts devoted to animal research. While there are significant differences between the various rodent models, this heterogeneity surely carries important lessons for understanding the potential heterogeneity underlying this complex disease in humans (Leiter et al. 2004). The NOD mouse and its related strain have contributed to our understanding of the genetics and pathogenesis of T1D and to the development of effective methods of disease prediction, prevention, and intervention. Pioneering studies using genome scanning and congenic and transgenic strategies were all performed with NOD mice. In addition, information generated with this strain has contributed and will continue to contribute greatly to elucidation of the genetics of multifactorial diseases and of the pathogenesis and molecular mechanisms of autoimmune diseases in general. In conclusion, there are, of course, discrepancies between animal models and the human disease; however, studies using animal models provide a "roadmap" to further investigation of pathogenesis, immunotherapy, and prevention of T1D in humans.

REFERENCES

Coleman DL, Kuzava JE, Leiter EH. (1990) Effect of diet on incidence of diabetes in nonobese diabetic mice. *Diabetes* 39:432–436.

Dietrich WF, Miller JC, Steen RG, et al. (1994) A genetic map of the mouse with 4,006 simple sequence length polymorphisms. *Nat Genet* 7:220–245.

Elliott RB, Reddy SN, Bibby NJ, et al. (1988) Dietary prevention of diabetes in the nonobese diabetic mouse. *Diabetologia* 31:62–64.

Fox H (1992) Androgen treatment prevents diabetes in nonobese diabetic mice. *J Exp Med* 175:1409–1412.

Ghosh S, Palmer SM, Rodrigues NR, et al. (1993) Polygenic control of autoimmune diabetes in nonobese diabetic mice. *Nat Genet* 4:404–409.

Giulietti A, Gysemans C, Stoffels K, et al. (2004) Vitamin D deficiency in early life accelerates Type 1 diabetes in nonobese diabetic mice. *Diabetologia* 47:451–462.

Hall RJ, Hollis-Moffatt JE, Merriman ME, et al. (2003) An autoimmune diabetes locus (*Idd21*) on mouse chromosome 18. *Mamm Genome* 14:335–339.

Hattori M, Buse JB, Jackson RA, et al. (1986) The NOD mouse: Recessive diabetogenic gene in the major histocompatibility complex. *Science* 231:733–735.

Hattori M, Hattori K, Fujisawa T. (2003) MHC class I Kd is diabetogenic in the NOD mouse. *Diabetes* 52:A248.

Ikegami H, Eisenbarth GS, Hattori M. (1990) Major histocompatibility complex-linked diabetogenic gene of the nonobese diabetic mouse. Analysis of genomic DNA amplified by the polymerase chain reaction. *J Clin Invest* 85:18–24.

Ikegami H, Makimo S. (1993) Genetic susceptibility to insulin-dependent diabetes mellitus: From the NOD mouse to man. In *Lessons from animal diabetes*, ed. E. Shafrir, Smith–Gordon, London, 4:39–50.

Ikegami H, Makimo S, Ogihara T. (1996) Molecular genetics of insulin-dependent diabetes mellitus: analysis of ongenic strains. In *Lessons from animal diabetes*, ed. E. Shafrir, Birkhauser, Boston, 6:33–46.

Ikegami H, Makino S, Yamato E, et al. (1995) Identification of a new susceptibility locus for insulin-dependent diabetes mellitus by ancestral haplotype congenic mapping. *J Clin Invest* 96:1936–1942.

Ikegami H, Yano N, Sato T, et al. (1989) Immunogenetics and immunopathogenesis of the NOD mouse. In *Immunotherapy of diabetes and selected autoimmune diseases*, CRC Press, Boca Raton, FL.

Inoue K, Ikegami H, Fujisawa T, et al. (2004) Allelic variation in class I K gene as candidate for a second component of MHC-linked susceptibility to T1D in nonobese diabetic mice. *Diabetologia* 47: 739–747.

Kagohashi Y, Udagawa J, Abiru N, et al. (2005) Maternal factors in a model of type 1 diabetes differentially affect the development of insulitis and overt diabetes in offspring. *Diabetes* 54:2026–2031.

Kagohashi Y, Udagawa J, Moriyama K, et al. (2005) Maternal environment affects endogenous virus induction in the offspring of type 1 diabetes model nonobese diabetic mice. *Congenit Anom* (Kyoto) 45:80–84.

Kaku K, Fiedorek FT Jr., Province M, et al. (1988) Genetic analysis of glucose tolerance in inbred mouse strains. Evidence for polygenic control. *Diabetes* 37:707–713.

Kissler S, Stern P, Takahashi K, et al. (2006) *In vivo* RNA interference demonstrates a role for Nramp1 in modifying susceptibility to type 1 diabetes. *Nat Genet* (in press).

Leiter EH, Serreze DV, Prochazka M. (1990) The genetics and epidemiology of diabetes in NOD mice. *Immunol Today* 11:147–149.

Leiter EH, Von Herrath M. (2004) Animal models have little to teach us about type 1 diabetes: 2. In opposition to this proposal. *Diabetologia* 47:1657–1660.

Makino S, Hayashi Y, Muraoka Y, et al. (1985) Establishment of the nonobese diabetic (NOD) mouse. In *Current topics in clinical and experimental aspects of diabetes mellitus*, ed. N Sakamoto, H Min, S Baba. Elsevier, Amsterdam, 25–32.

Makino S, Kunimoto K, Muraoka Y, et al. (1980) Breeding of nonobese, diabetic strain of mice. *Exp Anim* 29:1–13.

Makino S, Kunimoto K, Muraoka Y, et al. (1981) Effect of castration on the appearance of diabetes in NOD mouse. *Exp Anim* 30:137–140.

Makino S, Muraoka Y, Harada M, et al. (1988) Characteristics of the NOD mouse and its relatives. In *New lessons from diabetes in animals*, ed. R Larkins, P Zimmet, D Chisholm, Elsevier, Amsterdam, 747–750.

Makino S, Muraoka Y, Kishimoto Y, et al. (1985) Genetic analysis for insulitis in NOD mice. *Exp Anim* 34:425–432.

McAleer MA, Reifsnyder P, Palmer SM, et al. (1995) Crosses of NOD mice with the related NON strain. A polygenic model for IDDM. *Diabetes* 44:1186–1195.

Mestas J, Hughes CC. (2004) Of mice and not men: Differences between mouse and human immunology. *J Immunol* 172:2731–2738.

Mohan C, Morel L, Yang P, et al. (1999) Genetic dissection of lupus pathogenesis: A recipe for nephrophilic autoantibodies. *J Clin Invest* 103:1685–1695.

Muraoka Y, Matsui S, Watanabe H. (1992) Histopathological observation of the development of glomerular intracapillary deposits in the NON mouse. In *Current concepts of a new animal model: The NON mouse*, ed. N Sakamoto, N Hotta, K Uchida, Elsevier, Amsterdam, 107–120.

Ohsugi T, Kurosawa T. (1994) Increased incidence of diabetes mellitus in specific pathogen-eliminated offspring produced by embryo transfer in NOD mice with low incidence of the disease. *Lab Anim Sci* 44:386–388.

Ohtori H, Yoshida T, Inuma T. (1968) Small eye and cataract, a new dominant mutation in the mouse. *Exp Anim* 17:91–96.

Oldstone MB. (1988) Prevention of type I diabetes in nonobese diabetic mice by virus infection. *Science* 239:500–502.

Penha-Goncalves C, Moule C, Smink LJ, et al. (2003) Identification of a structurally distinct CD101 molecule encoded in the 950-kb *Idd10* region of NOD mice. *Diabetes* 52:1551–1556.

Podolin PL, Denny P, Armitage N, et al. (1998) Localization of two insulin-dependent diabetes (*Idd*) genes to the *Idd10* region on mouse chromosome 3. *Mamm Genome* 9:283–286.

Podolin PL, Denny P, Lord CJ, et al. (1997) Congenic mapping of the insulin-dependent diabetes (*Idd*) gene, d10, localizes two genes mediating the Idd10 effect and eliminates the candidate Fcgr1. *J Immunol* 159:1835–1843.

Podolin PL, Pressey A, Delarato NH, et al. (1993) I-E+ nonobese diabetic mice develop insulitis and diabetes. *J Exp Med* 178:793–803.

Pozzilli P, Signore A, Williams AJ, et al. (1993) NOD mouse colonies around the world—Recent facts and figures. *Immunol Today* 14:193–196.

Prochazka M, Serreze DV, Worthen SM, et al. (1989) Genetic control of diabetogenesis in NOD/Lt mice. Development and analysis of congenic stocks. *Diabetes* 38:1446–1455.

Roep BO, Atkinson M. (2004) Animal models have little to teach us about type 1 diabetes: 1. In support of this proposal. *Diabetologia* 47:1650–1656.

Rogner UC, Boitard C, Morin J, et al. (2001) Three loci on mouse chromosome 6 influence onset and final incidence of type I diabetes in NOD.C3H congenic strains. *Genomics* 74:163–171.

Saito T, Sato H, Kudo K, et al. (1989) Lipoprotein glomerulopathy: Glomerular lipoprotein thrombi in a patient with hyperlipoproteinemia. *Am J Kidney Dis* 13:148–153.

Suzuki T, Yamada T, Fujimura T, et al. (1987) Diabetogenic effects of lymphocyte transfusion on the NOD or NOD nude mouse. In *Immune-deficient animals in biochemical research*, ed. Rygaard, Brunner, Graem Thomsen, pp.112–116. Basel: Karger.

Todd JA. (1991) A protective role of the environment in the development of type 1 diabetes? *Diab Med* 8: 906–910.

Todd JA, Aitman TJ, Cornall RJ, et al. (1991) Genetic analysis of autoimmune type 1 diabetes mellitus in mice. *Nature* 351:542–547.

Ueda H, Howson JM, Esposito L, et al. (2003) Association of the T-cell regulatory gene CTLA4 with susceptibility to autoimmune disease. *Nature* 423:506–511.

Ueda H, Ikegami H, Yamato E, et al. (1995) The NSY mouse: A new animal model of spontaneous NIDDM with moderate obesity. *Diabetologia* 38:503–508.

Vijayakrishnan L, Slavik JM, Illes Z, et al. (2004) An autoimmune disease-associated CTLA-4 splice variant lacking the B7 binding domain signals negatively in T cells. *Immunity* 20:563–575.

Vijayakrishnan L, Slavik JM, Illes Z, et al. (2005) An autoimmune disease-associated CTLA4 splice variant lacking the B7 binding domain signals negatively in T cells. *Novartis Found Symp* 267:200–212; Discussion 212–218.

Wetzel JD, Barton ES, Chappell JD, et al. (2006) Reovirus delays diabetes onset but does not prevent insulitis in nonobese diabetic mice. *J Virol* 80:3078–3082.

Wicker LS, Appel MC, Dotta F, et al. (1992) Autoimmune syndromes in major histocompatibility complex (MHC) congenic strains of nonobese diabetic (NOD) mice. The NOD MHC is dominant for insulitis and cyclophosphamide-induced diabetes. *J Exp Med* 176:67–77.

Wicker LS, Miller BJ, Coker LZ, et al. (1987) Genetic control of diabetes and insulitis in the nonobese diabetic (NOD) mouse. *J Exp Med* 165:1639–1654.

Wicker LS, Todd JA, Prins JB, et al. (1994) Resistance alleles at two nonmajor histocompatibility complex-linked insulin-dependent diabetes loci on chromosome 3, *Idd3* and *Idd10*, protect nonobese diabetic mice from diabetes. *J Exp Med* 180:1705–1713.

Wilberz S, Partke HJ, Dagnaes-Hansen F, et al. (1991) Persistent MHV (mouse hepatitis virus) infection reduces the incidence of diabetes mellitus in non-obese diabetic mice. *Diabetologia* 34:2–5.

Williams AJ, Krug J, Lampeter EF, et al. (1990) Raised temperature reduces the incidence of diabetes in the NOD mouse. *Diabetologia* 33:635–637.

Zella JB, Deluca HF. (2003) Vitamin D and autoimmune diabetes. *J Cell Biochem* 88:216–222.

Zella JB, Mccary LC, Deluca HF. (2003) Oral administration of 1,25-dihydroxyvitamin D3 completely protects NOD mice from insulin-dependent diabetes mellitus. *Arch Biochem Biophys*. 417:77–80.

3 Obesity/Diabetes in Mice with Mutations in Leptin or Leptin Receptor Genes

Streamson Chua, Jr., Lieselotte Herberg, and Edward H. Leiter

CONTENTS

INTRODUCTION

This chapter represents an updating of a previously published review (Herberg and Leiter 2001) of two of the most intensively studied monogenic obesity mutations in the mouse, *obese* (*ob*) and *diabetes* (*db*). The previous observations that both mutations, although mapping to separate chromosomes, produced nearly identical obesity/diabetes syndromes when studied on a common inbred background, coupled with the results of parabiosis studies (Coleman 1973, 1978) suggested that the two mutations affected a common pathway. This has been confirmed by recent discoveries showing the *ob* mutation to be a defect in the gene encoding leptin (*Lep*) and the *db* mutation to be a defect in the leptin receptor gene (*Lepr*). This understanding has revolutionized our understanding of how the periphery, particularly fat cells, coordinates management of energy homeostasis with the brain (reviewed in Ahima and Flier 2000). Several other reviews contrast these two mutations with other monogenic obesity-producing genes in the mouse (Kim et al. 1998; McIntosh and Pederson 1999). In the rat, the fatty (*fa*) mutation on chromosome 5 (and its allele, corpulent, *cp*) are orthologs of the mouse *db* mutation. This chapter will help to integrate the extensive early literature describing the physiologic, biochemical, and behavioral effects of the mouse mutations with the more current information gained from the molecular information and the availability of recombinant leptin protein.

CURRENT NOMENCLATURE FOR MOUSE MUTATIONS

Since the discovery that *ob* is a mutation in the leptin structural gene and *db* is a mutation in the leptin receptor gene, the nomenclature for these mutations has been changed to reflect their molecular bases as noted in table 3.1. This new nomenclature will henceforth be used in preference to the inaccurate descriptors "ob-protein" and "ob-receptor". Gene symbols are italicized whereas gene products (leptin, leptin receptor) are not. Throughout this chapter, $Lep^{ob\text{-}1J}$ and $Lepr^{db\text{-}1J}$ will henceforth be denoted as Lep^{ob} and $Lepr^{db}$ in keeping with rules for nomenclature of mouse genes. The symbols Lep^{ob} and $Lepr^{db}$ will denote homozygous mutants in place of Lep^{ob}/Lep^{ob} and $Lepr^{db}/Lepr^{db}$. The wild-type alleles are denoted simply as *Lep* and *Lepr*, respectively. However, in descriptions of specific crosses entailing use of linked coat color mutations, any wild-type allele will be noted by "+". It should be noted that the gene symbols for the rat ortholog of the leptin receptor mutation have also been changed to reflect this new nomenclature. Hence, the *fatty* mutation is now referred to as $Lepr^{fa}$ and the corpulent mutation as $Lepr^{facp}$.

As noted in table 3.2, spontaneous mutations at the $Lepr^{db}$ locus are relatively frequent, whereas only two mutations at the Lep^{ob} locus have been described. In addition to the more recently discovered *Lepr* gene mutations listed in table 3.2, transgenic insertion within the *Lepr* gene has also generated a new mutant allele (Reichart et al. 2000). Gene targeting has also generated a series of *Lepr* alleles, certain of which have been used for tissue-specific and cell type-specific ablation of leptin receptor gene expression (McMinn et al. 2004; Bates et al. 2003; Cohen et al. 2001).

TABLE 3.1
Current Genetic Nomenclature for Mouse *ob* and *db* Mutations

Common Name	Gene Symbol Formal Designation	Earlier Symbol	Gene Product Current Name	Synonyms
Obese	$Lep^{ob\text{–}J}$	ob	Leptin	Ob-protein
			Leptin mRNA	Ob-mRNA
Diabetes	$Lepr^{db\text{–}1J}$	db	Leptin receptor	Ob-receptor
			Leptin receptor mRNA	Ob-receptor mRNA

TABLE 3.2
Additional Alleles of *Lep*^{ob} and *Lepr*^{db}

Gene Symbol	Spontaneous Occurrence of the Alleles in
$Lep^{ob\text{-}2J}$	The SM/J inbred strain at the Jackson Laboratory
$Lepr^{db\text{-}2J}$	A stock that has been discontinued
$Lepr^{db\text{-}3J}$	The 129/J strain at the Jackson Laboratory
$Lepr^{db\text{-}4J}$	The BXD-16 stock of Dr. Ben Taylor at the Jackson Laboratory
$Lepr^{db\text{-}Pas1}$	The DW inbred strain at the Institut Pasteur in Paris
$Lepr^{db\text{-}Pas2}$	The DW inbred strain at the Institut Pasteur in Paris
$Lepr^{db\text{-}ad}$	A heterogenous strain at the Institute of Genetics in Edinburgh
$Lepr^{db\text{-}dmpg\ (dumpling)}$	The B10.D2-H8b(57N)/SnJ congenic stock at the Jackson Laboratory
$Lepr^{db\text{-}rtnd\ (rotund)}$	The CBA/J strain at the Jackson Laboratory
$Lepr^{db\text{-}ripy\ (rolypoly))}$	The Nu/J strain at the Jackson Laboratory
$Lepr^{db\text{-}NCSU}$	The CD1 strain at North Carolina State University
$Lepr^{db\text{-}neo}$	129 ES cells and backcrossed to FVB/NJ
$Lepr^{db\text{-}flox}$	129 ES cells and backcrossed to FVB/NJ
$Lepr^{db\text{-}delta17}$	129 ES cells
$Lepr^{db\text{-}flox}$	129 ES cells
$Lepr^{db\text{-}delta}$	129 ES cells
$Lepr^{db\text{-}5J}$	NOD/LtJ strain at the Jackson Laboratory

Source: Dr. Ben Taylor, the Jackson Laboratory, personal communication (mutation no longer maintained).

MUTATION ORIGINS AND EFFECT OF INBRED STRAIN BACKGROUND ON SYNDROME DEVELOPMENT

The *Lep*^{ob} mutation on chromosome 6 was discovered at the Jackson Laboratory in a multiple recessive ("V") stock in 1949 (Ingalls et al. 1950). It was immediately recognized because of the marked obesity and hyperphagia exhibited by homozygous mutant mice. As was customary with most mutations developed at the laboratory, the *Lep*^{ob} mutation was subsequently transferred to the B6 inbred strain background, initially by standard backcrossing (Drasher et al. 1955). The *Lep*^{ob} mutation on the B6 background produces juvenile onset obesity, hyperinsulinemia with increasing insulin resistance, but a hyperglycemia that was relatively mild and transient. This remission from chronic hyperglycemia was correlated with a sustained hypertrophy of pancreatic islets primarily contributed by hyperplasia of the beta-cell mass.

The *Lepr*^{db} mutation is a recessive mutation on chromosome 4 that occurred spontaneously in the C57BLKS/J (BKS) inbred strain in 1966 (Hummel et al. 1966). This mutation, as well as a second mutation (*Lepr*$^{db\text{-}2J}$) arising in a mixed background stock (Hummel et al. 1972) and backcrossed to BKS was named "diabetes" because the obesity/diabetes syndrome did not remit as it did in B6-*Lep*^{ob} mice. Instead, an early hyperinsulinemia was not sustained; a progressively more severe hyperglycemia was correlated at the morphologic level with pancreatic beta-cell necrosis and islet atrophy. During this "end stage" period, a precipitous drop in insulin occurs concomitant with islet atrophy and beta-cell degeneration as well as a loss of body weight. Life span of the mutants is shortened compared to lean littermate controls.

DISCOVERY OF MOLECULAR NATURE OF "*ob*" AND "*db*" MUTATIONS

A key event in unraveling the molecular genetic basis of these two mutations initially came from the discovery that either mutation on the B6 inbred strain background produced identical obesity syndromes accompanied by a transient diabetes syndrome that remitted after puberty. In sharp contrast, both mutations on the BKS background produced identical diabetogenic obesity (diabesity) syndromes that failed to remit, leading to weight loss and severely shortened life span. The BKS strain is a recombinant congenic strain with a genome comprising ~80% B6-derived alleles and ~20% from a DBA/2-like strain (Naggert et al. 1995; Davis et al. 2005). The differential effect of the two backgrounds on diabesity development clearly indicated that, although the obesity was under monogenic control, the diabesity was under multigenic control. The nature of some these diabetogenic background modifiers is discussed later. The salient observation by Coleman and colleagues at the Jackson Laboratory (Coleman 1978) was that the *db* and *ob* mutations produced the same phenotype of extreme obesity and hyperphagia when studied on the same inbred background. This suggested that both mutations impaired genes encoding enzymes or factors in a common metabolic pathway, or else a receptor and its ligand.

Experimentally induced hypothalamic lesions in rodents induce the phenotypes of hyperphagia and extreme obesity. The observation by Hausberger (1958)—that parabiosis of "*ob*" mice with lean mice was followed by a suppressed or completely prevented weight gain in the "*ob*" partner of the pair—suggested existence of a humoral factor linking the fat cell with hypothalamic centers, although Hausberger could not conclude that adipose tissue was the source. However, the observation that donor fat cells took on the morphological characteristics of the host anticipated the hypothesis that a humoral factor present in only one partner of a lean/obese parabiosed pair would be able to regulate adipose tissue mass in the factor-deficient partner (Ashwell et al. 1977).

The first insight that the "*ob*" and "*db*" mutations affected a ligand–receptor pair involved in energy homeostasis came from the landmark parabiosis studies of Coleman and Hummel (1969a). These investigators reported that food intake was drastically reduced in BKS lean (+/+) partners parabiosed with a BKS-"*db*" mouse. In the lean parabiont, blood glucose and body weight dropped until it died of starvation; the "*db*" partner of the pair remained hyperglycemic and gained weight. In a second set of experiments, Coleman (1973) compared the effect of "*ob*" and "*db*" on a congenic (B6) background. Parabiosis of "*ob*" with "*db*" mice produced the same response in the "*ob*" partner as had been observed previously in the BKS lean partner parabiosed to the BKS-"*db*" mutant; namely, the "*ob*" parabiont stopped eating, developed hypoglycemia, and died of starvation. Parabiosis of a lean B6-+/+ mouse with an "*ob*" partner produced a drop in food intake and weight gain in the "*ob*" mouse. Coleman (1973) concluded that the "*db*" mouse produces a satiety factor that is missing in the "*ob*" mouse. Whereas the satiety center of the "*db*" mouse is insensitive to its own satiety factor, the satiety center of the "*ob*" mouse responds to the factor if it is provided by the "*db*" or the lean parabiont. This hypothesis was proven correct 21 years later.

POSITIONAL CLONING OF "*ob*" AND "*db*" IDENTIFIES LEPTIN AND ITS RECEPTOR

The Leptin Gene

Confirmation of the Coleman hypothesis concerning the relationship between the "*ob*" and "*db*" mutations came two decades later with the positional cloning of the "*ob*" gene and identification of its product, leptin, as a new protein hormone (Zhang et al. 1994). The wild-type mouse leptin gene contains three exons and two introns, with the coding sequence in exons 2 and 3 producing a 4.5-kb mRNA. The 167 amino acid-secreted protein made from this message was named leptin (Greek *leptos* = thin). Leptin is synthesized in white adipose tissue with different mRNA levels in different white fat depots. Other sites of leptin synthesis are brown adipose tissue, the placenta (Trayhurn et al. 1998), the stomach (Bado et al. 1998), and the ovary (Chehab et al. 1996). The "*ob*" mutation represents a nonsense mutation (C $\rightarrow$ T changing an arginine at position 105 to a stop codon). This results in the synthesis of a truncated, inactive protein that probably is rapidly degraded in the fat cell (Zhang et al. 1994).

A second spontaneous mouse mutation at the *Leptin* locus, named ob^{2J} (now $Lep^{ob\text{-}2J}$) was shown to be unable to synthesize mature leptin mRNA due to an insertion of a retroviral-like transposon into the first intron of the leptin gene (Moon and Friedman 1997). Thus, adipocytes from these obese and hyperphagic mutant mice (SM/Ckc-$Lep^{ob\text{-}2J}$) contained no leptin mRNA and there was no measurable leptin in serum. The discovery of the adipocyte as the major source of leptin led to a new appreciation of the adipocyte as a source of multiple endocrine factors ("adipokines") in addition to estrogen (Simpson et al. 1989) and angiotensinogen (Frederich et al. 1992). These factors, in addition to leptin, now include adiponectin, resistin, plasminogen activator inhibitor-1, nerve growth factor, vascular endothelial growth factor, and a variety of proinflammatory cytokines (Trayhurn and Wood 2005).

The Leptin Receptor Gene

Positional cloning of the leptin receptor gene from a mouse choroid plexus cDNA library followed rapidly behind the cloning of the leptin gene (Tartaglia et al. 1995). This receptor is frequently denoted as "Ob-R" in the literature, but LEPR is clearly a more accurate usage. The leptin receptor is a member of the IL-6 cytokine family of receptors. The leptin receptor gene (*Lepr*) encodes five alternatively spliced receptor isoforms (denoted as Ra, Rb, Rc, Rd, and Re). All isoforms share an extracellular domain of approximately 800 amino acids, a transmembrane domain of 34 amino acids, and the first 29 amino acids of the intracellular domain. However, only the Rb isoform encodes the full-length receptor containing approximately 300 amino acids (Tartaglia et al. 1995). This long form is required for signal transduction processes (Lee et al. 1996; Chen et al. 1996; Ghilardi et al. 1996) and is expressed at high levels in hypothalamic regions associated with neuroendocrine regulation of appetite and body weight. The spliced receptor variant identified in the mouse carrying the *Lepr*$^{db\text{-}1J}$ mutation shows a 106-bp insertion at the splice junction predicting a premature stop codon and changing Rb to Ra (Lee et al. 1996).

As noted previously, the ubiquitously expressed Ra isoform lacks a full-length intracellular signaling domain that functions through JAK/STAT activation. In addition to *Lepr*$^{db\text{-}1J}$, additional mutant alleles have been characterized at the molecular level. The *Lepr*$^{db\text{-}Ncsu}$ mutation represents a deletion of a G in exon 12, producing a frame shift introducing an 11 amino acid insertion followed by a premature termination of translation before the membrane spanning domain (Brown et al. 2000). The *Lepr*$^{db\text{-}3J}$ mutation results in a leptin receptor truncated at amino acid 625 (Lee et al. 1997) and the *Lepr*$^{db\text{-}Pas1}$ mutation results in a stop codon one residue after amino acid 281 (Li et al. 1998). *Lepr*$^{db\text{-}rtnd}$ contains a G deletion in exon 4, introducing a frameshift and premature transcript termination; *Lepr*$^{db\text{-}ripy}$ contains a large deletion, resulting in no transcript, and *Lepr*$^{db\text{-}dmpg}$ has a deletion in the extracellular domain (Kim et al. 2003). Thus, as a result of the splicing mutation in *Lepr*$^{db\text{-}1J}$ mice, the intracytoplasmic domain involved in signal transduction is missing, whereas *Lepr*$^{db\text{-}3J}$ and *Lepr*$^{db\text{-}Pas1}$ mice exhibit a truncated receptor in the extracellular domain. The NOD/LtJ-*Lepr*$^{db\text{-}5J}$ stock also expresses a full-length Rb receptor with a single amino acid change at residue 640 in the extracellular domain (Lee et al. 2005).

The multiplicity of intracellular signaling events activated by leptin is only now beginning to be understood (Fruhbeck 2006) and will not be reviewed here. Since the expression of other LEPR isoforms is unaffected in mice carrying *Lepr*$^{db\text{-}1J}$, the weight-reducing effect of leptin depends on the Rb form. The Re form appears to be a soluble leptin binding protein (Gavrilova et al. 1997). The manifold increase of leptin mRNA in adipocytes and leptin concentration in serum of *Lepr*^{db} mice reflects a dysfunctional feedback loop between the hypothalamic center that cannot bind leptin and the adipose tissue-enriched periphery.

Lepr^{db} AND *Lep*^{ob} MUTATIONS ON OTHER STRAIN BACKGROUNDS

The major effects of inbred strain background on development of the obesity/diabetes syndrome produced by mutations at the *Lep*^{ob} and *Lepr*^{db} loci have been well reviewed (Coleman, 1978; Leiter and Herberg 1997; McIntosh and Pederson 1999). The *Lepr* locus is proving to be a genetic "hot spot" in terms of the numbers of spontaneous monogenic obesities in mice that ultimately map to the locus. The expression of the *Lepr*^{db} mutation maintained in strains other than BKS and B6, either as newly arisen mutations or by congenic transfer of an older mutation (*Lepr*$^{db\text{-}1J}$ or *Lepr*$^{db\text{-}3J}$), is summarized in table 3.3. In addition, the *Lep*$^{ob\text{-}1J}$ mutation has been

TABLE 3.3
Diabetes-Susceptible/Resistant Strain Backgrounds

Susceptible	Gender	Resistant	Gender
(C57BLKS/J) BKS	Both sexes	(C57BL/6J) B6	Both sexes
CBA/Lt[a]	Males	V stock	Both sexes
DBA/2J	Males	129/J	Both sexes
SM/J[b]	Both sexes	MA/J	Both sexes
C3HeB/FeJ	Males	BALB/c	Both sexes
FVB	Both sexes	NOD/LtJ[c]	Both sexes
BTBR	Both sexes	Nu/J	Both sexes
CD1	Both sexes		Both sexes

[a]The *Lepr*$^{db\text{-}rtnd}$ mutation congenic on CBA/J was diabetogenic in both sexes, although males were more severely affected.
[b]Leiter, unpublished.
[c]The NOD/LtJ stock with the *Lepr*$^{db\text{-}5J}$ mutation (extracellular domain but not intracellular signaling domain affected) developed a type 2 diabetes that spontaneously remitted in mutant females and in one-third of mutant males without any reversion to type 1 diabetes. (Lee, C.H. et al. 2005. *Diabetes* 54:2525–2532; Lee, C.H. 2006. *Diabetes* 55:171–178.) This contrasts with a Japanese finding that NOD/Shi mice congenic for the *Lepr*$^{db\text{-}1J}$ mutation (no intact intracellular signaling domain) reportedly succumbed to type 1 diabetes/insulitis. (Nishimura, M. and Miyamoto, H. 1987. *J Immunogenet* 14:127–130.)

crossed into several backgrounds in addition to BKS. As would be anticipated from the preceding discussion regarding the different sets of diabesity modifiers in B6 versus BKS strain backgrounds, these more recently described stocks exhibit variations of the obesity/diabetes phenotype that are novel and could potentially yield insights into the genetic regulation of susceptibility to diabesity. Because the vast majority of the studies entail B6-*Lep*$^{ob\text{-}1J}$ or BKS-*Lepr*$^{db\text{-}1J}$, the bulk of this review will focus on those two models. The following brief discussion highlights some of the newer congenic stocks.

BALB/c-*Lep*^{ob}. This model is remarkable for the retention of reproductive competence in both sexes (Chehab et al. 2002; Qiu, Ogus, Mounzih, et al. 2001; Ewart-Toland et al. 1999). There is a modest reduction in the degree of obesity, although this is unlikely to be a confounder of fertility since calorically restricted B6-*Lep*^{ob} mice remain infertile. These mice have a reduction in insulin resistance, which might affect fertility. However, the females remain lactationally incompetent and have histologically immature mammary glands.

BTBR-*Lep*^{ob}. The BTBR strain has long been used for mutagenesis studies. Incidentally used in a cross with B6 wild type, the F1 mice developed a degree of insulin resistance that was not apparent in either of the parental strains. Subsequently, *Lep*^{ob} from B6 was backcrossed into BTBR to produce a new congenic strain. The BTBR-*Lep*^{ob} mice develop hyperglycemia with insulin resistance (Stoehr et al. 2000). Over time, beta cell mass and circulating insulin concentrations decrease in obese male and female mice. Mapping of phenotypic traits in the obese progeny of an F2 cross between B6 and BTBR has identified several diabetes modifier loci (Clee et al. 2005).

FVB-*Lepr*^{db} and FVB-*Lep*^{ob}. The FVB strain is often used for developing transgenic mice since the large size of the oocyte pronucleus facilitates microinjections of DNA. The strain was developed at NIH by inbreeding a Swiss albino colony originally founded at the Rockefeller University. Congenic FVB/N-*Lep*^{ob} and -*Lepr*^{db} mice develop hyperglycemia upon weaning to standard rodent chow, which has a high carbohydrate content (Chua et al. 2002). The hyperglycemia is sustained over several months. A novel phenotype that markedly distinguishes this background from the diabesity-susceptible BKS background is the persistence of elevated circulating insulin concentrations and increased beta cell mass despite chronic hyperglycemia. Genetic mapping studies show the presence of diabetes modifier loci on mouse chromosome 5 that regulate beta cell mass during hyperglycemia (Luo et al. 2006). It is likely that multiple loci throughout the genome contribute to the phenotype, although experimentally it is much more feasible to identify individual loci that produce observable effects.

NOD/LtJ-*Lepr*$^{db\text{-}5J}$. This is a spontaneous mutation that occurred at the Jackson Laboratory in the mouse strain that is nominally predisposed to develop a T cell-mediated autoimmune type 1 diabetes. The mutational site differs in this mutation from the other mouse *Lepr*^{db} mutations; it comprises a single base point mutation converting glycine at residue 640 of the extracellular domain to valine (Lee et al. 2005). Thus, a full-length LEPR-Rb signaling isoform is expressed. Nevertheless, mutants are obese, hyperinsulinemic, and hyperleptinemic, but retain some reproductive capacity. The remarkable observation pertaining to these mutant mice is that the induction of leptin resistance and hyperinsulinemia suppresses the development of type 1 diabetes in mutant mice of both sexes (Lee et al. 2005). This contrasts to an NOD stock in Japan made congenic for the *Lepr*$^{db\text{-}1J}$ mutation introgressed from BKS (Nishimura and Miyamoto 1987). Although this mutant stock also initially develops a type 2 diabetes syndrome, a type 1 diabetes syndrome characterized by insulitic destruction of the islet β-cells eventually prevails. Because the *Lepr*$^{db\text{-}1J}$ lacks the intracellular signaling domain present in the *Lepr*$^{db\text{-}5J}$ stock, the possibility is raised that the latter may maintain some signaling capacity.

With the identification of the exact nucleotide sequence changes in most of the obese and diabetes mutations, it is currently feasible to identify the alleles carried by any individual mouse directly. Indeed, this is actually necessary for the identification of mice in some transgenic rescue studies. Genotyping protocols based on the polymerase chain reaction have been published for the stocks listed in table 3.3 (Chua et al. 2004; Kowalski et al. 2001; Kim et al. 2003; Lee et al. 2005).

The exact nature of the differential BKS/B6 genetic modifiers contributing to the more severe diabetes syndrome on the BKS background has not yet been established. That these background modifiers comprise a large number of interactive quantitative trait loci (QTL) was indicated by results from a (BKS-+/*Lepr*^{db} × DBA/2)F2 genetic analysis (Togawa et al. 2006). Although BKS and B6 differ at *H2*, segregation analysis has eliminated the major histocompatibility complex as a diabetogenic determinant (Leiter, Le, et al. 1987). Rather, studies implicated differential strain metabolism of sex steroids, especially estrogen and dehydroepiandrosterone (DHEA) (Leiter and Chapman 1994). The importance of sex-limited gene expression effects is indicated by the marked sexual dimorphism in pathogenesis evidenced by some of the strain backgrounds shown in table 3.3 (e.g., DBA/2J, C3H/HeB, CBA/J).

Strain differences in β-cell proliferative capacity have also been suggested because the B6-+/+ islet proliferative rate *in vitro* was twice that observed in BKS-+/+ islets (Swenne and Andersson 1984). However, an intrinsic BKS background-mandated limitation in β-cell proliferative activity was questioned by the finding that another recessive mutation on the BKS background, the *fat* allele at the carboxypeptidase E (*Cpe*) locus is not associated with β-cell atrophy, but rather with hypertrophied/hyperplastic islets (Leiter et al. 1999). The difference in β-cell replicative ability may be a secondary response of the BKS genome to chronically elevated glucose in *Lepr*^{db} or *Lep*^{ob} mutations on the BKS background. Studies *in vitro* using wild-type BKS versus B6 islets exposed to high glucose show differential BKS induction of endogenous retroviral genomes that produce defective intracisternal type A particles within the rough endoplasmic reticulum (Leiter et al. 1986).

If hyperglycemia is controlled, as can be done with estrogen therapy (Prochazka et al. 1986), then continued islet hypertrophy/hyperplasia can be observed in BKS-*Lepr*^{db} mutants. In an F2 cross, higher hepatic malic enzyme activity level characteristic of grandparental B6--*Lepr*^{db} mice was associated with less severe hyperglycemia while the lower activity characteristic of BKS--*Lepr*^{db} grandparental mice was associated with greater diabetes severity (Coleman 1992). The locus controlling this phenotype was mapped to chromosome 12 and designated malic enzyme regulator (*Mod1r*). In a repetition of this F2 cross in which atherogenic responses to a high-fat diet were tested, an atherosclerosis susceptibility locus on chromosome 12 conferring higher susceptibility to BKS was indicated (Mu et al. 1999). However, this locus did not appear to control severity of diabetes under the dietary conditions used. Hence, the gene conferring higher atherogenic susceptibility to BKS mice did not appear to be identical to that conferring increased diabetes susceptibility. This latter gene may be the gene encoding lipin (Davis et al. 2005).

Given the strong diabetogenic responses of the BKS versus the B6 strain background to leptin or to leptin receptor gene mutations, it was anticipated that wild-type BKS males would be more responsive to diet-induced obesity than B6 males. In fact, the opposite was true; B6 males rather than BKS males proved to be more susceptible to "Western diet" induction of obesity accompanied by hyperglycemia and hyperinsulinemia (Surwit et al. 1994). This surprising result may, in part, be explained by the finding that islets from B6-+/+ mice exhibit lower insulin secretory responses to glucose when compared with islets from other strains (Wencel et al. 1995; Toye et al. 2005) such that the male B6 mice are less able to cope with diet-induced insulin resistance.

METABOLIC FEATURES OF *Lepr^db^* AND *Lep^ob^* MUTANT MICE

CENTRAL EFFECTS

*Lep^ob^* and *Lepr^db^* mice exhibit metabolic abnormalities found in rats with hypothalamic lesions, albeit in a less pronounced manner (Bray and York 1979). There are several lines of evidence suggesting that the majority of leptin's actions on metabolism, ingestive behavior, and reproduction are mediated by the nervous system. Replacement of LEPR-Rb expression within neurons of *Lepr^db^* mice leads to normalization of body weight, body composition, insulin sensitivity, and fertility (de Luca et al. 2005; Chua et al. 2004; Kowalski et al. 2001). It is noteworthy that the correction of hyperphagia was observed prior to the normalization of any other phenotypic trait characteristic of *Lepr^db^* mice. Deletion of LEPR-Rb expression in neurons leads to the recapitulation of the obesity/diabetes/infertility phenotype of leptin signaling deficiency (Cohen et al. 2001; McMinn et al. 2005).

The major focus of the central effects of leptin has been on hypothalamic neurons, principally the NPY/AGRP neurons (Morton and Schwartz 2001) and POMC/CART neurons (Xu et al. 2005; Cummings and Schwartz 2000) of the arcuate nucleus, the SF1 neurons of the ventromedial nucleus (Dhillon et al. 2006), and the MCH neurons of the lateral hypothalamic area (Shimada et al. 1998; Ludwig et al. 2001; Segal-Lieberman et al. 2003; Jo et al. 2005; Alon and Friedman 2006). In general, all of these neurons show leptin-mediated phosphorylation of STAT3 and nuclear accumulation of pSTAT3. Phosphorylation of STAT3 is the major event upon leptin binding to LEPR-Rb in neurons, so it is generally viewed that pSTAT3 is a marker for leptin activation. Careful analysis of the areas of the hypothalamus and other areas of the brain show that nuclear accumulation of pSTAT3 occurs in many areas outside the arcuate nucleus, VMN, and LHA (Munzberg et al. 2004).

When combined with information regarding the phenotypes of mice with neuron-type specific ablation of LEPR function (Dhillon et al. 2006; Balthasar et al. 2004), it is clear that many neuronal types in several areas of the brain contribute to the leptin signaling deficient phenotype. With the application of cell type-specific targeting to ablate gene expression, it is possible to evaluate the roles of individual neuronal types in leptin biology. The loss of LEPR function in POMC/CART cells leads to a mild obesity without the development of hyperphagia, insulin resistance, or infertility (Balthasar et al. 2004). Similarly, ablation of LEPR within SF1 neurons of the VMN leads to the development of mild obesity only (Dhillon et al. 2006). Further work remains to determine whether the defects in glucose metabolism, ingestive behavior, and reproduction are mediated by specific single cell types or combinations of cell types.

PERIPHERAL EFFECTS

Whereas leptin-deficient *Lep^ob^* mice are exquisitely sensitive to administration of recombinant leptin (Pelleymounter et al. 1995), hyperleptinemic *Lepr^db^* mice are very leptin resistant. This leptin resistance has a clear impact on metabolism in non-neuronal tissue sites. Site specific mutagenesis of a tyrosine to a serine residue in the intracellular domain associated with phosphorylation of STAT3 (T1138S), followed by a genetic "knock-in" into a B6 background, produced mice that were obese and hyperphagic, but not infertile (Bates et al. 2003). The finding that insulin resistance and glucose intolerance were less severe than in standard B6-$Lepr^{db\text{-}1J}$ mice indicates that STAT3-independent LEPR signaling functions must exist (Bates et al. 2005).

Indeed, evidence for direct effects of leptin on a multiplicity of non-neuronal cell types (including skeletal muscle, adipocytes, and β-cells) has been reviewed recently (Ceddia 2005; Fruhbeck 2006). These effects may entail tissue-dependent activation of PI3 kinase or 5′-AMP-activated protein kinase. There have been two transgenic mouse models of leptin overexpression (Ogawa and Nakao 2000; Qiu, Ogus, Lu et al. 2001; Ogus et al. 2003). Initially, both transgenic models are leaner than nontransgenic controls. However, over time, one model of leptin overexpressing mice develops mild obesity that is exacerbated by feeding of a high-fat diet. It is postulated that an interaction between chronically elevated leptin concentrations and high-fat feeding leads to fat accumulation. Presumably, peripheral actions of leptin are involved. Known peripheral leptin effects on various tissue/cell types are summarized next.

Immune System

B6-Lep^{ob} and BKS-$Lepr^{db}$ mice are immunodeficient. In B6-Lep^{ob} mice, reduced weight of the spleen and the thymus (Bray and York 1979), decreased number of IgG-producing lymphocytes (Chandra 1980), and impaired ability to reject skin grafts (Sheena and Meade 1978) reflect impaired immune function. In BKS-$Lepr^{db}$ mice, precocious and severe thymic involution due to a lymphocytic depletion and reduced levels of serum thymic factor (FST) (Dardenne et al. 1983), as well as T-cell lymphopenia and T-cell imbalance (Boillot et al. 1986), indicate immunodeficiency. That the antipancreatic β-cell immunity is secondary in the development of the hyperglycemic syndrome in these type 2 diabetes models has been shown in a study using double mutant mice homozygous for $Lepr^{db}$ in combination with one of several immune deficiency mutations (*Xid*, $Pkrdc^{scid}$, or $Foxn1^{nude}$) (Leiter, Prochazka, et al. 1987).

Chronic administration of recombinant leptin to preweaning NOD females (but not males) accelerates type 1 diabetes onset (Matarese et al. 2002). Reciprocally, the leptin-resistant NOD/LtJ-$Lep^{db\text{-}5J}$ stock is protected from type 1 diabetes due to peripheral suppression of autoreactive T cell activation (Lee et al. 2005, 2006). Other alterations in the immune system elicited by leptin (and insulin) resistance remain to be further elaborated; leptin increases in multiple sclerosis patients have been correlated with suppressed development of T regulatory cells (Matarese et al. 2005). Suppressed T cell activation in B6-Lep^{ob} mice also correlates with protection from autoimmune disease (La Cava et al. 2004). This protection is observed in several models of experimentally induced autoimmunity: experimental allergic encephalomyelitis (EAE) (Matarese, Di Giacomo, et al. 2001; Matarese, Sanna, et al. 2001) and antigen-induced arthritis (Busso et al. 2002). However, defects in leptin signaling also cause a chronic inflammatory response, primarily due to cytokine production from adipocytes and macrophages within adipose tissue. Thus, there is a necessary distinction of leptin's impact on autoimmunity and inflammation.

Platelets

The STAT3 signaling competent isoform of LEPR, LEPR-Rb, is present in platelets (Nakata et al. 1999; Konstantinides et al. 2001, 2004; Giandomenico et al. 2005). Leptin promotes platelet aggregation in the presence of known stimulators, such as ADP. In *Lep*^{ob} and *Lepr*^{db} mice, the thrombi that form in response to vascular injury are unstable and the time to complete occlusion is prolonged. Leptin's ability to promote platelet aggregation has been proposed as a potential factor in increased atherothrombotic events in obesity.

Skeletal Muscle

Leptin is proposed to act acutely on skeletal muscle to promote fatty acid oxidation by phosphorylation and activation of AMP kinase (Tanaka et al. 2005; Minokoshi et al. 2002). Activated AMP kinase phosphorylates and inhibits acetyl-CoA carboxylase (ACC), the enzyme that synthesizes malonyl CoA from acetyl-CoA. The drop in malonyl CoA disinhibits carnitine palmitoyl-CoA (CPT), leading to increased transport of fatty acids into mitochondria and promoting fatty acid oxidation. Over the long term, leptin also activates AMP kinase within skeletal muscle by sympathetically driven input into skeletal muscle.

Adipose Tissue

There is indirect but strong evidence that leptin directly regulates its synthesis in adipose tissue. The correlation of leptin concentrations with adipose tissue mass is strong (Guo et al. 2004). Increased leptin concentrations are observed in rodents that are heterozygous for leptin receptor mutations even after accounting for their increased fat mass (Zhang et al. 1997). Finally, partial loss of leptin receptors in adipose tissue and multiple peripheral tissues leads to elevated leptin concentrations (de Luca et al. 2005). This is with the preservation of leptin receptor function in the nervous system and maintenance of normal body mass. Increased adipose tissue mass is associated with increased numbers of macrophages within adipose tissue (Weisberg et al. 2003, 2006). These macrophages are activated and secrete cytokines that are proinflammatory and contribute to the insulin resistance associated with increased fat mass.

Endocrine Pancreas

A major component of the phenotype associated with leptin signaling deficiency is hyperinsulinemia. Leptin inhibits glucose-stimulated insulin secretion from rodent pancreas in a dose-dependent fashion, in part associated with central mechanisms (Muzumdar et al. 2003). However, direct effects may also be a factor since leptin has recently been colocalized with glucagon in pancreatic alpha cells (Reddy et al. 2004), a counter-regulatory islet cell type. Beta-cell mass expansion is also a major feature of these mouse models. It is likely that insulin resistance and ensuing hyperglycemia is the stimulus to β-cell mass expansion. Even in the BKS strain that exhibits eventual β-cell loss, there is a period of increased β-cell proliferation. In B6-*Lep*^{ob} mice, where the genetic background is permissive to β-cell expansion without eventual β-cell necrosis and islet atrophy, analysis of temporal changes provides convincing data that the expansion is due to proliferation of intraislet cells rather than from cells extrinsic to the islet (Bock et al. 2003, 2005). Strain differences in β-cell responses are an active research area and the problem is being attacked by the production of new congenic strains (Clee et al. 2005; Stoehr et al. 2004; Chua et al. 2002; Luo et al. 2006); the eventual goal is to identify the alleles of the modifier genes.

DYSREGULATED HYPOTHALAMIC–PITUITARY–ADRENAL (HPA) AXIS

Mice with defects in leptin or its receptor exhibit widespread metabolic abnormalities in the hypothalamic–pituitary–adrenal (HPA) axis (Bray and York 1979). Adrenal and thyroidal function has been studied in *Lep*^{ob} and *Lepr*^{db} mice by numerous groups. Hypercorticism, present in BKS-*Lepr*$^{db\text{-}1J}$ and B6-*Lep*^{ob} mutants, reflects dysregulated pro-opiomelanocortin (POMC) release and exacerbates the hyperglycemic syndrome. Leptin is known to stimulate corticotropin-releasing hormone (CRH) mRNA in the paraventricular nucleus (Friedman and Halaas 1998). This is consistent with the observation that CRH administration decreases feeding, oxygen consumption, and grooming activity in B6-*Lep*^{ob} mice (Drescher et al. 1994). Similarly, impaired thyroid function reflects the likely role of leptin as an activator of the thyroid releasing hormone promoter in PVH neurons (Guo et al. 2006).

Leptin mRNA in Tissues and Relative or Absolute Plasma Leptin Concentrations

Only fully developed white adipocytes express leptin mRNA (MacDougald et al. 1995). Interestingly, leptin mRNA in fat depots and plasma leptin are significantly higher in B6-*Lep*^{ob} and BKS-*Lepr*^{db} mice when compared with the respective lean +/+ litter mates, possibly indicating compensatory responses (Chung et al. 1998). Leptin mRNA expression correlates well with fat cell size rather than fat cell number: in *Lep*^{ob}mice of the Aston, United Kingdom, substrain and in BKS-*Lepr*^{db} mice, leptin mRNA expression was found to be higher in intra-abdominal fat pads with large fat cells when compared with subcutaneous fat pads with small cells (Trayhurn et al. 1995; Maffei, Halaas, et al. 1995). In brown adipose tissue, leptin mRNA is markedly lower when compared with white adipose tissue (Moinat et al. 1995; Maffei, Fei, et al. 1995) and differs according to ambient temperature and functional stage of brown adipose tissue (Cancello et al. 1998). Fully differentiated multilocular brown adipocytes expressing uncoupling protein-1 (UCP-1) are leptin negative, whereas unilocular, UCP-negative brown adipocytes localized mainly at the periphery of the intrascapular brown adipose tissue depot express leptin mRNA (Cinti et al. 1997).

Sensitive radioimmunoasay and ELISA kits for measurement of mouse leptin are available from several suppliers. Serum levels vary depending upon inbred strain, gender (females higher than males), and age. In lean mice, serum leptin levels are usually in a range of 5–20 ng/mL. Leptin is absent in plasma from *Lep*^{ob} mice. It is nine to ten times higher in plasma from BKS-*Lepr*^{db} mice when compared with lean litter mates (Frederich et al. 1995; Maffei, Halaas, et al. 1995). Plasma leptin concentrations correlate closely with BMI (body mass index).

Effect of Leptin Administration on *Lep*^{ob} and *Lepr*^{db} Mice

Earlier adipose tissue transplantation studies in the 1950s and 1970s (Hausberger 1958; Ashwell et al. 1977; Meade et al. 1979) failed to demonstrate the essential role of a fat-secreted hormone, but, in retrospect, this may have been a failure to transplant enough fat (Klebanov et al. 2005). The close link between central neuronal networks controlling food intake and the fat cell sending its feeding-inhibitory signal was demonstrated in B6-*Lep*^{ob} mice by various groups at the same time (Campfield et al. 1995; Halaas et al. 1995; Pelleymounter et al. 1995; Stephens et al. 1995; Weigle et al. 1995). The lack of endogenously produced leptin was compensated by the intraperitoneal or intracranial injection of recombinant leptin.

In these experiments, recombinant leptin injected into *Lep*^{ob} mice was followed by a dose-dependent decrease in food intake and a loss of body weight. In B6-*Lep*^{ob} mice a five- to six-hour i.v. infusion of leptin resulted in a 60% increase in glucose turnover when compared with PBS-treated *Lep*^{ob} controls. Hepatic glucose output as well as glucose-6-phosphatase increased, whereas hepatic phosphoenolpyruvate carboxykinase (PEPCK) decreased, suggesting that insulin sensitivity was at least partly restored by leptin. Moreover, leptin-induced glucose uptake was markedly increased in brown adipose tissue (Burcelin et al. 1999). Therefore, normalization of the increased metabolic efficiency after leptin administration (Pelleymounter et al. 1995) came about by a more complete oxidation of FFA in the presence of extra endogenous carbohydrate. In *Lepr*^{db} mice on either background, administration of recombinant leptin remains ineffective due to the defective leptin receptor.

PHYSIOLOGIC AND BIOCHEMICAL CHARACTERISTICS IN B6-*Lep*^{ob} MICE

Adipose Tissue Size

The abnormal adipose tissue enlargement in B6-*Lep*^{ob} mice is due to hyperphagia, increased food utilization, increased lipogenesis, and depressed lipolysis. Milk intake is identical in suckling Aston-*Lep*^{ob} and Aston-+/+ pups (Contaldo et al. 1981). Enlargement of adipose tissue, however, starts before weaning since carcass energy (kilojoules) of Aston-*Lep*^{ob} mice was recorded to be 12 and 40% higher at days 10 and 17 of life, respectively, when compared with lean mice (Thurlby and Trayhurn 1978). Total body fat content was 15% in 6- to 16-day-old B6-*Lep*^{ob} and increased to 60–75% in one-year-old *Lep*^{ob} mice, whereas in controls body fat content remains constantly around 15% on low-fat chow diet (Herberg, unpublished observation).

TABLE 3.4
Fat Cell Volume (× 10³ m³)

Mice	Days 6–16	>1 Year
B6-+/+	33.73	111.49
B6-*Lep*^{ob}/*Lep*^{ob}	198.70	1106.32

ADIPOSE TISSUE LOCALIZATION

In young B6-*Lep*^{ob} mice, the main portion of white adipose tissue is located subcutaneously in the inguinal and axillar region. In one-month-old mutants, the ratio of total/intraperitoneal body fat is 7.1/1 and decreases to 4.1/1 in nine-month-old B6-*Lep*^{ob} mice. The increase in intraperitoneal adipose tissue in older mice is mainly due to an enlargement of visceral adipose tissue (Herberg 1988). In older B6-*Lep*^{ob} mice, epididymal adipose tissue becomes less resilient and less pliable, and assumes a yellowish color (Herberg and Coleman 1977).

CELLULARITY OF ADIPOSE TISSUE

B6-Lep^{ob} mice exhibit the hypertrophic-hyperplastic type of adiposity. In 6- to 16-day-old mutants, mean fat cell volume is about six times higher than in the controls; mean fat cell volume increases with aging. In one-year-old B6-*Lep*^{ob} and lean controls, fat cell volume enlarges by 5.5-fold and 3-fold, respectively, when compared with 6- to 16-day-old pups. In one-year-old B6-*Lep*^{ob} mice, mean fat cell volume is about 10 times that observed in the controls (table 3.4). It is smallest in the retroperitoneal and largest in the subcutaneous pad (Herberg, unpublished). On testing the frequency of adipose cell size distribution in 3- to 18-week-old B6-*Lep*^{ob} and B6-+/+ mice, Kaplan et al. (1976) observed an increase in the frequency of small cells in *Lep*^{ob} mice only, suggesting that, in mutants, the number of adipocytes is not prefixed. Accordingly, preadipocytes isolated from the epididymal and retroperitoneal fat pads of B6-*Lep*^{ob} mice show a greater ability to proliferate than stromal-vascular cells from lean controls (Black and Begin-Heick 1995). Total fat cell number is highest in the subcutaneous and lowest in the gonadal fat pads (Johnson and Hirsch 1972).

THERMOGENESIS

A defective dietary and cold-induced thermogenesis contributes to adipose tissue enlargement. Thus, overfeeding Aston-*Lep*^{ob} and lean mice with a cafeteria type of diet revealed that *Lep*^{ob} mice deposit most of the extra ingested energy into fat, in contrast to lean controls (Trayhurn et al. 1982). About 70% of the heat production in nonshivering thermogenesis is produced in brown adipose tissue mitochondria. In B6-*Lep*^{ob} mice, the heat-producing proton conductance system is less active on cold exposure when compared with controls (Hogan and Himms-Hagen 1980). A diminished Na^+/K^+-ATPase (Bray and York 1971) as well as a masking of nucleotide binding in mutants (Hogan and Himms-Hagen 1980) contributes to the defective thermogenesis. A reduced binding of purine nucleotides (GDP, ADP) to the mitochondrial membrane of brown adipose tissue, as well as an attenuation of the cold-induced increase in uncoupling protein-1 (UCP-1) mRNA, is followed by reduced heat production and increased energy storage in B6-*Lep*^{ob} when compared with controls (Reichling et al. 1988).

Beta$_3$-adrenergic receptors activated by sympathetic stimulation are highly expressed on mouse brown adipocytes, but are dysfunctional in *Lep*^{ob} mice (Collins and Surwit 2001). The decreased GDP binding to brown adipose tissue mitochondria of 14-day-old Aston-*Lep*^{ob} (Goodbody and Trayhurn 1982) confirms the observation that adipose tissue enlargement precedes hyperphagia in mutants. Measurements of oxygen consumption at different environmental temperatures showed that Aston-*Lep*^{ob} mice spend less energy on thermoregulatory thermogenesis when compared with controls (Trayhurn and James 1978). With increasing environmental temperature from 17°C to thermoneutrality at 33°C, lean mice accumulated more energy, whereas Aston-*Lep*^{ob} showed little change in energy gain, reflecting their reduced energy expenditure (Thurlby and Trayhurn 1979). B6-*Lep*^{ob} mice kept at 4°C become hypothermic and die. However, if a 24-h period at 10°C precedes the 4°C period, the mice are able to adapt (Coleman 1982). The defect in thermoregulation, therefore, is only partial and, as shown, not severe enough to explain the increased metabolic efficiency characteristic of B6-*Lep*^{ob} mice (Coleman 1985).

Lipogenesis

De novo lipogenesis is markedly enhanced in B6-*Lep*^{ob} when compared with control mice, especially in young animals. In *Lep*^{ob} mice, hepatic fatty acid synthesis is increased 6-fold per total liver and 2.2-fold per total small intestine (Memon et al. 1994). Recently, fatty acid translocase mRNA and plasma membrane fatty acid binding protein mRNA were reported to be increased in hepatic and adipose tissue in B6-*Lep*^{ob} when compared with lean mice. The increase in hepatic microsomal acyl-CoA synthase activity suggests an increased esterification of FFA in *Lep*^{ob} hepatic tissue (Memon et al. 1999). Hepatic lipogenesis accounts for about 50% of total lipogenesis in B6-*Lep*^{ob} and for 20% in lean control mice. Hepatic and adipose tissue lipogenesis remains clearly enhanced under fasting conditions as shown in pair-feeding experiments. Although the high serum insulin levels favor adipose tissue enlargement, the rate of liver and adipose tissue lipogenesis does not show a strict proportionality to circulating serum insulin levels (Herberg and Coleman 1977). In B6-*Lep*^{ob} mice, re-esterification is also enhanced due to an increased glycerol kinase activity (Stern et al. 1983) that, theoretically, is capable of phosphorylating about 30% of glyceride-glycerol.

Lipolysis

Lipolysis in relation to adipose tissue enlargement has been discussed (Herberg and Coleman 1977). The marked increase in FFA and glycerol in serum from starved B6-*Lep*^{ob} mice reflects a high lipolytic rate as does the high basal and stimulated lipolysis per single fat cell. However, as noted for brown adipocytes described earlier,

β_3-adrenergic receptor function in white adipose tissue from B6-*Lep*^{ob} mice is also markedly decreased (Begin-Heick 1996). The lipolytic defect in adipose tissue, therefore, might be due to the reduced β_3–adrenergic receptor function.

Serum/Plasma Insulin

In mature lean adult controls, mean serum insulin level remained nearly constant throughout lifetime (<2 ng/mL). In B6-*Lep*^{ob} mice, moderately elevated serum insulin can be detected during the periweaning period with concentrations twice as high in 6- to 16-day-old *Lep*^{ob} mice when compared with controls. After weaning and shift to chow diets, circulating insulin concentrations increase as a function of increasing adiposity. In four- and seven-month-old *Lep*^{ob} mice, serum insulin levels were 2 to 3 times and 20 to 30 times, respectively, the level observed in controls. The highest serum insulin levels were observed in seven-month-old *Lep*^{ob} mice weighing about 75 g. Thereafter, serum insulin levels spontaneously declined and showed considerable variation among individual mutant mice. Multiple alterations in the early steps of intracellular insulin signaling in liver and muscle cells from *Lep*^{ob} mice are responsible for the insulin insensitivity of the target organs (Kerouz et al. 1997).

In general, highest insulin levels were observed in B6-*Lep*^{ob} males, indicating a more severe insulin resistance in mutant males as compared with mutant females. This is consistent with the transient hyperglycemia common in mutant males, but rare in mutant females. Presumably, the more pronounced insulin secretion in males is necessary for this remission from hyperglycemia. As reviewed previously (Friedman and Halaas 1998), high serum insulin levels are correctable by leptin in a dose-dependent manner. The effects of fasting or different composition of diets on serum or plasma insulin concentrations have been discussed earlier (Herberg and Coleman 1977). Basal and stimulated insulin secretion is increased in B6-*Lep*^{ob} mice *in vivo* and from islets *in vitro*. Arginine stimulation is followed by a prompt and highly pronounced increase in insulin levels.

After a glucose challenge, however, serum insulin initially drops and the subsequent increase is delayed. A 16-h prefasting period abolishes the delayed glucose-induced insulin response, whereas arginine-stimulated insulin release remains unaffected by fasting. Likewise, long-term food restriction keeping body weight of B6-*Lep*^{ob} mice within the normal range leads to a reduced glucose-induced insulin secretion known as "starvation diabetes" (Herberg and Coleman 1977). Various factors known to affect insulin secretion by different mechanism have been reviewed by Chan (1995).

The link between adipocyte and β-cell has been demonstrated by various groups. The findings of leptin receptor mRNA expression on pancreatic β-cells from normal rats (Kieffer et al. 1996) and of an insulin-induced up-regulation of leptin gene expression in adipose tissue (Mizuno et al. 1996) suggested an adipo-insular feedback loop or an adipo-insular axis. Thus, in normal and Aston-*Lep*^{ob} mice, leptin administration led to a marked decrease in plasma insulin levels, whereas blood glucose increased (Kulkarni et al. 1997). In isolated pancreatic islets from normal mice and rats, leptin failed to depress basal insulin secretion whereas insulin secretion stimulated by glucose plus isobutylmethylxanthine (IBMX) was significantly inhibited (Poitout et al. 1998). However, in isolated islets from Aston-*Lep*^{ob} mice (islets from *Lep*^{ob} mice are very sensitive to leptin), recombinant leptin acutely inhibited basal and glucose-stimulated insulin secretion (Emilsson et al. 1997). That leptin directly acts at the β-cell was demonstrated by the leptin-induced activation of ATP-sensitive K^+ channels, which was reversible by the specific K_{ATP} inhibitor tolbutamide (Kieffer et al. 1997), and the concomitant leptin-induced decrease in glucose-stimulated intracellular Ca^{2+} concentration in islets from *Lep*^{ob} mice and normal mice and rats (Kieffer et al. 1997; Fehmann et al. 1997).

Furthermore, a leptin-induced decrease in intracellular Ca^{2+} concentration was shown to be reversible by glucose plus GLP-1 and leptin suppression of insulin release was overcome by glucose plus GLP-1 (Kieffer et al. 1997). The maximal insulin secretion inhibiting effect of leptin, which occurred in normal rat islets after a prolonged exposure to leptin, suggested the leptin effect to be mediated through a change in proinsulin gene expression (Roduit and Thorens 1997). Recently, it was shown in isolated human pancreatic islets that leptin directly inhibits basal and stimulated insulin secretion by interfering with the secretory process and suppressing the GLP-1-stimulated expression of proinsulin mRNA (Seufert et al. 1999).

Blood Glucose

In B6-*Lep*^{ob} mice, hyperglycemia is transient. In one-month-old mutants of the Düsseldorf colony, nonfasted blood glucose was slightly lower than in controls (data not shown). Highest blood glucose levels occurred in four-month-old B6-*Lep*^{ob} mice, at which time serum insulin was still increasing. In B6-*Lep*^{ob} older than five months of age, blood glucose significantly dropped by 40% in either gender. From the fifth to sixth month of life until the thirteenth month, blood glucose remained nearly constant. The high circulating plasma glucagon levels as measured in B6-*Lep*^{ob} of the Düsseldorf colony (Herberg, unpublished) and by Flatt et al. (1980) in Aston-*Lep*^{ob} mice may contribute to the pronounced hyperglycemia in *Lep*^{ob} mice. The possible relationship between blood glucose levels and the activity of the gluconeogenic and glycolytic pathway has been discussed in detail (Herberg and Coleman 1977). Numerous studies on the relationship between blood glucose and the adrenal cortex revealed hypercorticosteronemia to be secondary in the development of the obese-hyperglycemic syndrome in *Lep*^{ob} mice (Herberg and Coleman 1977).

PHYSIOLOGIC AND BIOCHEMICAL CHARACTERISTICS OF BKS-*Lepr*^{db} MICE

BKS-*Lepr*^{db} mice resemble B6-*Lep*^{ob} mice in terms of a rapid development of juvenile obesity after weaning. However, the obesity-induced diabetes (diabesity) syndrome is considerably more severe and does not remit such that life span in most colonies is considerably shortened. Another major distinction between any *Lep*^{ob} and *Lepr*^{db} stock is that the latter is leptin resistant due to the defective leptin receptor (Friedman and Halaas 1998).

Obesity Development

BKS-*Lepr*^{db} mice, like B6-*Lep*^{ob} mice, are hyperphagic and three- to four-week-old mutants can be identified by their plump appearance. In BKS-*Lepr*^{db} mice, the pattern of body weight gain can differ from colony to colony. In principle, two developmental weight gain patterns have been observed. The one type described by Coleman and Hummel (1967) is characterized by a continuous increase in body weight up to 50–60 g in three- to four-month-old mice during which time circulating insulin is significantly elevated. With further aging, the mutants begin to lose weight and spontaneous death occurs at five to eight months of life. This "Bar Harbor" pattern of reduced survivorship correlating with increasing blood glucose, large declines in circulating insulin, declines in body weight to below 45 g, and unthrifty appearance after five months of chronic diabetes was considered to be representative (Coleman and Hummel 1967). In the colony maintained by one of us (L.H.) in Düsseldorf, the Bar Harbor pattern of unremitting, severe hyperglycemia was seen in most of the mice.

However, a second syndrome progression was observed in a subset of BKS-*Lepr*^{db} mice of the colony. This was characterized by continuous weight gains to 75–80 g associated with a more moderate hyperglycemia (~300 mg/dL), sustained high serum insulin concentrations, and extended survivorship (to 8–12 months). As discussed later, this extended survivorship very likely reflected more aggressive environmental management practice. Currently, the BKS-*Lepr*^{db} mice distributed by the Jackson Laboratory show less β-cell compensation during the early phase of syndrome, such that mutant mice are severely hyperglycemic by 16 weeks of age (plasma glucose > 500 mg/dL). These mice seldom attain weights > 45 g, becoming unthrifty and losing weight after this maximum weight is attained. This terminal stage is characterized by major declines in circulating insulin and widespread islet atrophy. These latter mice nevertheless maintain an adipose body composition during the period of weight loss (e.g., lean mass is diminished while fat tissue is conserved).

Adipose Tissue

In young BKS-*Lepr*^{db} mice, white adipose tissue is mainly localized in the inguinal and axillar subcutaneous region. Increasing body fat content in older BKS-*Lepr*^{db} mice is mainly due to visceral fat increase. Total body fat content is 16% in 6- to 16-day-old mice, 33% in 2-month-old mice, and 53% in 1-year-old BKS-*Lepr*^{db} mice (unpublished). During the first, dynamic phase, body fat mass increases mainly by hypertrophy (Herberg and Coleman 1977). Mean fat cell volume is smaller until weaning in BKS-*Lepr*^{db} when compared with B6-*Lep*^{ob} mice—that is, 20% of the value recorded in 6- to 16-day-old *Lep*^{ob} and 80% of the value recorded in 18- to 22-day-old *Lep*^{ob} mice. In 26-week-old male BKS-*Lepr*^{db} mice, wet weight, fat cell volume, and fat cell number of the retroperitoneal, gonadal, and subcutaneous adipose tissue are lower when compared with B6-*Lep*^{ob} mice (Johnson and Hirsch 1972). However, BMI is identical in B6-*Lep*^{ob} and BKS-*Lepr*^{db} mice.

Thermogenesis

Sympathetic stimulation of brown adipose, responsible for a major portion of non-shivering thermogenesis in mice, is reduced in *Lepr*^{db} mice as it is in *Lep*^{ob} mice (Collins and Surwit 1996). Although energy expenditure on thermoregulatory thermogenesis was reported to be reduced in BKS-*Lepr*^{db} mice (Trayhurn 1979), Coleman (1985) concluded that *Lepr*^{db} and *Lep*^{ob} mice (on the BKS or the B6 background) used equivalent amounts of energy for heat production compared to lean controls. In pair-fed mutants, the energy used by controls for thermoregulation (33%) was found insufficient to account for the increased metabolic efficiency in the mutants.

Lipogenesis

BKS-*Lepr*^{db} mice are characterized by an increased lipogenesis and high plasma triglyceride and FFA levels. Compared with lean controls, *Lepr*^{db} mice showed a 9-fold higher hepatic and a 10.5-fold higher small intestinal fatty acid synthesis when expressed on a per-organ basis (Memon et al. 1994). In BKS-*Lepr*^{db} and B6-*Lep*^{ob} mice, the hypothesis that adipsin, a serine protease homologue of complement factor D in humans, acts as a fat cell-derived regulatory molecule (Flier et al. 1987) could not be confirmed. The age- and obesity-related pattern of adipsin revealed that the impairment of adipsin expression is secondary to obesity (Dugail et al. 1990).

Blood Glucose

Insulin resistance-mediated alterations in hepatic enzymes associated with carbohydrate metabolism have been documented in BKS-*Lepr*^{db} mice (Coleman and Hummel 1967). Chronic and increasingly more severe hyperglycemia is the most prominent feature distinguishing BKS-*Lepr*^{db} and BKS-*Lep*^{ob} mice from the same mutations on the B6 inbred background. Currently, significant elevations in blood glucose are detected in BKS-*Lepr*^{db} and BKS-*Lep*^{ob} mice at the Jackson Laboratory within a week after weaning. By six to eight weeks of age, most individuals exhibit plasma glucose levels of >400 mg/dL, and by 12 weeks of age values range at 500–800 mg/dL, with males generally exhibiting higher values than females. An occasional mutant female will exhibit a B6-like pattern (e.g., mild, if any, hyperglycemia with compensatory sustained hyperinsulinemia). A carbohydrate-free diet initiated at weaning retarded or even prevented the development of the extreme hyperglycemia characteristic of chow-fed BKS-*Lepr*^{db} mice (Leiter et al. 1983).

Plasma and Pancreas Insulin

Normal plasma insulin level in BKS-+/+ mice is approximately 2 ng/mL. BKS-*Lepr*^{db} mice exhibit increased plasma insulin levels at two weeks of age when compared with lean controls. Increased plasma insulin levels persist until the end of the third month of age and then decrease. Maximum plasma insulin levels in mutant males in a research colony at the Jackson Laboratory rarely exceeded 10 ng/mL. When mutants become unthrifty and body weights decline, plasma insulin drops to ~2 ng/mL. Already in one-month-old mutants, pancreatic insulin content is lower when compared with controls. Whereas pancreatic insulin remains similar in one- to five-month-old BKS-+/+ mice, in BKS-*Lepr^db^* mice it declines progressively. In five-month-old mutants, pancreatic insulin is about 20% that measured in controls, thus reflecting increasing islet atrophy and β-cell degeneration (Coleman and Hummel 1969b).

Insulin Resistance

In BKS-*Lepr*^{db} mice, insulin administration up to 100 U/100 g body weight failed to normalize blood glucose concentrations of about 250 mg/dL (Coleman and Hummel 1967). The persistence of reduced numbers of hepatocyte plasma membrane insulin receptors in BKS-*Lepr*^{db} mice in which hyperinsulinemia was abolished after estrone treatment indicated that the reduced insulin binding capacity resulting from reduced numbers of insulin receptors was in some way closely marking the primary genetic defect (leptin resistance) rather than reflecting secondary ligand-mediated down-regulation (Prochazka et al. 1986).

It is also pertinent that administration of exogenous semilente insulin to BKS-*Lepr*^{db} mice resulted in reduction of liver PEPCK activity and PEPCK-mRNA content along with decrease almost to normal of plasma glucose levels. However, the attenuation of insulin resistance was achieved only after two-day treatment at plasma insulin levels of >1000 μU/mL, much higher than those effecting suppression of PEPCK in streptozotocin diabetic mice (Shafrir 1988).

BEHAVIORAL CHARACTERISTICS OF B6-*Lep*^{ob} AND BKS-*Lepr*^{db} MICE

Physical Activity and Demand for High Ambient Temperature

Male *Lep*^{ob} and *Lepr*^{db} mice do not exhibit any intrastrain aggressiveness as the lean controls do. B6-*Lep*^{ob} and BKS-*Lepr*^{db} mice exhibit considerably reduced physical activity secondary to obesity (Yen and Acton 1972). Due to a defective brown adipose tissue thermogenesis (Himms-Hagen 1985) B6-*Lep*^{ob} mice prefer higher (25–35°C) ambient temperature (Wilson and Sinha 1985) facilitating a mean core temperature of about 36.5°C (Carlisle and Dubuc 1984). *Lep*^{ob} and *Lepr*^{db} mice exhibit a higher rate of heat loss when compared with lean controls (Bellward and Dauncey 1988). Therefore, during resting periods, *Lep*^{ob} and *Lepr*^{db} mice huddle closely to each other to retain heat.

Food Intake

Hyperphagia, one of the main characteristics of *Lep*^{ob} and *Lepr*^{db} mice, is not a prerequisite of obesity since adipose tissue mass of mutants pair-fed with lean mice remains still higher and even increases due to an increased metabolic efficiency in mutants (Coleman 1985). The increase in meal size and the proportionally higher food intake in the light period of the light–dark cycle in male B6-*Lep*^{ob} mice when compared with controls is easy to understand in view of the lack of leptin in *Lep*^{ob} mice.

The equal number and periodicity of meals in *Lep*^{ob} and lean mice in either 12-hour period in the light–dark cycle points to a normal circadian feeding pattern in *Lep*^{ob} mice (Ho and Chin 1988). When either gender of *Lep*^{ob} mice was studied, it became obvious that female *Lep*^{ob} ate fewer meals in the dark when compared with male *Lep*^{ob} and that the number of meals ingested in the dark was higher in lean when compared with *Lep*^{ob} mice, respectively. The number of meals consumed during the light period was similar in *Lep*^{ob} and lean mice of either gender (Strohmayer and Smith 1987).

B6-*Lep*^{ob} and BKS-*Lepr*^{db} mice show a clear preference for sweet liquid diets (Sprott 1972). On a free-choice feeding, B6-*Lep*^{ob} mice prefer a high-carbohydrate diet (Mayer et al. 1951). In this study the diets offered to mice consisted of about 90% protein, 90% carbohydrate, or 90% fat. Therefore, the lesser intake (g) of the fat-enriched diet when compared with the slightly higher intake (g) of the high-carbohydrate diet resulted in a nearly 100% higher intake of calories from fat and reflects the poor adjustment to dietary diluted diets rather than a preference for high-fat diets, as concluded by the authors and frequently cited in the literature. Life span of B6-*Lep*^{ob} females can be extended by diet restriction, although the mutants retain their adipose body composition (Harrison et al. 1984). BKS-*Lepr*^{db} mice are highly sensitive to dietary carbohydrate (Leiter et al. 1983). A carbohydrate-free diet initiated at weaning retarded or even prevented the development of the extreme hyperglycemia characteristic of chow-fed BKS-*Lepr*^{db} mice.

Stress Responses

Enhanced sensitivity to various forms of stress is also a characteristic of *Lep*^{ob} and *Lepr*^{db} mice. The observation that food-deprived and physically restrained B6-*Lep*^{ob} mice developed gastric stress ulcers and became hypothermic (Greenberg and Ackerman 1984) points to a possible relationship between the autonomous nervous system and the leptin–leptin receptor loop. As discussed previously, various hypothalamic nuclei containing leptin receptors are known to modulate the activity of the sympathetic and the parasympathetic nervous systems (Friedman and Halaas 1998). Usually, sympathetic activity is reduced in genetic obesity (Bray 1991). Thus, blood glucose was increased by systemic administration of epinephrine to B6-*Lep*^{ob} mice or by social stress evoked by different grouping conditions (Surwit and Williams 1996). An exaggerated peripheral response to catecholamines is suggested to contribute to stress-induced hyperglycemia in B6-*Lep*^{ob} mice (Kuhn et al. 1987).

Central injection of epinephrine or the a-2 agonist clonidine into B6-*Lep*^{ob} mice was followed by an enhanced feeding response at the beginning of the dark period of the light–dark cycle (Currie and Wilson 1993). The hyperphagic effect of norepinephrine and clonidine as well as the anorectic effect of 5-hydroxytryptamine (serotonin) was shown to be dose dependent and nutrient selective in that B6-*Lep*^{ob} mice showed an increased preference for carbohydrate (Currie 1993).

Opioids modulate food intake as well as behavioral activation when administered to BKS-L*epr*^{db} (Levine et al. 1982) and B6-*Lep*^{ob} mice (Shimomura et al. 1982; Calcagnetti et al. 1987). Experiments on the effect of the opiate antagonist naxolone on grooming, rearing, and jumping following an immobilization or heat stress or a combination of both revealed possible links between endogenous opioids (endorphins) and behavioral responses in B6-*Lep*^{ob} and lean controls. Under the conditions tested, the naxolone effect was clearly stronger in *Lep*^{ob} when compared with the lean controls (Amir 1981).

SECONDARY COMPLICATIONS

VASCULOPATHY

Microvascular (Bohlen and Niggl 1979, 1980) and macrovascular (Kamata and Kojima 1997) lesions are present in BKS-*Lepr*^{db} mice. Increased protein glycation in BKS-*Lepr*^{db} mice is suggested to lead to a generalized endothelial damage (Cohen et al. 1996).

NEUROPATHY

In contrast to B6-*Lep*^{ob} or B6-*Lepr*^{db} mice, BKS-*Lepr*^{db} develop neuropathy (Hanker et al. 1980). Functional impairment of motor and sensory conduction velocity precedes degenerative changes. Thus, a decrease in MNCV (motor nerve conduction velocity) was observed in five-week-old BKS-*Lepr*^{db} mice. In seven-week-old *Lepr*^{db}, MNCV was reported to be significantly lower when compared with BKS-*Lepr*^{db} +/+ *m* mice. MNCV was significantly improved by insulin treatment of young mice (Robertson and Sima 1980) or ganglioside treatment of older BKS-*Lepr*^{db} mice (Norido et al. 1984). Therefore, the first phase of neuropathy was labeled "metabolic" and the second phase "neuronal."

Distal hind limb nerves are most affected by functional defects. This can be verified by a simple test: When lifted by the tail, BKS-*Lepr*^{db} mice adduct their hind limbs and keep them near the belly with the digits clenched. With increasing age and severity of the syndrome, the frequency of adduction increases. In contrast, BKS-+/+ and B6-*Lep*^{ob} extend their legs when held in this position (Carson et al. 1980; Hanker et al. 1980). Before any degenerative morphological changes become visible, axonal protein transport is markedly decreased (Vitadello et al. 1983; Calcutt et al. 1988). In myelinated and unmyelinated fibers, axons are swollen and contain conglomerates of membranous profiles. Honeycombed Schwann cell-axon networks develop and are followed by axonal atrophy (Sima and Robertson 1979). In Schwann cells, RER (rough endoplasmatic reticulum) is dilated; mitochondria are swollen and contain electron-lucent vacuoles (Carson et al. 1980). Loss and shrinkage of myelinated fibers occur and diameters become smaller in myelinated and unmyelinated fibers (Robertson and Sima 1980). The sequence of events suggests metabolic abnormalities as the primary cause of functional nerve defects rather than degenerative processes (Sima and Robertson 1979).

NEPHROPATHY

An early symptom of nephropathy is an increased GFR (glomerular filtration rate) determined by total clearance of ^{51}Cr-EDTA. GFR is increased in BKS-*Lepr*^{db} with extremely elevated blood glucose and in B6-*Lepr*^{db} with moderately elevated blood glucose. In older mutants, GFR decreases slowly and approaches the levels of controls, presumably because of advanced glomerular damage (Gartner 1978). Glomerular damage becomes evident in one-month-old BKS-*Lepr*^{db} mice by the deposition of immunoglobulins in the glomerular mesangium (Lee and Graham 1980), which leads to marked thickening of the mesangium. Immunoglobulins are deposited mainly at the glomerular hilus and in the juxtaglomerular mesangium of the juxtaglomerular apparatus (Bower et al. 1980). Immunoglobulin deposits in the distal tubuli are seen in BKS-*Lepr*^{db} of all ages and in hyperglycemic B6-*Lep*^{ob} mice (Meade et al. 1981) and are suggested to be derived from the mesangium of associated glomeruli. Albuminuria is present only in BKS-*Lepr*^{db} mice older than six months of age (Meade et al. 1981).

In BKS-*Lepr*^{db} mice, immunoglobulin deposition can be prevented by normalization of blood glucose by dietary means (Lee and Bressler 1981) or pharmacological intervention (Lee et al. 1982). Arginine and its metabolites reduce the accumulation of collagen in the glomerular basement membrane of BKS-*Lepr*^{db} mice by their inhibitory effect on advanced stage nonenzymatic glycosylation end products (AGEs) (Weninger et al. 1992; Lubec et al. 1994; Marx et al. 1995). Recently, treatment of BKS-*Lepr*^{db} mice with monoclonal antibodies specific for Amadori-modified glycated albumin prevented the formation of AGEs, reduced albuminuria, and decreased the elevated type IV collagen as well as the elevated fibronectin gene expression by about 50%. Concomitantly, mesangial enlargement of glomeruli was markedly reduced and the increase in serum creatinine as well as the decrease in creatinine clearance was prevented while the rise in blood urea nitrogen was attenuated (Cohen et al. 1996). The absence of receptors for advanced glycation end products (AGEs) in renal tissue (RAGE) from BKS-*Lepr*^{db} mice suggests that factors other than hyperglycemia affect RAGE expression (Ziyadeh et al. 1997). However, hyperglycemia may affect cell cycle arrest and hypertrophy of mesangial cells as demonstrated by the expression of p27^{Kip1} in nuclei of mesangial cells (Wolf et al.1998).

The chronically hyperglycemic FVB-*Lepr*^{db} mouse develops a diabetic nephropathy. Obese FVB mice develop albuminuria and histological lesions of the glomeruli and tubules that are highly reminiscent of human diabetic nephropathy (Chua et al. 2002; Wang et al. 2005). As diabetic mice of the FVB strain develop comparable pathology to that reported for OVE26 transgenic mice (Zheng et al. 2004) and A-ZIP transgenic mice (Suganami et al. 2005), it is evident that the FVB strain harbors alleles for developing severe diabetic nephropathy. Interestingly, in none of the models do the animals develop renal failure with elevated creatinine or BUN.

Myocardial Disease

Lipid droplets are scattered in myocytes from BKS-*Lepr*^{db} and B6-*Lep*^{ob} mice. Atrophy of cardiomyocytes increases with age in BKS-*Lepr*^{db} mice only (Giacomelli and Wiener 1979). Degenerative changes of mitochondria and abnormalities in various mitochondrial enzyme activities are also observed in BKS-*Lepr*^{db} mice only (Skoza et al. 1980; Kuo et al. 1983, 1985). Heart collagen accumulation can be reduced by arginine (Khaidar et al. 1994), presumably by inhibiting AGE formation.

Retinopathy

Studies on retinopathy in mutants are scarce. In BKS-*Lepr*^{db} mice, a significant decrease in retinal, intramural pericytes has been recorded (Midena et al. 1989). In lens tissue, sorbitol levels were found to be similar in B6-*Lep*^{ob} mice and lean litter mates and to be higher in BKS-*Lepr*^{db} when compared with lean litter mates. Feeding of an aldose reductase inhibitor resulted in a 70% decrease in lens sorbitol levels in BKS-*Lepr*^{db} mice (Vicario et al. 1989).

REPRODUCTIVE CHARACTERISTICS IN B6-*Lep*^{ob} MICE

Recently, it was shown that administration of recombinant leptin was followed by successful pregnancies in B6-*Lep*^{ob} mice (Chehab et al. 1996). Hence, leptin treatment of mutants provides an effective new method to directly propagate mutants. Depletion of adipose depots by food reduction did not restore the greatly reduced reproductive capacity of female *Lep*^{ob} mice (Drasher et al. 1955). However, elicitation of fertility by pituitary extracts (Runner 1954), hypothalamic extracts (Batt 1972), or gonadotropic hormones (Runner and Gates 1954) led to the suggestion of an alteration of the hypothalamic–pituitary system in female *Lep*^{ob} mice. In male *Lep*^{ob} mice, an impaired response to LH-RH (Swerdloff et al. 1978) and an inadequate release of LH-RH (Batt et al. 1982) led to the suggestion of a defect in the hypothalamic–pituitary axis.

In male *Lep*^{ob} mice, the descent of testes into the poorly developed scrotum is usually incomplete. A poor vascularization of the testes and the accessory glands as well as reduced weights of testes, epididymides, and seminal vesicles is common in *Lep*^{ob} mice (Jones and Harrison 1958). Leydig cell mass is reduced as is nuclear size of the frequently atrophic Leydig cells. In B6-*Lep*^{ob} males, an incomplete spermatogenesis has been described (Jones and Harrison 1958), whereas in V stock-*Lep*^{ob} mice, an intact spermatogenesis was reported (Hellman et al. 1963). Reproductive capability may be intact in food-restricted and food-nonrestricted males (Lane and Dickie 1954; Lane 1959). That hypogonadism is secondary was shown by the prompt LH-stimulated testicular production of testosterone *in vivo* (Swerdloff et al. 1976) and *in vitro* (Wilkinson and Moger 1981). Accordingly, administration of testosterone propionate was followed by nearly normal enlargement of the accessory glands (Jones and Harrison 1958).

In female B6-*Lep*^{ob} mice, ovaries contain a few Graffian follicles only, whereas corpora lutea are totally lacking. The uteri contain only 50% of the number of cells when compared with lean controls (Drasher et al. 1955). Female *Lep*^{ob} never come into estrus (Jones and Harrison 1958). The reproductive organs remain immature. Uterine tissue promptly responds to cessation of estrogen therapy by regression (Drasher et al. 1955). However, after induction of ovulation by gonadotropin, ova and ovaries transplanted into lean mice can produce viable offspring (Runner and Gates 1954). Implantation of fertilized ova from gonadotropin-stimulated obese females can be protected by daily injections of progesterone from the day of copulation until day 18 p.c. (Smithberg and Runner 1957).

REPRODUCTIVE CHARACTERISTICS IN BKS-*Lepr^db^* MICE

In BKS-*Lepr^db^* and *Lepr^db-ad^* mice kept on a heterogenous background (Falconer and Isaacson 1959), testes were described to be normal or slightly reduced in size. Numbers of Leydig cells, spermatozoa, or tailed spermatides are also reduced (Johnson and Sidman 1979; Batt and Harrison 1960). Ovaries are filled with Graffian follicles at all stages of development and lack any corpora lutea; the uteri are small and immature (Batt and Harrison 1963). Ovary transplantation into lean controls resulted in the successful production of litters (Johnson and Sidman 1979). Estrogen treatment resulted in prompt disappearance of the numerous intracellular uterine lipid vacuoles (Garris 1989). Measurements of gonadotropins before and after stimulation by releasing hormones in male and female *Lepr^db^* mice revealed the sterility to be of central origin (Johnson and Sidman 1979). However, a "knock-in" replacement of the tyrosine residue at position 1138 responsible for JAK2/STAT3 activation with an inactivating serine residue produced hyperphagia and obesity, but not sterility. This demonstrated distinct LEPR-Rb signaling pathways for neuroendocrine regulation of energy balance versus reproduction (Bates et al. 2003).

GESTATIONAL DIABETES IN BKS-*Lepr^db^* MICE

Gestastional diabetes and macrosomia in pups have been described in heterozygous *Lepr^db^*/+ mice (Kaufmann et al. 1987). A recent study revealed hepatic glucose production as well as plasma leptin levels to be markedly elevated in pregnant hyperglycemic *Lepr^db^*/+ mice. Since down-regulated insulin receptor β (IR-β), insulin receptor substrate 1 (IRS-1), and phosphoinositol (PI) 3-kinase could be stimulated in heterozygous, pregnant, transgenic mice overexpressing the human GLUT 4 gene, the authors suggested abnormalities in insulin receptor signaling in maternal muscle and a defect in insulin secretion as well as an enhanced nutrient availability to be the cause of gestational diabetes. A direct relationship between maternal hyperglycemia and fetal macrosomia was excluded (Ishizuka et al. 1999).

OTHER FEATURES OF BKS-*Lepr^db^* AND/OR B6-*Lep^ob^* MICE

KIDNEY PATHOLOGY

Polycystic kidneys and hydronephrosis occur in BKS-+/+ mice. The frequency increases with aging and can be as high as 63% in 11- to 15-week-old BKS-+/+ (Weide and Lacy 1991). In the former Düsseldorf colony, hydronephrosis or polycystic kidneys occurred in 25/57 and 20/36 BKS-+/+ at weeks 15–25 and 30–54, respectively. BKS-*Lepr^db^* mice were found to be free from kidney anomalies. However, 4/30 BKS-*Lep^ob^* mice had developed polycystic kidneys at week 25.

Resistance to Atherosclerosis

Plasma lipid and cholesterol levels are elevated in B6-*Lep*^{ob} and BKS-*Lepr*^{db} mice (Herberg and Coleman 1977). By fractionating total cholesterol into its constituents HDL-C and combined VLDL + LDL-C, Nishina, Lowe, et al. (1994) found the elevation of cholesterol levels to be due to an increase in the HDL subfraction. *Lep*^{ob} and *Lepr*^{db} mice on the B6 or BKS backgrounds exhibited higher HDL-C levels than homozygous or heterozygous lean controls. In B6-*Lepr*^{db} mice between the second and fourth weeks of life, HDL-C and blood glucose levels transiently increased. Since insulin precedes the early blood glucose increase in mutants, the authors determined plasma HDL-C levels in 14-week-old and 14-month-old female *Lepr*^{db} mice. B6-*Lepr*^{db} mice in which plasma insulin levels remain high throughout life showed similar HDL-C levels at either time. In BKS-*Lepr*^{db}, in which plasma insulin usually drops with aging, HDL-C levels were clearly higher in the 14-week-old when compared with the 14-month-old group. However, HDL-C levels were lower in BKS-*Lepr*^{db} when compared with B6-*Lepr*^{db} mice. VLDL + LDL-C levels were remarkably low in either age group. These results led the authors to speculate that cholesterogenesis is stimulated by insulin.

In a second set of experiments, Nishina, Naggert, et al. (1994) evaluated aortic lesions in *Lep*^{ob} and *Lepr*^{db} females on the B6 and the BKS backgrounds. Wild-type and mutant males were resistant to atherosclerosis under the dietary conditions used. After long-term (14 weeks) feeding of a high-fat, high-cholesterol diet, *Lep*^{ob} and *Lepr*^{db} mutant females exhibited reduced lesion area in comparison with normal females. Presumably, the greater resistance of the mutants to diet-induced atherosclerosis compared to the controls was due to higher levels of HDL-C in the former. In BKS controls, the lesions were twice as large when compared with B6 controls. In *Lep*^{ob} and *Lepr*^{db} mutants, aortic lesions were significantly larger if the mutations were carried on the BKS background. A BKS allele on proximal chromosome 12, designated *Ath6*, was found to be a significant contributor to the increased atherogenic susceptibility of this strain in a segregating F2 population produced by outcross of BKS-*Lepr*^{db} and B6-*Lepr*^{db} heterozygotes (Mu et al. 1999). Interestingly, a malic enzyme regulator gene (*Mod1r*) had been mapped into this region previously in the same type of F2 cross, with the allele conferring low malic enzyme activity (*Mod1r*^{b}) carried by BKS associated with more severe diabetes (Coleman and Kuzava 1991). However, the BKS-derived *Ath6* atherogenic allele did not correlate with glycemic state (Mu et al. 1999).

HUSBANDRY CONSIDERATIONS

As noted earlier, mutants are quite stress sensitive. Hence, access to the mouse room should be limited to the personnel involved in care of the mice and mice should not be disturbed during the dark period. Mouse rooms are customarily set for 14 h/10 h or 12 h/12 h light–dark cycles. Procedures for entry will depend upon the barrier level. A complete barrier facility is not necessary for maintaining a high frequency of diabetes in BKS-*Lepr*^{db} and *Lepr*^{ob} mice. However, because the mutant mice, particularly on the BKS background, become increasingly immunodeficient as they age, the facility should be free of the standardly encountered major murine pathogenic agents. This requires constant monitoring by veterinary care staff and diligence on the part of the investigator to prohibit introduction into the same room of other research animals or biologicals that have not been prescreened to be specific pathogen free (SPF).

To ensure a clean husbandry environment, caretakers should wear body covering. (A clean lab gown is a minimum requirement; donning of clean booties over shoes and a hair net are recommended.) Caretakers should wear gloves and handle mice by means of stainless steel forceps dipped in an iodine disinfectant solution. The room and associated clean supply area should be positively pressurized by HEPA-filtered, humidified air to remove particles of 0.3 µm or larger. Mouse cages may be kept on open shelves if covered by sterilizable filter bonnets, or they may be held in pressurized, individually ventilated (PIV) caging systems or in microisolator cages. Autoclaved or otherwise processed clean materials should be kept in a clean supply area and dirty materials removed from the mouse room into a separate dirty-materials corridor. HEPA-filtered cage changing stations are recommended.

Because of the rapid postweaning weight gains exhibited by mutants on BKS and B6 backgrounds, a diet with a lower fat content (4–6%) should be selected. Protein is generally between 18 and 21%. Diets should be autoclavable. Drinking water should be acidified (with hydrochloric acid to attain a pH of 2.8–3.2) to prevent growth of *Pseudomonas* sp. Alternatively, hyperchlorinated water (10 ppm sodium hypochlorite) may be used. Water bottles should be checked frequently; since the mutant mice are polydipsic, their water consumption is higher than that of lean controls. Bottles should be changed twice per week.

Bedding material can be autoclaved pine shavings of 2-mm length or other commercially available absorptive material. BKS-$Lepr^{db}$ and $Lepr^{ob}$ mice become increasingly polyuric as the severity of diabetes increases. Although the bedding material used absorbs the moisture quite well, the quantity of urine produced and the peculiarity of the mice to urinate into a corner of the cage requires a minimum of two cage changes per week. Daily changing of the cages to minimize exposure of the mice to a wet substratum greatly extends life span of BKS-$Lepr^{db}$ mice. Use of hygroscopic bedding is recommended. However, such material may be excessively dry for litters in breeding cages. Breeding pairs can be provided with nesting material (nestlets made from cellulose paper). Cages containing newborn litters are not changed until the seventh day postpartum; at that time, the nestlets containing pups can be transferred into the clean cages. Gloves should be worn when handling litters; if mating females show a tendency to eat their pups when disturbed, the pups can be sprinkled with urine from the dam or, alternatively, a small amount of a camphor-containing petroleum jelly like Vick's VapoRub can be deposited on the dam's nose.

Because mutant mice gain weight rapidly after weaning and are polyuric, fewer mutant mice can be maintained in the same cage as compared to lean controls, with more frequent cage changes required to prevent morbidity. This is an important consideration in studies designed to study development of diabetic complications or to implement long-term therapies. At the Jackson Laboratory, cages for weaned mice are transparent polycarbonate double pens measuring 28 (l) × 28 (w) × 13 (h) cm. One breeding pair can be maintained in each pen. For weaned mice being aged for experimental purposes, not more than three mutants would be caged per side. In the Düsseldorf colony, *Lepr*^{db} and *Lep*^{ob} offspring were grouped by sex at weaning with four mice housed in plastic cages measuring 26 × 20 cm and five to ten housed in larger cages of 42 × 26 cm, respectively. Lean mice are usually grouped by sex and housed separately. As mentioned earlier, mutant males are quite docile, so mice from different litters can be pooled in aging studies. However, lean control male litter mates should be caged together from weaning to reduce fighting behavior.

CONCLUDING REMARKS

As can be readily verified by a check of the current literature, B6-*Lep*^{ob} and BKS-*Lepr*^{db} mice continue to be the most utilized mouse genetic models of obesity and diabesity. Indeed, the obese-hyperglycemic syndrome displayed by those mice shows multiple similarities to human metabolic abnormalities present in the cardiometabolic syndrome and type 2 diabetes. Insulin resistance, inappropriate hyperglycemia, impaired glucose tolerance, and increased insulin secretion finally leading to β-cell exhaustion are seen in humans and mice. However, the neuroendocrine defects produced by either mutation are so serious as to elicit many pleiotropic changes in metabolism beginning in early life (often preweaning). Moreover, the obesity produced is extreme (if extrapolated to humans, it would be considered morbid obesity rather than the more moderate "garden variety" characterizing most maturity-onset obesities in humans). The severe neuroendocrine changes produced by either of these mutations and responsible for the hyperphagia and development of morbid obesity also elicit very early expression of metabolic abnormalities to an extreme degree not matched in most human diabesity. These include very high plasma insulin levels reflecting an extreme insulin resistance, very high leptin or no leptin, and hypercorticism and other changes associated with a dysregulated HPA axis.

Additional genus-unique features distinguish mice from humans, including a pronounced male sex bias in terms of susceptibility to diabesity, an ability to metabolize ketone bodies, resistance to atherosclerosis because of mutation-induced elevations in serum HDL concentrations, and a remarkable sensitivity to the diabetogenic effects of refined sugars. Nevertheless, research on leptin and leptin–leptin receptor loop in *Lep*^{ob} and *Lepr*^{db} mice has opened new fields of inquiry into previously unrecognized adipocyte–brain and adipocyte–islet interactions.

As important as these new interactions are to a full understanding of the complexity of mammalian metabolism regulation, study of these "models" has provided human medicine with something else of major importance: the realization of the importance of the genetic background in interfacing with the specific molecular defect in a mutant allele to determine phenotypic outcomes. As efforts to find good mouse models of diabetic complication intensify, it becomes critical to define which inbred strains carry the appropriate modifier genes for syndrome severity and secondary organ complications. For example, the *Lepr*$^{db\text{-}1J}$ mutation on the BKS and the *Lepr*$^{db\text{-}3J}$ on FVB/N inbred strain backgrounds produce severe hyperglycemia. But eventual declines in plasma insulin marking eventual beta cell loss, islet atrophy, and shortened life span characterize the former, whereas sustained hyperinsulinemia coupled with islet hyperplasia characterize the latter (Luo et al. 2006). Thus, while secondary complications associated with chronic hyperglycemia may develop in both mutant stocks, the extended longevity of mutants on the FVB background is clearly advantageous.

Others best summarize the utility of any single animal model of "diabesity" in terms of the complexity of obesity-associated diabetes in humans: "None is identical to any human syndrome; all establish that the mechanism found in one animal is a mechanism in other mammals and thus worth testing for" (Renold et al. 1988).

ACKNOWLEDGMENTS

This chapter is dedicated to our friend and colleague, Dr. Douglas L. Coleman, whose pioneering studies at the Jackson Laboratory pointed the way for molecular analyses to come. This work was supported by the Ministerium für Wissenschaft und Forschungs des Landes Nordrhein-Westfalen, the Deutsche Forschungsgemeinschaft (SFB 351) to L.H., and a grant from The American Diabetes Association (E.H.L.). Institutional shared services at The Jackson Laboratory were supported by National Cancer Institute Center Support Grant CA34196.

REFERENCES

Ahima RS, Flier JS. (2000) Leptin. *Annu Rev Physiol* 62:413–437.

Alon T, Friedman JM. (2006) Late-onset leanness in mice with targeted ablation of melanin concentrating hormone neurons. *J Neurosci* 26:389–397.

Amir S. (1981) Behavioral response of the genetically obese (*ob/ob*) mouse to heat stress: Effects of naloxone and prior exposure to immobilization stress. *Physiol Behav* 27:249–353.

Ashwell M, Meade CJ, Medawar P, Sowter C. (1977) Adipose tissue: Contributions of nature and nurture to the obesity of an obese mutant mouse (*ob/ob*). *Proc R Soc Lond B* 195:343–353.

Bado A, Levasseur S, Attoub S, et al. (1998) The stomach is a source of leptin. *Nature* 394:790–793.

Balthasar N, Coppari R, McMinn J, et al. (2004) Leptin receptor signaling in POMC neurons is required for normal body weight homeostasis. *Neuron* 42:983–991.

Bates SH, Kulkarni RN, Seifert M. Myers, MG. Jr. (2005) Roles for leptin receptor/STAT3-dependent and independent signals in the regulation of glucose homeostasis. *Cell Metab* 1:169–178.

Bates SH, Stearns WH, Dundon TA, et al. (2003) STAT3 signaling is required for leptin regulation of energy balance but not reproduction. *Nature* 421:856–859.

Batt RA. (1972) The response of the reproductive system in the female mutant mouse, obese (genotype *ob/ob*) to gonadotrophin-releasing hormones. *J Reprod Fertil* 31:496–497.

Batt RA, Everard DM, Gillies G, et al. (1982) Investigation into the hypogonadism of the obese mouse (genotype *ob/ob*). *J Reprod Fertil* 64:363–371.

Batt RA, Harrison GA. (1960) Features of the "adipose" mouse. *Heredity* 15:335–337.

Batt RA, Harrison GA. (1963) The reproductive system of the adipose mouse. *J Heredity* 54:135–138.

Begin-Heick N. (1996) Beta-adrenergic receptors and G-proteins in the *ob/ob* mouse. *Int J Obesity* 20: Suppl 3, S32–S35.

Bellward K, Dauncey MJ. (1988) Behavioral energy regulation in lean and genetically obese (*ob/ob*) mice. *Physiol Behav* 42:433–438.

Black MA, Begin-Heick N. (1995) Growth and maturation of primary-cultured adipocytes from lean and *ob/ob* mice. *J Cell Biochem* 58:455–463.

Bock T, Pakkenberg B, Buschard K. (2003) Increased islet volume but unchanged islet number in *ob/ob* mice. *Diabetes* 52:1716–1722.

Bock T, Pakkenberg B, Buschard K. (2005) Genetic background determines the size and structure of the endocrine pancreas. *Diabetes* 54:133.

Bohlen HG, Niggl BA. (1979) Arteriolar anatomical and functional abnormalities in juvenile mice with genetic or streptozotocin-induced diabetes mellitus. *Circ Res* 45:390–396.

Bohlen HG, Niggl BA. (1980) Early arteriolar disturbances following streptozotocin-induced diabetes mellitus in adult mice. *Microvasc Res* 20:19–29.

Boillot D, Assan R, Dardenne M, et al. (1986) T-lymphopenia and T-cell imbalance in diabetic *db/db* mice. *Diabetes* 35:198–203.

Bower G, Brown DM, Steffes MW, et al. (1980) Studies of the glomerular mesangium and the juxtaglomerular apparatus in the genetically diabetic mouse. *Lab Invest* 43:333–341.

Bray GA. (1991) Obesity, a disorder of nutrient partitioning: The MONA LISA hypothesis. *J Nutr* 121:1146–1162.

Bray GA, York DA. (1971) Genetically transmitted obesity in rodents. *Physiol Rev* 51:598–646.

Bray GA, York DA. (1979) Hypothalamic and genetic obesity in experimental animals: An autonomic and endocrine hypothesis. *Physiol Rev* 59:719–809.

Brown JA, Chua SC Jr, Liu SM, et al. (2000) Spontaneous mutation in the db gene results in obesity and diabetes in CD-1 outbred mice. *Am J Physiol* 278:R320–R330.

Burcelin R, Kamohara S, Li J, et al. (1999) Acute intravenous leptin infusion increases glucose turnover but not skeletal muscle glucose uptake in *ob/ob* mice. *Diabetes* 48:1264–1269.

Busso N, So A, Chobaz-Peclat V, et al. (2002) Leptin signaling deficiency impairs humoral and cellular immune responses and attenuates experimental arthritis. *J Immunol* 168:875–882.

Calcagnetti DJ, Flynn JJ, Margules DL. (1987) Opioid-induced linear running in obese (*ob/ob*) and lean mice. *Pharmacol Biochem Behav* 26:743–747.

Calcutt NA, Willars GB, Tomlinson DR. (1988) Axonal transport of choline acetyltransferase and 6-phosphofructokinase activities in genetically diabetic mice. *Muscle Nerve* 11:1206–1210.

Campfield LA, Smith FJ, Guisez Y, et al. (1995) Recombinant mouse OB protein: Evidence for a peripheral signal linking adiposity and central neural networks. *Science* 269:546–549.

Cancello R, Zingaretti MC, Sarzani R, et al. (1998) Leptin and UCP1 genes are reciprocally regulated in brown adipose tissue. *Endocrinology* 139:4747–4750.

Carlisle HJ, Dubuc PU. (1984) Temperature preference of genetically obese (*ob/ob*) mice. *Physiol Behav* 33:899–902.

Carson KA, Bossen EH, Hanker JS. (1980) Peripheral neuropathy in mouse hereditary diabetes mellitus. II. Ultrastructural correlates of degenerative and regenerative changes. *Neuropathol Appl Neurobiol* 6:361–374.

Ceddia RB. (2005) Direct metabolic regulation in skeletal muscle and fat tissue by leptin: Implications for glucose and fatty acids homeostasis. *Int J Obesity* 29:1175–1183.

Chan CB. (1995) Beta-cell stimulus—Secretion coupling defects in rodent models of obesity. *Can J Physiol Pharmacol* 73:1414–1424.

Chandra RK. (1980) Cell-mediated immunity in genetically obese C57BL/6J *ob/ob*) mice. *Am J Clin Nutr* 33:13–16.

Chehab FF, Lim ME, Lu R. (1996) Correction of the sterility defect in homozygous obese female mice by treatment with the human recombinant leptin. *Nat Genet* 12:318–320.

Chehab FF, Qiu J, Mounzih K, et al. (2002) Leptin and reproduction. *Nutr Rev* 60:S39–S46; discussion S68–S84, S85–S87.

Chen H, Charlat O, Tartaglia LA, et al. (1996) Evidence that the diabetes gene encodes the leptin receptor: Identification of a mutation in the leptin receptor gene in *db/db* mice. *Cell* 84:491–495.

Chua S Jr., Liu SM, Li Q, et al. (2002) Differential beta cell responses to hyperglycaemia and insulin resistance in two novel congenic strains of diabetes (FVB-*Lepr* (*db*)) and obese (DBA-Lep (*ob*)) mice. *Diabetologia* 45:976–990.

Chua SC Jr., Liu SM, Li Q, et al. (2004) Transgenic complementation of leptin receptor deficiency. II. Increased leptin receptor transgene dose effects on obesity/diabetes and fertility/lactation in *lepr-db/db* mice. *Am J Physiol* 286:E384–E392.

Chung WK, Belfi K, Chua M, et al. (1998) Heterozygosity for Lep(*ob*) or Lep(*rdb*) affects body composition and leptin homeostasis in adult mice. *Am J Physiol* 274:R985–R990.

Cinti S, Frederich RC, Zingaretti MC, et al. (1997) Immunohistochemical localization of leptin and uncoupling protein in white and brown adipose tissue. *Endocrinology* 138:797–804.

Clee SM, Nadler ST, Attie AD. (2005) Genetic and genomic studies of the BTBR *ob/ob* mouse model of type 2 diabetes. *Am J Ther* 12:491–498.

Cohen MP, Clements RS, Cohen JA, Shearman CW. (1996) Glycated albumin promotes a generalized vasculopathy in the *db/db* mouse. *Biochem Biophys Res Commun* 218:72–75.

Cohen P, Zhao C, Cai X, et al. (2001) Selective deletion of leptin receptor in neurons leads to obesity. *J Clin Invest* 108:1113–1121.

Coleman DL. (1973) Effects of parabiosis of obese with diabetes and normal mice. *Diabetologia* 9:294–298.

Coleman DL. (1978) Obese and diabetes: two mutant genes causing diabetes-obesity syndromes in mice. *Diabetologia* 14:141–148.

Coleman DL. (1982) Thermogenesis in diabetes-obesity syndromes in mutant mice. *Diabetologia* 22:205–211.

Coleman DL. (1985) Increased metabolic efficiency in obese mutant mice. *Int J Obes* 9:Suppl 2, 69–73.

Coleman DL. (1992) The influence of genetic background on the expression of mutations at the diabetes (*db*) locus in the mouse. VI: Hepatic malic enzyme activity is associated with diabetes severity. *Metabolism* 41:1134–1136.

Coleman DL, Hummel KP. (1967) Studies with the mutation, diabetes, in the mouse. *Diabetologia* 3:238–248.

Coleman DL, Hummel KP. (1969a) Effects of parabiosis of normal with genetically diabetic mice. *Am J Physiol* 217:1298–1304.

Coleman DL and Hummel KP. (1969b) The mutation, diabetes, in the mouse. In *Diabetes, Proc. VI. Congr. Internat. Diab. Fed. Stockholm, 1967, Internat. Congr. Ser. 172*, ed. Östman J and Milner RDG, Excerpta med: Amstrdam, 1969b, pp 813–820.

Coleman DL, Kuzava JE. (1991) Genetic regulation of malic enzyme activity in the mouse. *J Biol Chem* 266:21997–22002.

Collins S, Surwit RS. (1996) Pharmacologic manipulation of ob expression in a dietary model of obesity. *J Biol Chem* 271:9437–9440.

Collins S, Surwit RS. (2001) The beta-adrenergic receptors and the control of adipose tissue metabolism and thermogenesis. *Recent Prog Horm Res* 56:309–328.

Contaldo F, Gerber H, Coward WA, and Trayhurn P. (1981) Milk intake in pre-weanling genetically obese (ob/ob) mice. In: *Obesity: Pathogenesis and Treatment*, ed. Enzi G, Crepaldi G, Pozza G, and Renold AE, Academic Press: London and New York, pp. 319–322.

Cummings DE, Schwartz MW. (2000) Melanocortins and body weight: A tale of two receptors. *Nat Genet* 26:8–9.

Currie PJ. (1993) Differential effects of NE, CLON, and 5-HT on feeding and macronutrient selection in genetically obese (*ob/ob*) and lean mice. *Brain Res Bull* 32:133–142.

Currie PJ, Wilson LM. (1993) Potentiation of dark onset feeding in obese mice (genotype *ob/ob*) following central injection of norepinephrine and clonidine. *Eur J Pharmacol* 232:227–234.

Dardenne M, Savino W, Gastinel LN, et al. (1983) Thymic dysfunction in the mutant diabetic (*db/db*) mouse. *J Immunol* 130:1195–1199.

Davis RC, Schadt EE, Cervino AC, et al. (2005) Ultrafine mapping of SNPs from mouse strains C57BL/6J, DBA/2J, and C57BLKS/J for loci contributing to diabetes and atherosclerosis susceptibility. *Diabetes* 54:1191–1199.

de Luca C, Kowalski TJ, Zhang Y, et al. (2005) Complete rescue of obesity, diabetes, and infertility in *db/db* mice by neuron-specific LEPR-B transgenes. *J Clin Invest* 115:3484–3493.

Dhillon H, Zigman JM, Ye C, et al. (2006) Leptin directly activates SF1 neurons in the VMH, and this action by leptin is required for normal body-weight homeostasis. *Neuron* 49:191–203.

Drasher ML, Dickie MM, Lane WD. (1955) Physiological differences *in uteri* of obese stock mice. A comparison between obese mice and their thin sibs. *J Heredity* 46:209–212.

Drescher VS, Chen HL, Romsos DR. (1994) Corticotropin-releasing hormone decreases feeding, oxygen consumption and activity of genetically obese (*ob/ob*) and lean mice. *J Nutr* 124:524–530.

Dugail I, Quignard-Boulange A, Le Liepvre X, Lavau M. (1990) Impairment of adipsin expression is secondary to the onset of obesity in *db/db* mice. *J Biol Chem* 265:1831–1833.

Emilsson V, Liu YL, Cawthorne MA, et al. (1997) Expression of the functional leptin receptor mRNA in pancreatic islets and direct inhibitory action of leptin on insulin secretion. *Diabetes* 46:313–316.

Ewart-Toland A, Mounzih K, Qiu J, Chehab FF. (1999) Effect of the genetic background on the reproduction of leptin-deficient obese mice. *Endocrinology* 140:732–738.

Falconer DS, Isaacson JH. (1959) Adipose, a new inherited obesity of the mouse. *J Hered* 50:290–292.

Fehmann HC, Peiser C, Bode HP, et al. (1997) Leptin: A potent inhibitor of insulin secretion. *Peptides* 18:1267–1273.

Flatt PR, Buchanan KD, Bailey CJ. (1980) Glucagon and diabetes: Evidence for marked insensitivity to local regulation of A-cell function by endogenous insulin in obese-hyperglycaemic (*ob/ob*) mice. *Biochem Soc Trans* 8:58–59.

Flier JS, Cook KS, Usher P, Spiegelman BM. (1987) Severely impaired adipsin expression in genetic and acquired obesity. *Science* 237:405–408.

Frederich RC Jr., Kahn BB, Peach MJ, Flier JS. (1992) Tissue-specific n tritional regulation of angiotensinogen in adipose tissue. *Hypertension* 19:339–344.

Frederich RC, Lollmann B, Hamann A, et al. (1995) Expression of *ob* mRNA and its encoded protein in rodents. Impact of nutrition and obesity. *J Clin Invest* 96:658–663.

Friedman JM, Halaas JL. (1998) Leptin and the regulation of body weight in mammals. *Nature* 395:763–770.

Fruhbeck G. (2006) Intracellular signaling pathways activated by leptin. *Biochem J* 393:7–20.

Garris DR. (1989) Effects of estradiol and progesterone on diabetes-associated utero-ovarian atrophy in C57BL/KsJ (*db/db*) mutant mice. *Anat Rec* 225:310–317.

Gartner K. (1978) Glomerular hyperfiltration during the onset of diabetes mellitus in two strains of diabetic mice (c57bl/6j *db/db* and c57bl/ksj *db/db*). *Diabetologia* 15:59–63.

Gavrilova O, Barr V, Marcus-Samuels B, Reitman M. (1997) Hyperleptinemia of pregnancy associated with the appearance of a circulating form of the leptin receptor. *J Biol Chem* 272:30546–30551.

Ghilardi N, Ziegler S, Wiestner A, et al. (1996) Defective STAT signaling by the leptin receptor in diabetic mice. *Proc Natl Acad Sci USA* 93:6231–6235.

Giacomelli F, Wiener J. (1979) Primary myocardial disease in the diabetic mouse. An ultrastructural study. *Lab Invest* 40:460–473.

Giandomenico G, Dellas C, Czekay RP, et al. (2005) The leptin receptor system of human platelets. *J Thromb Haemost* 3:1042–1049.

Goodbody AE, Trayhurn P. (1982) Studies on the activity of brown adipose tissue in suckling, pre-obese, *ob/ob* mice. *Biochim Biophys Acta* 680:119–126.

Greenberg D, Ackerman SH. (1984) Genetically obese (*ob/ob*) mice are predisposed to gastric stress ulcers. *Behav Neurosci* 98:435–440.

Guo K, Lukacik P, Papagrigoriou E, et al. (2006) Characterization of human DHRS6, an orphan short chain dehydrogenase/reductase enzyme: A novel cytosolic type 2 r-beta-hydroxybutyrate dehydrogenase. *J Biol Chem* 281:10291–10297.

Guo KY, Halo P, Leibel RL, Zhang Y. (2004) Effects of obesity on the relationship of leptin mRNA expression and adipocyte size in anatomically distinct fat depots in mice. *Am J Physiol* 287:R112–R119.

Halaas JL, Gajiwala KS, Maffei M, et al. (1995) Weight-reducing effects of the plasma protein encoded by the obese gene. *Science* 269:543–546.

Hanker JS, Ambrose WW, Yates PE, et al. (1980) Peripheral neuropathy in mouse hereditary diabetes mellitus. I. Comparison of neurologic, histologic, and morphometric parameters with dystonic mice. *Acta Neuropathol* 51:145–153.

Harrison DE, Archer JR, Astle CM. (1984) Effects of food restriction on aging: Separation of food intake and adiposity. *Proc Natl Acad Sci USA* 81:1835–1838.

Hausberger FX. (1958) Parabiosis and transplantation experiments in hereditarily obese mice. *Anat Rec* 130:313.

Hellman B, Jacobsson L, Taeljedal IB. (1963) Endocrine activity of the testis in obese-hyperglycaemic mice. *Acta Endocrinol* 44:20–26.

Herberg L, Coleman DL. (1977) Laboratory animals exhibiting obesity and diabetes syndromes. *Metabolism* 26:59–99.

Herberg L. (1988) Insulin resistance in abdominal and subcutaneous obesity: comparison of C57BL/6J-ob/ob with New Zealand obese mice. In *Frontiers in Diabetes Research. Lessons from Animal Diabetes II.* ed. Shafrir E, Renold A. John Libbey: London, 367–373.

Herberg L, Leiter EH. (2001) Obesity/diabetes in mice with mutations in the leptin or leptin receptor genes. Animal Models of Diabetes: A Primer. ed. Sima A, Shafrir E. *Frontiers in Animal Diabetes Research.* Harwood Academic Publishers: Amsterdam, Netherlands, 63–107.

Himms-Hagen J. (1985) Defective brown adipose tissue thermogenesis in obese mice. *Int J Obesity* 9:Suppl 2, 17–24.

Ho A, Chin A. (1988) Circadian feeding and drinking patterns of genetically obese mice fed solid chow diet. *Physiol Behav* 43:651–656.

Hogan S, Himms-Hagen J. (1980) Abnormal brown adipose tissue in obese (*ob/ob*) mice: Response to acclimation to cold. *Am J Physiol* 239:E301–E309.

Hummel KP, Coleman DL, Lane PW. (1972) The influence of genetic background on expression of mutations at the diabetes locus in the mouse. I. C57BL-KsJ and C57BL-6J strains. *Biochem Genet* 7:1–13.

Hummel KP, Dickie MM, Coleman DL. (1966) Diabetes, a new mutation in the mouse. *Science* 153:1127–1128.

Ingalls AM, Dickie MM, Snell GD. (1950) Obese, a new mutation in the house mouse. *J Heredity* 41:317–318.

Ishizuka T, Klepcyk P, Liu S, et al. (1999) Effects of overexpression of human GLUT4 gene on maternal diabetes and fetal growth in spontaneous gestational diabetic C57BLKS/J Lepr(*db*/+) mice. *Diabetes* 48:1061–1069.

Jo YH, Wiedl D, Role LW. (2005) Cholinergic modulation of appetite-related synapses in mouse lateral hypothalamic slice. *J Neurosci* 25:1133–1144.

Johnson LM, Sidman, RL. (1979) A reproductive endocrine profile in the diabetes (*db*) mutant mouse. *Biol Reprod* 20:552–559.

Johnson PR, Hirsch J. (1972) Cellularity of adipose depots in six strains of genetically obese mice. *J Lipid Res* 13:2–11.

Jones N, Harrison GA. (1958) Genetically determined obesity and sterility in the mouse. *Stud Fert* 9:51–61.

Kamata K, Kojima S. (1997) Characteristics of contractile responses of aorta to norepinephrine in *db/db* mice. *Res Commun Mol Pathol Pharmacol* 96:319–328.

Kaplan ML, Trout JR, Leveille GA. (1976) Adipocyte size distribution in *ob/ob* mice during preobese and obese phases of development. *Proc Soc Exp Biol Med* 153:476–482.

Kaufmann RC, Amankwah KS, Colliver JA, Arbuthnot J. (1987) Diabetic pregnancy. The effect of genetic susceptibility for diabetes on fetal weight. *Am J Perinatol* 4:72–74.

Kerouz NJ, Horsch D, Pons S, Kahn CR. (1997) Differential regulation of insulin receptor substrates-1 and -2 (IRS-1 and IRS-2) and phosphatidylinositol 3-kinase isoforms in liver and muscle of the obese diabetic (*ob/ob*) mouse. *J Clin Invest* 100:3164–3172.

Khaidar A, Marx M, Lubec B, Lubec G. (1994) L-arginine reduces heart collagen accumulation in the diabetic *db/db* mouse. *Circulation* 90:479–483.

Kieffer TJ, Heller RS, Habener JF. (1996) Leptin receptors expressed on pancreatic beta-cells. *Biochem Biophys Res Commun* 224:522–527.

Kieffer TJ, Heller RS, Leech CA, et al. (1997) Leptin suppression of insulin secretion by the activation of ATP-sensitive K+ channels in pancreatic beta-cells. *Diabetes* 46:1087–1093.

Kim JH, Nishina PM, Naggert JK. (1998) Genetic models for non insulin dependent diabetes mellitus in rodents. *J Basic Clin Physiol Pharmacol* 9:325–345.

Kim JH, Taylor PN, Young D, et al. (2003) New leptin receptor mutations in mice: Lepr(db-rtnd), Lepr(db-dmpg) and Lepr(db-rlpy). *J Nutr* 133:1265–1271.

Klebanov S, Astle CM, DeSimone O, et al. (2005) Adipose tissue transplantation protects *ob/ob* mice from obesity, normalizes insulin sensitivity and restores fertility. *J Endocrinol* 186:203–211.

Konstantinides S, Schafer K, Koschnick S, Loskutoff DJ. (2001) Leptin-dependent platelet aggregation and arterial thrombosis suggests a mechanism for atherothrombotic disease in obesity. *J Clin Invest* 108:1533–1540.

Konstantinides S, Schafer K, Neels JG, et al. (2004) Inhibition of endogenous leptin protects mice from arterial and venous thrombosis. *Arterioscler Thromb Vasc Biol* 24:2196–2201.

Kowalski TJ, Liu SM, Leibel RL, Chua SC Jr. (2001) Transgenic complementation of leptin-receptor deficiency. I. Rescue of the obesity/diabetes phenotype of LEPR-null mice expressing a LEPR-B transgene. *Diabetes* 50:425–435.

Kuhn CM, Cochrane C, Feinglos MN. Surwit, RS. (1987) Exaggerated peripheral responses to catecholamines contribute to stress-induced hyperglycemia in the *ob/ob* mouse. *Pharmacol Biochem Behav* 26:491–495.

Kulkarni RN, Wang ZL, Wang RM, et al. (1997) Leptin rapidly suppresses insulin release from insulinoma cells, rat and human islets and, *in vivo*, in mice. *J Clin Invest* 100:2729–2736.

Kuo TH, Giacomelli F, Wiener J, Lapanowski-Netzel K. (1985) Pyruvate dehydrogenase activity in cardiac mitochondria from genetically diabetic mice. *Diabetes* 34:1075–1081.

Kuo TH, Moore KH, Giacomelli F, Wiener J. (1983) Defective oxidative metabolism of heart mitochondria from genetically diabetic mice. *Diabetes* 32:781–787.

La Cava A, Alviggi C, Matarese G. (2004) Unraveling the multiple roles of leptin in inflammation and autoimmunity. *J Mol Med* 82:4–11.

Lane PW. (1959) The pituitary–gonad response of genetically obese mice in parabiosis with thin and obese siblings. *Endocrinology* 65:863–868.

Lane PW, Dickie MM. (1954) Fertile, obese male mice. Relative sterility on obese males corrected by dietary restricion. *J Hered* 45:56–58.

Lee CH, Chen YG, Chen J, et al. (2006) Novel leptin receptor mutation in NOD/LtJ mice suppresses type 1 diabetes progression: II. Immunologic analysis. *Diabetes* 55:171–178.

Lee CH, Reifsnyder PC, Naggert JK, et al. (2005) Novel leptin receptor mutation in NOD/LtJ mice suppresses type 1 diabetes progression: Pathophysiological analysis. *Diabetes* 54:2525–2532.

Lee G, Li C, Montez J, et al. (1997) Leptin receptor mutations in 129 *db3J/db3J* mice and NIH *facp/facp* rats. *Mamm Genome* 8:445–447.

Lee GH, Proenca R, Montez JM, et al. (1996) Abnormal splicing of the leptin receptor in diabetic mice. *Nature* 379:632–635.

Lee SM, Bressler R. (1981) Prevention of diabetic nephropathy by diet control in the *db/db* mouse. *Diabetes* 30:106–111.

Lee SM, Graham A. (1980) Early immunopathologic events in experimental diabetic nephropathy: A study in *db/db* mice. *Exp Mol Pathol* 33:323–332.

Lee SM, Tutwiler G, Bressler R, Kircher CH. (1982) Metabolic control of prevention of nephropathy by 2-tetradecylglycidate in the diabetic mouse (*db/db*). *Diabetes* 31:12–18.

Leiter EH, Chapman HD. (1994) Obesity-induced diabetes (diabesity) in C57BL/KsJ mice produces aberrant transregulation of sex steroid sulfotransferase genes. *J Clin Invest* 93:2007–2013.

Leiter EH, Coleman DL, Ingram DK, Reynolds MA. (1983) Influence of dietary carbohydrate on the induction of diabetes in C57BL/KsJ-*db/db* diabetes mice. *J Nutr* 113:184–195.

Leiter EH, Fewell JW, Kuff EL. (1986) Glucose induces intracisternal type A retroviral gene transcription and translation in pancreatic beta cells. *J Exp Med* 163:87–100.

Leiter EH, Herberg L. (1997) The polygenetics of diabesity in mice. *Diabetes Rev* 5:131–148.

Leiter EH, Kintner J, Flurkey K, et al. (1999) Physiologic and endocrinologic characterization of male sex-biased diabetes in C57BLKS/J mice congenic for the fat mutation at the carboxypeptidease E locus. *Endocrine* 10:57–66.

Leiter EH, Le PH, Coleman DL. (1987) Susceptibility to db gene and streptozotocin-induced diabetes in C57BL mice: Control by gender-associated, MHC-unlinked traits. *Immunogenetics* 26:6–13.

Leiter EH, Prochazka M, Shultz LD. (1987) Effect of immunodeficiency on diabetogenesis in genetically diabetic (*db/db*) mice. *J Immunol* 138:3224–3229.

Levine AS, Morley JE, Brown DM, Handwerger BS. (1982) Extreme sensitivity of diabetic mice to naloxone-induced suppression of food intake. *Physiol Behav* 28:987–989.

Li C, Ioffe E, Fidahusein N, Connolly E, Friedman JM. (1998) Absence of soluble leptin receptor in plasma from dbPas/dbPas and other *db/db* mice. *J Biol Chem* 273:10078–10082.

Lubec B, Aufricht C, Herkner K, et al. (1994) Creatine reduces collagen accumulation in the kidneys of diabetic *db/db* mice. *Nephron* 67:214–217.

Ludwig DS, Tritos NA, Mastaitis JW, et al. (2001) Melanin-concentrating hormone overexpression in transgenic mice leads to obesity and insulin resistance. *J Clin Invest* 107:379–386.

Luo N, Liu SM, Liu H, et al. (2006) Allelic variation on chromosome 5 controls beta cell mass expansion during hyperglycemia in leptin receptor deficient diabetes mice. *Endocrinology*. In press.

MacDougald OA, Hwang CS, Fan H, Lane MD. (1995) Regulated expression of the obese gene product (leptin) in white adipose tissue and 3T3-L1 adipocytes. *Proc Natl Acad Sci USA* 92:9034–9037.

Maffei M, Fei H, Lee GH, et al. (1995) Increased expression in adipocytes of *ob* RNA in mice with lesions of the hypothalamus and with mutations at the *db* locus. *Proc Natl Acad Sci USA* 92:6957–6960.

Maffei M, Halaas J, Ravussin E, et al. (1995) Leptin levels in human and rodent: measurement of plasma leptin and *ob* RNA in obese and weight-reduced subjects. *Nat Med* 1:1155–1161.

Marx M, Trittenwein G, Aufricht C, et al. (1995) Agmatine and spermidine reduce collagen accumulation in kidneys of diabetic *db/db* mice. *Nephron* 69:155–158.

Matarese G, Carrieri PB, La Cava A, et al. (2005) Leptin increase in multiple sclerosis associates with reduced number of CD4(+)CD25+ regulatory T cells. *Proc Natl Acad Sci USA* 102:5150–5155.

Matarese G, Di Giacomo A, Sanna V, et al. (2001) Requirement for leptin in the induction and progression of autoimmune encephalomyelitis. *J Immunol* 166:5909–5916.

Matarese G, Sanna V, Di Giacomo A, et al. (2001) Leptin potentiates experimental autoimmune encephalomyelitis in SJL female mice and confers susceptibility to males. *Eur J Immunol* 31:1324–1332.

Matarese G, Sanna V, Lechler RI, et al. (2002) Leptin accelerates autoimmune diabetes in female NOD mice. *Diabetes* 51:1356–1361.

Mayer J, Dickie MM, Bates MW, Vitale JJ. (1951) Free selection of nutrients by hereditarily obese mice. *Science* 113:745–746.

McIntosh CHS, Pederson RA (1999) Noninsulin-dependent animal models of diabetes mellitus. In: *Experimental Models of Diabetes*. ed. McNeill JH, CRC Press LLC, Boca Raton, pp. 338–398.

McMinn JE, Liu SM, Dragatsis I, et al. (2004) An allelic series for the leptin receptor gene generated by CRE and FLP recombinase. *Mamm Genome* 15:677–685.

McMinn JE, Liu SM, Liu H, et al. (2005) Neuronal deletion of *Lepr* elicits diabesity in mice without affecting cold tolerance or fertility. *Am J Physiol* 289:E403–E411.

Meade CJ, Ashwell M, Sowter C. (1979) Is genetically transmitted obesity due to an adipose tissue defect? *Proc R Soc Lond B Biol Sci* 205:395–410.

Meade CJ, Brandon DR, Smith W, et al. (1981) The relationship between hyperglycaemia and renal immune complex deposition in mice with inherited diabetes. *Clin Exp Immunol* 43:109–120.

Memon RA, Fuller J, Moser AH, et al. (1999) Regulation of putative fatty acid transporters and Acyl–CoA synthetase in liver and adipose tissue in *ob/ob* mice. *Diabetes* 48:121–127.

Memon RA, Grunfeld C, Moser AH, Feingold KR. (1994) Fatty acid synthesis in obese insulin resistant diabetic mice. *Horm Metab Res* 26:85–87.

Midena E, Segato T, Radin S, et al. (1989) Studies on the retina of the diabetic *db/db* mouse. I. Endothelial cell-pericyte ratio. *Ophthalmic Res* 21:106–111.

Minokoshi Y, Kim YB, Peroni OD, et al. (2002) Leptin stimulates fatty-acid oxidation by activating AMP-activated protein kinase. *Nature* 415:339–343.

Mizuno TM, Bergen H, Funabashi T, et al. (1996) Obese gene expression: reduction by fasting and stimulation by insulin and glucose in lean mice, and persistent elevation in acquired (diet-induced) and genetic (yellow agouti) obesity. *Proc Natl Acad Sci USA* 93:3434–3438.

Moinat M, Deng C, Muzzin P, Assimacopoulos-Jeannet F, Seydoux J, Dulloo AG, et al. (1995) Modulation of obese gene expression in rat brown and white adipose tissues. *FEBS Lett* 373:131–134.

Moon BC, Friedman JM. (1997) The molecular basis of the obese mutation in *ob2J* mice. *Genomics* 42:152–156.

Morton GJ, Schwartz MW. (2001) The NPY/AgRP neuron and energy homeostasis. *Int J Obesity* 25:Suppl 5, S56–S62.

Mu JL, Naggert JK, Svenson KL, et al. (1999) Quantitative trait loci analysis for the differences in susceptibility to atherosclerosis and diabetes between inbred mouse strains C57BL/6J and C57BLKS/J. *J Lipid Res* 40:1328–1335.

Munzberg H, Flier JS, Bjorbaek C. (2004) Region-specific leptin resistance within the hypothalamus of diet-induced obese mice. *Endocrinology* 145:4880–4889.

Muzumdar R, Ma X, Yang X, et al. (2003) Physiologic effect of leptin on insulin secretion is mediated mainly through central mechanisms. *FASEB J* 17:1130–1132.

Naggert JK, Mu JL, Frankel W, et al. (1995) Genomic analysis of the C57BL/Ks mouse strain. *Mamm Genome* 6:131–133.

Nakata M, Yada T, Soejima N, Maruyama I. (1999) Leptin promotes aggregation of human platelets via the long form of its receptor. *Diabetes* 48:426–429.

Nishimura M, Miyamoto H. (1987) Immunopathological influence of the *Ay*, *db*, *ob* and *nu* genes placed on the inbred NOD background as murine models for human type I diabetes. *J Immunogenet* 14:127–130.

Nishina PM, Lowe S, Wang J, Paigen B. (1994) Characterization of plasma lipids in genetically obese mice: The mutants obese, diabetes, fat, tubby, and lethal yellow. *Metabolism* 43:549–553.

Nishina PM, Naggert JK, Verstuyft J, Paigen B. (1994) Atherosclerosis in genetically obese mice: The mutants obese, diabetes, fat, tubby, and lethal yellow. *Metabolism* 43:554–558.

Norido F, Canella R, Zanoni R, Gorio A. (1984) Development of diabetic neuropathy in the C57BL/Ks (*db/db*) mouse and its treatment with gangliosides. *Exp Neurol* 83:221–232.

Ogawa Y, Nakao K. (2000) [Transgenic skinny mice overexpressing leptin]. *Seikagaku* 72:554–558.

Ogus S, Ke Y, Qiu J, Wang B, Chehab FF. (2003) Hyperleptinemia precipitates diet-induced obesity in transgenic mice overexpressing leptin. *Endocrinology* 144:2865–2869.

Pelleymounter MA, Cullen MJ, Baker MB, et al. (1995) Effects of the obese gene product on body weight regulation in *ob/ob* mice. *Science* 269:540–543.

Poitout, V, Rouault C, Guerre-Millo M, et al. (1998) Inhibition of insulin secretion by leptin in normal rodent islets of Langerhans. *Endocrinology* 139:822–826.

Prochazka M, Premdas FH, Leiter EH, Lipson LG. (1986) Estrone treatment dissociates primary vs. secondary consequences of "diabetes" (*db*) gene expression in mice. *Diabetes* 35:725–728.

Qiu J, Ogus S, Lu R, Chehab FF. (2001) Transgenic mice overexpressing leptin accumulate adipose mass at an older, but not younger, age. *Endocrinology* 142:348–358.

Qiu J, Ogus S, Mounzih K, et al. (2001) Leptin-deficient mice backcrossed to the BALB/cJ genetic background have reduced adiposity, enhanced fertility, normal body temperature, and severe diabetes. *Endocrinology* 142:3421–3425.

Reddy S, Lau EM, Ross JM. (2004) Immunohistochemical demonstration of leptin in pancreatic islets of nonobese diabetic and CD-1 mice: Colocalization in glucagon cells and its attenuation at the onset of diabetes. *J Mol Histol* 35:511–519.

Reichart U, Kappler R, Scherthan H, et al. (2000) Partial leptin receptor gene deletion in transgenic mice prevents expression of the membrane-bound isoforms except for Ob-Rc. *Biochem Biophys Res Commun* 269:496–501.

Reichling S, Patel HV, Freeman KB, et al. (1988) Attenuated cold-induced increase in mRNA for uncoupling protein in brown adipose tissue of obese (*ob/ob*) mice. *Biochem Cell Biol* 66:193–198.

Renold AE, Porte D, Shafrir E. (1988) Definitions for diabetes types: Use and abuse of the concept "animal models of diabetes mellitus." In *Lessons from animal diabetes*, ed. Shafrir E, Renold AE. J. Libbey, London, 3–7.

Robertson DM, Sima AA. (1980) Diabetic neuropathy in the mutant mouse [C57BL/ks(*db/db*)]: A morphometric study. *Diabetes* 29:60–67.

Roduit R, Thorens B. (1997) Inhibition of glucose-induced insulin secretion by long-term pre-exposure of pancreatic islets to leptin. *FEBS Lett* 415:179–182.

Runner MN. (1954) Inherited hypofunction of the female pituitary in the sterile-obese syndrome in the mouse. *Rec Genet Soc Am* 23:63–64.

Runner MN, Gates A. (1954) Sterile, obese mothers. *J Heredity* 45:51–55.

Segal-Lieberman G, Bradley RL, Kokkotou E, et al. (2003) Melanin-concentrating hormone is a critical mediator of the leptin-deficient phenotype. *Proc Natl Acad Sci USA* 100:10085–10090.

Seufert J, Kieffer TJ, Leech CA, et al. (1999) Leptin suppression of insulin secretion and gene expression in human pancreatic islets: Implications for the development of adipogenic diabetes mellitus. *J Clin Endocrinol Metab* 84:670–676.

Shafrir E. (1992) Animal models of non-insulin dependent diabetes. *Diabetes/Metabolism Rev* 8:179–208.

Sheena J, Meade CJ. (1978) Mice bearing the *ob/ob* mutation have impaired immunity. *Int Arch Allergy Appl Immunol* 57:263–268.

Shimada M, Tritos NA, Lowell BB, et al. (1998) Mice lacking melanin-concentrating hormone are hypophagic and lean. *Nature* 396:670–674.

Shimomura Y, Oku J, Glick Z, Bray GA. (1982) Opiate receptors, food intake and obesity. *Physiol Behav* 28:441–445.

Sima AA, Robertson DM. (1979) Peripheral neuropathy in the diabetic mutant mouse. An ultrastructural study. *Lab Invest* 40:627–632.

Simpson ER, Merrill JC, Hollub AJ, et al. (1989) Regulation of estrogen biosynthesis by human adipose cells. *Endocr Rev* 10:136–148.

Skoza L, Giacomelli F, Wiener J. (1980) Lysosomal enzymes in the heart of the genetically diabetic mouse. *Lab Invest* 43:443–448.

Smithberg M, Runner MN. (1957) Pregnancy induced in genetically sterile mice. *J Heredity* 48:97–100.

Sprott RL. (1972) Long-term studies of feeding behavior of obese, diabetic, and viable yellow mutant mice under *ad lib* and operant conditions. *Psychol Rep* 30:991–1003.

Stephens TW, Basinski M, Bristow PK, et al. (1995) The role of neuropeptide Y in the antiobesity action of the obese gene product. *Nature* 377:530–532.

Stern JS, Hirsch J, Drewnowski A, et al. (1983) Glycerol kinase activity in adipose tissue of obese rats and mice: Effects of diet composition. *J Nutr* 113:714–720.

Stoehr JP, Byers JE, Clee SM, et al. (2004) Identification of major quantitative trait loci controling body weight variation in *ob/ob* mice. *Diabetes* 53:245–249.

Stoehr JP, Nadler ST, Schueler KL, et al. (2000) Genetic obesity unmasks nonlinear interactions between murine type 2 diabetes susceptibility loci. *Diabetes* 49:1946–1954.

Strohmayer AJ, Smith GP. (1987) The meal pattern of genetically obese (*ob/ob*) mice. *Appetite* 8:111–123.

Suganami T, Mukoyama M, Mor, K, et al. (2005) Prevention and reversal of renal injury by leptin in a new mouse model of diabetic nephropathy. *FASEB J* 19:127–129.

Surwit R, Seldin M, Kuhn C, et al. (1994) Diet-induced obesity and diabetes in C57BL/6J and C57BL/KsJ mice. *Mouse Genome* 92:523–525.

Surwit RS, Williams PG. (1996) Animal models provide insight into psychosomatic factors in diabetes. *Psychosom Med* 58:582–589.

Swenne I, Andersson A. (1984) Effect of genetic background on the capacity for islet cell replication in mice. *Diabetologia* 27:464–467.

Swerdloff RS, Batt RA, Bray GA. (1976) Reproductive hormonal function in the genetically obese (*ob/ob*) mouse. *Endocrinology* 98:1359–1364.

Swerdloff RS, Peterson M, Vera A, et al. (1978) The hypothalamic-pituitary axis in genetically obese (*ob/ob*) mice: Response to luteinizing hormone-releasing hormone. *Endocrinology* 103:542–547.

Tanaka T, Hidaka S, Masuzaki H, et al. (2005) Skeletal muscle AMP-activated protein kinase phosphorylation parallels metabolic phenotype in leptin transgenic mice under dietary modification. *Diabetes* 54:2365–2374.

Tartaglia LA, Dembski M, Weng X, et al. (1995) Identification and expression cloning of a leptin receptor, OB-R. *Cell* 83:1263–1271.

Thurlby PL, Trayhurn P. (1978) The development of obesity in preweaning *ob/ob* mice. *Br J Nutr* 39:397–402.

Thurlby PL, Trayhurn P. (1979) The role of thermoregulatory thermogenesis in the development of obesity in genetically obese (*ob/ob*) mice pair-fed with lean siblings. *Br J Nutr* 42:377–385.

Togawa K, Moritani M, Yaguchi H, Itakura M. (2006) Multidimensional genome scans identify the combinations of genetic loci linked to diabetes-related phenotypes in mice. *Hum Mol Genet* 15:113–128.

Toye AA, Lippiat JD, Proks P, et al. (2005) A genetic and physiological study of impaired glucose homeostasis control in C57BL/6J mice. *Diabetologia* 48:675–686.

Trayhurn P. (1979) Thermoregulation in the diabetic-obese (*db/db*) mouse. The role of non-shivering thermogenesis in energy balance. *Pflugers Arch* 380:227–232.

Trayhurn P, Duncan JS, Hoggard N, Rayner DV. (1998) Regulation of leptin production: A dominant role for the sympathetic nervous system? *Proc Nutr Soc* 57:413–439.

Trayhurn P, James WP. (1978) Thermoregulation and non-shivering thermogenesis in the genetically obese (*ob/ob*) mouse. *Pflugers Arch* 373:189–193.

Trayhurn P, Jones PM, McGuckin MM, Goodbody AE. (1982) Effects of overfeeding on energy balance and brown fat thermogenesis in obese (*ob/ob*) mice. *Nature* 295:323–325.
Trayhurn P, Thomas ME, Duncan JS, Rayner DV. (1995) Effects of fasting and refeeding on *ob* gene expression in white adipose tissue of lean and obese (*ob/ob*) mice. *FEBS Lett* 368:488–490.
Trayhurn P, Wood IS. (2005) Signaling role of adipose tissue: Adipokines and inflammation in obesity. *Biochem Soc Trans* 33:1078–1081.
Vicario PP, Slater EE, Saperstein R. (1989) The effect of ponalrestat on sorbitol levels in the lens of obese and diabetic mice. *Biochem Int* 19:553–561.
Vitadello M, Couraud JY, Hassig R, et al. (1983) Axonal transport of acetylcholinesterase in the diabetic mutant mouse. *Exp Neurol* 82:143–147.
Wang Z, Jiang T, Li J, et al. (2005) Regulation of renal lipid metabolism, lipid accumulation, and glomerulosclerosis in FVB *db/db* mice with type 2 diabetes. *Diabetes* 54:2328–2335.
Weide LG, Lacy PE. (1991) Hereditary hydronephrosis in C57BL/KsJ mice. *Lab Anim* 41:415–418.
Weigle DS, Bukowski TR, Foster DC, et al. (1995) Recombinant ob protein reduces feeding and body weight in the *ob/ob* mouse. *J Clin Invest* 96:2065–2070.
Weisberg SP, Hunter D, Huber R, et al. (2006) CCR2 modulates inflammatory and metabolic effects of high-fat feeding. *J Clin Invest* 116:115–124.
Weisberg SP, McCann D, Desai M, et al. (2003) Obesity is associated with macrophage accumulation in adipose tissue. *J Clin Invest* 112:1796–1808.
Wencel HE, Smothers C, Opara EC, et al. (1995) Impaired second phase insulin response of diabetes-prone C57BL/6J mouse islets. *Physiol Behav* 57:1215–1220.
Weninger M, Xi Z, Lubec B, et al. (1992) L-arginine reduces glomerular basement membrane collagen N epsilon-carboxymethyllysine in the diabetic *db/db* mouse. *Nephron* 62:80–83.
Wilkinson M, Moger WH. (1981) Transpubertal modulation of pituitary and testicular function in the *ob/ob* mouse. *Horm Res* 14:95–103.
Wilson LM, Sinha HL. (1985) Thermal preference behavior of genetically obese (*ob/ob*) and genetically lean (+/?) mice. *Physiol Behav* 35:545–548.
Wolf G, Schroeder R, Thaiss F, et al. (1998) Glomerular expression of p27Kip1 in diabetic *db/db* mouse: Role of hyperglycemia. *Kidney Int* 53:869–879.
Xu AW, Kaelin CB, Takeda K, et al. (2005) PI3K integrates the action of insulin and leptin on hypothalamic neurons. *J Clin Invest* 115:951–958.
Yen TT, Acton JM. (1972) Locomotor activity of various types of genetically obese mice. *Proc Soc Exp Biol Med* 140:647–650.
Zhang Y, Olbort M, Schwarzer K, et al. (1997) The leptin receptor mediates apparent autocrine regulation of leptin gene expression. *Biochem Biophys Res Commun* 240:492–495.
Zhang Y, Proenca R, Maffei M, et al. (1994) Positional cloning of the mouse obese gene and its human homologue. *Nature* 372:425–432.
Zheng S, Noonan WT, Metreveli NS, et al. (2004) Development of late-stage diabetic nephropathy in OVE26 diabetic mice. *Diabetes* 53:248–257.
Ziyadeh FN, Cohen MP, Guo J, Jin Y. (1997) RAGE mRNA expression in the diabetic mouse kidney. *Mol Cell Biochem* 170:147–152.

4 The Zucker Diabetic Fatty (ZDF) Rat—Lessons from a Leptin Receptor Defect Diabetic Model

Richard G. Peterson

CONTENTS

INTRODUCTION

The obese male Zucker diabetic fatty (ZDF) rat has been a significant diabetic model for studying the mechanisms of onset and treatment of type 2 diabetes (T2D). The obese ZDF female has also been developed into a diabetic model by means of dietary manipulation. This chapter will review the history and development of the model and some of the ways that the ZDF has been used and update the information included in the previous edition of this book (Peterson 2001). The ZDF model, like most other obese rat models, exhibits leptin receptor defects. Although these defects cause obesity, the type of obesity expressed is unlike common forms of human obesity. The genetic mutation responsible for the obesity and insulin resistance in this animal is the underlying substrate for expression of the diabetic condition. Perhaps due to the lack of alternative models, the leptin mutation does not seem to inhibit the use of this model in diabetes studies; however, there may be excessive insulin resistance and the obesity cannot be controlled by drugs developed to work through hypothalamic mechanisms. Looking toward the future, diabetic models developed with polygenetic obesity might become an improvement over the currently available models.

BACKGROUND AND HISTORY OF THE ZDF RAT

The ZDF rat was developed from a Zucker colony that had demonstrated diabetic propensity into a reproducible type 2 diabetic model at Indiana University during the 1980s. All "Zucker" founder rats were originally from a colony of rats that contained the obesity mutation [*Lepr*^{fa} (*fa*)] that was originally observed (Zucker and Zucker 1961). Over the years, animals from this outbred colony were distributed to various academic and industrial laboratories; significant diversity was and continues to be observed between closed colonies of Zucker rats as traits are inadvertently selected for.

The Zucker colony maintained by Walter Shaw at Eli Lilly began to demonstrate more diabetic potential than other colonies. As a result these rats became useful in the testing of drugs under development for diabetes and metabolic syndrome. Animals from the Lilly colony were moved to the DRTC Animal Core at Indiana University School of Medicine. Additional male rats demonstrating diabetes were identified in both colonies and the initial paper describing this condition was published (Clark et al. 1983). Shortly after this publication, the DRTC Animal core was turned over to another investigator and the cesarean derivation and development of the model continued (Peterson et al. 1990; Peterson 1995, 2001). As the model traits became fixed, requests for the model from investigators increased and, in 1991, the rat was licensed to Genetic Models, Inc. for breeding and distribution. After 10 years of market development, breeding and distribution of the ZDF rats were transferred to Charles River Laboratories where they are now available for purchase.

GENERAL CHARACTERISTICS OF THE ZDF RAT

The obese male ZDF rat is characterized by hyperinsulinemia and hyperglycemia beginning at six to seven weeks of age. Glucose levels increase over the next three to four weeks, when the average glycemia levels reach about 500 mg/dL. Over this period the insulin levels peak and then decrease over the next four to six weeks to a level of about 1 ng/mL; the insulin levels continue to drop below this level over time. Complications of diabetes become evident as the animals age. This general pattern of the development of diabetes has been reported in multiple publications (Peterson et al. 1990; Peterson, 1995, 2001; Etgen and Oldham 2000). The diet and other conditions are very important for the consistent development of a homogeneous expression of diabetes; the recommended diet is Purina 5008. This trait is viewed as desirable by the drug development industry today. Manipulation of the animals with protocols that require excessive bleeding and fasting for glucose tolerance testing can lead to unpredictable results. These aggressive protocols can result in a delay in the development of diabetes and inconsistent expression of the diabetic condition.

The obese ZDF female does not become reliably diabetic on standard chow diets. Initially, attempts to induce diabetes in females were frustrating. They included ovariectomy and feeding a variety of chow diets. The sex-hormone modulation by ovariectomy did not result in diabetes. However, a small number of the obese females became diabetic on Purina chow #5015. Over time, a number of other synthetic diets were developed and tested. Feeding these diets to obese female ZDF rats did not result in producing diabetes. However, since some obese females did get diabetic on 5015, the formulation was closely examined and was found to have a fairly unique ingredient that contained high levels of pork fat. This discovery led to the addition of more of this product (30% by weight) to the 5015 formulation. This diet became identified as C13004 (research diets).

Additional defined diets were subsequently formulated. A number of these reproduced and refined the diabetic potential of the original gain-based diet. Feeding the original C13004 formula or these defined diets to obese females has resulted in consistent development of diabetes in most obese ZDF females. It was also observed that these diets could be started at a variety of ages with similar results until the obese females reached about six months of age, when this regimen was no longer effective. The series of studies also demonstrated that if the diet were continued for about four weeks, the diabetic state remained when the rats were put back on normal chow diet (Purina 5008). Several publications have documented some of the usefulness of this model (Corsetti et al. 2000; Macchia et al. 2002; Liang et al. 2005; Zhou et al. 2005; Li et al. 2005; Schrijvers et al. 2006). A significant difference between the diabetic male and female is that the female retains higher insulin levels for a longer period of time after the diabetic state is initiated. This seems to make this model more desirable in testing certain drugs that depend on some insulin action and beta cell function.

The development of diabetes and the changes in beta cell function in the male obese ZDF rats have been studied by numerous investigators. Obesity, generally associated with the genetic condition, accompanied with concurrent insulin resistance appears to be the initiating factor for diabetes. If obesity is controlled by diet and caloric restriction, the diabetic condition can be delayed or prevented (Ohneda et al. 1995; Corsetti et al. 2000; Belury et al. 2003; Banz et al. 2005). Various nutriceuticals and pharmaceuticals can also delay or prevent the diabetic condition (Peterson et al. 1993; Peterson 1994; Sreenan et al. 1996; Pickavance et al. 1998, 2005; Smith et al. 2000; Byrne and Bradlow 2001; Finegood et al. 2001; Brand et al. 2003; Anis et al. 2004; Gedulin et al. 2005; Huang et al. 2005; Liang et al. 2005; Wargent et al. 2005; Li et al. 2005).

The obese ZDF rat and many of the other obese and/or diabetic rat models have the *Lepr^fa^* (*fa*) or the *Lepr^cp^* (*cp* or *k*) mutation of the leptin receptor. When these recessive mutations are homozygous or one of each of these mutations is present, obesity ensues; this is due to the inactivity of leptin in the hypothalamus. Obesity then results in insulin resistance and metabolic syndrome in some rat strains while other strains develop overt diabetes. The associated complications that develop vary significantly with the background strain of the animals. Although leptin receptor defect obesity is associated with the development of these conditions and as such has value, the obesity does not reflect the common polygenetic obesity usually found in humans and in other animal species. These polygenetic obesity conditions have been observed in many animal species, including humans, and are typically expressed with the excess availability of calorie-dense food. However, although excessive and calorie-dense food is certainly a contributing factor, genetics remains as a very important factor. With respect to these issues, each available animal model, irrespective of the mechanism behind the obese condition, can contribute significantly to the study of metabolic disease conditions if it is used properly. New and refined models should allow for new approaches to study disease conditions.

The specific genetic factor that allows the ZDF rat to become overtly diabetic has not been specifically identified. Nevertheless, Griffen and colleagues (2001) have demonstrated that the insulin promoter seems to be dysfunctional in the ZDF rat. These investigators examined beta cell function in homozygous lean ZDF fetuses. They demonstrated that there was no difference in beta-cell number and insulin content between homozygous and heterozygous ZDF-lean fetuses. However, insulin promoter activity was reduced 30–50% and insulin mRNA levels were reduced by 45% in homozygous ZDF-lean fetal islets. The islet amyloid polypeptide promoter was unaffected. This study demonstrated that the ZDF rat carries a genetic defect in beta-cell transcription that is independent of the leptin receptor mutation that causes the obesity and insulin resistance. The defect in beta-cell transcription is likely to be the inherited defect in the beta-cell gene that contributes to the development of diabetes in the setting of obesity and insulin resistance. The relationship of this gene to human diabetes has not been determined.

Despite the issues related to leptin receptor defect, the ZDF rat has contributed significantly to the study of development of diabetes and to the development of drugs to treat this condition. Changes in beta-cell structure and function in the obese ZDF male are well documented in the literature. One of the original observations was the loss of GLUT-2 transporter (Orci et al. 1990; Johnson et al. 1990; Unger 1991; Milburn et al. 1993). The loss of pancreatic beta cells and insulin release has also been well documented. Beta cell apoptosis was described in 1988 by Pick and colleagues (1998). Metabolic changes have been described that precede or accompany the development of diabetes in this model.

The conditions suggested as contributing to the beta-cell dysfunction are glucotoxicity (Harmon et al. 2001; Poitout and Robertson 2002; Girard 2005) and lipotoxicity (Unger 1995; Seufert et al. 1998; Higa et al. 1999; Kakuma et al. 2000; Girard 2000, 2005). Other studies have associated this loss with lipid accumulation in beta cells (Lee et al. 1994; Unger 1995; Zhou et al. 1998; Shimabukuro et al. 1998; Unger and Orci 2001; Cnop et al. 2001). The lipotoxicity hypothesis for beta-cell failure has been primarily supported by Unger's group at Southwestern University. The balance between these two hypotheses has been reviewed (Poitout and Robertson 2002; Girard 2005). Poitout's conclusion was that the glucose is the main factor causing the loss of beta-cell function.

The diabetic condition in obese ZDF male rat can be treated or prevented by a number of compounds that increase insulin sensitivity, decrease weight gains, and/or reduce glucose levels. Some of the compounds have been effective in preventing hyperglycemia and the beta-cell loss. The ZDF rat is very sensitive to diet; certain high-fat diets will increase the level of diabetes while other diets will reduce the numbers of animals getting diabetic or prevent it completely (Slieker et al. 1992; Peterson et al. 1993; Peterson 1994; Corsetti et al. 2000). A number of recent studies have demonstrated that thiazolidinediones (TZD) are effective in preventing and treating diabetes in the ZDF (Shimabukuro et al. 1998; Smith et al. 2000; McCarthy et al. 2000; Finegood et al. 2001; Corpe et al. 2001; Yue et al. 2005; Girard 2005; Pelzer et al. 2005). Other compounds and materials that have been tested and are effective in the ZDF include:

non-TZD PPAR agonists (Cobb et al. 1998; Brown et al. 1999; Dana et al. 2001; Corpe et al. 2001; Etgen et al. 2002; Baylis et al. 2003; Brand et al. 2003; Chen et al. 2005; Huang et al. 2005, 2006; Li et al. 2005)
a variety of natural products and food additives (Peterson 1994; Balon et al. 1995; Houseknecht et al. 1998; Poucheret et al. 1998; Prasad 2001; Wang et al. 2001; Nagao et al. 2003; Belury et al. 2003; Banz et al. 2005; Fu et al. 2005; Wasan et al. 2006)
incretins and related compounds (Hargrove et al. 1995; Berghofer et al. 1997; Shen et al. 1998; Sreenan et al. 2000; Hui et al. 2002; Sudre et al. 2002; Sturis et al. 2003; Oh et al. 2003; Gedulin et al. 2005)
diazoxide (Alemzadeh and Tushaus 2004, 2005)
vanadium-containing compounds (Poucheret et al. 1998; Yuen et al. 1999, 2003; Wang et al. 2001; Winter et al. 2005; Wasan et al. 2006)
a variety of other compounds (Ishikawa et al. 1998; Liu et al. 1998; Mest et al. 2001; Byrne and Bradlow 2001; Fujinami et al. 2001; Sudre et al. 2002; Ring et al. 2003; Henriksen et al. 2003; Wilkes et al. 2005; Konno et al. 2005; Liang et al. 2005; Pold et al. 2005; Wargent et al. 2005)

OTHER DIABETIC RAT MODELS

Several other rat strains have demonstrated diabetic characteristics and have been used as models for the diabetic condition and its complications. Several of these animals are obese due to leptin receptor defects while some better established but less available models are free of leptin mutations. Many of these animals are reviewed in the chapters in this book and are therefore mentioned solely for the purpose of comparison with the ZDF in this chapter.

Metabolic Syndrome/Prediabetic Models

A number of rat models demonstrate obesity and other prediabetic characteristics without ever becoming frankly diabetic. Included in this category are several obese strains that have leptin receptor defects $Lepr^{fa}$(*fa*) or $Lepr^{cp}$(*cp* or *k*) and include but are not limited to Zucker rats-*fa*, LA/N-*cp*, SHR/N-*cp*, and SHROB-*cp*.

The obese females of the strains where the obese males become frankly diabetic also usually fall in this category along with polygenic obese models and DIO models.

Each of the animals in this category has some of the characteristics of metabolic syndrome. This chapter will not comment on the specific characteristics of the individual models.

Leptin Receptor Defect Diabetic Models

The primary rat leptin receptor defect models that are diabetic are the ZDF, SHHF, ZSF1, BBZDR, and VDF. The leptin defects are *Lepr^fa^*(*fa*) or *Lepr^cp^*(*cp* or *k*). The ZDF, BBZDR, and VDF have the *fa* mutation while the SHHF has the *cp* mutation; the obese ZSF1 rat is a cross between the ZDF and SHHF and thus has one of each of the mutations. These models will only be briefly mentioned so that they are recognized as important models with individual characteristics that make them important for specific applications.

The ZDF rat has been well described earlier and therefore it will only be referenced for comparison in this section. The SHHF rat was developed by Sylvia McCune (Meier et al. 1995; McCune et al. 1999). Although the obese males do develop a degree of hyperglycemia, they do not become fasting hyperglycemic and they never express the low insulin levels that the ZDF rat demonstrates when it becomes overtly diabetic (Gao et al. 1994; Friedman et al. 1997; McCune et al. 1999). The main feature of the SHHF model is that it develops heart failure at various ages depending on gender and obesity (Park et al. 1997, 2004; Sharkey et al. 1998; McCune et al. 1999; Heyen et al. 2002). Myocyte remodeling is a significant feature of the failing heart and is expressed in the SHHF rat (Gerdes et al. 1996; Onodera et al. 1998; Tamura et al. 1998).

The *ZSF1 rat* is produced by crossing a heterozygous (*fa*/+) ZDF with a heterozygous (*cp*/+) SHHF rat. The obese animals that result from these breedings are heavier than obese ZDF rats, have elevated glucose levels that tend to moderate as the animals age, and have high lipid levels as they age (Dominguez et al. 2006). Some of these changes occur as the model begins to develop complications. The main unique complication is kidney failure, which is not a prominent feature of either of the parental strains, although the SHHF rat does have significant kidney pathology (Radin et al. 1992; Jackson et al. 2001). This condition develops slowly at first, but by about six to eight months of age, the glomerular changes are significant and they progress over the next few months to kidney failure. This model has been extensively studied in Tofovic's laboratory (Tofovic et al. 2000, 2002; Tofovic, Dubey, et al. 2001; Tofovic, Kusaka, et al. 2001). These rats are also quite sensitive to dietary modulation (Davis et al. 2005).

The *BBZDR/Wor rat* is a cross between the Zucker rat and BBDR/Wor (diabetes-resistant) strain (Guberski et al. 1993; Tirabassi et al. 2004). This model develops T2D with obesity and insulin resistance at about 10 weeks of age. The publications demonstrate that this model also expresses typical complications of diabetes such as cardiovascular disease, nephropathy, neuropathy, and retinopathy (Guberski et al. 1993; Ellis et al. 1998, 2000, 2002; Tirabassi et al. 2004; Ito et al. 2006).

The *VDF rat* (Vancouver diabetic fatty) was characterized in Pederson's laboratory in the Department of Physiology, University of British Columbia, Vancouver, Canada, and demonstrates hyperinsulinemia, marked obesity, and mild diabetes

(Pospisilik et al. 2002a, 2002b; Carr et al. 2003). This model is maintained at the university and has not been bred for distribution.

Non-Leptin-Receptor Defect Diabetic Models

The models that do not have leptin receptor defects are the OLETF and GK rats. Both models were developed by selective breeding for traits that were recognized in outbred rat colonies. The OLETF rat (Kawano et al. 1992, 1994) is reviewed in another chapter in this book. Although the GK rat does not have all of the characteristics of T2D, this model does have beta-cell problems that might be associated with beta-cell failure of T2D. It could be interesting to breed this specific genetic beta-cell defect onto a monogenic or polygenetic obesity animal model. Such a cross could result in a new animal model that exhibits more of the characteristics of typical human T2D. Howard Jacobs's laboratory has recently placed the beta-cell defect from the GK rat on a faun-hooded background (Nobrega et al. 2004). The faun-hooded rat has a propensity for hypertension and kidney problems. When they introgressed the mitochondria and some passenger loci from the FHH/EurMcwi rat into the genetic background of diabetic GK rats, they created a new rat strain, T2DN (T2DN/Mcwi), that has a nephropathy that closely resembles morphological and physiological characteristics of human diabetic nephropathy.

LIMITATIONS OF CURRENT MODELS

No animal model can completely duplicate the human diabetic condition. Yet all of the models contribute in their individual ways. A significant issue is that there is no polygenic obesity model that has the typical characteristics of metabolic syndrome and goes on to develop overt diabetes. The ideal model would develop obesity on chow diets or on synthetic "Western" diets. Along with obesity, the animal should express insulin resistance, hypertension, and hyperlipidemia. Over time, the "ideal model" should develop hyperglycemia with beta-cell failure and loss. To be most useful in experimental protocols, the aforementioned characteristics should appear in every animal over a relatively short period. The model should also develop the typical complications of the diabetic state, which include cardiovascular, kidney, and nerve pathology.

Although this animal might be ideal, it would not completely represent the human population. The typical individual patient does not express all of the features described in this "ideal" animal. A better solution would be to have a series of animal models that have individual characteristics that would have combinations of the preceding characteristics that could be used to study the specific disease conditions that accompany diabetes. Although the research community appears to have many of these characteristics in the available animal models, we do have some characteristics that are not like the human condition. Among such conditions are the mechanisms behind the development of obesity and the conditions that would result in ideal cardiovascular and other complications. Such model issues could be addressed and resolved over time as we more completely understand the genetics and mechanisms behind the human condition. Then we can naturally select and genetically modify animals to obtain the "ideal models."

NEW MODELS UNDER DEVELOPMENT

Our laboratory has begun a process of crossing appropriate animals and attempting to identify more novel models by purposefully selecting for the phenotypic characteristics of metabolic syndrome. The first model we are developing is based on crossing a polygenetic DIO rat, developed and selected by Levin, with the homozygous lean ZDF rat. The purpose of this selection process was to develop a polygenic obesity model that develops type 2 diabetes.

DIO rats have been used by a number of laboratories. The strains of rats used have varied from setting to setting but the general principle is to feed rats a palatable high-energy diet and thus cause increased energy intake and increased fat mass. Levin's laboratory has characterized and used specific Sprague–Dawley rats from the Charles River CD colony for a number of years (Levin et al. 1989, 1997; Ricci and Levin 2003). Levin has been able to produce diet-prone (DP) and diet-resistant (DR) substrains by selective breeding of the colony (Levin et al. 1997). The Levin DIO has become a useful model for investigation of obesity mechanisms. Heiman's laboratory at Lilly has characterized and used a Long–Evans model of obesity and used it to test compounds (Hsiung, Hertel, et al. 2005; Hsiung, Kusaka, et al. 2005). A chapter in this book is dedicated to a review of this and other models.

Again, this new model is produced by initially crossing the lean ZDF with a polygenic DIO rat selected by Levin. The breeding process to develop a new model is based on the hypothesis that the obesity and diabetic traits were inherited independently. If this hypothesis is true, the genetic problem with the beta cell, which leads to diabetes when the animal is obese, should be present in the lean animal. The work of Griffen's laboratory (Griffen et al. 2001) demonstrated that this was likely to be true. The data so far have demonstrated that this is possible and several generations of obese, insulin-resistant rats have developed a level of type 2 diabetes that appears to be similar to the diabetes seen in the ZDF rat. It appears likely that this experiment will result in a new, potentially improved diabetes model that has significant advantages over the ZDF and other animal models for T2D.

This model is being referred to as ZDSD. There are no publications yet on this new model but there have been several meetings where the characteristics have been described. These include rapid weight gain on rat chow, high insulin levels, indicating insulin resistance before hyperglycemia develops, and decreased insulin levels as diabetes develops, indicating beta-cell dysfunction. These general characteristics are similar to the ZDF diabetic traits and diabetic patients but this new model does not have the leptin receptor defects found in the ZDF rat. The data on this model support the hypothesis presented previously and appear to support it as a viable T2D model. Although this model will be useful in the type of obesity that precedes insulin resistance and diabetes, the presence of the complications of diabetes remains to be determined.

As suggested earlier, there are other genetic models of type 2 diabetes with traits that might be placed on other genetic backgrounds to enhance the models that are currently available. The example of the GK rat previously mentioned is only one possibility. Similar breeding experiments could be done with mouse models that had desirable traits that could be combined.

SUMMARY

There are many rat models that range the spectrum from mildly obese to overt diabetes. Each of these models has advantages and disadvantages but they all have value related to the diseases they represent. The obvious disadvantages are in those whose genetic characteristics do not resemble the human condition. With a better understanding of the human condition and the specific characteristics of current animal models, specific new animal models can be developed by cross-breeding strains and selecting for desired characteristics. Although not discussed in this chapter, insertion of human genes could also help in making these models more closely resemble the human metabolic state and thus make them more valuable for studying the diseases and testing target-specific drugs.

REFERENCES

Alemzadeh R, Tushaus KM. (2004) Modulation of adipoinsular axis in prediabetic Zucker diabetic fatty rats by diazoxide. *Endocrinology* 145:5476–5484.

Alemzadeh R, Tushaus K. (2005) Diazoxide attenuates insulin secretion and hepatic lipogenesis in Zucker diabetic fatty rats. *Med Sci Monit* 11:BR439–BR448.

Anis Y, Leshem O, Reuveni H, et al. (2004) Antidiabetic effect of novel modulating peptides of G-protein-coupled kinase in experimental models of diabetes. *Diabetologia* 47:1232–1244.

Balon TW, Gu JL, Tokuyama Y, et al. (1995) Magnesium supplementation reduces development of diabetes in a rat model of spontaneous NIDDM. *Am J Physiol* 269:E745–E752.

Banz, WJ, Davis J, Steinle JJ, et al. (2005) (+)-Z-bisdehydrodoisynolic acid ameliorates obesity and the metabolic syndrome in female ZDF rats. *Obesity Res* 13:1915–1924.

Baylis CE. Atzpodien A, Freshour G, et al. (2003) Peroxisome proliferator-activated receptor [gamma] agonist provides superior renal protection versus angiotensin-converting enzyme inhibition in a rat model of type 2 diabetes with obesity. *J Pharmacol Exp Ther* 307:54–60.

Belury MA, Mahon A, Banni S. (2003) The conjugated linoleic acid (CLA) isomer, t10c12-CLA, is inversely associated with changes in body weight and serum leptin in subjects with type 2 diabetes mellitus. *J Nutr* 133:257S–260S.

Berghofer P, Peterson RG, Schneider K, et al. (1997) Incretin hormone expression in the gut of diabetic mice and rats. *Metabolism* 46:261–267.

Brand CL, Sturis L, Gotfredsen CF, et al. (2003) Dual PPARalpha/gamma activation provides enhanced improvement of insulin sensitivity and glycemic control in ZDF rats. *Am J Physiol* 284:E841–E854.

Brown KK, Henke BR, Blanchard SG, et al. (1999) A novel N-aryl tyrosine activator of peroxisome proliferator-activated receptor-gamma reverses the diabetic phenotype of the Zucker diabetic fatty rat. *Diabetes* 48:1415–1424.

Byrne JJ, Bradlow HL. (2001) DHEA-PC slows the progression of type 2 diabetes (non-insulin-dependent diabetes mellitus) in the ZDF/Gmi-*fa/fa* rat. *Diabetes Technol Ther* 3:211–219.

Carr RD, Brand CL, Bodvarsdottir TB, et al. (2003) NN414, a SUR1/Kir6.2-selective potassium channel opener, reduces blood glucose and improves glucose tolerance in the VDF Zucker rat. *Diabetes* 52:2513–2518.

Chen XM, Osborne C, Rybczynski PJ, et al. (2005) Pharmacological profile of a novel, non-TZD PPARgamma agonist. *Diabetes Obesity Metab* 7:536–546.

Clark JB, Palmer CJ, Shaw WN. (1983) The diabetic Zucker fatty rat. *Proc Soc Exp Biol Med* 173:68–75.

Cnop M, Hannaert JC, Hoorens A, et al. (2001) Inverse relationship between cytotoxicity of free fatty acids in pancreatic islet cells and cellular triglyceride accumulation. *Diabetes* 50:1771–1777.

Cobb JE, Blanchard SG, Boswell EG, et al. (1998) N-(2-benzoylphenyl)-L-tyrosine PPAR-gamma agonists. 3. Structure–activity relationship and optimization of the N-aryl substituent. *J Med Chem* 41:5055–5069.

Corpe C, Sreenan S, Burant C. (2001) Effects of type-2 diabetes and troglitazone on the expression patterns of small intestinal sugar transporters and PPAR-gamma in the Zucker diabetic fatty rat. *Digestion* 63:116–123.

Corsetti JP, Sparks JD, Peterson RG, et al. (2000) Effect of dietary fat on the development of non-insulin dependent diabetes mellitus in obese Zucker diabetic fatty male and female rats. *Atherosclerosis* 148:231–241.

Dana SL, Hoener PA, Bilakovics JM, et al. (2001) Peroxisome proliferator-activated receptor subtype-specific regulation of hepatic and peripheral gene expression in the Zucker diabetic fatty rat. *Metabolism* 50:963–971.

Davis JM, Iqbal J, Steinle J, et al. (2005) Soy protein influences the development of the metabolic syndrome in male obese ZDF × SHHF rats. *Horm Metab Res* 37:316–325.

Dominguez JH, Wu P, Hawes JW, et al. (2006) Renal injury: Similarities and differences in male and female rats with the metabolic syndrome. *Kidney Int*, in press.

Ellis EA, Grant MB, Murray FM, et al. (1998) Increased NADH oxidase activity in the retina of the BBZ/Wor diabetic rat. *Free Radical Biol Med* 24:111–120.

Ellis EA, Guberski DL, Hutson D, et al. (2002) Time course of NADH oxidase, inducible nitric oxide synthase and peroxynitrite in diabetic retinopathy in the BBZ/WOR rat. *Nitric Oxide* 6:295–304.

Ellis EA, Guberski DL, Somogyi-Mann L, et al. (2000) Increased H2O2, vascular endothelial growth factor and receptors in the retina of the BBZ/Wor diabetic rat. *Free Radical Biol Med* 28:91–101.

Etgen GJ, Oldham BA. (2000) Profiling of Zucker diabetic fatty rats in their progression to the overt diabetic state. *Metabolism* 49:684–688.

Etgen GJ, Oldham BA, Johnson WA, et al. (2002) A tailored therapy for the metabolic syndrome: The dual peroxisome proliferator-activated receptor-alpha/gamma agonist LY465608 ameliorates insulin resistance and diabetic hyperglycemia while improving cardiovascular risk factors in preclinical models. *Diabetes* 51:1083–1087.

Finegood DT, McArthur MD, Kojwang D, et al. (2001) Beta-cell mass dynamics in Zucker diabetic fatty rats. Rosiglitazone prevents the rise in net cell death. *Diabetes* 50:1021–1029.

Friedman JE, Ferrara CM, Aulak KS, et al. (1997) Exercise training down-regulates *ob* gene expression in the genetically obese SHHF/Mcc-*fa*(*cp*) rat. *Horm Metab Res* 29:214–219.

Fu WJ, Haynes TE, Kohli R, et al. (2005) Dietary L-arginine supplementation reduces fat mass in Zucker diabetic fatty rats. *J Nutr* 135:714–721.

Fujinami K, Kojima K, Aragane K, et al, (2001) Postprandial hyperlipidemia in Zucker diabetic fatty *fa*/*fa* rats, an animal model of type II diabetes, and its amelioration by acyl-CoA:cholesterol acyltransferase inhibition. *Jpn J Pharmacol* 86:127–129.

Gao JW, Sherman M, McCune SA, et al. (1994) Effects of acute running exercise on whole body insulin action in obese male SHHF/Mcc-*facp* rats. *J Appl Physiol* 77:534–541.

Gedulin BR, Smith P, Prickett KS, et al. (2005) Dose–response for glycemic and metabolic changes 28 days after single injection of long-acting release exenatide in diabetic fatty Zucker rats. *Diabetologia* 48:1380–1385.

Gerdes AM, Onodera T, Wang X, et al. (1996) Myocyte remodeling during the progression to failure in rats with hypertension. *Hypertension* 28:609–614.

Girard J. (2000) Fatty acids and beta cells. *Diabete Metab* (Paris) 26:Suppl 3, 6–9.

Girard J. (2005) Glitazones and pancreatic function. *Ann Endocrinol (Paris)* 66:1S18–1S23.

Griffen SC, Wang J, German MS. (2001) A genetic defect in beta-cell gene expression segregates independently from the *fa* locus in the ZDF rat. *Diabetes* 50:63–68.

Guberski DL, Butler L, Manzi SM, et al. (1993) The BBZ/Wor rat: Clinical characteristics of the diabetic syndrome. *Diabetologia* 36:912–919.

Hargrove DM, Nardone NA, Persson LM, et al. (1995) Glucose-dependent action of glucagon-like peptide-1 (7-37) *in vivo* during short- or long-term administration. *Metabolism* 44:1231–1237.

Harmon JS, Gleason CE, Tanaka Y, et al. (2001) Antecedent hyperglycemia, not hyperlipidemia, is associated with increased islet triacylglycerol content and decreased insulin gene mRNA level in Zucker diabetic fatty rats. *Diabetes* 50:2481–2486.

Henriksen EJ, Kinnick TR, Teachey MK, et al. (2003) Modulation of muscle insulin resistance by selective inhibition of GSK-3 in Zucker diabetic fatty rats. *Am J Physiol* 284:E892–E900.

Heyen JR, Blasi ER, Nikula K, et al. (2002) Structural, functional, and molecular characterization of the SHHF model of heart failure. *Am J Physiol* 283:H1775–H1784.

Higa M, Zhou YT, Ravazzola M, et al. (1999) Troglitazone prevents mitochondrial alterations, beta cell destruction, and diabetes in obese prediabetic rats. *Proc Natl Acad Sci USA* 96:11513–11518.

Houseknecht KL, Vanden Heuvel JP, Moya-Camarena SY, et al. (1998) Dietary conjugated linoleic acid normalizes impaired glucose tolerance in the Zucker diabetic fatty *fa/fa* rat. *Biochem Biophys Commun* 244:678–682.

Hsiung HM, Hertel J, Zhang XY, et al. (2005) A novel and selective beta-melanocyte-stimulating hormone-derived peptide agonist for melanocortin 4 receptor potently decreased food intake and body weight gain in diet-induced obese rats. *Endocrinology* 146:5257–5266.

Hsiung HM, Smiley DL, Zhang XY, et al. (2005) Potent peptide agonists for human melanocortin 3 and 4 receptors derived from enzymatic cleavages of human beta-MSH(5-22) by dipeptidyl peptidase I and dipeptidyl peptidase IV. *Peptides* 26:1988–1996.

Huang TH, Peng G, Kota BP, et al. (2005) Anti-diabetic action of *Punica granatum* flower extract: Activation of PPAR-gamma and identification of an active component. *Toxicol Appl Pharmacol* 207:160–169.

Huang TH, Peng G, Li GQ, et al. (2006) *Salacia oblonga* root improves postprandial hyperlipidemia and hepatic steatosis in Zucker diabetic fatty rats: Activation of PPAR-alpha. *Toxicol Appl Pharmacol* 210:225–235.

Hui H, Farilla L, Merkel P, et al. (2002) The short half-life of glucagon-like peptide-1 in plasma does not reflect its long-lasting beneficial effects. *Eur J Endocrinol* 146:863–869.

Ishikawa Y, Saito MN, Ikemoto T, et al. (1998) Actions of the novel oral antidiabetic agent HQL-975 in insulin-resistant non-insulin-dependent diabetes mellitus model animals. *Diabetes Res Clin Pract* 41:101–111.

Ito I, Jarajapu YP, Guberski DL, et al. (2006) Myogenic tone and reactivity of rat ophthalmic artery in acute exposure to high glucose and in a type II diabetic model. *Invest Ophthalmol Vis Sci* 47:683–692.

Jackson EK, Kost CK Jr., Herzer WA, et al. (2001) A(1) receptor blockade induces natriuresis with a favorable renal hemodynamic profile in SHHF/Mcc-*fa(cp)* rats chronically treated with salt and furosemide. *J Pharmacol Exp Ther* 299:978–987.

Johnson JH, Ogawa A, Chen L, et al. (1990) Underexpression of β cell high Km glucose transporters in noninsulin-dependent diabetes. *Science* 250:546–549.

Kakuma T, Lee Y, Higa M, et al. (2000) Leptin, troglitazone, and the expression of sterol regulatory element binding proteins in liver and pancreatic islets. *Proc Natl Acad Sci USA* 97:8536–8541.

Kawano K, Hirashima T, Mori S, et al. (1994) OLETF (Otsuka Long-Evans Tokushima Fatty) rat: a new NIDDM rat strain. *Diabetes Res Clin Pract* 24:S317–S320.

Kawano K, Hirashima T, Mori S, et al. (1992) Spontaneous long-term hyperglycemic rat with diabetic complications. Otsuka Long-Evans Tokushima Fatty (OLETF) strain. *Diabetes* 41:1422–1428.

Konno R, Kaneko Y, Suzuki K, et al. (2005) Effect of 5-campestenone (24-methylcholest-5-en-3-one) on Zucker diabetic fatty rats as a type 2 diabetes mellitus model. *Horm Metab Res* 37:79–83.

Lee Y, Hirose H, Ohneda M, et al. (1994) Beta-cell lipotoxicity in the pathogenesis of non-insulin-dependent diabetes mellitus of obese rats: Impairment in adipocyte–beta-cell relationships. *Proc Natl Acad Sci USA* 91:10878–10882.

Levin BE, Dunn-Meynell AA, Balkan B, et al. (1997) Selective breeding for diet-induced obesity and resistance in Sprague–Dawley rats. *Am J Physiol* 273:R725–R730.

Levin BE, Hogan S, Sullivan AC. (1989) Initiation and perpetuation of obesity and obesity resistance in rats. *Am J Physiol* 256:R766–R771.

Li X, Hansen PA, Xi L, et al. (2005) Distinct mechanisms of glucose lowering by specific agonists for peroxisomal proliferator activated receptor gamma and retinoic acid X receptors. *J Biol Chem* 280:38317–38327.

Liang Y, Chen X, Osborne M, et al. (2005) Topiramate ameliorates hyperglycaemia and improves glucose-stimulated insulin release in ZDF rats and *db/db* mice. *Diabetes Obesity Metab* 7:360–369.

Liu X, Perusse F, Bukowiecki J. (1998) Mechanisms of the antidiabetic effects of the beta 3-adrenergic agonist CL-316243 in obese Zucker-ZDF rats. *Am J Physiol* 274:R1212–R1219.

Macchia PE, Jiang P, Yuan YD, et al. (2002) RXR receptor agonist suppression of thyroid function: Central effects in the absence of thyroid hormone receptor. *Am J Physiol* 283:E326–E331.

McCarthy KJ, Routh RJ, Shaw E, et al. (2000) Troglitazone halts diabetic glomerulosclerosis by blockade of mesangial expansion. *Kidney Int* 58:2341–2350.

McCune SA, Baker PB, Stills HF. (1999) SHHF/Mcc-*cp* rat: Model of obesity, non-insulin-dependent diabetes, and congestive heart failure. *ILAR News* 32:23–27.

Meier DA, Hennes MM, McCune SA, et al. (1995) Effects of obesity and gender on insulin receptor expression in liver of SHHF/Mcc-FA*cp* rats. *Obesity Res* 3:465–470.

Mest HJ, Raap A, Schloos J, et al. (2001) Glucose-induced insulin secretion is potentiated by a new imidazoline compound. *Naunyn Schmiedebergs Arch Pharmacol* 364:47–52.

Milburn JLJ, Ohneda M, Johnson JH, et al. (1993) Beta-cell GLUT-2 loss and non-insulin-dependent diabetes mellitus: Current status of the hypothesis. *Diabetes Metab Rev* 9:231–236.

Nagao K, Inoue N, Wang YM, et al. (2003) Conjugated linoleic acid enhances plasma adiponectin level and alleviates hyperinsulinemia and hypertension in Zucker diabetic fatty (*fa/fa*) rats. *Biochem Biophys Res Commun* 310:562–566.

Nobrega MA, Fleming S, Roman RJ, et al. (2004) Initial characterization of a rat model of diabetic nephropathy. *Diabetes* 53:735–742.

Oh S, Lee M, Ko KS, et al. (2003) GLP-1 gene delivery for the treatment of type 2 diabetes. *Mol Ther* 7:478–483.

Ohneda M, Inman LR, Unger RH. (1995) Caloric restriction in obese pre-diabetic rats prevents beta-cell depletion, loss of beta-cell GLUT 2 and glucose incompetence. *Diabetologia* 38:173–179.

Onodera T, Tamura T, Said S, et al. (1998) Maladaptive remodeling of cardiac myocyte shape begins long before failure in hypertension. *Hypertension* 32:753–757.

Orci L, Ravazzola M, Baetens D, et al. (1990) Evidence that down-regulation of beta-cell glucose transporters in non-insulin-dependent diabetes may be the cause of diabetic hyperglycemia. *Proc Natl Acad Sci USA* 87:9953–9957.

Park S, McCune SA, Radin MJ, et al. (1997) Verapamil accelerates the transition to heart failure in obese, hypertensive, female SHHF/Mcc-*fa*(*cp*) rats. *J Cardiovasc Pharmacol* 29:726–733.

Park SC, Liu-Stratton Y, Medeiros LC, et al. (2004) Effect of male sex and obesity on platelet arachidonic acid in spontaneous hypertensive heart failure rats. *Exp Biol Med* 229:657–664.

Pelzer T, Jazbutyte V, Arias-Loza PA, et al. (2005) Pioglitazone reverses down-regulation of cardiac PPARgamma expression in Zucker diabetic fatty rats. *Biochem Biophys Res Commun* 329:726–732.

Peterson RG. (1994) Alpha-glucosidase inhibitors in diabetes: Lessons from animal studies. *Eur J Clin Invest* 24:Suppl 3, 11–18.

Peterson RG. (1995) The Zucker diabetic fatty (ZDF) rat. In *Lessons from Animal Diabetes*, ed. E Shafrir. Smith–Gordon, London, 5:225–230.

Peterson RG. (2001) *The Zucker Diabetic Fatty (ZDF) Rat: Animal Models of Diabetes. A Primer.* Harwood Academic Publishers, Amsterdam, 109–128.

Peterson, RG, Doss DI, Neel MA, et al. (1993) The effectiveness of acarbose in treating Zucker diabetic fatty rats (ZDF/Drt-fa). In *Drugs in Development*, vol. I. *α-Glucosidase Inhibition: Potential Use in Diabetes*, ed. JR Vasselli, CA Maggio, A Scriabine. Neva Press, Branford, CT, 167–172.

Peterson RG, Shaw WN, Neel MA, et al. (1990) Zucker diabetic fatty rat as a model for non-insulin-dependent diabetes mellitus. *ILAR News* 32:16–19.

Pick A, Clark J, Kubstrup C, et al. (1998) Role of apoptosis in failure of beta-cell mass compensation for insulin resistance and beta-cell defects in the male Zucker diabetic fatty rat. *Diabetes* 47:358–364.

Pickavance L, Widdowson PS, King P, et al. (1998) The development of overt diabetes in young Zucker diabetic fatty (ZDF) rats and the effects of chronic MCC-555 treatment. *Br J Pharmacol* 125:767–770.

Pickavance LC, Brand CL, Wassermann K, et al. (2005) The dual PPARalpha/gamma agonist, ragaglitazar, improves insulin sensitivity and metabolic profile equally with pioglitazone in diabetic and dietary obese ZDF rats. *Br J Pharmacol* 144:308–316.

Poitout V, Robertson RP. (2002) Minireview: Secondary beta-cell failure in type 2 diabetes—A convergence of glucotoxicity and lipotoxicity. *Endocrinology* 143:339–342.

Pold R, Jensen LS, Jessen N, et al. (2005) Long-term AICAR administration and exercise prevents diabetes in ZDF rats. *Diabetes* 54:928–934.

Pospisilik JA, Stafford SG, Demuth HU, et al. (2002a) Long-term treatment with the dipeptidyl peptidase IV inhibitor P32/98 causes sustained improvements in glucose tolerance, insulin sensitivity, hyperinsulinemia, and beta-cell glucose responsiveness in VDF (*fa/fa*) Zucker rats. *Diabetes* 51:943–950.

Pospisilik JA, Stafford SG, Demuth HU, et al. (2002b) Long-term treatment with dipeptidyl peptidase IV inhibitor improves hepatic and peripheral insulin sensitivity in the VDF Zucker rat: A euglycemic-hyperinsulinemic clamp study. *Diabetes* 51:2677–2683.

Poucheret P, Verma S, Grynpas MD, et al. (1998) Vanadium and diabetes. *Mol Cell Biochem* 188:73–80.

Prasad K. (2001) Secoisolariciresinol diglucoside from flaxseed delays the development of type 2 diabetes in Zucker rat. *J Lab Clin Med* 138:32–39.

Radin MJ, Jenkins MJ, McCune SA, et al. (1992) Effects of enalapril and clonidine on glomerular structure, function, and atrial natriuretic peptide receptors in SHHF/Mcc-*cp* rats. *J Cardiovasc Pharmacol* 19:464–472.

Ricci MR, Levin BE. (2003) Ontogeny of diet-induced obesity in selectively bred Sprague–Dawley rats. *Am J Physiol* 285:R610–R618.

Ring DB, Johnson KW, Henriksen KJ, et al. (2003) Selective glycogen synthase kinase 3 inhibitors potentiate insulin activation of glucose transport and utilization *in vitro* and *in vivo*. *Diabetes* 52:588–595.

Schrijvers BF, Flyvbjerg A, Tiltonm RG, et al. (2006) A neutralizing VEGF antibody prevents glomerular hypertrophy in a model of obese type 2 diabetes, the Zucker diabetic fatty rat. *Nephrol Dial Transplant* 21:324–329.

Seufert J, Weir GC, Habaneras JF. (1998) Differential expression of the insulin gene transcriptional repressor CCAAT/enhancer-binding protein beta and transactivator islet duodenum homeobox-1 in rat pancreatic beta cells during the development of diabetes mellitus. *J Clin Invest* 101:2528–2539.

Sharkey LC, Holycross BJ, Park S, et al. (1998) Effect of ovariectomy in heart failure-prone SHHF/Mcc-*facp* rats. *Am J Physiol* 275:R1968–R1976.

Shen HQ, Roth MD, Peterson RG. (1998) The effect of glucose and glucagon-like peptide-1 stimulation on insulin release in the perfused pancreas in a non-insulin-dependent diabetes mellitus animal model. *Metabolism* 47:1042–1047.

Shimabukuro M, Zhou YT, Lee Y, et al. (1998) Troglitazone lowers islet fat and restores beta cell function of Zucker diabetic fatty rats. *J Biol Chem* 273:3547–3550.

Slieker LJ, Sundell KL, Heath WF, et al. (1992) Glucose transporter levels in tissues of spontaneously diabetic Zucker *fa/fa* rat (ZDF/drt) and viable yellow mouse (Avy/a). *Diabetes* 41:187–193.

Smith SA, Lister CA, Toseland CD, et al. (2000) Rosiglitazone prevents the onset of hyperglycemia and proteinuria in the Zucker diabetic fatty rat. *Diabetes Obesity Metab* 2:363–372.

Sreenan S, Sturis J, Pugh W, et al. (1996) Prevention of hyperglycemia in the Zucker diabetic fatty rat by treatment with metformin or troglitazone. *Am J Physiol* 271:E742–E747.

Sreenan SK, Mittal AA, Dralyuk F, et al. (2000) Glucagon-like peptide-1 stimulates insulin secretion by a Ca2+-independent mechanism in Zucker diabetic fatty rat islets of Langerhans. *Metabolism* 49:1579–1587.

Sturis J, Gotfredsen CF, Romer J, et al. (2003) GLP-1 derivative liraglutide in rats with beta-cell deficiencies: Influence of metabolic state on beta-cell mass dynamics. *Br J Pharmacol* 140:123–132.

Sudre B, Broqua P, White RB, et al. (2002) Chronic inhibition of circulating dipeptidyl peptidase IV by FE 999011 delays the occurrence of diabetes in male Zucker diabetic fatty rats. *Diabetes* 51:1461–1469.

Tamura T, Onodera T, Said S, et al. (1998) Correlation of myocyte lengthening to chamber dilation in the spontaneously hypertensive heart failure (SHHF) rat. *J Mol Cell Cardiol* 30:2175–2181.

Tirabassi RS, Flanagan JF, Wu T, et al. (2004) The BBZDR/Wor rat model for investigating the complications of type 2 diabetes mellitus. *ILAR J* 45:292–302.

Tofovic SP, Dubey RK, Jackson RK. (2001) 2-Hydroxyestradiol attenuates the development of obesity, the metabolic syndrome, and vascular and renal dysfunction in obese ZSF1 rats. *J Pharmacol Exp Ther* 299:973–977.

Tofovic SP, Kost CK Jr., Jackson EK, et al. (2002) Long-term caffeine consumption exacerbates renal failure in obese, diabetic, ZSF1 (*fa-fa*(*cp*)) rats. *Kidney Int* 61:1433–1444.

Tofovic SP, Kusaka H, Jackson EK, et al. (2001) Renal and metabolic effects of caffeine in obese (*fa/fa*(*cp*)), diabetic, hypertensive ZSF1 rats. *Renal Failure* 23:159–173.

Tofovic SP, Kusaka H, Kost CK Jr., et al. (2000) Renal function and structure in diabetic, hypertensive, obese ZDF × SHHF-hybrid rats. *Renal Failure* 22:387–406.

Unger RH. (1991) Diabetic hyperglycemia: Link to impaired glucose transport in pancreatic beta cells. *Science* 251:1200–1205.

Unger RH. (1995) Lipotoxicity in the pathogenesis of obesity-dependent NIDDM. Genetic and clinical implications. *Diabetes* 44:863–870.

Unger RH, Orci L. (2001) Diseases of liporegulation: New perspective on obesity and related disorders. *FASEB J* 15:312–321.

Wang J, Yuen VG, McNeill JH. (2001) Effect of vanadium on insulin and leptin in Zucker diabetic fatty rats. *Mol Cell Biochem* 218:93–96.

Wargent E, Stocker C, Augstein P, et al. (2005) Improvement of glucose tolerance in Zucker diabetic fatty rats by long-term treatment with the dipeptidyl peptidase inhibitor P32/98: Comparison with and combination with rosiglitazone. *Diabetes Obesity Metab* 7:170–181.

Wasan KM, Risovic V, Yuen VG, et al. (2006) Differences in plasma homocysteine levels between Zucker fatty and Zucker diabetic fatty rats following 3 weeks oral administration of organic vanadium compounds. *J Trace Elem Med Biol* 19:251–258.

Wilkes JJ, Nelson E, Osborne M, et al. (2005) Topiramate is an insulin-sensitizing compound *in vivo* with direct effects on adipocytes in female ZDF rats. *Am J Physiol* 288:E617–E624.

Winter CL, Lange JS, Davis MG, et al. (2005) A nonspecific phosphotyrosine phosphatase inhibitor, bis(maltolato)oxovanadium(IV), improves glucose tolerance and prevents diabetes in Zucker diabetic fatty rats. *Exp Biol Med* 230:207–216.

Yue TL, Bao W, Gu JL, et al. (2005) Rosiglitazone treatment in Zucker diabetic fatty rats is associated with ameliorated cardiac insulin resistance and protection from ischemia/reperfusion-induced myocardial injury. *Diabetes* 54:554–562.

Yuen VG, Bhanot S, Battell ML, et al. (2003) Chronic glucose-lowering effects of rosiglitazone and bis(ethylmaltolato)oxovanadium(IV) in ZDF rats. *Can J Physiol Pharmacol* 81:1049–1055.

Yuen VG, Vera E, Battell ML, et al. (1999) Acute and chronic oral administration of bis(maltolato)oxovanadium(IV) in Zucker diabetic fatty (ZDF) rats. *Diab Res Clin Pract* 43:9–19.

Zhou YP, Madjidi A, Wilson ME, et al. (2005) Matrix metalloproteinases contribute to insulin insufficiency in Zucker diabetic fatty rats. *Diabetes* 54:2612–2619.
Zhou YT, Shimabukuro M, Lee Y, et al. (1998) Enhanced *de novo* lipogenesis in the leptin-unresponsive pancreatic islets of prediabetic Zucker diabetic fatty rats: Role in the pathogenesis of lipotoxic diabetes. *Diabetes* 47:1904–1908.
Zucker LM, Zucker TF. (1961) Fatty, a new mutation in the rat. *J Heredity* 52:275–287.

5 The Goto–Kakizaki Rat

Claes-Göran Östenson

CONTENTS

INTRODUCTION

The Goto–Kakizaki (GK) rat is a nonobese substrain of Wistar rat origin, developing non-insulin-dependent diabetes mellitus early in life. Glucose intolerance is most likely primarily due to impaired B-cell mass and function on the background of a polygenic inheritance. In addition, secondary defects in B-cell function and insulin action may superimpose (e.g., due to chronic hyperglycemia [glucotoxicity]). Since the GK rat can be regarded as one of the best available rodent strains for the study of inherited type 2 diabetes (T2D), it has been extensively used in experimental diabetes research reported in hundreds of publications.

ORIGIN AND BREEDING OF THE GK RAT

In the 1970s, Goto and coworkers in Sendai, Japan, initiated the GK substrain. From 211 healthy Wistar rats, they selected nine animals of each gender with the highest blood glucose levels within the normal range during an oral glucose tolerance test (OGTT) (Goto et al. 1975, 1988; Suzuki et al. 1992). Repeated selection of rats with tendency to enhanced OGTT response for breeding resulted in clear-cut glucose intolerance after five generations (F_6). Since the ninth generation (F_8), sister–brother mating was consequently performed implying that the GK rat today—after more than 100 generations—is a highly inbred substrain. This has been verified by inter-animal skin transplantation (unpublished data), as well as microsatellite mapping of the GK rat genome (Galli et al. 1996).

In the F_8 generation of GK rat, impaired insulin response to glucose was demonstrated in the perfused isolated pancreas (Kimura et al. 1982). Thereafter, glucose intolerance and impairment of glucose-induced insulin secretion have been constant features of the GK rat, also, when bred in colonies outside Japan (Suzuki et al. 1992; Östenson et al. 1996). However, despite these similarities, other characteristics such as islet cell morphology and islet metabolism seem to differ between some of the GK rat colonies (see following discussion).

GK Rat Colonies

Until the end of the 1980s, GK rats were bred only in Sendai. Colonies were then initiated with breeding pairs from Japan in Paris (Portha et al. 1991), Stockholm (Östenson, Khan, et al. 1993), Cardiff, Wales (Lewis et al. 1996), and Coimbra, Portugal (Duarte et al. 2004), and still exist. Some other colonies existed for shorter periods during the 1990s in London (Hughes et al. 1994), Aarhus, Denmark, and Seattle, Washington (Metz et al. 1999). At present (2006), there are also GK rat colonies derived from Paris (e.g., Tampa [Villar-Palasi and Farese 1994] and Brussels [Sener et al. 1996]) and in some Japanese universities. Also, GK rats are available commercially from a Danish producer of laboratory animals, M&B Animal Models A/S (Ry, Denmark), and several Japanese breeders, including Charles River Japan (Yokohama), Oriental Yeast (Tokyo), Clea (Osaka), and Takeda Labix Ltd (Osaka).

Maintenance and Breeding of GK Rat

Although the Japanese originally suggested that GK rats ideally should be fed oriental laboratory chow, it appears that the rats feed and breed perfectly well also on conventional lab rodent diet and tap water. Experience from the Stockholm GK rat colony, initiated in 1989 with F_{40} breeding pairs from Sendai, has indicated the importance of regulated environmental properties for optimal breeding. Thus, a humidity of 40–60% and temperature of 25–26°C in the breeding room seem to be advantageous for pregnancy and minimize the number of stillbirths, which were reported to be high in GK rats (Malaisse-Lagae et al. 1997). It is likely that a too dry environment may reduce fertility by impairing the tubar ciliar activity. Insulin treatment of pregnant GK rats as well as foster mothers of Wistar rat strain has not improved the litter size considerably (Östenson, unpublished data). In the Stockholm GK rat colony, the litter size ranges between 2 and 12 pups, with a mean of 8–9.

It should be emphasized that all GK rats display diabetic features, which are present at an early age and even in the fetuses (Abdel-Halim et al. 1994; Movassat et al. 1997; Serradas et al. 1998). Consequently, outbred Wistar rats have been used as control animals by most investigators. In most studies, the body weights of GK rats (in the age range of one week to four months) have been 10–30% lower than those of age-matched control animals (Sener et al. 1993; Hughes et al. 1994; Movassat et al. 1997; Giroix et al. 1999).

Glucose Tolerance and Plasma Insulin Levels

In the original GK rat colony, oral glucose tolerance tests were performed by administering 2 g/kg body weight to overnight fasted animals. It has been reported that glucose tolerance in the Sendai GK colony continued to decrease until the 35th generation; it has since been maintained at a stable level (Suzuki et al. 1992). In other laboratories, glucose tolerance has been assessed after i.p. injection of glucose solution (2 g/kg) (Östenson, Khan, et al. 1993; Abdel-Halim et al. 1994; Hughes et al. 1994) or i.v. (0.5–1.0 g/kg) (Portha et al. 1991; Gauguier et al. 1994). In male GK rats, irrespective of Stockholm or Paris origin, fasting blood glucose levels have been typically 7–9 m*M* compared with 3–5 m*M* in age-matched Wistar controls (Abdel-Halim et al. 1994) and in the fed state 10–18 and 6–8 m*M*, respectively (Sener et al. 1993; Ling et al. 1998; Giroix et al. 1999). In female GK rats, somewhat lower blood glucose concentrations have been noted.

Nonfasting plasma insulin levels in GK rats from all colonies have been similar or somewhat increased as compared with age-matched control Wistar rats (Portha et al. 1991; Suzuki et al. 1992; Östenson, Abdel-Halim, et al. 1993; Sener et al. 1993; Tsuura et al. 1993; Okamoto et al. 1995; Movassat et al. 1997; Salehi et al. 1999; Metz et al. 1999), while fasting plasma insulin levels have been lower relative to control animals (Galli et al. 1996). Fetal plasma insulin concentrations have been reported significantly lower in GK rats from the Paris colony compared with control fetuses (Serradas et al. 1995).

GENETIC CONSIDERATIONS

The hereditary nature of diabetes in the GK rat is obvious from the way the strain was produced (Goto et al. 1975, 1988; Suzuki et al. 1992). Studies on the offspring in crosses between GK and Wistar rats demonstrated that these F1 hybrid rats, regardless of whether the mother was a GK or a Wistar rat, exhibit glucose intolerance and glucose-induced insulin secretion intermediate to their parents (Abdel-Halim et al. 1994; Gauguier et al. 1994; Serradas et al. 1998). These findings indicate that conjunction of GK genes from both parents is required for defective insulin response and glucose tolerance to be fully expressed. They furthermore lead to the conclusion that maternal hyperglycemia *in utero* plays no or minimal role in the development of defects in the offspring, which is supported by a more recent study (Gill-Randall et al. 2004). In this context, it is noteworthy that no mitochondrial DNA deletion or polymorphism, which may be responsible for maternal inheritance, was detected in GK rats (Gauguier et al. 1994; Serradas et al. 1995). However, mitochondrial DNA content was decreased in pancreatic islets of adult, but not fetal, GK rats, probably reflecting the effect of metabolic dysfunction in the rat.

Repeated backcrossing of F1 hybrids of GK and Wistar rats with Wistar rats successively improved glucose tolerance, suggesting that several additive genes are responsible for glucose intolerance in the GK rat (Abdel-Halim et al. 1994). This observation was strongly supported by two genetic studies employing linkage analysis (Galli et al. 1996; Gauguier et al. 1996). Thus, with a combination of phenotypic and genotypic studies on F2 offspring following initial crosses between diabetic GK and healthy Fischer or brown Norway rats, respectively, at least six different independently segregating loci predisposing to glucose intolerance and impaired insulin secretion were found. In both studies, a major locus, *Niddm1* or *Nidd/gk1*, was found on chromosome 1 and estimated to contribute mainly to postprandial and less to fasting hyperglycemia. The somewhat less important *Niddm2* or *Nidd/gk2*, influencing postprandial and fasting glycemia, was localized to chromosome 2; another locus linked to body weight, *Weight1* or *bw/gk1,* was shown on chromosome 7 of the rat.

Although congenic strains of GK rats have been produced (Wallace et al. 2004; Wallis et al. 2004) identifying these and other putative diabetes loci, so far no obvious candidate genes of diabetes have been assigned to the described chromosomal loci. Interestingly, however, the genes coding for two uncoupling proteins, Ucp2 and Ucp3, were positioned to a region of chromosome 1 linked to glucose intolerance (Kaisaki et al. 1998). In addition, the mitochondrial glycerol-3-phosphate dehydrogenase gene was mapped to a part of chromosome 3 that contains a region linked to glucose intolerance in the GK rat (Koike et al. 1996), and the gene coding for insulin-degrading enzyme (IDE) was also suggested as a candidate susceptibility gene in GK rats of the Swedish colony (Fakhrai-Rad et al. 2000).

ISLET MORPHOLOGY AND HORMONE CONTENT

Islet Structure and Composition

One striking morphologic feature of GK rat pancreatic islets is the occurrence of so-called starfish-shaped islets (Goto et al. 1988; Suzuki et al. 1992; Guenifi et al. 1995). These islets are characterized by disrupted structure with cords of pronounced fibrosis—that is, connective tissue separating strands of endocrine cells, thereby resembling the appearance of a starfish (figs. 5.1 and 5.2). Accordingly, the mantle of glucagon and somatostatin cells is disrupted and these cells are found intermingled between beta-cells. These changes are rare or absent in the pancreas of young GK rats, but increase in prevalence with ageing (Suzuki et al. 1992). The parallel increase in IGF-2 expression, presumably inappropriately processed, with age in Swedish GK rat islets but not in control rat islets (Höög et al. 1996) may at least partly account for these changes in islet morphology. Partly at variance, in fetal GK rats of the Paris colony, IGF-2 production appeared impaired and this may contribute to retarded beta-cell growth (Serradas et al. 2002).

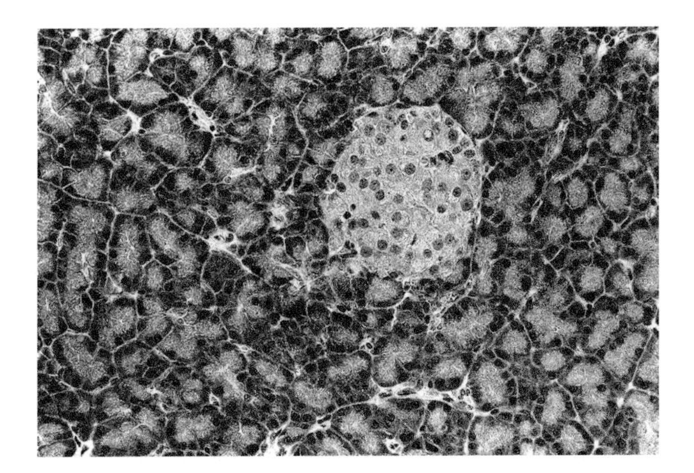

FIG. 5.1 An islet from a three-month-old GK rat. The islet is round in shape with clearly defined borders. (Magnification × 180; Dr. A. Höög, Stockholm.)

Early observations in the Japanese colony (Suzuki et al. 1992) suggested degeneration and paucity of beta-cells in the ageing GK rat. Beta-cell replication rates studied as labeling index after administration of ^{3}H-thymidine were, however, similar in GK and Wistar rats at one, three, and six months of age, indicating that the changes with ageing in GK rat islets cannot be accounted for by altered beta-cell replication (Östenson et al. 1996). This conclusion was, at least partly, supported by a study in the Paris GK rat colony (Movassat and Portha 1999).

Immunohistochemical methods have been used to assess the distribution of various islet endocrine cell types in GK rats. From such studies, it appears that this

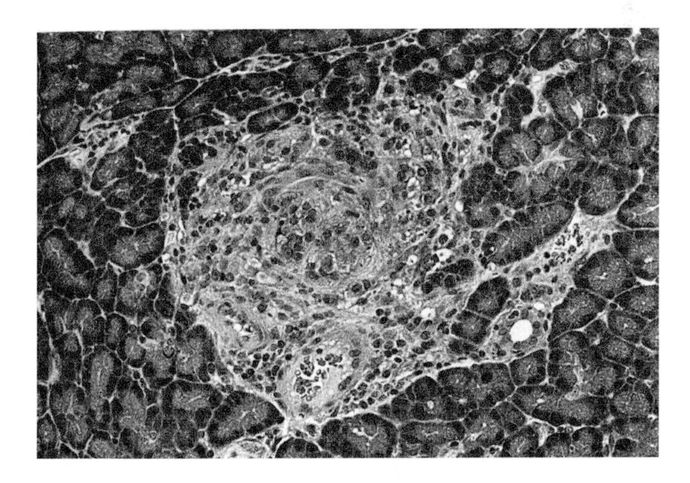

FIG. 5.2 A starfish-shaped islet from the same GK rat as in figure 5.1. Note the irregular shape with ill-defined borders and fibrous strands traversing the islet, as well as the typically enlarged size. (Magnification × 180; Dr. A. Höög, Stockholm.)

distribution differs between some of the GK rat colonies. Thus, in the Stockholm colony beta-cell density and relative volume of islet endocrine cells were alike in two- to three-month-old GK rats and control Wistar rats (Östenson, Abdel-Halim, et al. 1993; Guenifi et al. 1995). Similar results were reported in a study in Dallas with GK rats obtained directly from Sendai (Ohneda et al. 1993). In contrast, GK rats from the Paris colony reportedly display already at fetal stage a >50% reduction of the beta-cell mass, which is maintained throughout the adult life and apparently precedes onset of hyperglycemia at about the third week after birth (Movassat et al. 1997). Incidentally, in the Stockholm colony, GK pups were hyperglycemic already at the first week (Abdel-Halim et al. 1994).

Extensive studies in Paris GK rats have suggested that the defective beta-cell mass and function reflect a complex interaction of at least three pathogenic factors (Plachot et al. 2001; Miralles and Portha 2001; Portha 2005), including (1) several genes causing impaired insulin secretion; (2) decreased beta-cell neogenesis due to gestational metabolic programming; and (3) secondary or acquired loss of beta-cell differentiation in response to chronic exposure to hyperglycemia (glucotoxicity). Accordingly, GK rat beta-cell mass was shown to be further reduced by exposure to a carbohydrate-rich diet (Koyama et al. 1998). Reduction of beta-cell mass was protected by treatment of GK rats with an α-glucosidase inhibitor, voglibose (Koyama et al. 2000), and a similar effect in parallel with less severe hyperglycemia was achieved by treatment with GLP-1 or exendin-4 in postnatal GK rats (Tourrel et al. 2002).

Concerning the glucagon-producing A-cell, in the Paris GK rat colony their relative volume was reported to be normal during fetal and early neonatal period, while reduced by approximately one-third at older ages (Movassat et al. 1997). No corresponding study has been performed regarding the somatostatin-producing D-cells or other islet cell types.

Pancreatic and Islet Hormonal Content

As for islet cellular composition, there is some controversy regarding the content of pancreatic hormones in GK rats. In Paris animals, markedly reduced pancreatic and islet insulin, expressed per islet DNA (<40% of control), has been consistently reported also in the fetuses (Sener et al. 1996; Movassat et al. 1997; Giroix et al. 1999). In other GK rat colonies, corresponding insulin levels have been similar or more moderately decreased, compared with control rats (Östenson, Abdel-Halim, et al. 1993; Abdel-Halim et al. 1993; Suzuki et al. 1997; Salehi et al. 1999). Incidentally, islet insulin content was normal in GK rats of the Seattle colony, which originally was derived from Paris via Tampa, Florida (Metz et al. 1999). Apparently, male GK rats are more prone than females to develop reduced islet insulin content, which may lead to a more marked hyperglycemia in males (unpublished observations). F1 hybrids of GK and Wistar rats did not show reduced pancreatic insulin (Abdel-Halim et al. 1994). The biosynthesis of insulin in isolated islets of Paris and Stockholm colonies has been grossly normal (Giroix, Vesco, et al. 1993; Guest et al. 2002).

No major alteration in pancreatic glucagon content, expressed per pancreatic weight, has been demonstrated in GK rats (Abdel-Halim et al. 1993), although the A-cell mass was decreased by about 35% in adult GK rats of the Paris colony (Movassat et al. 1997). Pancreatic somatostatin content was slightly but significantly increased in Stockholm GK rats (Abdel-Halim et al. 1993). No studies on glucagon and somatostatin biosynthesis have been published.

ISOLATION OF PANCREATIC ISLETS

Pancreatic islets can be easily isolated by collagenase digestion from the pancreas of young GK rats; however, the increasing prevalence of starfish-shaped islets (fig. 5.2) may reduce the yield of islets per pancreas from animals older than four months. In fact, GK rat islets seem to resist the action of collagenase well and may need slightly higher enzyme concentration or longer exposure to the enzyme than Wistar rat islets (Östenson, unpublished observation).

ISLET AND BETA CELL FUNCTION

Glucose-Induced Insulin Secretion

Impaired glucose-stimulated insulin secretion in GK rats has been demonstrated *in vivo* (fig. 5.3) (Portha et al. 1991; Gauguier et al. 1994, 1996; Galli et al. 1996; Salehi et al. 1999), in the perfused isolated pancreas (Portha et al. 1991; Östenson, Khan, et al. 1993; Abdel-Halim et al. 1993, 1994, 1996), and in isolated pancreatic islets (Östenson, Khan, et al. 1993; Giroix, Vesco, et al. 1993; Giroix, Sener, et al. 1993; Hughes et al. 1994).

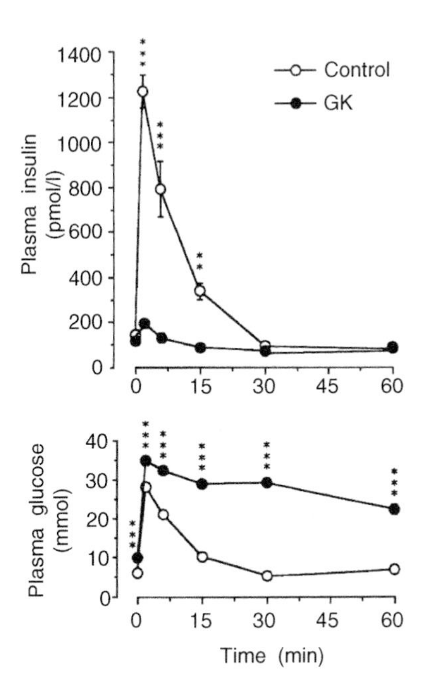

FIG. 5.3 Glucose and insulin responses *in vivo* to an i.v. glucose challenge in GK rats (filled circles) and control Wistar rats (unfilled circles). ***: $p < 0.001$ and **: $p < 0.01$ between GK and control. (Modified after Salehi A. et al. 1999. *Endocrinology* 140:3045–3053.)

It is generally accepted that glucose stimulates insulin secretion through its metabolism. After rapid transport over the beta-cell membrane, aided by the glucose transporter GLUT-2, the hexose is phosphorylated by glucokinase/hexokinase to glucose-6-phosphate and further metabolized by glycolysis and Krebs cycle, resulting in a yield of ATP. An increased cytosolic ATP/ADP ratio closes the ATP-regulated K^+-channels, which depolarizes the cell membrane and opens voltage-dependent L-type Ca^{2+}-channels. The resulting increase in cytoplasmic calcium ions stimulates exocytosis of insulin granules (Efendic et al. 1991).

A number of alterations or defects have been shown in the stimulus-secretion coupling for glucose in GK rat islets. GLUT-2 is underexpressed, but not likely to the extent that it could explain the impairment of insulin release (Ohneda et al. 1993). This assumption is supported by the fact that glucokinase/hexokinase activities were normal in GK rat islets (Östenson, Abdel-Halim, et al. 1993b; Tsuura et al. 1993). In addition, glycolysis rates in GK rat islets are unchanged or increased compared with control islets (Östenson, Khan, et al. 1993; Giroix, Vesco, et al. 1993; Giroix, Sener, et al. 1993; Giroix, Sener, Portha, et al. 1993; Hughes et al. 1994, 1998; Ling et al. 1998). Furthermore, oxidation of glucose has been reported decreased (Giroix, Vesco, et al. 1993), unchanged (Östenson, Khan, et al. 1993; Hughes et al. 1994, 1998; Giroix, Sener, et al. 1993), or even enhanced (Ling et al. 1998). Also, lactate production has been shown to be increased in GK rat islets (Ling et al. 1998). These somewhat contradictory results may be partly due to differences among the different GK rat colonies.

Although it may be hypothesized that the defective insulin response to glucose in GK rat beta-cells is accounted for by an impaired ATP production and closure of the ATP-regulated K^+-channels (Tsuura et al. 1993), the rate of ATP production in islet mitochondria was similar in Stockholm GK and control rats (Ling et al. 1998). Also, ATP/ADP levels were found to be normal in GK rats of Paris and Seattle overall (Giroix, Sener, et al. 1993; Metz et al. 1999). Other defects in GK islet glucose metabolism include increased glucose cycling, due to increased glucose-6-phosphatase activity (Östenson, Khan, et al. 1993; Ling et al. 1998); impaired glycerol phosphate shuttle, due to markedly reduced activity of the FAD-linked glycerol phosphate dehydrogenase (Östenson, Abdel-Halim, et al. 1993; MacDonald et al. 1996); reduced pyruvate dehydrogenase activity (Zhou et al. 1995); and decreased pyruvate carboxylase activity (MacDonald et al. 1996). It is possible that these alterations may affect ATP concentrations locally. However, the enzyme dysfunctions were restored by normalization of glycemia in GK rats (MacDonald et al. 1996; Ling et al. unpublished observations), but with only partial improvement of glucose-induced insulin release. Hence, it is likely that these altered enzyme activities result from a glucotoxic effect rather than being primary causes behind the impaired secretion. Also, lipotoxic effects leading to defective insulin release have been observed in GK rats on high-fat diet (Shang et al. 2002; Briaud et al. 2002), possibly mediated by a mechanism partly involving modulation of UCP-2 expression.

Abnormalities in the function of the ATP-regulated K^+-channels and L-type Ca^{2+} channels (Kato et al. 1996) do not seem to account for the major impairment in insulin release in the GK rat (Hughes et al. 1998; Marie et al. 2001). Indeed, glucose-stimulated insulin secretion was markedly impaired in GK rat islets also when the islets were depolarized by a high concentration of potassium chloride and the ATP-regulated K^+-channels kept open by diazoxide (Abdel-Halim et al. 1996). Similar results were obtained when insulin release was induced by exogenous calcium in electrically permeabilized GK rat islets (Okamoto et al. 1995). These findings, together with a study on animals from the Seattle colony (Metz et al. 1999), indicate that important defects reside late in signal transduction (i.e., in the exocytotic machinery). In fact, markedly reduced expression of several exocytotic SNARE complex proteins, important for the docking and fusion between insulin granules and beta-cell membrane, have been demonstrated in GK rat islets (Nagamatsu et al. 1999; Gaisano et al. 2002; Zhang et al. 2002). Thus, a reduced number of docking granules accounts for impaired beta-cell secretion (Ohara-Imaizumi et al. 2004), and this defect may partly be primary and partly due to glucotoxicity (Gaisano et al. 2002).

Other intriguing aspects of possible mechanisms behind defective glucose-induced insulin release in GK rat islets are the findings of dysfunction of islet lysosomal glycogenolytic enzymes (Salehi et al. 1999), as well as marked impairment of the glucose-heme oxygenase-carbon monoxide signaling pathway (Mosén et al. 2005).

In the light of a normal ATP generation by glucose, it could be proposed that a primary defect in glucose metabolism, with a key role in impaired insulin response to the hexose, resides in glycolysis. It still remains, however, to find that link. An attractive possibility to be explored is that increased activity of adenylyl cyclase III (AC-III), due to functional mutations in the promoter region of the AC-III gene (Abdel-Halim et al. 1998), results in increased cAMP production and thereby enhanced activity of lactate dehydrogenase (Derda et al. 1980). This would in turn lead to increased glucose utilization and reduced cytoplasmic pool of NADH. It is well documented that altered NADH/NAD ratio is associated with impaired glucose-induced insulin release (Eto et al. 1999). The increased cAMP production has also offered the possibility to fully restore insulin response to glucose in the presence of an AC-III stimulator such as forskolin (Abdel-Halim et al. 1996, 1998), a phosphodiesterase inhibitor, or GLP-1 (Guenifi et al. 1998).

Also, cholinergic stimulation has been demonstrated to restore glucose-induced insulin secretion from GK rat islets (Guenifi et al. 2001), and it was proposed that such a stimulation is not mediated mainly through activation of the PKC pathway, but via a paradoxical activation of the cAMP/PKA pathway to enhance Ca^{2+}-stimulated insulin release in the GK rat beta-cell (Dolz et al. 2005). However, diminished levels and/or abnormal activation of several PKC isoenzymes in GK rat islets could also account for the defective signals downstream to glucose metabolism, responsible for impaired insulin secretion (Warwar et al. 2006).

Peroxovanadium is an inhibitor of islet protein-tyrosine phosphatase (PTP) activity that was shown to enhance glucose-stimulated insulin secretion from GK rat islets (Abella et al. 2003; Chen and Ostenson, 2005). One possible target for this effect could be PTP sigma that is overexpressed in GK rat islets (Ostenson et al. 2002); indeed, exposure of GK islets to PTP sigma antisense increased the insulin response to glucose. In addition, defects in islet protein histidine phosphorylation have been proposed to contribute to impaired insulin release in GK rat islets (Kowluru 2003).

Insulin Response to Nonglucose Secretagogues

It is of interest that the insulin response to sulfonylurea was decreased in GK rats compared with control rat islets (Giroix, Vesco, et al. 1993). Among other nonglucose insulin stimulators, arginine has been shown to induce a normal or even augmented insulin response from the GK rat pancreas (Kimura et al. 1982; Portha et al. 1991; Abdel-Halim et al. 1994; Hughes et al. 1994). Since preperfusion for 50–90 min in the absence of glucose reduced the insulin response to arginine in the GK but not in the control pancreas (Portha et al. 1991), it is likely that previous exposure to glucose (*in vivo* or during the perfusion experiment) potentiates arginine-induced insulin secretion. Therefore, arginine-induced insulin release may also be impaired in GK rats like in T2D patients, as was shown after normalization of glycemia (Porte 1991).

Interestingly, using islets from F1 hybrids of GK and Wistar rats, impaired potentiation by glucose of arginine-stimulated insulin release was shown (Guenifi et al. 1998). Insulin responses to another amino acid, leucine, and its metabolite, α-ketoisocaproate (KIC) were diminished in GK rats from the Paris and Stockholm colonies (Giroix, Vesco, et al. 1993; Giroix et al. 1999; Guenifi et al. 1998). This was attributed to defective catabolism of the amino acids (i.e., reduced generation of acetyl-CoA [Giroix et al. 1999]). However, in GK rat islets from London and Kyoto colonies, KIC induced normal insulin responses (Hughes et al. 1994; Tsuura et al. 1993). Also, in islets from mildly diabetic F1 hybrids of GK and Wistar rats, where glucotoxicity was shown to play a minor role, if any, for defective insulin release (Abdel-Halim et al. 1995), KIC stimulated insulin release normally in the absence, but not in the presence, of glucose (Guenifi et al. 1998). Furthermore, KIC but not glucose exerted similar stimulation in GK and control islets that were depolarized by potassium chloride. The latter findings further support the view that an abnormal B-cell metabolism of glucose proximal to the Krebs cycle is likely to account for the impairment of insulin release.

ISLET MICROCIRCULATION

Pancreatic and islet blood flow, estimated with microsphere technique, were shown to be increased in GK rats (Svensson, Ostenson, et al. 1994; Atef et al. 1994) and in glucose-intolerant F1 hybrids of GK and Wistar rats (Svensson, Abdel-Halim, et al. 1994). Similar enhancement of islet microcirculation has been noted in other conditions with increased functional demand on the B-cells—for example, pregnancy and hyperglycemia induced by glucose infusion (Svensson 1994). The increased islet blood flow in GK rats may be accounted for by an altered vagal nerve regulation mediated by nitric oxide (NO), since vagotomy as well as inhibition of NO synthase normalized GK islet flow (Svensson, Ostenson, et al. 1994; Svensson, Abdel-Halim, et al. 1994). In addition, islet capillary pressure was increased in GK rats (Carlsson et al. 1996); however, this defect was restored after two weeks of normalization of glycemia by phlorizin treatment. Hypoglycemia induced decreased islet perfusion, presumably mediated by the central nervous system (Carlsson et al. 2003). The precise relationship between islet microcirculation and B-cell secretory function remains to be established, however. In this context, it is of interest that adipose tissue blood flow was shown to be increased in normal-weight GK rats, but not in obese Zucker rats (Kampf et al. 2005), leading to a speculated mechanism for prevention of obesity as well as for increasing islet blood flow in GK rats due to marked release of FFA from the fat cells.

INSULIN SENSITIVITY

Studies of insulin sensitivity in the GK rat have been performed *in vivo*, using clamp techniques, and *in vitro* in muscle, liver, and adipose tissue. Thus, in hyperinsulinemic-euglycemic clamp studies combined with tracer determination of hepatic glucose production (HGP) in Sendai rats, the main defect responsible for a rather mild insulin resistance in GK rats was increased HGP in connection with dysregulation of hepatíc fructose-2,6-bisphosphate (Suzuki et al. 1992). These findings were later confirmed in clamp studies in animals of the Paris colony; however, in addition to hepatic insulin resistance at basal and hyperinsulinemic states, decreased insulin sensitivity in extrahepatic tissues was demonstrated (Bisbis et al. 1993). The latter defect in liver was characterized by decreased receptor number but normal tyrosine kinase activity. In addition, the antagonistic effect of insulin on glucagon-induced HGP was reported attenuated (Doi et al. 2001); furthermore, vanadium improved liver insulin receptor function in GK rat liver (Kato et al. 2004). Another study in GK rats demonstrated that improvement of glucose tolerance by pioglitazone and eicosapentaenoic acid may be exerted by directly increasing hepatic insulin sensitivity, while similar effect by fibrates might be due to effects on hepatic glycogen metabolism (Matsuura et al. 2004).

In extrahepatic tissues, defective activation of glucose accumulation into glycogen by skeletal muscle, perhaps due to chronic activation of protein kinase C, has been suggested to contribute to insulin resistance and hyperglycemia in GK rats (Villar-Palasi and Farese 1994; Avignon et al. 1996). Also, in insulin receptor signaling in adipocytes and skeletal muscle of GK rats, specific defects have been demonstrated. Thus, impaired insulin-stimulated tyrosine phosphorylation of insulin-receptor substrate-1 (IRS-1) and inhibited effect of insulin on MAP kinase activation, probably due to altered serine/threonine phosphatase regulation, could account partly for the impaired insulin effect on glucose uptake and glycogen synthesis as seen in GK rat adipocytes (Begum and Ragolia, 1998).

Furthermore, in skeletal muscle of GK rats, defective postreceptor signaling was characterized by impaired insulin-stimulated glucose transport as well as PI-3-K activated Akt kinase (protein kinase B) (Krook et al. 1997). However, the latter changes were fully restored when glycemia was nearly normalized in GK rats by phlorizin treatment (reducing hyperglycemia by inhibiting renal glucose reabsorption). These observations strongly suggest that impaired insulin action in the GK rat, at least in extrahepatic tissues, is secondary to hyperglycemia. Such a glucotoxic effect is also supported by studies in F1 hybrids of GK and Wistar rats in which insulin-mediated muscle glucose transport at one month of age was normal, while that of two-month-old hybrid rats was moderately decreased compared with age-matched Wistar rats (Nolte et al. 1995). Altogether, these observations indicate the primacy of a B-cell secretory defect in the pathogenesis of diabetes in the GK rat.

DIABETIC COMPLICATIONS

At an early stage, GK rats were used in Sendai for studies of late diabetic complications. Signs of neuropathy were noted as reduced motor nerve conduction velocity (MNCV) in the tail nerve as early as in two-month-old GK relative to Wistar rats (Goto et al. 1982, 1988; Suzuki et al. 1992). In addition, morphological alterations such as axonal degeneration and segmental demyelination as well as increased sorbitol and decreased myoinositol levels were observed in sciatic nerves in six-month-old GK rats. Some of these defects were prevented by treatment with aldose reductase inhibitor (Goto et al. 1982). Reduced MNCV in the femoral nerve was also reported in eight-month-old GK rats of the Stockholm colony (Östenson et al. 1997) and more detailed studies of peripheral nerve morphology and function were performed in mildly diabetic 6- and 12-month-old GK rats (Murakawa et al. 2002). In these animals, signs of diabetic osteopathy were also noted (Östenson et al. 1997; Ahmad et al. 2003, 2004). In GK rats of the Coimbra colony, studies suggested that insulin modulates synaptosomal GABA and/or glutamate transport, thus exerting a neuroprotective role under oxidative stress or diabetic conditions (Duarte et al. 2004).

Morphological changes indicating retinopathy seem to develop rather late in GK rats. Thus, an altered retinal endothelial cell/pericyte ratio was demonstrated in animals more than one year old, but not in eight-month-old GK rats (Agardh et al. 1997, 1998). Biochemical and functional alterations in the GK rat retina have been described as earlier phenomena, such as reduced glutathion levels (Agardh et al. 1998), increased tissue levels of vascular endothelial growth factor (VEGF) (Sone et al. 1997), and impaired retinal blood flow (Miyamoto et al. 1996). An increased production of nitric oxide in GK rat retinas has been suggested to account for increased blood–retinal barrier permeability and breakdown (Carmo et al. 2000).

Regarding studies of experimental nephropathy, a gradual increase in glomerular basal membrane with ageing has been shown in GK rats from three months of age (Suzuki et al. 1992). GK rats have also been found to develop impaired renal function (by 50–75%), reflected as increased serum creatinine and urea nitrogen levels as compared with Wistar rats (Vesely et al. 1999). GK rats have been used in several studies of diabetic nephropathy (Phillips et al. 2001; Janssen et al. 2003; Schrijvers et al. 2004).

Ventromedial hypothalamic (VMH) lesions in GK rats induced accentuated hyperglycemia and hypertriglyceridemia with visceral fat accumulation and reduced pancreatic insulin content (Yoshida et al. 1996). In addition, the VMH lesion enhanced proteinuria and glomerular basal membrane thickening as well as induced morphological changes in the aortal intima characteristic of an early stage of atherosclerosis. Thus, the VMH-lesioned GK rat has been suggested to be a model of microangiopathy and macroangiopathy. Normal GK rats have been shown to display mild cardiomyopathy, evident as exaggerated diastolic dysfunction during hypoxia (El-Omar et al. 2004). Impaired insulin response in GK rat heart muscle may be due to decreased levels of insulin receptor beta, IRS-1, and GLUT-4 proteins (Desrois et al. 2004a), and these changes appeared more advanced in female than male animals (Desrois et al. 2004b). Lack of antioxidant coenzyme Q9 has been suggested to be responsible for an increased susceptibility of GK heart mitochondria to oxidative damage and subsequent impaired myocardial function (Santos et al. 2003).

CONCLUDING REMARKS

In conclusion, an increasing amount of studies has demonstrated that the GK rat is a useful animal model of nonobese T2D, in which the primary defect most likely resides in the beta-cell. In addition, the nature of diabetes heredity in the GK rat (i.e., the additive action of several genes) may well resemble the polygenic basis for disease in the majority of T2D patients. During the long-term inbreeding of GK rats—almost 30 years—the animals appear to have maintained rather stable levels of glucose intolerance and impairment of glucose-induced insulin response and also when they were studied in substrains of GK rats in various laboratories. However, other characteristics such as islet morphology and insulin content, as well as islet metabolism, have been shown to differ considerably between different substrain colonies, suggesting that newly introduced genetic mutations account for contrasting phenotypic properties.

REFERENCES

Abdel-Halim SM, Guenifi A, Efendic S, Östenson C-G. (1993) Both somatostatin and insulin responses to glucose are impaired in the perfused pancreas of the spontaneously non-insulin dependent diabetic GK (Goto–Kakizaki) rat. *Acta Physiol Scand* 148:219–226.

Abdel-Halim SM, Guenifi A, Grill V, et al. (1994) Impact of diabetic inheritance on glucose tolerance and insulin secretion in spontaneously diabetic GK-Wistar rats. *Diabetes* 43:281–288.

Abdel-Halim SM, Guenifi A, He B, et al. (1998) Mutations in the promoter of adenylyl cyclase (AC)-III gene, overexpression of AC-III mRNA, and enhanced cAMP generation in islets from spontaneously diabetic GK rat model of type-2 diabetes. *Diabetes* 47:498–504.

Abdel-Halim SM, Guenifi A, Jansson L, et al. (1995) A defective stimulus-secretion coupling rather than glucotoxicity mediates the impaired insulin secretion in the mildly diabetic F1 hybrids of GK-Wistar rats. *Diabetes* 44:1280–1284.

Abdel-Halim SM, Guenifi A, Khan A, et al. (1996) Impaired coupling of glucose signal to the exocytotic machinery in diabetic GK rats; a defect ameliorated by cAMP. *Diabetes* 45:934–940.

Abella A, Marti L, Camps M, et al. (2003) Semicarbazide-sensitive amine oxidase/vascular adhesion protein-1 activity exerts an antidiabetic action in Goto–Kakizaki rats. *Diabetes* 52:1004–1013.

Agardh C-D, Agardh E, Hultberg B, et al. (1998) Glutathion levels are reduced in Goto–Kakizaki rat retina, but are not influenced by aminoguanidine treatment. *Curr Eye Res* 17:251–256.

Agardh C-D, Agardh E, Zhang H, Östenson C-G. (1997) Altered endothelial/pericyte ratio in Goto–Kakizaki rat retina. *J Diab Complications* 11:158–162.

Ahmad T, Ohlsson C, Sääf M, et al. (2003) Skeletal changes in the type 2-diabetic Goto–Kakizaki rat. *J Endocrinol* 178:111–116.

Ahmad T, Ugarph-Morawski A, Li J, et al. (2004) Bone and joint neuropathy in rats with type 2-diabetes. *Regul Peptides* 119:61–67.

Atef N, Portha B, Pénicaud L. (1994) Changes in islet blood flow in rats with NIDDM. *Diabetologia* 37:677–680.

Avignon A, Yamada K, Zhou X, et al. (1996) Chronic activation of protein kinase C in soleus muscles and other tissues of insulin-resistant type II diabetic Goto–Kakizaki (GK), obese/aged, and obese/Zucker rats. A mechanism for inhibiting glycogen synthesis. *Diabetes* 45:1396–1404.

Begum N, Ragolia L. (1998) Altered regulation of insulin signaling components in adipocytes of insulin-resistant type II diabetic Goto–Kakizaki rats. *Metabolism* 47:54–62.

Bisbis S, Bailbe D, Tormo M-A, et al. (1993) Insulin resistance in the GK rat: Decreased receptor number but normal kinase activity in liver. *Am J Physiol* 265:E807–E813.

Briaud I, Kelpe CL, Johnson LM, et al. (2002) Differential effects of hyperlipidemia on insulin secretion in islets of Langerhans from hyperglycemic versus normoglycemic rats. *Diabetes* 51:662–668.

Carlsson PO, Berne C, Östenson CG, et al. (2003) Hypoglycemia induces decreased islet blood perfusion mediated by the central nervous system in normal and type 2 diabetic GK rats. *Diabetologia* 46:1124–1130.

Carlsson PO, Jansson L, Östenson C-G, Källskog Ö. (1997) Islet capillary blood pressure increase mediated by hyperglycemia in NIDDM GK rats. *Diabetes* 46:947–952.

Carmo A, Cunha-Vaz JG, Carvalho AP, Lopes MC. (2000) Nitric oxide synthase activity in retinas from non-insulin-dependent diabetic Goto–Kakizaki rats: Correlation with blood-retinal barrier permeability. *Nitric Oxide* 4:590–596.

Chen J, Östenson CG. (2005) Inhibition of protein-tyrosine phosphatases stimulates insulin secretion in pancreatic islets of diabetic Goto–Kakizaki rats. *Pancreas* 30:314–317.

Derda DF, Miles MF, Schweppe JS, Jungmann RA. (1980) Cyclic AMP regulation of lactate dehydrogenase. *J Biol Chem* 225:11112–11121.

Desrois M, Sidell RJ, Gauguier D, et al. (2004a) Initial steps of insulin signaling and glucose transport are defective in the type 2 diabetic rat heart. *Cardiovasc Res* 61:288–296.

Desrois M, Sidell RJ, Gauguier D, et al. (2004b) Gender difference in hypertrophy, insulin resistance and ischemic injury in the aging type 2 diabetic rat heart. *J Mol Cell Cardiol* 37:547–555.

Doi Y, Iwai M, Matssura B, Onji M. (2001) Glucagon attenuates the action of insulin on glucose output in the liver of Goto–Kakizaki rat perfused *in situ*. *Eur J Physiol* 442:537–541.

Dolz M, Bailbé D, Giroix MH, et al. (2005) Restitution of defective glucose-stimulated insulin secretion in diabetic GK rat by acetylcholine uncovers paradoxical stimulatory effect of β-cell muscarinic receptor activation on cAMP production. *Diabetes* 54:3229–3237.

Duarte AI, Santos MS, Seica R, Oliveira CR. (2004) Oxidative stress affects synaptosomal γ-aminobutyric acid and glutamate transport in diabetic rats. The role of insulin. *Diabetes* 53:2110–2116.

Efendic S, Kindmark H, Berggren P-O. (1991) Mechanisms involved in the regulation of the insulin secretory process. *J Intern Med* 229:9–22.

El-Omar MM, Yang ZK, Phillips AO, Shah AM. (2004) Cardiac dysfunction in the Goto–Kakizaki rat. A model of type II diabetes mellitus. *Basic Res Cardiol* 99:133–141.

Eto K, Tsubamoto Y, Terauchi Y, et al. (1999) Role of NADH shuttle system in glucose-induced activation of mitochondrial metabolism and insulin secretion. *Science* 283:981–985.

Fakhrai-Rad H, Nikoshkov A, Kamel A, et al. (2000) Insulin-degrading enzyme identified as a candidate diabetes susceptibility gene in GK rats. *Hum Mol Genet* 9:2149–2158.

Gaisano HY, Östenson CG, Sheu L, et al. (2002) Abnormal expression of pancreatic islet exocytotic soluble N-ethylmaleimide-sensitive factor attachment protein receptors in Goto–Kakizaki rats is partially restored by phlorizin treatment and accentuated by high glucose treatment. *Endocrinology* 143:4218–4226.

Galli J, Li L-S, Glaser A, et al. (1996) Genetic analysis of non-insulin dependent diabetes mellitus in the GK rat. *Nat Genet* 12:31–37.

Gauguier D, Froguel P, Parent V, et al. (1996) Chromosomal mapping of genetic loci associated with non-insulin-dependent diabetes in the GK rat. *Nat Genet* 12:38–43.

Gauguier D, Nelson I, Bernard C, et al. (1994) Higher maternal than paternal inheritance of diabetes in GK rats. *Diabetes* 43:220–224.

Gill-Randall RJ, Adams D, Ollerton RL, Alcolado JC. (2004) Is human type 2 diabetes maternally inherited? Insights from an animal model. *Diabetic Med* 21:759–762.

Giroix M-H, Saulnier C, Portha B. (1999) Decreased pancreatic islet response to L-leucine in the spontaneously diabetic GK rat: enzymatic, metabolic and secretory data. *Diabetologia* 42:965–977.

Giroix M-H, Sener A, Bailbe D, et al. (1993) Metabolic, ionic, and secretory response to D-glucose in islets from rats with acquired or inherited non-insulin-dependent diabetes. *Biochem. Med Metab Biol* 50:301–321.

Giroix M-H, Sener A, Portha B, Malaisse WJ. (1993) Preferential alteration of oxidative relative to total glycolysis in pancreatic islets of two rat models of inherited or acquired type 2 diabetes mellitus. *Diabetologia* 36:305–309.

Giroix M-H, Vesco L, Portha B. (1993) Functional and metabolic perturbations in isolated pancreatic islets from the GK rat, a genetic model of noninsulin-dependent diabetes. *Endocrinology* 132:815–822.

Goto Y, Kakizaki M, Masaki N. (1975) Spontaneous diabetes produced by selective breeding of normal Wistar rats. *Proc Jpn Acad* 51:80–85.

Goto Y, Kakizaki M, Yagihashi S. (1982) Neurological findings in spontaneously diabetic rats. *Excerpta Medica ICS* 581:26–38.

Goto Y, Suzuki K-I, Sasaki M, et al. (1988) GK rat as a model of nonobese, noninsulin-dependent diabetes. Selective breeding over 35 generations. In *Lessons from Animal Diabetes*, ed. E. Shafrir, A.E. Renold. Libbey, London, 2:301–303.

Guenifi A, Abdel-Halim SM, Efendic S, Östenson C-G. (1998) Preserved initiatory and potentiatory effect of α-ketoisocaproate on insulin release in islets of glucose intolerant rats. *Diabetologia* 41:1368–1373.

Guenifi A, Abdel-Halim SM, Höög A, et al. (1995) Preserved β-cell density in the endocrine pancreas of young, spontaneously diabetic Goto–Kakizaki (GK) rats. *Pancreas* 10:148–153.

Guenifi A, Simonsson E, Karlsson S, et al. (2001) Carbachol restores insulin release in diabetic GK rat islets by mechanisms largely involving hydrolysis of diacylglycerol and direct interaction with the exocytotic machinery. *Pancreas* 22:164–171.

Guest PC, Abdel-Halim SM, Gross DJ, et al. (2002) Proinsulin processing in the diabetic Goto–Kakizaki rat. *J Endocrinol* 175:637–647.

Höög A, Sandberg-Nordqvist A-C, Abdel-Halim SM, et al. (1996) Increased amounts of a high-molecular-weight insulin-like growth factor-II (IGF-II) peptide and IGF-II messenger ribonucleic acid in pancreatic islets of diabetic GK rats. *Endocrinology* 137:2415–2423.

Hughes SJ, Faehling M, Thorneley CW, et al. (1998) Electrophysiological and metabolic characterization of single β-cells and islets from diabetic GK rats. *Diabetes* 47:73–81.

Hughes SJ, Suzuki K, Goto Y. (1994) The role of islet secretory function in the development of diabetes in the GK Wistar rat. *Diabetologia* 37:863–870.

Janssen U, Vassiliadou A, Riley SG, et al. (2004) The quest for a model of type II diabetes with nephropathy: The Goto–Kakizaki rat. *J Nephrol* 17:769–773.

Kaisaki PJ, Woon PY, Wallis RH, et al. (1998) Localization of tub and uncoupling proteins (Ucp) 2 and 3 to a region of rat chromosome 1 linked to glucose intolerance and adiposity in the Goto–Kakizaki (GK) type 2 diabetic rat. *Mamm Genome* 9:910–912.

Kampf C, Bodin B, Kallskog O, et al. (2005) Marked increase in white adipose tissue blood perfusion in the type 2 diabetic GK rats. *Diabetes* 54:2620–2627.

Kato S, Ishida H, Tsuura Y, et al. (1996) Alterations in basal and glucose-stimulated voltage-dependent Ca^{2+} channel activities in pancreatic β-cells of non-insulin-dependent diabetes mellitus GK rat. *J Clin Invest* 97:2417–2425.

Kato K, Yamada S, Ohmori Y, et al. (2004) Natural vanadium-containing Mt. Fuji ground water improves hypo-activity of liver insulin receptor in Goto–Kakizaki rats. *Molec Cell Biochem* 267:203–207.

Kimura K, Toyota T, Kakizaki M, et al. (1982) Impaired insulin secretion in the spontaneous diabetes rats. *Tohoku J Exp Med* 137:453–459.

Koike G, Van Vooren P, Shiozawa M, et al. (1996) Genetic mapping and chromosome localization of the rat mitochondrial glycerol-3-phosphate dehydrogenase gene, a candidate for non-insulin-dependent diabetes mellitus. *Genomics* 38:96–99.

Kowluru A. (2003) Defective protein histidine phosphorylation in islets from the Goto–Kakizaki diabetic rat. *Am J Physiol* 285:E498–E503.

Koyama M, Wada R, Mizukami H, et al. (2000) Inhibition of progressive reduction of islet beta-cell mass in spontaneously diabetic Goto–Kakizaki rats by alpha-glucosidase inhibitor. *Metabolism* 49:347–352.

Koyama M, Wada R, Sakuraba H, et al. (1998) Accelerated loss of islet β-cells in sucrose-fed Goto–Kakizaki rats, a genetic model of non-insulin-dependent diabetes mellitus. *Am J Pathol* 153:537–545.

Krook A, Kawano Y, Song XM, et al. (1997) Improved glucose tolerance restores insulin-stimulated Akt kinase activity and glucose transport in skeletal muscle from diabetic Goto–Kakizaki rats. *Diabetes* 46:2110–2114.

Lewis BM, Ismail IS, Issa B, et al. (1996) Desensitization of somatostatin, TRH and GHRH responses to glucose in the diabetic GK rat hypothalamus. *J Endocrinol* 151:13–17.

Ling Z-C, Efendic S, Wibom R, et al. (1998) Glucose metabolism in Goto–Kakizaki rat islets. *Endocrinology* 139:2670–2675.

MacDonald MJ, Efendic S, Östenson C-G. (1996) Normalization by insulin treatment of low mitochondrial glycerol phosphate dehydrogenase and pyruvate carboxylase in pancreatic islets of the GK rat. *Diabetes* 45:886–890.

Malaisse-Lagae F, Vanhoutte C, Rypens F, et al. (1997) Anomalies of fetal development in GK rats. *Acta Diabetol* 34:55–60.

Marie JC, Bailbé D, Gylfe E, Portha B. (2001) Defective glucose-dependent cytosolic Ca^{2+} handling in islets of GK and nSTZ rat models of type 2 diabetes. *J Endocrinol* 169:169–176.

Matsuura B, Kanno S, Minami H, et al. (2004) Effects of antihyperlipidemic agents on hepatic insulin sensitivity in perfused Goto-Kakizaki rat liver. *J Gastroenterol* 39:339–345.

Metz SA, Meredith M, Vadakekalam J, et al. (1999) A defect late in stimulus-secretion coupling impairs insulin secretion in Goto–Kakizaki diabetic rats. *Diabetes* 48:1754–1762.

Miralles F, Portha B. (2001) Early development of β-cells is impaired in the GK rat model of type 2 diabetes. *Diabetes* 50:S84–S88.

Miyamoto K, Ogura Y, Nishiwaki H, et al. (1996) Evaluation of retinal microcirculatory alterations in the Goto–Kakizaki rat. *Invest Ophtalmol Vis Sci* 37:898–905.

Mosén H, Salehi A, Alm P, et al. (2005) Defective glucose-stimulated insulin release in the diabetic Goto–Kakizaki (GK) rat coincides with reduced activity of the islet carbon monoxide signaling pathway. *Endocrinology* 146:1553–1558.

Movassat J, Portha B. (1999) Beta-cell growth in the neonatal Goto–Kakizaki rat and regeneration after treatment with streptozotocin at birth. *Diabetologia* 42:1098–1106.

Movassat J, Saulnier C, Serradas P, Portha B. (1997) Impaired development of pancreatic beta-cell mass is a primary event during the progression to diabetes in the GK rat. *Diabetologia* 40:916–925.

Murakawa Y, Zhang W, Pierson CR, et al. (2002) Impaired glucose tolerance and insulinopenia in the GK rat causes peripheral neuropathy. *Diabetes/Metab Res Rev* 18:473–483.

Nagamatsu S, Nakamichi Y, Yamamura C, et al. (1999) Decreased expression of t-SNARE, syntaxin 1, and SNAP-25 in pancreatic beta-cells is involved in impaired insulin secretion from diabetic GK rat islets: Restoration of decreased t-SNARE proteins improves impaired insulin secretion. *Diabetes* 48:2367–2373.

Nolte LA, Abdel-Halim SM, Martin IK, et al. (1995) Development of decreased insulin-induced glucose transport in skeletal muscle of glucose intolerant hybrids of diabetic GK rats. *Clin Sci* 88:301–306.

Ohara-Imaizumi M, Nishiwaki C, Kikuta T, et al. (2004) TIRF imaging of docking and fusion of single insulin granule motion in primary rat pancreatic β-cells: Different behavior of granule motion between normal and Goto–Kakizaki diabetic rat β-cells. *Biochem J* 381:13–18.

Ohneda M, Johnson JH, Inman LR, et al. (1993) GLUT2 expression and function in β-cells of GK rats with NIDDM. *Diabetes* 42:1065–1072.

Okamoto Y, Ishida H, Tsuura Y, et al. (1995) Hyperresponse in calcium-induced insulin release from electrically permeabilized pancreatic islets of diabetic GK rats and its defective augmentation by glucose. *Diabetologia* 38:772–778.

Östenson C-G, Abdel-Halim SM, Andersson A, Efendic S. (1996) Studies on the pathogenesis of NIDDM in the GK (Goto–Kakizaki) rat. In *Lessons from Animal Diabetes*, ed. E. Shafrir. Birkhäuser, Boston, 6:299–315.

Östenson C-G, Abdel-Halim SM, Rasschaert J, et al. (1993) Deficient activity of FAD-linked glycerophosphate dehydrogenase in islets of GK rats. *Diabetologia* 36:722–726.

Östenson C-G, Fière V, Ahmed M, et al. (1997) Decreased cortical bone thickness in spontaneously non-insulin-dependent diabetic GK rats. *J Diab Complic* 11:319–322.

Östenson C-G, Khan A, Abdel-Halim SM, et al. (1993) Abnormal insulin secretion and glucose metabolism in pancreatic islets from the spontaneously diabetic GK rat. *Diabetologia* 36:3–8.

Östenson CG, Sandberg-Nordqvist AC, Chen J, et al. (2002) Overexpression of protein-tyrosine phosphatase PTP sigma is linked to impaired glucose-induced insulin secretion in hereditary diabetic Goto–Kakizaki rats. *Biochem Biophys Res Commun* 291:945–950.

Phillips AO, Baboolai K, Riley S, et al. (2001) Association of prolonged hyperglycemia with glomerular hypertrophy and renal basement membrane thickening in the Goto–Kakizaki model of non-insulin-dependent diabetes mellitus. *Am J Kidney Dis* 37:400–410.

Plachot C, Movassat J, Portha B. (2001) Impaired β-cell regeneration after partial pancreatectomy in the adult Goto–Kakizaki rat, a spontaneous model of type II diabetes. *Histochem Cell Biol* 116:131–139.

Porte D, Jr. (1991) β-Cells in type II diabetes mellitus. *Diabetes* 40:166–180.

Portha B. (2005) Programmed disorders of β-cell development and function as one cause for type 2 diabetes? The GK rat paradigm. *Diabetes/Metab Res Rev* 21:495–504.

Portha B, Serradas P, Bailbé D, et al. (1991) β-Cell insensitivity to glucose in the GK rat, a spontaneous nonobese model for type II diabetes. *Diabetes* 40:486–491.

Salehi A, Henningsson R, Mosén H, et al. (1999) Dysfunction of the islet lysosomal system conveys impairment of glucose-induced insulin release in the diabetic GK rat. *Endocrinology* 140:3045–3053.

Santos DL, Palmeira CM, Seica R, et al. (2003) Diabetes and mitochondrial oxidative stress: A study using heart mitochondria from the diabetic Goto–Kakizaki rat. *Mol Cell Biochem* 246:163–170.

Schrijvers BF, De Vriese AS, Van de Voorde J, et al. (2004) Long-term renal changes in Goto–Kakizaki rat, a model of lean type 2 diabetes. *Nephrol Dial Transplant* 19:1092–1097.

Sener A, Malaisse-Lagae F, Östenson C-G, Malaisse WJ. (1993) Metabolism of endogenous nutrients in islets of Goto–Kakizaki (GK) rats. *Biochem J* 296:329–334.

Sener A, Malaisse-Lagae F, Ulusoy S, et al. (1996) Contrasting secretory behavior of pancreatic islets from old rat in two models of non-insulin dependent diabetes. *Diabetes Res* 31:67–76.

Serradas P, Gangnerau MN, Giroix MH, et al. (1998) Impaired pancreatic beta cell function in the fetal GK rat. Impact of diabetic inheritance. *J Clin Invest* 101:899–904.

Serradas P, Giroix M-H, Saulnier C, et al. (1995) Mitochondrial deoxyribonucleic acid content is specifically decreased in adult, but not fetal, pancreatic islets of the Goto–Kakizaki rat, a genetic model of noninsulin-dependent diabetes. *Endocrinology* 136:5623–5631.

Serradas P, Goya L, Lacorne M, et al. (2002) Fetal insulin-like growth factor-2 production is impaired in the GK rat model of type 2 diabetes. *Diabetes* 51:392–397.

Shang W, Yasuda K, Takahashi A, et al. (2002) Effect of high dietary fat on insulin secretion in genetically diabetic Goto–Kakizaki rats. *Pancreas* 25:393–399.

Sone H, Kawakami Y, Okuda Y, et al. (1997) Ocular vascular endothelial growth factor levels in diabetic rats are elevated before observable retinal proliferative changes. *Diabetologia* 40:726–730.

Suzuki K-I, Goto Y, Toyota T. (1992) Spontaneously diabetic GK (Goto–Kakizaki) rats. In *Lessons from Animal Diabetes*, ed. E. Shafrir. Smith–Gordon, London, 4:107–116.

Suzuki N, Aizawa T, Asanuma N, et al. (1997) An early insulin intervention accelerates pancreatic β-cell dysfunction in young Goto–Kakizaki rats, a model of naturally occurring noninsulin-dependent diabetes. *Endocrinology* 138:1106–1110.

Svensson AM. (1994) Pancreatic islet blood flow in the rat. Thesis, *Acta Universitatis Upsaliensis,* vol. 483, Uppsala.

Svensson AM, Abdel-Halim SM, Efendic S, et al. (1994) Pancreatic and islet blood flow in F1-hybrids of the non-insulin-dependent diabetic GK-Wistar rat. *Eur J Endocrinol* 130:612–616.

Svensson AM, Östenson C-G, Sandler S, et al. (1994) Inhibition of nitric oxide synthase by N^G-nitro-L-arginine causes a preferential decrease in pancreatic islet blood flow in normal rats and spontaneously diabetic GK rats. *Endocrinology* 135:849–853.

Tourrel C, Bailbe D, Lacorne M, et al. (2002) Persistent improvement of type 2 diabetes in the Goto–Kakizaki rat model by expansion of the β-cell mass during the prediabetic period with glucagon-like peptide-1 or exendin-4. *Diabetes* 51:1443–1452.

Tsuura Y, Ishida H, Okamoto Y, et al. (1993) Glucose sensitivity of ATP-sensitive K^+ channels is impaired in β-cells of the GK rat. *Diabetes* 42:1446–1453.

Vesely DL, Gower WR Jr., Dietz JR, et al. (1999) Elevated atrial natriuretic peptides and early renal failure in type 2 diabetic Goto–Kakizaki rats. *Metabolism* 48:771–778.

Villar-Palasi C, Farese RV. (1994) Impaired skeletal muscle glycogen synthase activation by insulin in the Goto–Kakizaki (GK) rat. *Diabetologia* 37:885–888.

Wallace KJ, Wallis RH, Collins SC, et al. (2004) Quantitative trait locus dissection in congenic strains of the Goto–Kakizaki rat identifies a region conserved with diabetes loci in human chromosome 1q. *Physiol Genomics* 19:1–10.

Wallis RH, Wallace KJ, Collins SC, et al. (2004) Enhanced insulin secretion and cholesterol metabolism in congenic strains of the spontaneously diabetic (type 2) Goto–Kakizaki rat are controlled by independent genetic loci in rat chromosome 8. *Diabetologia* 47:1096–1106.

Warwar N, Efendic S, Östenson CG, et al. (2006) Dynamics of glucose-induced localization of PKC isoenzymes in pancreatic β-cells. Diabetes-related changes in the GK rat. *Diabetes* 55:590–599.

Yoshida S, Yamashita S, Tokunaga K, et al. (1996) Visceral fat accumulation and vascular complications associated with VMH lesioning of spontaneously non-insulin-dependent diabetic GK rat. *Int J Obesity Relat Metab Disord* 20:909–916.

Zhang W, Khan A, Östenson CG, et al. (2002) Down-regulated expression of exocytotic proteins in pancreatic islets of diabetic GK rats. *Biochem Biophys Res Commun* 291:1038–1044.

Zhou Y-P, Östenson C-G, Ling Z-C, Grill V. (1995) Deficiency of pyruvate dehydrogenase activity in pancreatic islets of diabetic GK rats. *Endocrinology* 136:3546–3551.

6 The New Zealand Obese Mouse: Polygenic Model of Obesity, Glucose Intolerance, and the Metabolic Syndrome

Barbara C. Fam and Sofianos Andrikopoulos

CONTENTS

ABSTRACT

The New Zealand obese (NZO) mouse is an excellent and well established model of obesity, glucose intolerance, and the metabolic syndrome as it exhibits classical characteristics of these diseases, such as increased weight gain and food intake, leptin insensitivity, insulin resistance, impaired glucose-mediated insulin secretion, hypercholesterolemia, and hypertension. The polygenic nature of the NZO mouse makes it an attractive model to unravel the genetic mechanisms responsible for these defects since human obesity, glucose intolerance, and the metabolic syndrome also appear to be polygenic. This review aims to detail all the metabolic abnormalities that are associated with the NZO mouse and the current results of studies investigating the genetic causes of these defects. We believe that determining the cause of obesity and glucose intolerance will be the key to understanding these aberrations in the NZO mouse and, possibly, in humans.

BACKGROUND

The origins of the NZO mouse can be traced back to the 1930s when the original mixed colony of mice was bred and maintained in London. In the 1940s this colony was brought over to New Zealand where selective inbreeding was conducted on the basis of coat color. This continual breeding eventually led to mice becoming obese and, based on this phenotype, a new strain was selectively developed, the NZO/B1. Other lines based on different coat colors were also developed from this mixed population such as the New Zealand black (NZB/B1), NZY/B1, and the New Zealand chocolate (NZC) lines (for more detailed reviews, see Proietto and Larkins, 1993, and Andrikopoulos et al. 2001).

Traditionally, the NZO mouse is viewed as a model of obesity and glucose intolerance since it displays characteristics representative of these two disorders, such as excessive body weight, increased adiposity, leptin insensitivity (Thorburn et al. 2000), impaired glucose-stimulated insulin secretion, and insulin resistance in muscle, fat, and liver (Proietto and Larkins 1993). There is also evidence that these mice present with additional defects that are characteristic of the human metabolic syndrome, such as dyslipidemia, hypercholesterolemia, and hypertension (Ortlepp et al. 2000). Unlike other animal models of obesity and glucose intolerance, such as the *ob/ob*, *db/db*, and A^y mice and the *fa/fa* (Zucker) rat, which have single-point genetic mutations, the syndrome in the NZO mouse is likely to be the result of defects in multiple genes. In this regard, the NZO mouse is an excellent model of human obesity, glucose intolerance, and the overall metabolic syndrome.

As seems to be the case with many obese animal models, the NZO mouse exhibits poor breeding patterns. In the early days this became so much of a concern that Bielschowsky and Bielschowsky reported in 1956 their fears of nearly losing the line. This is highlighted by the finding that only 16 of 50 NZO females have their second litter 19–25 days after the first, compared to 45 of 50 NZC female mice (Bielschowsky and Bielschowsky 1956). Research performed many years later discovered that the breeding difficulties were due not only to the strains' early onset of obesity but also to defects in the transport of the adipocyte derived cytokine, leptin (a satiety factor integral in the regulation of food intake), and signaling of its receptor (Halaas et al. 1997; Igel et al. 1997). The use of diet restriction in the female breeders may be one effective technique to improve the rate of fertility. An alternative approach may be through treatment with a β3 adrenergic receptor agonist (CL316,243) at a concentration of 0.001% w/w in the diet for a month from weaning to retard weight gain (Koza et al. 2004).

The fact that the NZO mouse is an inbred model makes finding a suitable lean control line difficult. Researchers over the years have used a number of lean mice as controls, including the C57BL/6, BALB/c, and the albino ICR mouse. Our laboratory has previously used the lean NZC mouse as the control strain because it has similar metabolic characteristics to other lean controls (Andrikopoulos and Proietto 1995; Andrikopoulos et al. 1993, 1996; Veroni et al. 1991).

Obesity and Metabolic Characteristics of the NZO Mouse

NZO mice have birth weights similar to those of control mice (Crofford and Davis 1965). By four to six weeks of age the mice are markedly heavier and have increased body lipid composition and elevated free fatty acid levels compared to lean NZC mice. These differences persist into adulthood as the NZO mouse develops marked juvenile-onset obesity (Andrikopoulos et al. 1996; Crofford and Davis 1965). Studies have shown that this is due, in part, to early hyperphagia (Larkins and Martin 1972; Jurgens et al. 2006) and more recently to reduced total energy expenditure and voluntary locomotor activity but not spontaneous movement (Jurgens et al. 2006). Interestingly, NZO mice display functional adaptive thermogenesis on exposure to cold (Jurgens et al. 2006), unlike other obese strains such as the *ob/ob* mouse that have a reduced thermogenic capacity (Davis and Mayer 1954).

Leptin and its receptor have long been regarded as the fundamental components of the feedback mechanism between adipose tissue and the brain to regulate food intake and body weight (Thorburn et al. 2000). NZO mice have elevated serum leptin concentrations similar to other murine models of obesity, yet they are resistant to the peripheral effects of leptin to lower food intake (Igel et al. 1997). In contrast, central administration of leptin in NZO mice resulted in normal reductions in food intake (Halaas et al. 1997), implying that leptin resistance is not the primary cause of hyperphagia, even though levels of neuropeptide Y (NPY), a hypothalamic signal that causes an increase in food intake, has been reported to be overexpressed in these mice (Rizk et al. 1998).

Studies examining the sequence of the hypothalamic leptin receptor (long form, ObRb) reported a variant of this receptor in the NZO mouse compared to the wild-type receptor that had three amino acid substitutions (Igel et al. 1997). Despite this, there were no differences in the mRNA expression levels of the receptor, but there was a slight reduction in its signaling abilities. Interestingly, studies in the related NZB strain, which is lean, showed that it also displayed this same variant but did not become obese. It therefore appears that the polymorphic receptor alone cannot produce obesity in the NZO mouse but may, to some small extent, contribute to it—possibly through alterations in signaling (Igel et al. 1997). These studies tell us that the peripheral leptin resistance and resulting hyperphagia in the NZO mouse may possibly be due to defects in leptin transport across the blood–brain barrier, through the transport receptor (ObRe) or the blood–brain barrier receptor (short form, ObRa) or possibly through defects located distal to the hypothalamic leptin receptor. Figure 6.1 illustrates the possible defects that may contribute to the obesity phenotype of the NZO mouse.

Hyperglycemia appears to be an early defect in NZO mice, presenting at four to six weeks of age (Veroni et al. 1991). Hyperglycemia in four-week-old fed mice appears to be more pronounced than in overnight fasted mice. In adulthood, fed and

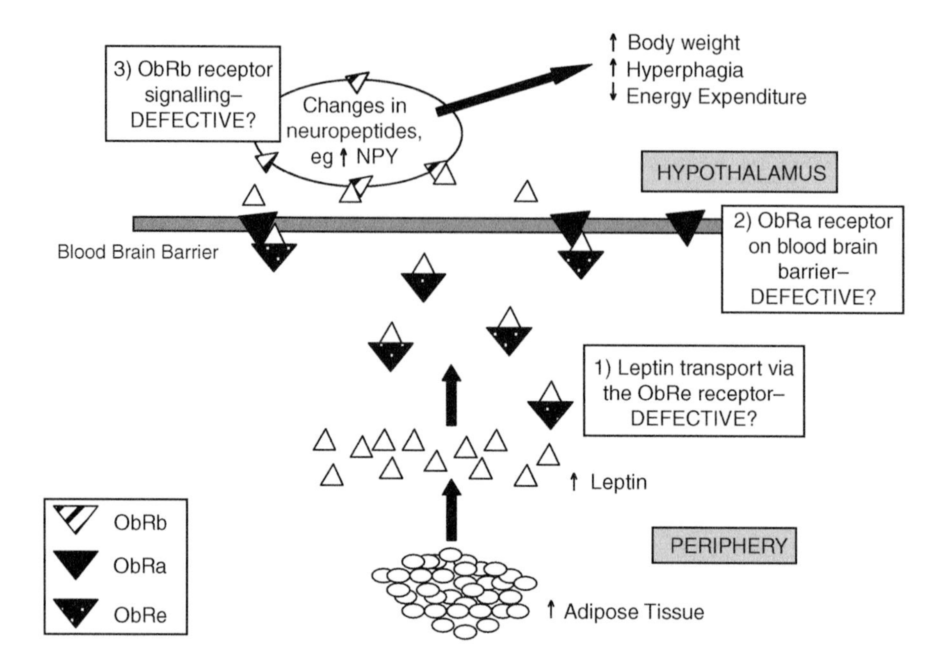

FIG. 6.1 Diagrammatic representation of the potential defects within the hypothalamus and periphery of the NZO mouse leading to increased body weight, hyperphagia, and reduced energy expenditure. In the normal state, leptin is produced within adipose tissue and is secreted into the circulation in proportion to adipose tissue mass. Leptin is transported to the blood–brain barrier via leptin transport receptors, ObRe. Leptin enters the hypothalamus from the periphery via binding to the short-form leptin receptor, ObRa. Once within, leptin binds to the long-form leptin receptor, ObRb, located on neurones that produce neuropeptides involved in the regulation of body weight. Leptin then signals via this receptor to activate or suppress expression of these neuropeptides. For example, leptin acts to lower NPY to lower food intake and increase energy expenditure. In the obese state, like that of the NZO mouse, there is an increased amount of adipose tissue leading to increased plasma leptin levels. As discussed in the review, the defects leading to obesity could occur at (1) defective transport via ObRe; (2) defective transport across the blood–brain barrier via ObRa; or (3) defective signaling at ObRb.

fasted plasma glucose concentrations are elevated (Veroni et al. 1991; Plum et al. 2000; Ortlepp et al. 2000; Kluge et al. 2000; Giesen et al. 2003). On the other hand, wide variations in blood glucose have been reported and not all accounts in the literature show elevated levels for the NZO mouse (Bielschowsky and Bielschowsky 1956; Upton et al. 1980). This may be partly due to the method of blood sampling and glucose measurement as well as the choice of control mice. Studies by Leiter and colleagues (1998) from the Jackson Laboratory in the United States have shown that approximately 60% of male NZO animals developed elevated plasma glucose levels with a small percentage (approximately 5%) developing severe hyperglycemia in the presence of hypoinsulinemia, suggesting that this obese model is a threshold model for diabetes.

In contrast, male mice maintained in our colony at the Walter and Eliza Hall Institute (WEHI) in Melbourne, Australia, show increased plasma glucose levels compared to NZC mice (Andrikopoulos and Proietto 1995; Andrikopoulos et al. 1993, 1996; Veroni et al. 1991), but very rarely become severely hyperglycemic. Even though both facilities report hyperglycemia, the degree to which it occurs is different between the colonies of mice and could explain why some studies have not seen this phenomenon. Female mice have also been reported to be hyperglycemic during adulthood (Plum et al. 2000; Ortlepp et al. 2000; Kluge et al. 2000; Giesen et al. 2003), but little is known about whether it is present early on. Glucose levels in NZO mice have also been shown to be weight dependent, with heavier animals displaying higher blood glucose concentrations than lighter mice (Herberg et al. 1970). Therefore, while hyperglycemia is present in the NZO mouse, it does not reach a level that would cause the polyuria, polydypsia, or weight loss typical of a diabetic situation.

NZO mice have also been reported to be hyperinsulinemic (Larkins 1971; Veroni et al. 1991; Ortlepp et al. 2000; Kluge et al. 2000; Plum et al. 2000); however, as with glucose levels, the results can be variable (Upton et al. 1980). Interestingly, NZO mice exhibit hypertrophy of pancreatic islets with an increase in number and size. The increase in size is mainly due to an increase in the number of β-cells that occupy the whole islet; a peripheral zone formed by α and δ cells is often not present (Bielschowsky and Bielschowsky 1956). This appears to be a compensatory mechanism of the strain that occurs in response to the obesity and hyperglycemic demand.

Glucose tolerance has been assessed by intraperitoneal injection of a glucose bolus (2 g/kg body weight). NZO mice had increased basal blood glucose concentrations and their postbolus glucose excursion was also significantly higher compared to C57BL/6 control mice (Larkins 1971). Basal plasma insulin concentrations were approximately twofold higher in NZO compared to control mice, whereas insulin secretion in response to glucose was significantly impaired. In contrast, an intraperitoneal bolus of arginine induced an insulin secretory response that was several fold higher than that for control mice (Larkins 1972), suggesting selective loss of the glucose recognition system for insulin secretion.

This impairment in glucose tolerance appears to be the result of obesity in the NZO mouse. This was shown by calorie restriction studies where the body weight of the NZO mice was not different from control C57BL/6 mice (Larkins 1973b). This resulted in improvement in glucose tolerance with stimulation of insulin secretion in response to a glucose bolus.

Hyperglucagonemia has also been reported in the NZO mouse, which may be an important contributor to their metabolic disturbances (Upton et al. 1980). In contrast, plasma growth hormone levels are not different in NZO compared to C57BL/6 control mice (Larkins 1971).

INSULIN SECRETION IN NZO MICE

Studies using the intraperitoneal glucose tolerance test suggested that NZO mice may have impaired insulin secretion. However, it was the use of the intravenous glucose tolerance test and *in vitro* pancreatic islet cultures that confirmed defective insulin secretion in NZO mice. The following sections summarize *in vivo* and *in vitro* studies characterizing impaired insulin release in the NZO mouse.

In Vivo Studies

Cameron and colleagues (1974) were the first to determine the rate of insulin secretion in response to various secretagogues administered intravenously. NZO mice displayed a markedly blunted early phase secretion of insulin in response to a 1 g/kg glucose bolus, while the late phase of insulin release was depressed compared to randomly bred white control mice. This was later confirmed by Veroni and colleagues in 1991, who also reported a defect in early and late phase insulin secretion in response to an intravenous glucose (0.6 g/kg) bolus. Furthermore, Veroni and colleagues showed that this defect was present in young, four- to five-week-old NZO mice compared to control NZC mice—suggesting that this is an early perturbation in the syndrome. The rate of insulin secretion was also defective in response to other nonglucose secretagogues, such as glucagon and aminophylline, whereas it was greatly exaggerated in response to arginine (Cameron et al. 1974).

In Vitro Studies

Basal insulin release from NZO islets cultured *in vitro* was five times higher compared to islets from C57BL/6 control mice (Larkins 1973b), thus corroborating the basal hyperinsulinemia seen *in vivo*. In response to 8.4 mmol/L glucose, islets from control mice showed a significant increase in insulin secretion whereas NZO mouse islets did not respond. Furthermore, although higher glucose concentrations (16.7 mmol/L) caused NZO mouse islets to secrete significant amounts of insulin, the fold stimulation was less compared to control islets (twofold vs. fivefold) (Larkins et al. 1980). In contrast, NZO mouse islets secreted more insulin in response to arginine compared to islets from C57BL/6 mice, confirming the *in vivo* data (Cameron et al. 1974). Moreover, insulin secretion in response to the glycolytic intermediate glyceraldehyde elicited a large response from NZO mouse islets, at least equal to that of control mouse islets (Larkins et al. 1980).

From these experiments, it appears that defects in insulin secretion in the NZO mouse lie within glucose metabolism, somewhere between glucose transport into the islet β-cell and the triose-phosphate step, where glyceraldehyde enters the glycolytic pathway. It is therefore possible that defects in GLUT2 or the β-cell glucose sensor glucokinase are responsible for defective insulin secretion in the NZO mouse. However, on further examination of the Larkins (1973a) study, the effects of another nonglucose secretagogue, tolbutamide (a sulphonylurea), revealed a significant response in control islets to secrete insulin but had no effect in the NZO mouse. Sulphonylureas act on the sulphonylurea receptor to elicit a response. Studies have shown that this receptor is in actual fact an ATP-sensitive potassium channel expressed in a variety of cells including the β-cell (Proks et al. 2002). Within this cell, the receptor plays a crucial role in glucose-mediated insulin secretion as it serves as a link between glucose metabolism and electrical activity of the cell (Seino

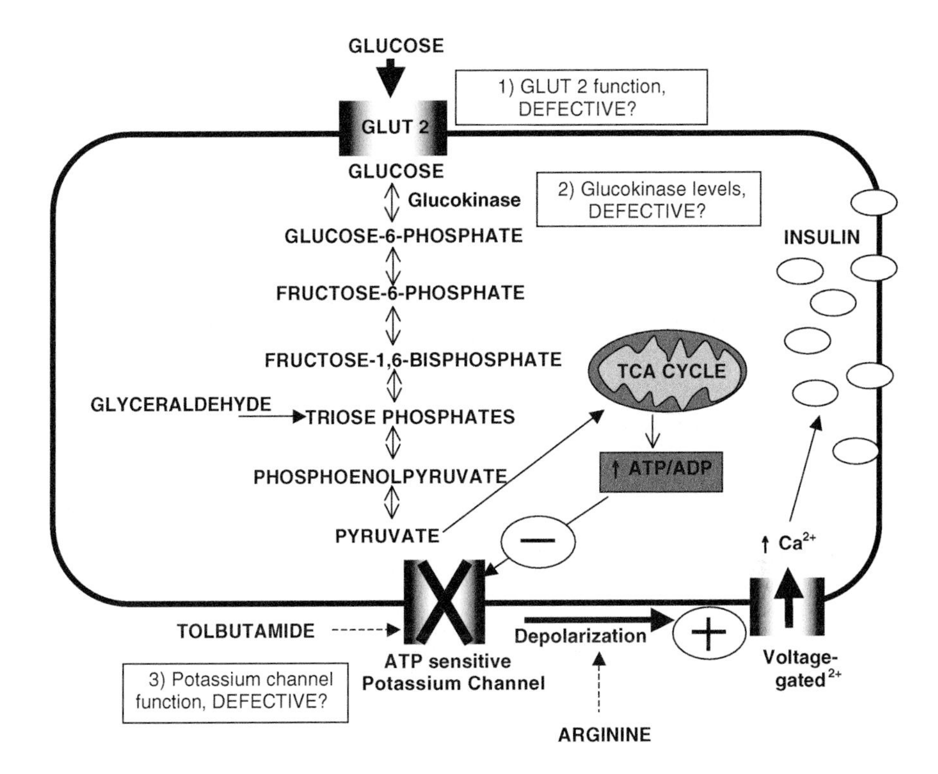

FIG. 6.2 Diagrammatic representation of the potential defects associated with glucose-mediated insulin secretion from the β-cell of the NZO mouse. In the normal state, glucose enters the β-cell via binding to the GLUT2 receptor. Glucose then enters the glycolytic cycle to be utilized in the production of pyruvate. This end product enters the TCA cycle and results in an increase in the ratio of ATP:ADP. This increase closes the ATP sensitive potassium channels (or sulphonylurea receptor); causes depolarization of the membrane, opening of the calcium channel; and allows an influx of calcium into the cytosol to cause insulin secretion via exocytosis of insulin granules. Nonglucose secretagogues such as arginine can stimulate insulin secretion by direct depolarization of the membrane, while tolbutamide acts directly on the sulphonylurea receptor to close the receptor. As discussed in the review, literature has suggested that the defects could be at the level of (1) GLUT2 receptor; (2) reduced levels of glucokinase; or (3) defective ATP sensitive potassium channels.

et al. 2000). Given that the NZO islets do not secrete insulin in response to tolbutamide, it is possible that defects in the sulphonylurea receptor may also contribute to the disturbances in insulin secretory function in the NZO mouse. This concept will need to be explored further. Figure 6.2 illustrates the glucose-mediated insulin secretory response and where the potential defects may lie in the NZO mouse. Overall, the NZO mouse displays impaired glucose-mediated insulin secretion and in this regard is a good animal model in determining the molecular mechanisms of this defect.

INSULIN ACTION IN NZO MICE

Insulin resistance in the NZO mouse was reported with the earliest characterization of the obesity syndrome in 1953 and performed with the relatively crude method of the insulin tolerance test (Bielschowsky and Bielschowsky 1953). This test showed that NZO mice tolerated an intraperitoneal injection of 2–8 units/kg of insulin whereas injections of 0.4 units/kg in NZC control mice invariably proved to be fatal (Bielschowsky and Bielschowsky 1956). This was confirmed by Crofford and Davis (1965) who, using the insulin tolerance test, showed that NZO mice were four to five times more insulin resistant than randomly bred albino mice. Although this test showed that NZO mice were insulin resistant, it did not describe the site of resistance (e.g., hepatic or peripheral) and did not give any information on the intracellular perturbations responsible for the insulin resistance.

MUSCLE AND FAT INSULIN RESISTANCE

In 1991 we adapted the euglycemic/hyperinsulinemic clamp technique to study glucose turnover in NZO and NZC control mice (Veroni et al. 1991). Results from this study showed that under basal conditions, glucose uptake in white adipose tissue, heart, diaphragm, white quadriceps, and white gastrocnemius was increased in 20-week-old NZO mice, whereas there was no difference in brown adipose tissue, soleus, red quadriceps, and red gastrocnemius muscle. This is similar to what has been described in patients with type 2 diabetes—that is, increased postabsorptive glucose disposal, probably as a result of the mass action effect of glucose (Yki-Jarvinen 1990). Most tissues assayed from NZO mice displayed a defect in glucose uptake in response to insulin. There was no stimulation over basal in white adipose tissue, whereas brown adipose tissue, soleus, red and white quadriceps, and white gastrocnemius showed a smaller increment compared to NZC control mice. In contrast, at four weeks of age, white adipose tissue and white quadriceps responded normally, while other tissues from NZO mice showed impaired sensitivity to insulin. This suggests progressive deterioration of peripheral insulin resistance in the NZO mouse.

To determine whether the defect in glucose uptake in NZO mice was due to a decrease in glucose transport, we measured the level of the insulin-stimulatable glucose transporter GLUT4 by immunoblotting (Ferreras et al. 1994). At 20 weeks of age, all tissues that showed a decrease in 2-deoxyglucose uptake also had decreased levels of GLUT4, except for soleus muscle. At four weeks of age, there was a decrease in brown adipose tissue GLUT4 but no difference in other tissues tested, supporting the idea that impaired insulin-stimulated glucose uptake in peripheral tissues is secondary to other metabolic perturbations in the NZO mouse. In *in vitro* experiments, 2-deoxyglucose uptake in isolated soleus muscle was defective at all insulin concentrations tested in NZO mice compared to lean BALB/c mice, as was glucose utilization and glycogen synthesis (Veroni and Larkins 1986). This suggests that although total GLUT4 levels were not decreased in NZO mouse soleus muscle, there may be a defect in transporter intrinsic activity or other distal step in glucose metabolism.

We also determined the activity of glycogen synthase and phosphorylase in quadriceps muscle of 1-day- and 20-week-old NZO and lean NZC mice. Active glycogen synthase activity was decreased in NZO mice compared to NZC mice at both ages tested, whereas there was no difference in glycogen phosphorylase activity (Thorburn et al. 1995). A defect in muscle glycogen synthesis has previously been reported by Stauffacher and Renold (1969), who showed that diaphragm and adipose tissue from NZO mice incorporated less glucose into glycogen in response to the same insulin levels compared to lean albino Swiss mice. Defects in muscle glycogen synthase have been described in patients with type 2 diabetes and these have been proposed to be of primary importance to the syndrome (Groop et al. 1993).

Sneyd and colleagues characterized adipose tissue insulin resistance in a series of somewhat contradictory experiments in the 1970s. Adipocytes isolated from NZO mice showed a markedly reduced rate of lipolysis (as measured by glycerol release) in response to 0.1 μg/mL isoprenaline compared to adipocytes from NZY control mice (Lovell-Smith and Sneyd 1973). In contrast, the opposite response was observed in response to 0.55 μmol/L epinephrine (Upton et al. 1979). This is most likely due to the fact that isoprenaline is a specific β-adrenergic agonist, whereas epinephrine is an α- as well as a β-adrenergic agonist. Defective lipolysis was present only in young mice, suggesting that this may be a primary cause of the development of obesity in NZO mice. This defect in lipolysis was attributed to diminished cellular cAMP concentrations as a result of increased phosphodiesterase activity (Lovell-Smith and Sneyd 1974).

A more comprehensive study of adipocyte insulin resistance showing decreased insulin-stimulated glucose transport and utilization in NZO compared to lean NZC mice was conducted by Macaulay and Larkins (1988). The key glucose oxidation enzyme, pyruvate dehydrogenase, was also found to be unresponsive to the stimulatory effects of insulin in NZO mice. Importantly, this study determined that an intracellular mediator, identified as an inositol-containing membrane glycophospholipid, that was produced by insulin in control NZC mice was absent in NZO adipocytes. Thus, muscle and fat insulin resistance in the NZO mouse seems to be due to defects in multiple sites, including a decrease in the expression of GLUT4, decreased activity of key glucose metabolism enzymes such as glycogen synthase and pyruvate dehydrogenase, and the production of insulin mediators.

Hepatic Insulin Resistance

The euglycemic/hyperinsulinemic clamp showed that NZO mice had increased endogenous glucose production compared to NZC mice in the basal state at both 4 and 20 weeks of age. Furthermore, insulin significantly inhibited endogenous glucose production in NZC control mice at both ages tested, whereas it did not have this effect in NZO mice (Veroni et al. 1991). To determine the substrates responsible for increased endogenous glucose production in the NZO mouse, we examined the *in vivo* rate of alanine and glycerol gluconeogenesis. The rate of conversion of both these substrates to glucose was increased in NZO mice compared to lean NZC control mice (Andrikopoulos and Proietto 1995). In the case of glycerol gluconeogenesis, this was due to an increase in substrate availability as well as an intrahepatic mechanism, whereas with alanine gluconeogenesis this was largely due to an increased rate of an intrahepatic mechanism (Andrikopoulos and Proietto 1995).

Increased gluconeogenesis from alanine and glycerol has been described in patients with type 2 diabetes (Puhakainen et al. 1992; Nurjhan et al. 1992). An earlier report from Rudorff et al. (1970) showed that basal alanine gluconeogenesis was similar between NZO and control white mice and that, whereas insulin inhibited gluconeogenesis in control mice, it had no effect in NZO mice. The discrepancy in basal alanine gluconeogenesis with our results may be because we performed the studies *in vivo* so that the hepatic glucose and insulin concentrations and neural input were not disturbed. Rudorff et al. (1970) used the liver perfusion technique in which the animals were sacrificed prior to the experiment and the concentrations of hormones and metabolites in the perfusion medium may not be physiologically relevant.

To identify the intrahepatic mechanisms responsible for accelerated alanine and glycerol gluconeogenesis, we measured the activity of key regulatory enzymes in the gluconeogenic and glycolytic pathways (Andrikopoulos et al. 1993). The activity of the key hepatic glycolytic enzymes glucokinase and pyruvate kinase was enhanced in NZO mice compared to NZC control mice. This is an appropriate response to the hyperinsulinemia. The activity of the hepatic gluconeogenic enzyme phosphoenolpyruvate carboxykinase (PEPCK) and glucose-6-phosphatase was depressed in NZO compared to NZC control mice. Moreover, since PEPCK is primarily regulated at the transcriptional level, we determined PEPCK mRNA levels and found that these were also diminished in NZO mice. This decreased PEPCK activity is surprising since it has been shown that this enzyme is rate limiting (Rongstad 1979) and transgenic animals overexpressing PEPCK display hepatic insulin resistance and elevated plasma glucose concentrations (Lamont et al. 2006; Rosella et al. 1995; Valera et al. 1994).

The reduced PEPCK activity and mRNA levels is the appropriate response to the hyperinsulinemic environment in the NZO mouse. In contrast, the activity of the gluconeogenic enzymes pyruvate carboxylase and fructose-1,6-bisphosphatase was increased in NZO mice. The increase in pyruvate carboxylase activity is not surprising since this enzyme is allosterically activated by free fatty acids and we have shown that plasma free fatty acids are increased in NZO mice (Andrikopoulos et al. 1996). The increase in pyruvate carboxylase activity is probably responsible for the increased rate of alanine conversion to glucose in NZO mice. The increase in fructose-1,6-bisphosphatase activity was surprising and was investigated further, since it explained the accelerated rate of glycerol conversion to glucose. Furthermore, elevated fructose-1,6-bisphosphatase activity has been proposed to account for increased glycerol gluconeogenesis in patients with type 2 diabetes (Nurjhan et al. 1992).

This is supported by an early report showing increased fructose-1,6-bisphosphatase activity in liver biopsied from patients with type 2 diabetes (Willms et al. 1970). Western and northern blot experiments revealed that while fructose-1,6-bisphosphatase protein levels were increased in NZO mice, mRNA levels were similar to NZC control mice (Andrikopoulos et al. 1996). Furthermore, the activity and protein levels of this enzyme were decreased in one-day-old NZO compared to NZC mice, suggesting that the defect is acquired in response to the disturbances in lipid metabolism. This is corroborated by experiments showing that feeding NZC control mice a high-fat diet for 12 days resulted in increased activity and protein levels of hepatic fructose-1,6-bisphosphatase (Andrikopoulos and Proietto 1995); this has also been confirmed in rats fed a high-fat diet (Song et al. 2001).

These results suggest that under obese and high-fat feeding circumstances, fructose-1,6-bisphosphatase may contribute significantly to increased hepatic glucose production. However, in a recent study from our group, we showed that transgenic mice with a specific upregulation of human hepatic fructose-1,6-bisphosphatase in the liver had increased glycerol gluconeogenesis but no change in hepatic glucose output or glucose tolerance (Lamont et al. 2006). Thus, how important this enzyme is to contributing to increased glucose production in obesity and type 2 diabetes is still to be determined.

The Metabolic Syndrome in NZO Mice

The metabolic syndrome is characterized by obesity, insulin resistance, hyperglycemia, hypertension, and dyslipidemia (reviewed in Zimmet et al. 1999, 2005). A number of rodent models have been used to study the disorder; however, not all models present with all the traits. A study by Ortlepp and colleagues in 2000 tested the suitability of the NZO mouse as a strain for metabolic syndrome, since some of these traits are common in the strain. The study found significant increases in diastolic and systolic blood pressure, serum cholesterol and triglyceride levels in the NZO mice compared to the lean Swiss Jackson Laboratory (SJL) mouse. Therefore, this strain presents with all components of the metabolic syndrome and is a good model for the identification of the genetic basis of this syndrome (Ortlepp et al. 2000).

Immune Abnormalities in NZO mice

Antibodies against the insulin receptor (Melez et al. 1980; Harrison and Itin 1979), as well as against native DNA and denatured, single-stranded DNA (Melez et al. 1980), have been described in NZO mice as having an association with the development of diabetes. However, it appears unlikely that immune abnormalities are responsible for the obesity, insulin resistance, and impaired insulin secretion in these mice for two reasons. First, antibody levels are higher in female NZO mice and the levels decline after six months of age to normal levels (Melez et al. 1980; Harrison and Itin 1979). Second, while glucose tolerance is impaired in NZO mice, it is normal in the related NZB/W strain despite severe immunological abnormalities in these mice (Upton 1984). A more recent study has further confirmed this by showing that anti-insulin receptor autoantibodies are not essential to the development of insulin resistance and transition to type 2 diabetes in the NZO mouse (McInerney et al. 2004).

GENETICS OF OBESITY AND DIABETES IN NZO MICE

The development of obesity and glucose intolerance in the NZO mouse is governed by multiple genes rather than as a single defect in a single gene. Despite all that is known about the metabolic disturbances in the NZO mouse, few clues remain as to the actual genes involved in its physiological and biochemical abnormalities. The use of quantitative trait loci (QTL) analysis and the generation of congenic mouse lines is one of the techniques employed to decipher the underlying genetics of such polygenic diseases. This involves crossing phenotypically divergent inbred lines, in which all genetic loci are homozygous, and then backcrossing or intercrossing the F1 population. In the resultant progeny (backcross or F2), different alleles at a QTL influencing the polygenic trait are shuffled in many combinations (Lander and Botstein 1989). The specific QTL allele combinations inherited by an individual backcross or F2 mouse can then be inferred by genotyping each mouse at markers spaced at regular intervals across the whole genome, known as genome-wide scanning.

Since 1998, a number of genome-wide scans have been conducted to investigate the genetic basis of obesity and diabetes using the obese NZO mouse (Plum et al. 2000, 2002; Taylor et al. 2001; Reifsnyder and Leiter 2002; Giesen et al. 2003; Reifsnyder et al. 2000; Kluge et al. 2000; Leiter et al. 1998). These studies vary from each other in the type of lean strains used in the initial cross; the SJL mouse (Becker et al. 2004; Giesen et al. 2003; Plum et al. 2000, 2002; Kluge et al. 2000), the small mouse (SM) (Taylor et al. 2001), and the nonobese nondiabetic (NON) mouse strains were used (Reifsnyder et al. 2000; Reifsnyder and Leiter 2002; Leiter et al. 1998). Table 6.1 summarizes the current literature of obesity- and diabetes-associated QTLs with more detailed review of these studies given later.

Leiter and colleagues in 1998 were the first group to use genome-wide scans to identify diabetes susceptibility QTLs in the NZO mouse. The study used F2 populations derived from crossing the NZO mouse with the NON strain to model how polygenic thresholds for the transition from impaired glucose tolerance (IGT) to diabetes were reached. Male NON mice showed increased levels of fasting blood glucose and IGT, yet they did not make the transition to overt diabetes. The NZO mice were obese with some males becoming hyperglycemic. The F1 male progeny (from both cross-directions) consistently developed type 2 diabetes, implicating the contribution of both genomes to diabetes susceptibility. Genetic analysis from these F2 mice identified three mouse loci with complex relationships to diabetes-related subphenotypes: *Nidd 1*, *Nidd 2*, and *Nidd 3*. Each of these loci represented a different phenotypic effect. *Nidd 1*, on chromosome 4, appeared to be associated with reduced plasma insulin, increased fed blood glucose, and lower body weight. *Nidd 2*, on chromosome 18, contributed to blood glucose, and *Nidd 3*, on chromosome 11, cosegregated with a reduced capacity to sustain elevated plasma insulin and reduced body weight over time (Leiter et al. 1998).

In 2000, Plum and colleagues published a study in which they crossed the NZO mouse with the lean and atherosclerosis-resistant SJL strain. The F1 progeny were backcrossed to female NZO mice to segregate the genes influencing hyperglycemia and hypoinsulinemia (the reciprocal backcross to female SJL mice exhibited unstable obesity-related traits). Genome-wide scanning of these backcross mice revealed a different locus from that of Leiter and colleagues. *Nidd/SJL* was identified on chromosome 4 and appeared to be responsible for the ~60% prevalence of hyperglycemia detected in the male backcross mice. Moreover, this locus was located 10–28 cm distal to the *Nidd1* locus identified by Leiter and colleagues. It was concluded that the *Nidd/SJL* loci represented a diabetes gene that lowered the obesity threshold for the development of hyperglycemia and hypoinsulinemia (Plum et al. 2000).

Closer examination of the two studies, however, revealed that three of the four identified loci were not NZO derived; *Nidd1* and *Nidd2* were contributed to by the NON genome while *Nidd/SJL* was contributed to by the SJL genome. The *Nidd3* locus was the

TABLE 6.1
Summary Table of Obesity and Diabetes QTLs Identified in NZO Mice

Strain	Progeny and Sex	Chromosome	QTL Name	Phenotype	Genome Derived	Ref.
NZO × NON	F2 males	4	Nidd1	↓ PI , ↑ BG, ↓ bw	NON	(Leiter et al. 1998)
		18	Nidd2	BG	NON	
		11	Nidd3	Reduced capacity to sustain ↑ PI and ↓ bw	NZO	
NZO × SJL	NZO backcross males	4	Nidd/SJL	Lowering obesity threshold for development of ↓ PI, ↑ BG	SJL	(Plum et al. 2000)
NZO × NON	NZO backcross males	1	—	bw, ↑ BG and ↑ PI	NZO	(Reifsnyder et al. 2000)
		15	—	BG	NZO	
		12	—	Adiposity, BMI, leptin	NZO	
		5	—	Adiposity, leptin	NZO	
		8 (weak)	—	Adiposity, BMI	NZO	
		3 (weak)	—	Insulin	NZO	
		13 (weak)	—	BMI	NZO	
		15 (weak)	—	bw, BMI, leptin	NZO	
NZO × SJL	NZO backcross females	5	Nob1	↑ bw	NZO	(Kluge et al. 2000)
NZO × SJL	NZO backcross males	5	Chol/NZO	Hypercholesterolemia	NZO	(Giesen et al. 2003)
NZO × SM	F1 hybrid males and females and F2 males and females	1, 2, 5, 6, 7, and 17	—	Adiposity index (sum of four white adipose tissue weight divided by bw)	NZO	(Taylor et al. 2001)

Notes: PI = plasma insulin; BG = blood glucose; bw = body weight; BMI = body mass index.

only one derived from the NZO genome. These results suggest that other inbred nonobese genomes can also contribute obesity and diabetes genes. This provides evidence that otherwise unaffected mice can harbor susceptibility loci and that the phenotypes of obesity and diabetes are the result of complex gene–gene and gene–environment interactions.

It is important to note that some studies have shown a contribution to obesity/diabetes by QTL from the NZO strain. Reifsnyder and colleagues, in 2000, intercrossed the NZO to the NON mouse and backcrossed the F1 progeny to the NZO. Using male progeny, the authors identified strong links between obesity and diabetes traits on chromosomes 1 (body weight, hyperglycemia, and hyperinsulinemia), 15 (plasma glucose), 12 (adiposity, body mass index [BMI], and leptin), and 5 (adiposity and leptin). All these linkages were contributed to by the NZO genome. The authors also found suggestive weak links with chromosomes 18 (adiposity and BMI), 3 (insulin), 13 (BMI), and two separate loci on 15 (body weight, BMI, and leptin), also derived from NZO alleles.

Another study in 2000 intercrossed the NZO with the lean SJL mice, backcrossed to the NZO, and studied the phenotypes in the female progeny (Kluge et al. 2000). The authors in this study found strong linkages with increased body weight on chromosome 5 and hyperinsulinemia on chromosome 19 and denoted these QTLs *Nob1* and *Nob2*, respectively. On further analysis, the data revealed a striking difference between the contributions of *Nob1* and *Nob2* to body weight and hyperinsulinemia. Whereas *Nob1* did not seem to exert a primary effect on insulin levels, *Nob2* increased them independently of body fat. Thus, these data appeared to be consistent with the idea that hyperinsulinemia in the NZO mouse is not solely due to obesity genes, but also to genes that affect insulin action independent of body weight. Furthermore, these authors also found that the leptin receptor variant of the NZO mouse enhanced the effect of *Nob1* on body weight and *Nob2* on insulin levels.

Using progeny from an intercross between NZO and another strain of lean mouse, the SM mouse, Taylor and colleagues (2001) identified 10 significant adiposity QTLs that were derived from NZO alleles. These were present on chromosomes 1 (three loci), 2, 5 (two loci), 6 (two loci), 7, and 17. More recently, Giesen and colleagues (2003) intercrossed the NZO with the SJL strains, backcrossed to the NZO, and found that, in the male backcross progeny, hypercholesterolemia was significantly linked to chromosome 5 (*Chol1/NZO*). Interestingly, this QTL was positioned 40 cm distal to the *Nob1* loci previously discovered (Kluge et al. 2000), but no increases in body weight, insulin or hyperglycemia were linked to this distal QTL (Giesen et al. 2003).

These studies have highlighted the complexity of the obesity and diabetes phenotypes of the NZO mouse and have provided valuable information on the genetics behind these abnormalities. However, this classic approach of breeding congenics to isolate the critical chromosomal regions may in essence fail since the individual gene variants contribute small effects that may not be detectable on a different, lean background (Jurgens et al. 2006; Reifsnyder and Leiter 2002). To better analyze such polygenic syndromes, researchers have adopted the use of recombinant congenic strains (RCS) (Reifsnyder and Leiter 2002; Leiter and Reifsnyder 2004); this has previously proven to be successful in dissecting the complex genetic interactions contributing to cancer (van Wezel et al. 1999) and type 1 diabetes (Serreze et al. 1994).

These mice are generated by backcrossing twice to a recipient strain of mouse, such as the NON lean mouse, and then inbreeding to fix all alleles to homozygosity (Leiter and Reifsnyder 2004). At the second backcross, the genome would have, on average, ~87.5% recipient alleles (NON) and ~12.5% of the donor alleles (NZO), for example. This increases the chance of bringing together QTL-causing phenotypes in different combinations while limiting the genetics of the donor strain to a small percentage (Leiter and Reifsnyder 2004), thus allowing for more rigorous assessment of the contribution of the genomic effects from the NZO mouse.

Selection for specific genomic intervals in the RCS allows testing of interactions predicted by previous crosses and allows unknown effects to be uncovered. In 2002, Reifsnyder and Leiter constructed 10 RCSs on the NON background. Each strain possessed different levels of NZO derived-genomic regions that presented with differing levels of obesity and diabetes susceptibility. RCS-2, -1, and -10 showed increasing levels of diabetes susceptibility, with RCS-10 the most affected strain. RCS-6, -7, -8, and -9 represent diabetes-prone strains with different combinations of diabetogenic QTLs, and RCS-3, -4, and -5 represented obese strains that did not progress to overt diabetes.

Recently, two studies have been conducted making use of these new strains of mice in determining the underlying biochemical aspects of thiazolidinedione-induced hepatosteatosis in F1 progeny of an NZO × NON intercross (Pan et al. 2005; Leiter et al. 2006). Thiazolidinediones such as rosiglitazone are potent therapeutic agents that have been used in the treatment of type 2 diabetes in rodent models and humans (Aston-Mourney et al. 2005; Spiegelman 1998); however, the actions have also been associated with lipid accumulation in tissues such as liver (Tiikkainen et al. 2004). Chronic treatment with rosiglitazone in F1 progeny from an NZO × NON intercross effectively suppressed hyperglycemia, hyperinsulinemia, and hyperlipidemia compared to untreated F1 mice (Watkins et al. 2002). However, there was a worsening of an underlying hepatosteatosis in these mice.

Using four RCSs (RCS-1, -2, -8, and -10), Pan and colleagues in 2005 reported that RCS-8 exhibited the same hepatosteatotic response to chronic treatment of rosiglitazone as that of the F1 progeny. This was due to further impairment of the hepatic phosphatidylcholine biosynthetic enzymes that were already present. In a study published recently, Leiter and colleagues (2006) used the same four lines of RCSs to further decipher the longitudinal changes in various plasma analytes and found that plasminogen activator inhibitor-1 (PAI-1) was the most effective predictor of this adverse drug response in F1 progeny and RCS-8 mice.

Therefore, the differing levels of obesity and diabetes susceptibility in these new mouse strains may be more representative of what is considered "typical polygenic" type 2 diabetes in humans and as a result provide better models to determine the underlying genes causing these defects.

CONCLUDING REMARKS

The NZO mouse displays all the classic features of obesity and glucose intolerance, including excessive body weight, hyperphagia, reduced energy expenditure, leptin insensitivity, impaired glucose-mediated insulin secretion, and hepatic as well as peripheral insulin resistance. We have been able to show that the disturbances in glucose metabolism in these mice are secondary to its obesity and defects in lipid metabolism. These abnormalities, in combination with the polygenic nature of the NZO mouse, make it an excellent model for the study of obesity-related glucose intolerance and the overall metabolic syndrome. The exact causes of these aberrations have not been fully uncovered, but thanks to the advancement in gene technology and through use of congenics and genome-wide scanning, we are inching closer to identifying the underlying genetic causes. The clues that we gain from this mouse model will likely bring us closer to determining the cause of human obesity and type 2 diabetes, leading to the development of therapeutic targets that can delay or prevent the onset of these diseases in humans.

AVAILABILITY

NZO mice are commercially available from the Jackson Laboratory, Bar Harbor, Maine, the Walter and Eliza Hall Institute (WEHI), Parkville, Victoria, Australia, and Bomholtgard, Aarhus, Denmark.

ACKNOWLEDGMENTS

This work has been supported by a grant from the National Health and Medical Research Council of Australia and Diabetes Australia Research Trust Fund. SA is supported by an RD Wright Biomedical Fellowship from the National Health and Medical Research Council of Australia.

REFERENCES

Andrikopoulos S, Proietto J. (1995) The biochemical basis of increased hepatic glucose production in a mouse model of type 2 (non-insulin-dependent) diabetes mellitus. *Diabetologia* 38:1389–1396.

Andrikopoulos S, Rosella G, Gaskin E, et al. (1993) Impaired regulation of hepatic fructose-1,6-bisphosphatase in the New Zealand obese mouse model of NIDDM. *Diabetes* 42:1731–1736.

Andrikopoulos S, Rosella G, Kaczmarczyk SJ, et al. (1996) Impaired regulation of hepatic fructose-1,6-bisphosphatase in the New Zealand obese mouse: An acquired defect. *Metabolism* 45:622–626.

Andrikopoulos S, Thorburn A, Proietto J. (2001) The New Zealand obese mouse: a polygenic model of Type 2 diabetes. In *Animal Models of Diabetes: A Primer*, ed. A Sima, E Shafrir, Harwood Academic, Amsterdam, 171–184.

Aston-Mourney K, Proietto J, Andrikopoulos S. (2005) Investigational agents that protect pancreatic islet beta-cells from failure. *Expert Opin Invest Drugs* 14:1241–1250.

Becker W, Kluge R, Kantner T, et al. (2004) Differential hepatic gene expression in a polygenic mouse model with insulin resistance and hyperglycemia: Evidence for a combined transcriptional dysregulation of gluconeogenesis and fatty acid synthesis. *J Mol Endocrinol* 32:195–208.

Bielschowsky M, Bielschowsky F. (1953) A new strain of mice with hereditary obesity. *Proc Univ Otago Med School* 31:29–31.

Bielschowsky M, Bielchowsky F. (1956) The New Zealand strain of obese mice. Their response to stilboestrol and to insulin. *Austral J Exp Biol* 34:181–1898.

Cameron DP, Opat F, Insch S. (1974) Studies of immunoreactive insulin secretion in NZO mice *in vivo*. *Diabetologia* 10:649–654.

Crofford OB, Davis CKJ. (1965) Growth characteristics, glucose tolerance and insulin sensitivity of New Zealand obese mice. *Metabolism* 14:271–280.

Davis TR, Mayer J. (1954) Imperfect homeothermia in the hereditary obese-hyperglycemic syndrome of mice. *Am J Physiol* 177:222–226.

Ferreras L, Kelada ASMK, McCoy M, Proietto J. (1994) Early decrease in GLUT4 protein levels in brown adipose tissue of New Zealand obese mice. *Int J Obesity* 18:760–765.

Giesen K, Plum L, Kluge R, et al. (2003) Diet-dependent obesity and hypercholesterolemia in the New Zealand obese mouse: Identification of a quantitative trait locus for elevated serum cholesterol on the distal mouse chromosome 5. *Biochem Biophys Res Commun* 304:812–817.

Groop LC, Kankuri M, Schalin-Janti C, et al. (1993) Association between polymorphism of the glycogen synthase gene and non-insulin-dependent diabetes mellitus. *N Engl J Med* 328:10–14.

Halaas JL, Boozer C, Blair-West J, et al. (1997) Physiological response to long-term peripheral and central leptin infusion in lean and obese mice. *Proc Natl Acad Sci USA* 94:8878–8883.

Harrison LC, Itin A.(1979) A possible mechanism for insulin resistance and hyperglycemia in NZO mice. *Nature* 279:334–336.

Herberg L, Major E, Hennigs U, et al. (1970) Differences in the development of the obese-hyperglycemic syndrome in *ob/ob* and NZO mice. *Diabetologia* 6:292–299.

Igel M, Becker W, Herberg L, Joost HG. (1997) Hyperleptinemia, leptin resistance, and polymorphic leptin receptor in the New Zealand obese mouse. *Endocrinology* 138:4234–4239.

Jurgens HS, Schurmann A, Kluge R, et al. (2006) Hyperphagia, lower body temperature, and reduced running wheel activity precede development of morbid obesity in New Zealand obese mice. *Physiol Genomics* 25:234–241.

Kluge R, Giesen K, Bahrenberg G, et al. (2000) Quantitative trait loci for obesity and insulin resistance (*Nob1*, *Nob2*) and their interaction with the leptin receptor allele (LeprA720T/T1044I) in New Zealand obese mice. *Diabetologia* 43:1565–1572.

Koza RA, Flurkey K, Graunke DM, et al. (2004) Contributions of dysregulated energy metabolism to type 2 diabetes development in NZO/H1Lt mice with polygenic obesity. *Metabolism* 53:799–808.

Lamont BJ, Visinoni S, Fam BC, et al. (2006) Expression of human fructose-1,6-bisphosphatase in the liver of transgenic mice results in increased glycerol gluconeogenesis. *Endocrinology* 147:2764–2772.

Lander ES, Botstein D. (1989) Mapping Mendelian factors underlying quantitative traits using RFLP linkage maps. *Genetics* 121:185–199.

Larkins RG. (1971) Plasma growth hormone in the New Zealand obese mouse. *Diabetologia* 7:302–307.

Larkins RG. (1972) Endocrine abnormalities in the NZO mouse (PhD thesis). Department of Medicine, Royal Melbourne Hospital, University of Melbourne, Melbourne.

Larkins RG. (1973a) Defective insulin secretion in the NZO mouse: *In vitro* studies. *Endocrinology* 93:1052–1056.

Larkins RG. (1973b) Defective insulin secretory response to glucose in the New Zealand obese mouse. Improvement with restricted diet. *Diabetes* 22:251–255.

Larkins RG, Martin FIR. (1972) Selective defect in insulin release in one form of spontaneous laboratory diabetes. *Nature* 235:86–88.

Larkins RG, Simeonova L, Veroni MC. (1980) Glucose utilization in relation to insulin secretion in NZO and C57Bl mouse islets. *Endocrinology* 107:1634–1638.

Leiter EH, Reifsnyder PC. (2004) Differential levels of diabetogenic stress in two new mouse models of obesity and type 2 diabetes. *Diabetes* 53:S4–11.

Leiter EH, Reifsnuder PC, Flurkey K, et al. (1998) NIDDM genes in mice. Deleterious synergism by both parental genomes contributes to diabetogenic thresholds. *Diabetes* 47:1287–1295.

Leiter EH, Reifsnyder PC, Zhang W, et al. (2006) Differential endocrine responses to rosiglitazone therapy in new mouse models of type 2 diabetes. *Endocrinology* 147:919–926.

Lovell-Smith CJ, Sneyd JGT. (1973) Lipolysis and adenosine 3′,5′-cyclic monophosphate in adipose tissue of the New Zealand obese mouse. *J Endocrinol* 56:1–11.

Lovell-Smith CJ, Sneyd JGT. (1974) Lipolysis and adenosine 3′,5′-cyclic monophosphate in adipose tissue of the New Zealand obese mouse; the activities of adipose tissue adenyl cyclase and phosphodiesterase. *Diabetologia*, 10:655–659.

Macaulay SL, Larkins RG. (1988) Impaired insulin action in adipocytes of New Zealand obese mice: A role for postbinding defects in pyruvate dehydrogenase and insulin mediator activity. *Metabolism* 37:958–965.

McInerney MF, Najjar SM, Brickley D, et al. (2004) Anti-insulin receptor autoantibodies are not required for type 2 diabetes pathogenesis in NZL/Lt mice, a New Zealand obese (NZO)-derived mouse strain. *Exp Diabesity Res* 5:177–185.

Melez KA, Harrison LC, Gilliam JN, Steinberg AD. (1980) Diabetes is associated with autoimmunity in the New Zealand obese (NZO) mouse. *Diabetes* 29:835–840.

Nurjhan N, Consoli A, Gerich J. (1992) Increased lipolysis and its consequence on gluconeogenesis in non-insulin-dependent diabetes mellitus. *J Clin Invest* 89:169–175.

Ortlepp JR, Kluge R, Giesen K, et al. (2000) A metabolic syndrome of hypertension, hyperinsulinaemia and hypercholesterolaemia in the New Zealand obese mouse. *Eur J Clin Invest* 30:195–202.

Pan HJ, Reifsnyder P, Vance DE, et al. (2005) Pharmacogenetic analysis of rosiglitazone-induced hepatosteatosis in new mouse models of type 2 diabetes. *Diabetes* 54:1854–62.

Plum L, Giesen K, Kluge R, et al. (2002) Characterization of the mouse diabetes susceptibilty locus Nidd/SJL: Islet cell destruction, interaction with the obesity QTL *Nob1*, and effect of dietary fat. *Diabetologia* 45:823–30.

Plum L, Kluge R, Giesen K, et al. (2000) Type 2 diabetes-like hyperglycemia in a backcross model of NZO and SJL mice: Characterization of a susceptibility locus on chromosome 4 and its relation with obesity. *Diabetes* 49:1590–1596.

Proietto J, Larkins RG. (1993) A perspective on the New Zealand obese mouse. In *Lessons from Animal Diabetes*, ed. Shafrir, E. Smith– Gordon, London, 4:65–74.

Proks P, Reimann F, Green N, et al. (2002) Sulfonylurea stimulation of insulin secretion. *Diabetes* 51:S368–376.

Puhakainen I, Koivisto VA, Uki-Jarvinen H. (1992) Lipolysis and gluconeogenesis from glycerol are increased in patients with noninsulin-dependent diabetes mellitus. *J Clin Endocrinol Metab* 75:789–794.

Reifsnyder PC, Churchill G, Leiter EH. (2000) Maternal environment and genotype interact to establish diabesity in mice. *Genome Res* 10:1568–1578.

Reifsnyder PC, Leiter EH. (2002) Deconstructing and reconstructing obesity-induced diabetes (diabesity) in mice. *Diabetes* 51:825–832.

Rizk NM, Liu LS, Eckel J. (1998) Hypothalamic expression of neuropeptide-Y in the New Zealand obese mouse. *Int J Obesity* 22:1172–1177.

Rongstad, R. (1979) Rate-limiting steps in metabolic pathways. *J Biol Chem* 254:1875–1878.

Rosella G, Zajac JD, Baker L, et al. (1995) Impaired glucose tolerance and increased weight gain in transgenic rats overexpressing a non-insulin-responsive phosphoenolpyruvate carboxykinase gene. *Mol Endocrinol* 9:1396–1404.

Rudorff KH, Huchzermeyer H, Windeck R, Staib W. (1970) Uber den Einfluss von insulin auf die Alaningluconeogenese in der isoliert perfundierten Leberber von New Zealand obese mice. *Eur J Biochem* 16:481–486.

Seino S, Iwanaga T, Nagashima K, Miki T. (2000) Diverse roles of K(ATP) channels learned from Kir6.2 genetically engineered mice. *Diabetes* 49:311–318.

Serreze DV, Prochazka M, Reifsnyder PC, et al. (1994) Use of recombinant congenic and congenic strains of NOD mice to identify a new insulin-dependent diabetes resistance gene. *J Exp Med* 180:1553–1558.

Song S, Andrikopoulos S, Filippis C, et al. (2001) Mechanism of fat-induced hepatic gluconeogenesis: Effect of metformin. *Am J Physiol* 281:E275–E282.

Spiegelman BM. (1998) PPAR-gamma: Adipogenic regulator and thiazolidinedione receptor. *Diabetes* 47:507–514.

Stauffacher W, Renold AE. (1969) Effect of insulin *in vivo* on diaphragm and adipose tissue of obese mice. *Am J Physiol* 216:98–105.

Taylor BA, Wnek C, Schroeder D, Phillips SJ. (2001) Multiple obesity QTLs identified in an intercross between the NZO (New Zealand obese) and the SM (small) mouse strains. *Mamm Genome* 12:95–103.

Thorburn A, Litchfield A, Fabris S, Proietto J. (1995) Abnormal transient rise in hepatic glucose production after oral glucose in non-insulin-dependent diabetic subjects. *Diabetes Res Clin Prac* 28:127–135.

Thorburn AW, Holdsworth A, Proietto J, Morahan G. (2000) Differential and genetically separable associations of leptin with obesity-related traits. *Int J Obesity* 24:742–750.

Tiikkainen M, Hakkinen AM, Korsheninnikova E, et al. (2004) Effects of rosiglitazone and metformin on liver fat content, hepatic insulin resistance, insulin clearance, and gene expression in adipose tissue in patients with type 2 diabetes. *Diabetes* 53:2169–2176.

Upton JD. (1984) Intravenous glucose tolerance tests in the New Zealand strains of mice. *Horm Metab Res* 16:290–292.

Upton JD, Sneyd JGT, Livesey J. (1980) Blood glucose, plasma insulin and plasma glucagon in NZO mice. *Horm Metab Res* 12:173–174.

Upton JD, Sneyd JGT, Rennie PIC. (1979) Insulin resistance in the New Zealand obese mouse (NZO): Lipolysis and lipogenesis in isolated adipocytes. *Arch Biochem Biophys* 197:139–148.

Valera A, Pujol A, Pelegrin M, Bosch F. (1994) Transgenic mice overexpressing phosphoenolpyruvate carboxykinase develop non-insulin-dependent diabetes mellitus. *Proc Natl Acad Sci USA* 91:9151–9154.

van Wezel T, Ruivenkamp CA, Stassen AP, et al. (1999) Four new colon cancer susceptibility loci, Scc6 to Scc9 in the mouse. *Cancer Res* 59:4216–4218.

Veroni MC, Larkins RG. (1986) Evolution of insulin resistance in isolated soleus muscle of the NZO mouse. *Horm Metab Res* 18:299–302.

Veroni MC, Proietto J, Larkins, RG. (1991) Evolution of insulin resistance in New Zealand obese mice. *Diabetes* 40:1480–1487.

Watkins SM, Reifsnyder PR, Pan HJ, et al. (2002) Lipid metabolome-wide effects of the PPARgamma agonist rosiglitazone. *J Lipid Res* 43:1809–1817.

Willms B, Ben-Ami P, Söling HD. (1970) Hepatic enzyme activities of glycolysis and gluconeogenesis in diabetes of man and laboratory animals. *Horm Metab Res* 2:135–141.

Yki-Jarvinen H. (1990) Acute and chronic effects of hyperglycaemia on glucose metabolism. *Diabetologia* 33:579–585.

Zimmet P, Boyko EJ, Collier GR, de Courten M. (1999) Etiology of the metabolic syndrome: Potential role of insulin resistance, leptin resistance, and other players. *Ann NY Acad Sci* 892:25–44.

Zimmet P, Magliano D, Matsuzawa Y, Alberti G, Shaw J. (2005) The metabolic syndrome: A global public health problem and a new definition. *J Atheroscler Thromb* 12:295–300.

7 The JCR:LA-*cp* Rat: Animal Model of the Metabolic Syndrome Exhibiting Micro- and Macrovascular Disease

James C. Russell, Sandra E. Kelly, and Spencer D. Proctor

CONTENTS

INTRODUCTION

The metabolic syndrome, a prediabetic state characterized by abdominal obesity, hypertriglyceridemia and insulin resistance, has become an established and intractable health problem in all prosperous societies. In humans, there is often, but not uniformly, a progression from the metabolic syndrome to pancreatic failure and type 2 diabetes (T2D) (i.e., hyperglycemic and hypoinsulinemic status). The sequelae of the syndrome, particularly the very high risk for cardiovascular disease (CVD) (Steiner 1994), has prompted growing concern worldwide. Individuals show a significant variation in susceptibility to obesity and the metabolic syndrome (Anand et al. 2000) that is undoubtedly, at least in part, genetic in origin (Hegele et al. 2003; Gupta et al. 2002). The metabolic syndrome is a particularly insidious disease state due to the asymptomatic character of its early stages. The clinical history typically involves a long period of increasing abdominal obesity without obvious underlying disease, followed by a rapid development of frank T2D. As the clinical diabetes becomes recognized and treated, medical follow-up reveals established CVD.

The significant damage to the vascular system has largely occurred during the prediabetic period, a process that is only now becoming widely appreciated. Consequently, a crucial feature of the metabolic syndrome is widespread vascular dysfunction and atherosclerosis that develops silently during the prediabetic and early diabetic phases of T2D. Consequently, the metabolic syndrome is a major contributor to the widespread burden of ischemic disease of the heart and brain and renal failure in prosperous societies worldwide. Recently, we have begun to appreciate that the metabolic syndrome and its pathophysiological complications are modulated by complex interactions between the environment (in the broadest sense) and the genome (Hegele et al. 2003). However, prevention and even amelioration of this disease burden will require greater understanding of the underlying physiological mechanisms.

A growing awareness of the importance of the obesity/insulin resistance syndrome as a primary cause of vascular disease has led to a heightened interest in animal models that mimic the human syndrome. In particular, there is a need for small-animal models for research into the underlying pathophysiological mechanisms and the assessment of pharmaceutical interventions. Such models should not only be metabolically similar to humans with insulin resistance or overt T2D, but also exhibit the associated pathological sequelae: atherosclerosis, vasculopathy, and ischemic end stage disease. We have developed a unique strain that meets these criteria—the JCR:LA-*cp* rat—and, along with collaborators, have addressed some of the issues of the mechanisms underlying the metabolic syndrome. We have also used this animal model to screen putative new treatment approaches to the metabolic syndrome and the associated end stage vascular and renal disease.

THE *cp* GENE AND STRAINS

Koletsky and SHROB

An autosomal recessive gene was initially isolated by Simon Koletsky (1973) in a cross between rats of the Sprague–Dawley strain (a standard albino strain) and the SHR strain. The SHR is a spontaneously hypertensive rat, due to a mutation that arose in the Wistar rat colony at Kyoto University in Japan. Rats that were homozygous for the mutant gene isolated by Koletsky were not only obese, but also hypertensive and exhibited a malignant atherosclerosis, accompanied by the development of aortic aneurysms, and a life span of only some 10 months (Koletsky 1975). Rats heterozygous or homozygous normal were lean with absence of disease and a normal life span. Koletsky designated the strain of rats as obese SHR or SHROB. Animals of this strain were sent to Hansen at the National Institutes of Health in Bethesda (NIH) in 1972.

Congenic Strains

Hansen proceeded to create a defined and stabilized strain incorporating the Koletsky gene, which he designated as *cp* (corpulent). This proved difficult, due to the extensive load of deleterious genes in the complex background genome. Hansen therefore crossed the SHROB with two in-house inbred strains at NIH—the SHR/N and the LA/N—and backcrossed repeatedly to the parent strains to create (after 12 backcrosses) two congenic strains: the SHR/N-*cp* and the LA/N-*cp* (Hansen 1983; Greenhouse et al. 1988). The congenic strains are inbred and retain only the *cp* gene from the original SHROB. As with the original SHROB rat, animals homozygous for the *cp* gene (*cp/cp*) are obese and exhibit metabolic and pathological sequelae, while rats that are heterozygous (*cp*/+) or homozygous normal (+/+) are lean and normal in all respects.

Rat Mutations

The *cp* mutation causes an absence of the transmembrane portion of the ObR leptin receptor due to a T2349A transversion resulting in a Tyr763Stop nonsense mutation in the gene (Wu-Peng et al. 1997). This results in an absence of any apparent leptin action in the *cp/cp* rats, an extreme hyperleptinemia, and fundamental changes in metabolism (McClelland et al. 2004). In contrast, the other rat mutation leading to obesity, the *fa* gene of the fatty Zucker rat (Zucker and Zucker 1961), a glycine to proline substitution at position 269 of the ObR, results in a 10-fold reduction in binding affinity for leptin (Chua et al. 1996). The *fa/fa* Zucker rat develops a less severe variant of the metabolic syndrome, with no progression to diabetes, has lower circulating levels of insulin, and does not exhibit any cardiovascular complications (Amy et al. 1988). A variant of the Zucker rat, the Zucker diabetic fatty (ZDF), has been developed more recently by Peterson and is described in this book.

JCR:LA-*cp* Strain

In 1978, at the fifth backcross to the LA/N strain, initial breeding stock was sent from NIH to the laboratory of one of the authors (JCR) at the University of Alberta. These animals were the founders of a colony that retains approximately 3% of the genome derived from the SHROB. The susceptibility to vascular disease in the JCR:LA-*cp* rat is multifactorial and polygenetic in origin. The other strains incorporating the *cp* gene, including the LA/N-*cp*, do not show this trait and have lost the severity of the metabolic dysfunction in the course of inbreeding to create the congenic status. Unlike the NIH colonies that were maintained inbred and congenic, the JCR:LA-*cp* strain has been maintained as a closed outbred colony to retain the unknown genetic elements that lead to cardiovascular disease; therefore, this strain was designated JCR:LA-*cp* in 1989. Recently, the JCR:LA-*cp* colony was rederived and established at Charles River Laboratories (Wilmington, Massachusetts), designated as Crl:JCR(LA)-$Lepr^{cp}$.

SHHF/Mcc Strain

At the seventh backcross to the SHR/N strain, breeding stock was sent to the G. D. Searle Company in Indianapolis, where rats bred in the colony were found to develop a cardiomyopathy spontaneously (Ruben et al. 1984). The Searle colony was transferred to McCune, who ultimately relocated it to Ohio State University and it was designated as SHHF/Mcc. This strain has also been rederived and established at Charles River Laboratories, designated as SHHF/MccCrl-$Lepr^{cp}$. The two noncongenic strains exhibit pathophysiological characteristics not evident in the related congenic strains. Importantly, whereas the JCR:LA-*cp* rats develop atherosclerosis and ischemic myocardial lesions, the SHHF/Mcc-*cp* rats develop a fatal cardiomyopathy, apparently due to a single gene. The SHHF/Mcc-*cp* strain has been inbred by McCune for the cardiomyopathic trait and now exhibits a very high frequency of cardiomyopathy with ultimate congestive heart failure (McCune et al. 1995; Haas et al. 1995).

LA/N-*cp* and SHR/N-*cp* Strains

The congenic LA/N-*cp* strain is phenotypically indistinguishable from the JCR:LA-*cp* strain and has less severe hyperinsulinemia (Recant et al. 1989; Pederson et al. 1991; Russell et al. 1999) than the equivalent rats of the JCR:LA-*cp* strain. In addition, the LA/N-*cp* rats do not exhibit vascular disease and myocardial lesions (Michaelis and Russell, unpublished observations) evident in the JCR:LA-*cp* strain. Similarly the SHR/N-*cp* rats, while hyperinsulinemic, do not develop ischemic myocardial lesions, showing only a perivascular fibrosis consistent with their hypertensive status (Michaelis and Russell, unpublished observations).

WKY/N-*cp* Strain

The WKY/N is the NIH variant of the normotensive parent strain of the SHR rat and used as a control animal. Hansen also created a congenic WKY/N-*cp* strain which does not develop myocardial lesions, but is very sensitive to high sugar intake, which results in a diffuse myocardial fibrosis. The Veterinary Resources Branch at the NIH that maintained the congenic *cp* rat strains, along with many other rodent models, has been effectively dissolved. Some strains of rats have been preserved as embryos and a few others transferred to other organizations, particularly the Rat Resource Center at the University of Missouri, Columbia. The status of other strains, including the WKY/N-*cp*, remains obscure.

PATHOPHYSIOLOGY OF THE JCR:LA-*cp* RAT

Genotype, Hyperphagia, Body Weight, and Phenotype

JCR:LA-*cp* rats that are homozygous normal (+/+) or heterozygous for the *cp* gene (*cp*/+) are lean and metabolically normal, while rats that are homozygous (*cp*/*cp*) develop the metabolic syndrome with obesity and the range of metabolic and physiological abnormalities. On weaning at three weeks of age, the *cp*/*cp* rats are no heavier than their lean +/? and +/+ litter mates. Nonetheless, they have a detectably more rounded body phenotype and can be unequivocally identified. The food intake of the *cp*/*cp* rats is greater than that of the +/? at three weeks and increases rapidly as the animals mature. The body weight of the *cp*/*cp* animals also increases more rapidly than that of the +/? rats; by eight weeks of age, they are already markedly obese. By nine months of age, the *cp*/*cp* male rats normally weigh 800–900 g compared with approximately 400–450 g for the lean male rats. Beyond 12 months of age, these rats have reduced food intake and show deterioration in their condition and a decline in weight. Extensive data on food intake and body weight under various conditions have been published (Russell, Amy, Manickavel, et al. 1989; Russell, Amy, Manickavel, and Dolphin 1989; Russell et al. 1995; Russell, Dolphin, et al. 1998; Russell et al. 2000). Although a reduction of food intake to that of the *cp*/+ or +/+ animals (20 g/day at 12 weeks) results in a lower body weight, it does not normalize the weight to the +/? rats and the *cp*/*cp* retains a rounded and obese phenotype being simply smaller.

Leptin Status

The absence of any of the isoforms of the ObR receptor in *cp*/*cp* rats in the plasma membrane is the core defect in the JCR:LA-*cp* strain. It is a necessary, but not sufficient, condition for the development of end stage disease. The absence of vascular pathology in the other congenic strains incorporating the *cp* gene confirms that the presence of additional genetic contributors is also essential to the disease processes in the JCR:LA-*cp* rat. Although the functions of leptin are still not entirely clear, a primary element is the inhibition of the hypothalamic neuropeptide Y (NPY) mechanism that is the single most powerful mediator of eating. The *cp*/*cp* rat has elevated levels of NPY in the arcuate nucleus and median eminence, the hypothalamic areas involved in the control of eating (Williams et al. 1990, 1992); the rats are significantly hyperphagic (Russell et al. 1990). Leptin also inhibits the release of insulin from pancreatic beta-cells (Emilsson et al. 1997). The *cp/cp* rat hypersecretes insulin, both basal and in response to secretagogues, that is greatly in excess of that of the *fa/fa* Zucker rat (Pederson et al. 1991), which exhibits a reduced binding affinity of leptin to the ObR receptor and is thus merely leptin resistant.

Insulin Resistance and Hyperlipidemia

The primary metabolic abnormalities of the *cp/cp* rat are a profound insulin resistance and hypertriglyceridemia. A major question is: Which comes first, the insulin resistance and hyperinsulinemia or the hyperlipidemia? Plasma triglyceride (TG) concentrations are elevated in male *cp/cp* rats at 3 weeks of age and continue to rise rapidly until 12 weeks of age. This is accompanied by an elevated intracellular TG content and precedes the rapid, substantive rise in plasma insulin levels that occurs at about five and one-half weeks of age (Russell, Bar-Tana, et al. 1998). The hyperlipidemia is due to a marked hypersecretion of very low-density lipoprotein (VLDL), resulting in high VLDL levels and elevated TG concentrations (Dolphin et al. 1987; Russell, Koeslag, et al. 1989). In 12-week-old *cp/cp* rats, these are typically 280 mg/100 mL in males and 500 mg/100 mL in females. More modest increases occur in cholesterol and phospholipid concentrations, probably due to VLDL requirements for particle stability. There is a prominent flow of the cholesterol pool through to the high-density lipoprotein (HDL) fraction in these rats; however, lipoprotein lipase activity and VLDL clearance do not appear to be affected.

The hyperphagic *cp/cp* rats absorb large amounts of glucose from a carbohydrate diet and the elevated VLDL secretion in the animals is essentially a compensatory response, by the liver, to dispose of the glucose (Brindley and Russell 2002). In the absence of any insulin-mediated glucose uptake (Russell et al. 1994), the *cp/cp* rat can only divert excess glucose to the liver, where high insulin levels promote upregulated fatty acid and TG synthesis, leading to increased VLDL assembly and secretion (Brindley and Russell 2002). VLDL-associated hyperlipidemia is well recognized as an independent risk factor for CVD, consistent with the lipid effects observed in the *cp/cp* rat.

Other cholesterol-rich remnant fractions such as postprandial chylomicrons (CM) and/or their remnants (CM-r) (which contain apolipoprotein B48) are also important in the development of atherosclerosis. We have demonstrated that CM-r readily permeate the arterial wall and are preferentially retained, leading to the focal accumulation of cholesterol, a hallmark of atherogenesis (Proctor et al. 2002, 2004). In addition, there is evidence that postprandial lipemia is also a significant risk factor for CVD in individuals with visceral and/or central obesity (Mamo et al. 2001; Schaefer et al. 2002). In this context, we have shown that the *cp/cp* rat has significant postprandial lipemia as measured by oral fat challenge experiments and assessment of plasma apoB48 kinetics (Vine et al. 2005). Thus, the insulin-resistant state in the *cp/cp* rat is associated with enhanced circulating levels of two different classes of atherogenic lipid particles. In this respect, the *cp/cp* rat has a similar hyperlipidemic profile to that of prediabetic humans and offers an unusual opportunity to study atherogenic mechanisms and putative antiatherogenic treatments.

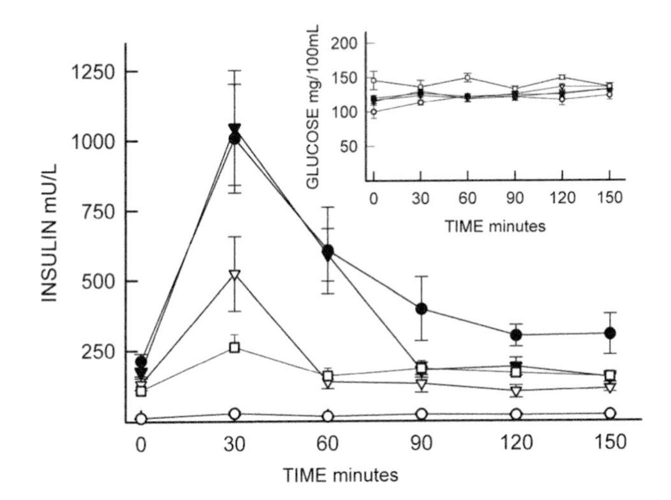

FIG. 7.1 Plasma glucose and insulin response of male rats at 12 weeks of age to a meal tolerance test consisting of 5 g of rat show. ○ = +/? control rats; ● = *cp/cp* control; □ = *cp/cp* S15262-treated 38 mg/kg/d; ◀ = *cp/cp* treated with metformin at 100 mg/kg/d; Δ = *cp/cp* treated with troglitazone at 100 mg/kg/d.

The profound insulin resistance, fully developed at eight weeks of age in the *cp/cp* rat, results in a complete absence of insulin-mediated glucose uptake (Russell et al. 1994; Russell, Bar-Tana, et al. 1998). The *cp/cp* rats are essentially normoglycemic (fasting plasma glucose levels are somewhat higher than in other strains of rat in all genotypes), but do have an impaired glucose tolerance under a high glucose load (Russell et al. 1987). The commonly used indices of abnormal glucose and insulin metabolism are the intravenous and oral glucose tolerance tests, which are not sensitive to changes in the sensitivity to insulin in these animals (Russell et al. 1999). We have found that a standardized meal tolerance test is the most sensitive and efficient method of assessing insulin sensitivity in these animals. This consists of a test meal (a 5-g pellet of rat chow) given after an overnight fast and subsequent measurement of plasma insulin and glucose over time. As seen in figure 7.1, the rats maintain euglycemia over the test period while the insulin response at 30 min postprandial provides a sensitive index of insulin sensitivity. As also shown in figure 7.1, the response to treatment with insulin-sensitizing agents can be differentiated, with the insulin response varying from the hyperinsulinemic response seen in the control animals to negligible insulin response similar to the +/? rat.

MACROVASCULAR DISEASE

Atherosclerosis

In metabolic terms, the *cp/cp* rats strongly mimic the human syndrome of abdominal obesity, with a concomitant insulin resistance, hyperinsulinemia, and hypertriglyceridemia. Despite the very high rate of insulin secretion (Pederson et al. 1991), the *cp/cp* rat does not progress to pancreatic beta-cell failure and resultant development

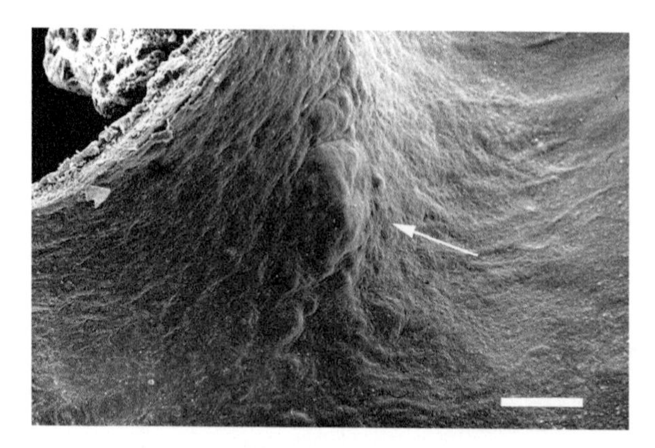

FIG. 7.2 Scanning electron micrograph of the aortic arch of a nine-month-old male *cp/cp* rat. There is a large raised intimal lesion over the lesser curve of the arch, with intact overlying endothelium. Bar = 0.25 mm.

of insulin-dependent diabetes. However, the prediabetic, insulin-resistant state is strongly associated with CVD in humans, and the *cp/cp* rats similarly spontaneously develop marked vasculopathy, atherosclerosis, and ischemic lesions. Early atherosclerotic changes are evident in young adult rats (12 weeks of age) and increase in severity with age so that, by nine months, essentially all *cp/cp* males have advanced intimal lesions of the aortic arch. A typical example of such a lesion, visualized by scanning electron microscopy, is shown in figure 7.2. Thrombi of various sizes and ages are frequently found on the arterial surface (fig. 7.3) and occasionally found

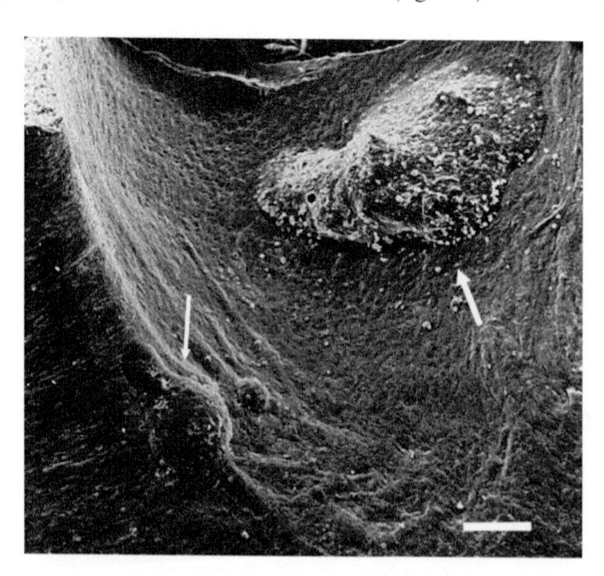

FIG. 7.3 Scanning electron micrograph of one of the branches of the arch of the aorta of a nine-month-old male *cp/cp* rat showing a large organized thrombus in the lumen of the branch. There is also another old thrombus on the flow divider at the orifice of the branch. The endothelium is abnormal, with adherent macrophages and underlying intimal lesions. Bar =0.25 mm.

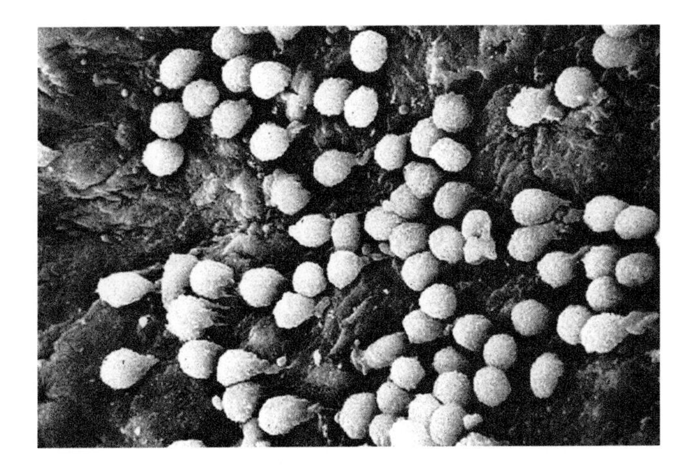

FIG. 7.4 Scanning electron micrograph of an area of the aorta of a nine-month-old *cp/cp* male rat with extensive adherent and activated macrophages.

occluding coronary arteries. The presence of thrombi is probably related to the increased levels of plasminogen activator inhibitor-1 (PAI-1) in the *cp/cp* rat (Schneider et al. 1998).

As shown in figure 7.4, adherent macrophages are often present, particularly over areas of abnormal endothelium. Transmission electron microscopy shows the intimal lesions as being very similar to atherosclerotic lesions in humans, encompassing lipids, proteoglycan, collagen, macrophages, or vascular smooth muscle cells (VSMC), and cellular debris in the intimal space (fig. 7.5). The endothelium overlying the lesions is usually intact and appears more or less normal. The VSMC of the media appear to be activated and migratory, moving through breaks in the internal elastic lamina into the intimal space. This is consistent with the observation that aortic VSMC of the male *cp/cp* rat are hyperproliferative and hyper-responsive to various cytokines, particularly insulin (Absher et al. 1997). The hyperactivity of the VSMC is essentially prevented by experimental manipulations that significantly reduce the hyperinsulinemia (Absher et al. 1999).

Vasculopathy

The vascular lesions and medial SMC abnormalities of the *cp/cp* rat are accompanied by vascular dysfunction in the form of enhanced contractile response to noradrenaline or phenylephrine (PE) and impaired nitric oxide-mediated relaxation (O'Brien et al. 1997, 1998). These effects are evident in aortae and mesenteric resistance vessels, but are not evident at all in *+/?* rats and less severe in *cp/cp* female rats (O'Brien et al. 2000). The noradrenergic-mediated hypercontractility has two components: impaired endothelial nitric oxide (NO) metabolism and release, accompanied by a vascular smooth muscle hypercontractility, that is evident even in the presence of inhibition of nitric oxide synthase (NOS) by L-NAME (N^G-nitro-L-arginine methyl ester) (see, for instance, data in Proctor, Kelly, et al. 2006). There is no

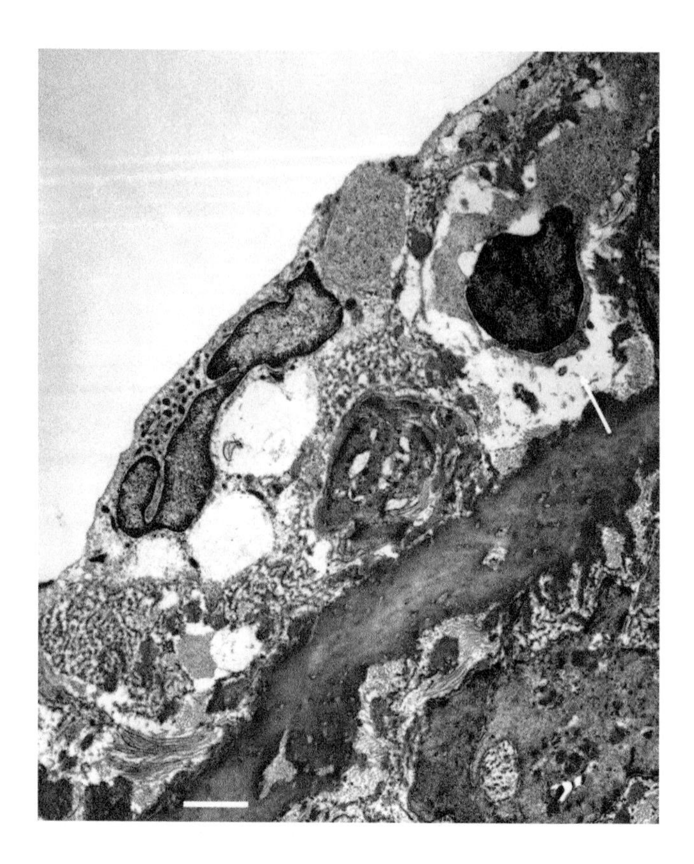

FIG. 7.5 Transmission electron micrograph of the aorta of a six-month-old male *cp/cp* rat showing an advanced atherosclerotic lesion with the lumen on the upper left. The internal elastic lumina are intact, visible as a light band across the lower portion, with an underlying necrotic smooth muscle/foam cell (arrow). The intimal space contains collagen, proteoglycan, and lipid inclusions, There is a highly abnormal smooth muscle cell in the medial space on the lower right. Bar = 1 μm.

impairment in relaxation of PE precontracted vessels in response to the direct NO donor sodium nitroprusside, indicating that the response of the VSMC per se is not compromised.

Vascular dysfunction in the *cp/cp* male rat is also evident in the coronary circulation (Russell et al. 2004), consistent with the development of ischemic lesions of the heart, especially in older rats. We have found that coronary artery dysfunction in these animals is accompanied by significant alterations in myocardial metabolism (Lopaschuk et al. 1991). When perfused *in vitro*, hearts from *cp/cp* rats require high concentrations of insulin (2000 mU/L) and Ca^{++} ≤1.75 m*M* in order to maintain mechanical function. In addition, the *cp/cp* heart has a specific impairment of NO-mediated relaxation of coronary resistance vessels that is not associated with impaired baseline myocardial contractility (Brunner et al. 2000). We have not been able to detect discrepancies in the level of NOS activity within the aorta or the left ventricle, or in plasma levels of biopterin between *cp/cp* and *+/?* male rats. However, we have been able to reverse the defect in endothelium-dependent relaxation in the coronary vascular system with exogenous tetrahydrobiopterin (H_4biopterin), suggesting that NO production is impaired in the *cp/cp* rat, but secondary to a defect in H_4biopterin metabolism (Wascher et al. 2000). This opens up the possibility that suboptimal levels of H_4biopterin may lead to an NOS-dependent generation of superoxide anions that, in turn, could result in physiologically significant endothelial damage.

Sexual Dimorphism of Disease

The vascular lesions and the vascular dysfunction are less severe in *cp/cp* females than in males, despite VLDL levels approximately twice those of the *cp/cp* males, but probably reflecting insulin levels approximately half those of males. Atherosclerotic lesions, intravascular thrombi, and other manifestations of CVD do develop in *cp/cp* females, but only at advanced ages. The dimorphism in the severity of cardiovascular disease is not affected by early castration in either sex, although plasma lipid levels do show significant changes (Russell et al. 1993). The mechanisms underlying the sexual dimorphism in CVD remain obscure, although they are not due to direct action of the steroid sex hormones.

MICROVASCULAR DISEASE

Glomerular Sclerosis

Microvascular damage occurring at the arteriolar level is a prominent feature of type 1 and T2D (Molitch et al. 2003). Microalbuminuria is a marker for early microvascular damage that leads to retinal damage, glomerular sclerosis, and end-stage renal failure (Savage et al. 1996). T2D is present at a much greater incidence in the human population than type 1 and is the primary contributor to the burden of microvascular complications of diabetes. If microvascular damage in T2D is related to hyperinsulinemia, the damage should occur in the early prediabetic, hyperinsulinemic stages and should be evident in the *cp/cp* rat. In fact, urinary albumin excretion in the *cp/cp* rat is significantly increased as early as 12 weeks of age, when hyperinsulinemia is well established (Proctor, Kelly et al. 2006). Accompanying the elevated urinary albumin concentrations is a significant incidence of glomerular sclerosis, as illustrated in figure 7.6.

Polycystic Ovary Disease

Polycystic ovary disease (PCOS) is the leading cause of infertility in humans and the etiology remains poorly understood. There is a strong association between PCOS and the metabolic syndrome, consistent with microvascular dysfunction/damage as a contributing factor. However, effective studies of this relationship remain inhibited by the lack of an established animal model (Cussons et al. 2005). The *cp/cp* rats of the JCR:LA-*cp* strain, of both sexes, are functionally sterile. While the *cp/cp* males produce viable sperm, they do not breed, for reasons that are not known. The *cp/cp* females do not conceive with *+/?* males and we have found that they have small

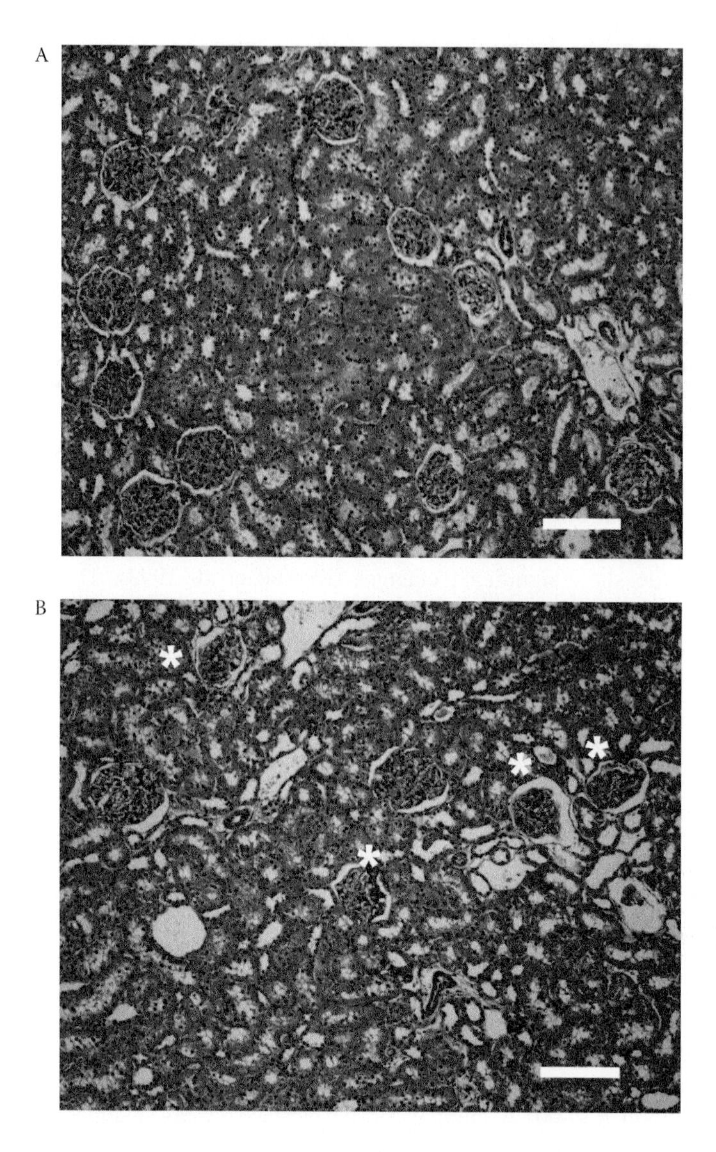

FIG. 7.6 Micrographs of kidneys of 12-week-old male rats: panel A, +/?; panel B, *cp/cp*. H&E stain, bars = 100 μm. Sclerotic glomeruli are indicated by asterisks.

ovaries with few follicles and ova (fig. 7.7). Thus, the *cp/cp* female rat offers a model that is a close analog of the human PCOS and the opportunity to explore underlying mechanisms and putative treatments.

Impaired Wound Healing

Impaired wound healing has long been recognized as a serious complication of diabetes. The underlying mechanism leading to the impaired healing and ulcer formation

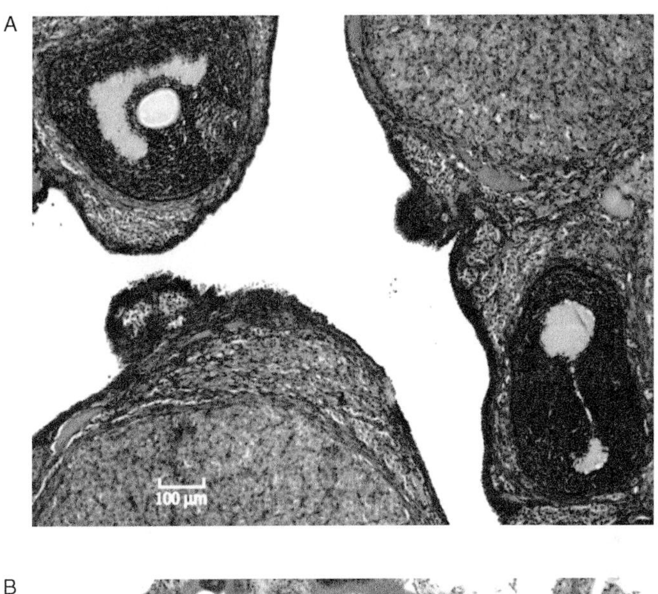

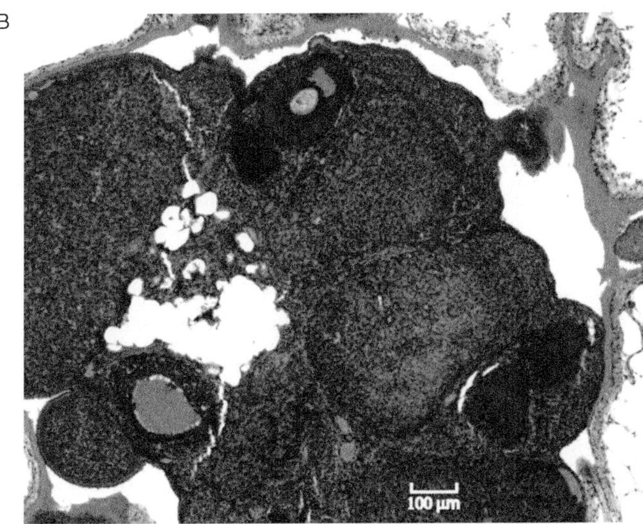

FIG. 7.7 Representative micrographs of ovaries of 12-week-old female rats; +/?, panel A, bar = 200 µm, and *cp*/*cp*, panel B, bar = 100 µm. The ovaries of the *cp*/*cp* rats were smaller and disrupted with only rare follicles or ova visible. The ovaries of the +/? rats were variable, but showed frequent follicles and ova.

remains an open question, with lack of an established animal model for the condition a major impediment to studies. Our work on the JCR:LA-*cp* rat suggests that the origin may lie in vascular dysfunction. We have recently shown that the *cp*/*cp* rat has a markedly reduced rate of healing of punch biopsy skin wounds (Bauer et al. 2004). This is accompanied by increased tissue levels of TGFβ-1, a cytokine previously found to be elevated in the plasma of the *cp/cp* rat that elicits an exaggerated growth response from VSMC of the *cp/cp* rat (Absher et al. 1997). We concluded that TGFβ-1 may not be effectively regulated in the *cp/cp* rat, leading to autoinduction of TGFβ-1 and increased plasma and tissue concentrations, with major effects on a variety of cell types and tissues (Bauer et al. 2004). The data, to date, indicate that the *cp/cp* rat offers a model that will permit definitive experimental studies to clarify the origin of impaired wound healing of diabetes and development of effective clinical treatments.

Dietary Effects

We have recently found that the JCR:LA-*cp* rat is very sensitive to dietary cholesterol. Normal rat chow is essentially cholesterol free. The addition of 0.25% cholesterol, from weaning to six months of age, leads to exacerbated intimal lesions in male *cp/cp*, but not in *+/?* rats. A diet containing 1% cholesterol leads to advanced atherosclerotic lesions at 12 weeks of age. The dietary cholesterol results in increased VLDL cholesterol ester concentrations that appear to be pathogenic, leading to larger and more lipid-laden atherosclerotic lesions.

Dietary supplementation with ω-3 fatty acid-rich oils, such as fish oils, has been an attractive concept for the amelioration of risk of CVD. Unfortunately, we found that a simple fish oil-supplemented diet conferred no protection against myocardial lesions in the *cp/cp* rat (Russell, Amy, and Dolphin 1991). In contrast, a phytosterol preparation (85% stigmasterol esterified with fish oil, 51% EPA + DHA) at a dose of 2.6 g/kg per day reduced plasma TG by 50% and cholesterol esters by almost 20% (Russell et al. 2002). At a low dose of 86 mg/kg, hypercontractility of aortic rings in response to PE was normalized and the relaxant response to the endothelial NO metabolism was significantly improved. Thus, dietary fatty acid and/or phytosterol intake can modify the abnormal lipid metabolism and favorably affect vascular disease in this model.

Recent studies in our laboratory have shown that longer term dietary intake of ω-3 fatty acids can lead to a powerful metabolic response in the *cp/cp* rat. For example, male *cp/cp* rats treated with a lipid-balanced (nutritionally adequate) 5% fish oil diet for six months showed reduced body-weight gain and improved insulin sensitivity as well as striking effects on fasting plasma lipids and metabolism. Mechanistically, olive oil and menhaden oil supplementation in *cp/cp* rats has been shown to reduce expression of SREBP-1c, with concomitant reductions in hepatic TG content, lipogenesis, and expression of enzymes related to lipid synthesis (Elam et al. 2001; Deng et al. 2004). These results confirm that polyunsaturated fatty acids reduce *de novo* lipogenesis in the *cp/cp* rat. The effects are mediated, at least in part, through reduction of hepatic expression for SREBP-1c.

Similarly, we have shown that mixed isoforms of conjugated linoleic acid (CLA), a natural constituent of dairy products, have beneficial effects on the micro- and macrovasculopathy associated with hyperinsulinemia and prediabetic state in the *cp/cp* rat (Proctor, Kelly et al. 2006). CLA reduced food intake, body weight, and fasting insulin in *cp/cp* rats. Striking reductions were also observed in the elevated urinary albumin excretion and the severity of glomerular sclerosis in rats treated with CLA.

Environmental Effects

There is strong evidence linking the metabolic syndrome, T2D and CVD to exposure to airborne fine particulate pollutants with aerodynamic diameters ≤ 2.5 μm ($PM_{2.5}$) (O'Neill et al. 2005). While the increased risk of exposure to $PM_{2.5}$ in humans is quite clear, the underlying mechanisms are obscure and difficult to study. We have recently used the JCR:LA-*cp* rat to study the vascular effects of residual oil fly ash (ROFA) from a large U.S. power plant (Proctor, Dreher, et al. 2006). The bioavailable constituents of the ROFA, at microgram concentrations, increased PE-mediated contraction in *cp/cp*, but not in *+/?* rat aortae, and the effect was exacerbated by L-NAME. Acetylcholine-mediated relaxation of *cp/cp* and *+/?* aortae was reduced. Initial exposure of aortae to ROFA caused a small contractile response, which was markedly greater on second exposure in the *cp/cp*, but slight in *+/?* aortae. The data demonstrate that oil combustion particles enhance noradrenergic-mediated vascular contraction, impair endothelium-mediated relaxation, and induce direct vasocontraction in the presence of the metabolic syndrome. Our observations provide the first direct evidence of the causal properties of $PM_{2.5}$ and confirm the pathophysiological role of the insulin-resistant or early prediabetic state in susceptibility to environmentally induced cardiovascular events.

RESPONSES TO PHARMACEUTICALS AND OTHER INTERVENTIONS

An animal model of type 2 diabetes, such as the JCR:LA-*cp* rat, would have limited usefulness if its metabolism and pathological sequelae were invariant. One of the values of animal models is to enable the exploration of underlying mechanisms of disease and use of the insights gained to develop new preventative or therapeutic treatments. To this end, we have conducted numerous studies of the effects of pharmaceutical and other treatments on the metabolism and cardiovascular disease of the *cp/cp* rat. Fortunately, the JCR:LA-*cp* rat strongly mimics the obese insulin-resistant human in these studies, strengthening the justification for use of the strain in screening or experimental verification of new antiobesity and insulin-sensitizing agents. The effectiveness of some agents in prevention of end-stage cardiovascular disease in this animal model is highly encouraging in terms of our ultimate ability to discover effective treatments for the human population.

The simplest treatments for the obesity/insulin-resistance syndrome are food restriction and substantive levels of physical activity. Severely restricting the food of the *cp/cp* male rat (to ~60% of the intake of a *+/?* control, or 12 g/day) is effective in reducing plasma insulin and TG levels and is cardioprotective (Russell et al. 1990). However, this is a quite unrealistic and severe approach in humans who, in any event, have greater choice in their behavior. It is possible to induce rats to run voluntarily in running wheels through mild restriction of their food supply (Morse et al. 1995), and we have been able to achieve running of 6000–8000 m/day for prolonged periods under this condition. Rats that run such distances or are weight paired to runners show prevention of myocardial lesions (Russell, Amy, Manickavel, et al. 1989) and normalization of the vascular VSMC (Absher et al. 1999). These results parallel reports of cardioprotection through exercise in humans, and our data suggest that the effects are related to a reduction in the hyperinsulinemia. The rats provide a model to further test these concepts experimentally.

Alteration of the diet with the aim of reducing CVD remains an attractive notion, but as discussed previously, simple alteration of the lipid composition of the diet was not been shown to change insulin levels or cardiovascular outcomes (Dolphin et al. 1988; Russell, Amy, and Dolphin 1991). In contrast, rats that consumed 4% ethanol in drinking water exhibited a 50% reduction in plasma insulin concentrations, greatly reduced hyperplasia of the islets of Langerhans, and a dramatic reduction in myocardial lesions (Russell, Amy, Manickavel, and Dolphin 1989). These results are consistent with the well-established inverse correlation between moderate consumption of ethyl alcohol and CVD in the human population (Lucas et al. 2005).

A number of pharmacological agents have been found to have beneficial metabolic effects or antiatherosclerotic and cardioprotective effects in the *cp/cp* rat. The anorectic drugs benfluorex and D–fenfluramine reduce food intake and body weight, increase insulin sensitivity, reduce the severity of vascular lesions and dysfunction, and are cardioprotective (Russell et al. 1997; Russell, Dolphin, et al. 1998). The highly effective inhibitor of TG synthesis, MEDICA 16, strongly inhibits the development of insulin resistance and is antiatherogenic and cardioprotective (Russell et al. 1995; Russell, Bar-Tana, et al. 1998). Other compounds, such as captopril and probucol, are cardioprotective in the absence of any improvement in metabolism or reduction in vascular lesions (Russell et al. 1998a, 1998b). These effects reaffirm that this animal model responds to a variety of interventions in ways that parallel effects seen in the human disease state.

Comparisons with Other Obese Rat Strains

Comparisons between the two genetically determined rat models of obesity are inevitable and valuable. The oldest and best known of the rat obesity mutations is the *fa* or fatty Zucker. Because the *fa/fa* rat is only leptin resistant, the fatty rats are less hyperinsulinemic and, consistent with our findings of the importance of high insulin levels in the vascular disease, do not develop intimal lesions or myocardial ischemic damage (Amy et al. 1988). Zucker rat colonies frequently are infected with mycoplasma and this appears to be associated with abnormalities of the endothelium (Amy et al. 1988). This makes use of Zucker rats in cardiovascular studies problematic, unless the animals can be shown to be mycoplasma free.

The rat strains with the *cp* gene provide an interesting demonstration of the multifactoral and polygenetic basis of the development of T2D and related CVD. The fully congenic LA/N-*cp* and SHR/N-*cp* strains have been backcrossed many times ($N > 16$) and have lost all genetic contribution from the original Koletsky strain. As a consequence, neither strain exhibits the end-stage CVD that is prominent in the JCR:LA-*cp* rat (Russell, Amy, Michaelis, et al. 1991). The congenic strains have been valuable for studies of the pathophysiology of insulin resistance and T2D, especially in the hands of O. E. Michaelis IV and his collaborators (Recant et al. 1989; Triana et al. 1991; Velasquez et al. 1995; Yamini et al. 1992).

MAINTENANCE OF THE STRAIN

With the establishment of a commercial breeding program for the JCR:LA-*cp* strain by Charles River Laboratories, it is no longer necessary or economical to breed the rats, except under special circumstances. Any such breeding will require a clear plan and allowance for the inevitable tendency for genetic drift, which will result in the loss of some of the unique genetic characteristics of a colony. In the case of other strains with the *cp* gene, this has led to the reduced severity of a metabolic dysfunction. Most importantly, the inbreeding needed to create the congenic strains also bred out the propensity for the development of the severe cardiovascular disease seen in the original Koletsy obese SHR rat (Koletsky 1975). In maintaining the JCR:LA-*cp* rat, we have ensured genetic stability of the colony through a formal assortive breeding program with 10 separate lines and using a technique described by Poiley (1960).

Our remaining breeding colony is maintained solely by members of our research group, who of necessity avoid all contact with other rodents. All our staff wear only facility clothing and footwear in the animal unit as well as hats, masks, and gloves. Food and other items are autoclaved before being brought into the unit and caging is subjected to high-temperature wash. These barrier precautions have been successful in preventing infection with viral or bacterial pathogens over many years. Because of the significant metabolic and physiological responses of the rats to any infectious process, it is most important that experimental animals be maintained using similar protocols. Failure to do so can compromise the quality of the results.

Rats are maintained on aspen wood chip bedding in polycarbonate cages, although we are converting our housing to a controlled isolated caging system (Techniplast, Slim Line™, Techniplast S.p.a., Buguggiate, Italy). Upon weaning, young rats are normally housed in a group for a week to reduce stress. Older rats are normally housed in pairs because they are social animals and this can reduce stress levels. However, it is not good practice to house a lean male with a *cp/cp* male beyond six weeks of age because the lean rat will be aggressive toward the obese animal. Careful humidity control is essential in the animal rooms and is a significant problem in areas, such as western Canada, where humidity levels are low.

EXPERIMENTAL TECHNIQUES

Animal Handling and Drug Administration

The *cp/cp* rats are more stress sensitive than normal rats and are metabolically compromised. Even simple relocation from one room to another can have significant metabolic effects on the animals, and transport on a noisy metal cart can induce ischemic damage to the heart. We have developed specific experimental approaches to minimize stress and to improve the quality of our data. For instance, we have found that drugs and other agents are best administered to the rats in their feed or drinking water, rather than by gavage. Agents that are soluble and stable can be placed in the water bottle, which is weighed daily or twice weekly and the consumption calculated. Because the rats are typically weighed on the same schedule, the concentration of the agent in the water can easily be adjusted to achieve a desired dosage. In our laboratory, drugs are incorporated into powdered (unpelleted) rat chow, available from the manufacturer. The calculated amount (based on body weight and expected or actual food intake) is mixed into the feed using a small commercial pasta maker (Bottene, 36035 Marano Vic., Italy), usually with premixing in a mortar and pestle. The feed and drug mixture is then moistened with water and extruded through a custom-made die. The resulting pellets are air dried without heat in a convection oven or on an open rack.

ANESTHESIA

Conventional barbiturate and other injection anesthesia agents do not work well in the *cp/cp* rats. The large fat stores result in rapid clearance of the drugs so that the animals tend to be too lightly anesthetized for surgery. Higher doses of anesthetic, however, typically result in anesthesia that is too deep and cannot be reversed, and subsequent respiratory arrest. Inhalation anesthesia with the volatile agents isofluorane or halothane avoids these problems. With young rats, anesthetic induction can be accomplished by holding the rat on its back in the hand, with the fingers under the animal's jaw, and placing a cone over the muzzle. Alternatively, the rat can be placed in a large jar or even left in a cage and isofluorane in oxygen blown in. Induction is performed using 3.5% halothane in oxygen, reduced to 2.5% once the animal is unconscious, with approximately 2% required to maintain a surgical plane. This concentration is adjusted from time to time to give the minimum level of anesthesia required during long procedures such as euglycemic insulin clamps or VLDL secretion rate studies. During the anesthetic period, the rats need to be placed on a warmed table to prevent hypothermia.

When animals are euthanized at the end of a protocol, blood and tissues are frequently collected for assay. It is important to ensure that the anesthesia and accompanying procedures are performed so as to minimize any stress on the rats and negative effects on metabolic parameters. We perform all anesthetic administration and procedures in a room located well away from the other rats in order to prevent distress in any animals who will subsequently undergo manipulation. It needs to be borne in mind that rats communicate widely in the ultrasonic sound range above human hearing, and any upset and fright or even simple new occurrences spread rapidly throughout the group. Our research team makes ongoing efforts to ensure that the rats are never handled roughly or given adverse experiences. Because of restriction of animal handling in the breeding and maintenance units to members of the research group, the rats are familiar with the technical staff, resulting in sociable and unstressed behavior. For example, experimenters are able to pick up a naive six-month-old rat and take a blood sample from its tail without struggle by the animal or the need for restraint.

DIURNAL CYCLE

Rats are nocturnal, sleeping during the light phase of their diurnal cycle. This activity pattern is reflected in their metabolism, and it is thus preferable to conduct metabolic studies during the dark period of the rats' cycle. Animals subjected to such studies are normally maintained on a reversed light cycle, with lights off from 0600 to 1800 h. The timing of lights is often adjusted around the experimenter's day for convenience in performing procedures. All manipulations of the rats are performed under a dim light in a dedicated procedure room.

Assessment of Insulin and Glucose Metabolism

The conventional method of assessing insulin and glucose metabolism in the rat has been the intravenous glucose tolerance test (IVGTT); the euglycemic insulin clamp is the ultimate approach. Both methods present serious problems in the *cp/cp* rat. Because the IVGTT shows relatively small differences in the rate of clearance of an injected glucose load (typically 0.5 g/kg) (Russell et al. 1999), it is difficult to detect experimental changes in the severity of insulin resistance using this test. A euglycemic insulin clamp potentially gives detailed information on the glucose clearance and hepatic output and is practical in the *cp/cp* rat. However, the procedure is complex and lengthy and requires that the rat be anesthetized throughout or that indwelling cannulae be surgically implanted in advance. The relatively lengthy anesthesia required for the IVGTT and the euglycemic insulin clamp leads to very significant stress in these rats. The *cp/cp* rats respond to stress with significant variations in insulin and glucose metabolism as well as central nervous system changes (Leza et al. 1998; McArthur et al. 1998), all of which can obscure experimental effects.

In contrast to the IVGTT, which bypasses the complex gut hormone responses to food intake, a meal tolerance test (MTT) is much more sensitive to changes in insulin and glucose metabolism. We have developed a standardized MTT that minimizes stress on the rats and can be administered repeatedly to the same animal (Russell et al. 1999). The procedure uses a tail bleed method to obtain blood samples without restraint and with minimal disturbance. Rats are accustomed to the procedure through an initial sham procedure, usually done one week in advance. The animal is deprived of food over the light period (normally overnight if animals are kept on a reversed light/dark cycle). Two hours into the dark period, the rat is placed on a plate warmed to 37°C for 10 min to increase circulation through the tail. The animal is then held lightly by the base of the tail with one hand (the rat usually will try to crawl under the upper arm of the experimenter). The conical tip of the tail (0.5 mm) is cut with a pair of fine scissors (the tail is pinched lightly in the sham procedure) and blood is milked from the tail into micro-blood-collecting tubes (Microtainer, Becton Dickinson, Franklin Lakes, New Jersey). A 0.8-mL blood sample is taken during each repeated sampling, and up to 1.5 mL can be collected at one time without distress. The rat is then returned to its cage and given a 5-g pellet of its regular diet. The pellet is usually eaten within 15 min, and the clock is started when it is half consumed. Further tail bleed samples are taken at intervals up to 150 min, and all are assayed for insulin and glucose.

As shown in figure 7.1, the rats maintain euglycemia under this protocol, but at the expense of a very large postprandial insulin response that can reach 1000 mU/L. The insulin response is short lived and the plasma concentrations at 30 min provide an excellent index of insulin sensitivity, which is essentially the plasma insulin level required to control the glucose concentration. Treatments that lower insulin resistance result in a reduced 30-min insulin concentration and can in some cases completely prevent the postprandial insulin response (Russell, Dolphin, et al. 1998; Russell et al. 1999).

There is a well-established practice in metabolic studies to measure plasma lipid and insulin/glucose concentrations in the fasted state (in the rat, after a

16-h starvation period). This results in relatively consistent values, but does not reflect the metabolic status of an animal in its normal day-to-day situation. In contrast, measurements on blood samples taken 2–3 h into the light period, when the rat has eaten its first meal of the day, are much more representative of the animal's ongoing metabolism. Plasma TG and insulin levels will be significantly higher in the *cp/cp* rat, but glucose will not show any increase. The development of the hyperinsulinemia and hypertriglyceridemia as the juvenile *cp/cp* rat matures can be followed accurately in the fed state (Russell, Dolphin, et al. 1998), but is obscured in the fasted state.

SUMMARY

The underlying mechanisms leading to the insulin-resistant state remain elusive, although it is clear that, in the *cp/cp* rat, the insulin resistance is acquired after weaning and is not in itself directly genetically determined. The processes leading to insulin resistance in the *cp/cp* rat are amenable to study and offer the prospect of understanding the slower sequences leading to insulin resistance in humans. Our group and others are pursuing this approach. Vasculopathy and atherosclerosis in the *cp/cp* rats are unique in and make the strain complementary to other models of cardiovascular disease, many of which are focused on hypercholesterolemic mechanisms. The JCR:LA-*cp* strain provides a stable prediabetic model complementary to the ZDF rat that rapidly progresses to a full-blown T2D. The origins of the lesions in the JCR:LA-*cp* rat are clearly polygenetic and related to some unknown components of the genome derived from the Koletsky obese SHR. The identification of the genes responsible would provide an important indication of the currently unknown genetic factors that lead to cardiovascular disease in humans.

Regardless of the mechanisms responsible for the insulin resistance, there is a real interest in developing new pharmaceutical agents with insulin-sensitizing and/or antiatherosclerotic or cardioprotective properties. The assessment of new candidate drugs requires an animal model that mimics the human disease state. The JCR:LA-*cp* rat is a unique model of the obesity/insulin-resistance syndrome, possessing all of the disease characteristics seen in humans, including artherosclerosis, glomerular sclerosis, and ischemic lesions of the heart. There is a growing interest in using this economical small-animal model in new drug screening and development programs for insulin-sensitizing and antiatherosclerotic compounds.

AVAILABILITY AND SHIPPING

At the time of writing, rats are available from the authors, in limited quantity for collaborative research, from the original breeding colony at the University of Alberta. The address is:

Alberta Institute for Human Nutrition
4-10 Agriculture Forestry Centre
University of Alberta, Edmonton, Alberta T6G 2P5

Canada
e mail: Jim.Russell@ualberta.ca

Rats are available in commercial quantities from:

Charles River Laboratories International, Inc.
Attn: Research Models and Services
251 Ballardvale Street
Wilmington, MA 01887-1000, USA
www.criver.com/research_models_and_services/research_models/rats_a_c.html

Long-distance shipping of the rats requires careful attention to detail. Air freight operations vary widely between airlines. Some are unwilling to carry live animals or take them in transit from a connecting carrier. Many aircraft, including some recent types from Airbus Industries, do not have sufficient installed heater capacity to allow for the winter shipment of live animals. Subsequent to events of September 11, 2001, interlined air freight of live animals (i.e., routing involving transfer between carriers) has become impractical due to a required 48-hour embargo on all such shipments within the United States. Shipments across international borders are best handled through specialized air freight companies who act as their own brokers and can expedite Customs and veterinary clearance. All well-run laboratory animal centers have their own rules regarding the importation of animals.

REFERENCES

Absher PM, Schneider DJ, Baldor LC, et al. (1999) The retardation of vasculopathy induced by attenuation of insulin resistance in the corpulent JCR:LA-*cp* rat is reflected by decreased vascular smooth muscle cell proliferation *in vivo*. *Atherosclerosis* 143:245–251.

Absher PM, Schneider DJ, Russell JC, Sobel BE. (1997) Increased proliferation of explanted vascular smooth muscle cells: A marker presaging atherogenesis. *Atherosclerosis* 131:87–194.

Amy RM, Dolphin PJ, Pederson RA, Russel, JC. (1988) Atherogenesis in two strains of obese rats: The fatty Zucker and LA/N-corpulent. *Atherosclerosis* 69:199–209.

Anand SS, Yusuf S, Vuksan V, et al. (2002) Differences in risk factors, atherosclerosis, and cardiovascular disease between ethnic groups in Canada: The Study of Health Assessment and Risk in Ethnic Groups (SHARE). *Lancet* 356:279–284.

Bauer BS, Ghahary A, Scott PG, et al. (2004) The JCR:LA:*cp* rat: A novel model for impaired wound healing. *Wound Repair Regen* 12:86–92.

Brindley DN, Russell JC. (2002) Animal models of insulin resistance and cardiovascular disease: Some therapeutic approaches using the JCR:LA-*cp* rat. *Diabetes Obesity Metab* 4:1–10.

Brunner F, Wölkart G, Russell JC, Wascher T. (2000) Vascular dysfunction and myocardial contractility in the JCR:LA-corpulent rat. *Cardiovasc Res* 47:150–158.

Chua SC Jr, White DW, Wu-Peng XS, *et al.* (1996) Phenotype of fatty due to Gln269Pro mutation in the leptin receptor (*Lepr*). *Diabetes* 45:1141–1143.

Cussons AJ, Stuckey BGA, Watts GF. (2006) Cardiovascular disease in the polycystic ovary syndrome: New insights and perspectives. *Atherosclerosis* 185:227–239.

Deng X, Elam MB, Wilcox HG, et al. (2004) Dietary olive oil and menhaden oil mitigate induction of lipogenesis in hyperinsulinemic corpulent JCR:LA-cp rats: Microarray analysis of lipid-related gene expression. *Endocrinology* 145:5847–5861.

Dolphin PJ, Amy RM, Koeslag DG, et al. (1988) Reduction of hyperlipidemia in the LA/N-corpulent rat by dietary fish oil containing omega-3 fatty acids. *Biochim Biophys Acta* 962:317–329.

Dolphin PJ, Stewart B, Amy RM, Russell JC. (1987) Serum lipids and lipoproteins in the atherosclerosis prone LA/N-corpulent rat. *Biochim Biophys Acta* 919:140–148.

Elam MB, Wilcox HG, Cagen LM, et al. (2001) Increased hepatic VLDL secretion, lipogenesis, and SREBP-1 expression in the corpulent JCR:LA-*cp* rat. *J Lipid Res* 42:2039–2048.

Emilsson V, Lim YL, Cawthorne MA, et al. (1997) Expression of the functional leptin receptor in RNA in pancreatic islets and directly inhibitory action of leptin on insulin secretion. *Diabetes* 46:313–316.

Greenhouse DD, Michaelis OE IV, Peterson RG. (1988) The development of fatty and corpulent rat strains. In *New Models of Genetically Obese Rats for Studies in Diabetes, Heart Disease and Complications of Obesity*, ed. Hansen CT, Michaelis OE IV, National Institutes of Health, Bethesda, MD, 3–6.

Gupta M, Doobay AV, Singh N, et al. (2002) Risk factors, hospital management and outcomes after acute myocardial infarction in South Asian Canadians and matched control subjects. *CMAJ* 166:717–722.

Haas GJ, McCune SA, Brown DM, Cody RJ. (1995) Echocardiographic characterization of left ventricular adaptation in a genetically determined heart failure rat model. *Am Heart J* 130:806–811.

Hansen CT. (1983) Two new congenic rat strains for nutrition and obesity research. *Fed Proc* 42:537.

Hegele RA, Zinman B, Hanley AJG, et al. (2003) Genes, environment and Oji-Cree type 2 diabetes. *Clin Biochem* 36:63–170.

Koletsky S. (1973) Obese spontaneously hypertensive rats: A model for the study of atherosclerosis. *Exp Mol Pathol* 19:52–60.

Koletsky S. (1975) Pathological findings and laboratory data in a new strain of obese hypertensive rats. *Am J Pathol* 80:29–142.

Leza JC, Salas E, Sawicki G, et al. (1998) The effect of stress on homeostasis in JCR:LA-*cp* rats. Role of nitric oxide. *J Pharmacol Exp Ther* 28:1397–1403.

Lopaschuk GD, Russell JC. (1991) Myocardial function and energy substrate metabolism in the insulin-resistant JCR:LA-corpulent rat. *J Appl Physiol* 71:1302–1308.

Lucas DL, Brown RA, Wassef M, et al. (2005) Alcohol and the cardiovascular system research challenges and opportunities. *J Am Coll Cardiol* 45:1916–1924.

Mamo JC, Watts GF, Barrett PH, et al. (2001) Postprandial dyslipidemia in men with visceral obesity: An effect of reduced LDL receptor expression? *Am J Physiol* 281:E626–E632.

McArthur MD, Graham SE, Russell JC, Brindley DN. (1998) Exaggerated stress-induced release of nonesterified fatty acids in JCR:LA-corpulent rats. *Metabolism* 47:1383–1390.

McClelland GB, Kraft CS, Michaud D, et al. (2004) Leptin and the control of respiratory gene expression in muscle. *Biochim Biophys Acta* 1688:86–93.

McCune S, Park S, Radin MJ, Jurin RR. (1995) The SHHF/Mcc-fa^{cp}: A genetic model of congestive heart failure. In *Mechanisms of Heart Failure*, ed. Singal PK, Beamish RE, Dhalla NS. Kluwer Academic, Boston, 91–106.

Molitch ME, DeFronzo RA, Franz MJ, et al. (2004) Nephropathy in diabetes. *Diabetes Care* 27:(Suppl 1), S79–S83.

Morse AD, Hunt TWM, Wood GO, Russell JC. (1995) Diurnal variation of intensive running in food-deprived rats. *Can J Physiol Pharmacol* 73:1519–1523.

O'Brien SF, McKendrick JD, Radomski MW, et al. (1998) Vascular wall reactivity in conductance and resistance arteries: Differential effects of insulin resistance. *Can J Physiol Pharmacol* 76:72–76.

O'Brien SF, Russell JC. (1997) Insulin resistance and vascular wall function: Lessons from animal models. *Endocrin Metab* 4:155–162.

O'Brien SF, Russell JC, Dolphin PJ, Davidge ST. (2000) Vascular wall function in insulin-resistant JCR:LA-cp rats: Role of male and female sex. *J Cardiov Pharmacol* 36:176–181.

O'Neill MS, Veves A, Zanobetti A, et al. (2005) Diabetes enhances vulnerability to particulate air pollution-associated impairment in vascular reactivity and endothelial function. *Circulation* 111:2913–2920.

Pederson RA, Campos RV, Buchan AMJ, et al. (1991) Comparison of the enteroinsular axis in two strains of obese rats: The fatty Zucker and JCR:LA-corpulent. *Int J Obes* 15:461–470.

Poiley SM. (1960) A systematic method of breeder rotation for noninbred laboratory animal colonies. *Anim Care Panel* 10:59–161.

Proctor SD, Dreher KL, Kelly SE, Russell JC. (2006) Hypersensitivity of prediabetic JCR:LA-*cp* rats to fine airborne combustion particle-induced direct and noradrenergic-mediated vascular contraction. *Toxicol Sci* 90:385–391.

Proctor SD, Kelly SE, Stanhope KL, et al. (2006) Synergistic effects of conjugated linoleic acid and chromium picolinate improve vascular function and renal pathophysiology in the insulin-resistant JCR:LA-*cp* rat. *Diabetes Obesity Metab*, in press.

Proctor SD, Vine DF, Mamo JCL. (2002) Arterial retention of apolipoprotein-B48 and B100-containing lipoproteins in atherogenesis. *Curr Opin Lipidol* 13:461–470.

Proctor SD, Vine D.F, Mamo JC. (2004) Arterial permeability and efflux of apolipoprotein B-containing lipoproteins assessed by *in situ* perfusion and three-dimensional quantitative confocal microscopy. *Arterioscler Thromb Vasc Biol* 24:2162–2167.

Recant L, Voyles NR, Timmers KI, et al. (1989) Comparison of insulin secretory patterns in obese nondiabetic LA/N-*cp* and obese diabetic SHR/N-*cp* rats. Role of hyperglycemia. *Diabetes* 38:691–697.

Ruben Z, Miller JE, Rohrbacher E, Walsh GM. (1984) A potential model for a human disease: spontaneous cardiomyopathy-congestive heart failure in SHR/N-*cp* rats. *Hum Pathol* 15:902–903.

Russell JC, Amy RM, Dolphin PJ. (1991) Effect of n-3 fatty acids on atherosclerosis prone JCR:LA-corpulent rats. *Exp Mol Pathol* 55:285–293.

Russell JC, Amy RM, Graham S, Dolphin PJ. (1993) Effect of castration on hyperlipidemic, insulin resistant JCR:LA-corpulent rats. *Atherosclerosis*100:113–122.

Russell JC, Amy RM, Graham SE, et al. (1995) Inhibition of atherosclerosis and myocardial lesions in the JCR:LA-*cp* rat by β-β′tetramethylhexadecanedioic acid (MEDICA 16). *Arterioscler Thromb Vasc Biol* 15:918–923.

Russell JC, Amy RM, Manickavel V, Dolphin PJ. (1989) Effects of chronic ethanol consumption in atherosclerosis-prone JCR:LA-corpulent rat. *Arteriosclerosis* 9:122–128.

Russell JC, Amy RM, Manickavel V, et al. (1987) Insulin resistance and impaired glucose tolerance in the atherosclerosis prone LA/N-corpulent rat. *Arteriosclerosis* 7:620–626.

Russell JC, Amy RM, Manickavel V, et al. (1989) Prevention of myocardial disease in the JCR:LA-corpulent rat by running. *J Appl Physiol* 66:1649–1655.

Russell JC, Amy RM, Michaelis OE IV, et al. (1991) Myocardial disease in corpulent strains of rats. In *Lessons from Animal Diabetes*, ed. Shafrir, E. Smith–Gordon, London, 3:402–407.

Russell JC, Bar-Tana J, Shillabeer G, et al. (1998) Development of insulin resistance in the JCR:LA-*cp* rat: Role of triacylglycerols and effects of MEDICA 16. *Diabetes* 47:770–778.

Russell JC, Dolphin PJ, Graham SE, et al. (1998) Improvement of insulin sensitivity and cardiovascular outcomes in the JCR:LA-*cp* rat by D-fenfluramine. *Diabetologia* 41:380–389.

Russell JC, Ewart HS, Kelly SE, et al. (2002) Improvement of vascular dysfunction in insulin resistance by a marine oil-based phytosterol compound. *Lipids* 37:147–152.

Russell JC, Graham SE, Amy RM, Dolphin PJ. (1998a) Cardioprotective effect of probucol in the atherosclerosis-prone JCR:LA-*cp* rat. *Eur J Pharmacol* 350:203–210.

Russell JC, Graham SE, Amy RM, Dolphin PJ. (1998b) Inhibition of myocardial lesions in the JCR:LA-corpulent rat by captopril. *J Cardiovasc Pharmacol* 31:971–977.

Russell JC, Graham SE, Dolphin PJ, et al. (1997) Antiatherogenic effects of long-term benfluorex treatment in male insulin resistant JCR:LA-*cp* rats. *Atherosclerosis* 132:187–197.

Russell JC, Graham SE, Dolphin PJ. (1999) Glucose tolerance and insulin resistance in the JCR:LA-*cp* rat: Effect of miglitol (Bay m1099). *Metabolism* 48:701–706.

Russell JC, Graham S, Hameed M. (1994) Abnormal insulin and glucose metabolism in the JCR:LA-corpulent rat. *Metabolism* 43:538–543.

Russell JC, Kelly SE, Schäfer S. (2004) Vasopeptidase inhibition improves insulin sensitivity and endothelial function in the JCR:LA-*cp* rat. *J Cardiovasc Pharmacol* 44:258–265.

Russell JC, Koeslag DG, Amy RM, Dolphin PJ. (1989) Plasma lipid secretion and clearance in the hyperlipidemic JCR:LA-corpulent rat. *Arteriosclerosis* 9:869–876.

Russell JC, Manickavel V, Koeslag DG, Amy RM. (1990) Effects of advancing age and severe food restriction on pathological processes in the insulin resistant JCR:LA-corpulent rat. *Diabetes Res* 15:53–62.

Russell JC, Ravel D, Pégorier J-P, et al. (2000) Beneficial insulin-sensitizing and vascular effects of S15261 in the insulin-resistant JCR:LA-*cp* rat. *J Pharm Exp Ther* 295:753–760.

Savage S, Estacio RO, Jeffers B, Schrier RW. (1996) Urinary albumin excretion as a predictor of diabetic retinopathy, neuropathy, and cardiovascular disease in NIDDM. *Diabetes Care* 19:1243–1248.

Schaefer EJ, McNamara JR, Shah PK, et al. (2002) Elevated remnant-like particle cholesterol and triglyceride levels in diabetic men and women in the Framingham Offspring Study. *Diabetes Care* 25:989–994.

Schneider DJ, Absher PM, Neimane D, et al. (1998) Fibrinolysis and atherogenesis in the JCR:LA-*cp* rat in relation to insulin and triglyceride concentrations in blood. *Diabetologia* 41:141–147.

Steiner G. (1994) Hyperinsulinemia and hypertriglyceridemia. *J Int Med* 736(suppl):23–26.

Triana RJ, Suits GW, Garrison S, et al. (1991) Inner ear damage secondary to diabetes mellitus. I. Changes in adolescent SHR/N-cp rats. *Arch Otolarynol* 117:635–640.

Velasquez MT, Abraham AA, Kimmel PL, et al. (1995) Diabetic glomerulopathy in the SHR/N-corpulent rat: Role of dietary carbohydrate in a model of NIDDM. *Diabetologia* 38:31–38.

Vine DF, Russell JC, Proctor SD. (2005) Impaired metabolism of chylomicrons and corresponding postprandial response can contribute to accelerated vascular disease in the JCR:LA *cp/cp* rat model of obesity and insulin resistance. *Can J Cardiol* 21(Suppl. C):A677.

Wascher TC, Wölkart G, Russell JC, Brunner F. (2000) Delayed insulin transport across endothelium in insulin resistant JCR:LA-*cp* rats. *Diabetes* 49:803–809.

Williams G, Cardoso H, Domin J, et al. (1990) Disturbances of regulatory peptides in the hypothalamus of the JCR:LA-corpulent rat. *Diabetes Res* 15:1–7.

Williams G, Shellard L, Lewis DA, et al. (1992) Hypothalamic neuropeptide Y disturbances in the obese *cp/cp* JCR:LA-corpulent rat. *Peptides* 13:537–540.

Wu-Peng XS, Chua SC Jr., Okada N, et al. (1997) Phenotype of the obese Koletsky (*f*) rat due to Tyr763Stop mutation in the extracellular domain of the leptin receptor (*Lepr*). *Diabetes* 46:513–518.

Yamini S, Carswell N, Michaelis OE IV, Szepesi B. (1992) Adaptation in enzyme (metabolic) pathways to obesity, carbohydrate diet and to the occurence of NIDDM in male and female SHR/N-*cp* rats. *Int J Obesity* 16:765–774.

Zucker LM, Zucker TF. (1961) Fatty, a new mutation in the rat. *J Heredity* 52:275–278.

8 The SHROB (Koletsky) Rat as a Model for Metabolic Syndrome

Richard J. Koletsky, Rodney A. Velliquette, and Paul Ernsberger

CONTENTS

SUMMARY

Human metabolic syndrome is characterized by insulin resistance, hypertension, abdominal obesity, hyperlipidemia (especially hypertriglyceridemia), and glucose intolerance in the absence of diabetes. The spontaneously hypertensive obese (SHROB) rat is a unique nondiabetic strain with all of the primary and many of the secondary characteristics associated with human metabolic syndrome. The obese phenotype results from a single recessive trait, a nonsense mutation affecting all forms of the leptin receptor, designated fa^k. The absence of hypothalamic leptin receptors leads to changes in neuropeptides that favor the development of obesity.

While only moderately hyperphagic, SHROB show enhanced lipogenesis and develop massive adipose depots that are equally distributed between subcutaneous and abdominal regions. Fasting glucose is unchanged relative to lean littermates, but recovery from a glucose challenge is delayed, suggesting reduced disposal of glucose. SHROB have fasting insulin levels 44-fold greater and glucagon levels 2-fold higher than SHR (lean) siblings. Insulin resistance in SHROB is due partly to excess glucagon, as well as to diminished expressions of insulin receptor and IRS-1 proteins with impairment of subsequent steps in the insulin signaling cascade, including the activation of protein kinase B (Akt). Despite multiple metabolic derangements, hypertension is not exacerbated in SHROB compared to SHR. Dietary manipulations and pharmacological interventions, including treatments of hypertension, diabetes, and hyperlipidemia, can modify the expressions of the various phenotypic components of metabolic syndrome. The SHROB serves as a useful model to test dietary and pharmacotherapeutic interventions in metabolic syndrome.

HISTORICAL BACKGROUND

The SHROB rat, originally designated the Koletsky rat strain, arose spontaneously in 1969 (Koletsky 1972). The founder was a female spontaneously hypertensive rat (SHR), descended from early breeding stock provided in 1968 to Case Western Reserve University in Cleveland, Ohio, by the National Heart Lung and Blood Institute (NHLBI) of the National Institutes of Health (NIH) in Bethesda, Maryland. The NHLBI colony of SHR was obtained as an inbred strain from Kyoto University around 1967, where the strain was first characterized beginning in 1963 (Okamoto and Aoki 1963). Since commercial breeders first obtained SHR breeders from the NIH in 1972 (Kurtz et al. 1989), the SHR/Kol colony from which the SHROB/Kol derived was genetically closer to the original Kyoto inbred strain than other commercially available SHR.

The founder female had been mated with a male Sprague–Dawley rat. The resulting hybrid offspring that remained hypertensive (systolic blood pressure > 150 mmHg) were inbred, and after several generations an abnormal phenotype was noted among some of the litters. These obese rats were not only hypertensive, like their lean littermates, but also showed additional phenotypic characteristics, including hyperlipidemia, hyperinsulinemia, and proteinuria with fulminant kidney disease (Koletsky and Ernsberger 1996).

Lean rats from the original litter of SHR/Sprague–Dawley hybrids that contained obese offspring then bred by brother–sister mating. Lean heterozygotes were identified by the production of obese offspring. Inbreeding within the same closed facility in the Animal Resource Center of Case Western Reserve University School of Medicine has continued from 1971 to the present day, comprising at least 60 generations of continuous inbreeding.

SHROB/Kol breeders were sent to the NIH in 1972, where they were backcrossed with other SHR strains and two normotensive strains (WKY/N and LA/N) in a complex scheme, resulting in at least five distinct strains (Greenhouse et al. 1990). The recessive trait was for some time designated as *cp* (corpulent). The hypertensive phenotype was lost from the backcrossed SHR/N-*cp*, with obese animals showing systolic blood pressures <140 mmHg. The other *cp* strains are also normotensive (Brindley and Russell 2002). The relationship of the genotypes and phenotypes of various hybrid strains to the parent SHROB/Kol colony is unknown (Koletsky 1975a).

The obese genotype involves both sexes and represents a homozygous recessive trait originally designated as *f* (Koletsky 1972) but later reclassified as fa^k (Koletsky and Ernsberger 1996; Takaya et al. 1996). Male and female SHROB are infertile, and the recessive fa^k/fa^k genotype can only be inherited when both parents are heterozygous and each carries the same recessive allele (Fa^k/fa^k). Lean SHR heterozygous carriers (Fa^k/Fa^k) are indistinguishable from homozygous wild-type littermates except through breeding of obese offspring. However, the consequences of heterozygosity for the fa^k trait have not been systematically investigated.

The fa^k mutation in the SHROB is a nonsense mutation of the leptin receptor gene, resulting in a premature stop codon in the leptin receptor extracellular domain at position 763 (Takaya et al. 1996). As a result of this additional stop codon, none of the leptin receptor isoforms in SHROB include a membrane-bound segment (Ishizuka et al. 1998). In contrast, the Zucker fatty (*fa*) rat has a missense mutation at position 269, which reduces its functionality (Yamashita et al. 1997). The mutation fa^k truncates all forms of the leptin receptor and eliminates all possible downstream events triggered by leptin.

Heterozygous Fa^k/fa^k breeders have been crossed with heterozygous *Fa/fa* breeders from a Zucker fatty rat colony. The resulting obese offspring (fa^k/fa) were slightly (3%) lighter than an SHROB and slightly (14%) heavier than a Zucker fatty rat (Yen et al. 1977). Blood pressure, lipid levels, and other metabolic abnormalities were never reported. The success of this SHROB/Zucker cross led some authors to suggest that the recessive traits causing obesity in the Zucker fatty rat and the SHROB were due to mutations at the same allelic site. The trait in the Zucker rat is known as *fa* while the SHROB became known as fa^k (Yen et al. 1977).

Differences have been reported between the Zucker fatty rat and the SHROB. The most notable is blood pressure. The SHROB is consistently hypertensive while the Zucker animal, according to most reports, is normotensive or at best has only slightly elevated blood pressure (Buñag and Barringer 1988). SHROB (Koletsky et al. 1995; Abramowsky et al. 1984; Koletsky 1975b) and Zucker fatty rats (Kasiske et al. 1985) develop renal disease. However, the renal disease is more rapid, severe, and extensive in SHROB, leading to mortality in SHROB but not in Zucker rats. A further difference is the lack of circadian rhythms in food and water intake and corticosterone excretion in Zucker rats (White et al. 1989) contrasted to the normal day–night variations in all these variables in SHROB rats (Koletsky and Ernsberger 1992).

Furthermore, SHROB are somewhat heavier than Zucker fatty rats, particularly females (17%) (Yen et al. 1977). These differences between Zucker fatty rats and SHROB may reflect similar mutations on differing genetic backgrounds (SHR vs. Sherman and Merck stock rats) or the effects of a nonsense mutation of the leptin receptor in SHROB versus a missense mutation in the Zucker. The Zucker diabetic fatty (ZDF) differs from standard Zucker fatty rats with regard to blood glucose; the ZDF (but not the Zucker) develops hyperglycemia and attendant complications (Friedman et al. 1991). Diabetes in the ZDF develops from a combination of obesity and susceptibility genes, possibly related to glucose transport (Friedman et al. 1991). The SHROB, despite being more obese than the ZDF or the standard Zucker, is highly resistant to the development of diabetes even when fed a diabetogenic diet (Ernsberger, Koletsky, Kline, et al. 1999). The JCR:LA-*cp* rat descended from the SHROB is also normoglycemic and insulin resistant (Brindley et al. 1992).

PHENOTYPIC FEATURES OF THE SHROB STRAIN

Obesity

SHROB weigh nearly the same as their lean littermates up through eight weeks of age (Koletsky et al. 2001; Koletsky and Ernsberger 1996). Two weeks earlier, SHROB show a rounded contour of the lower trunk, resembling the outline of a light bulb; this trait is the earliest to distinguish SHROB from lean SHR. From 8 to 16 weeks of age, SHROB gain weight rapidly (3–5 g/day). Mature SHROB routinely reach peak weights between 750 and 900 g. At high but stable weights, SHROB overeat relative to lean SHR littermates by only about 20%, despite a doubling or tripling of body weight (Velliquette et al. 2005). Thus, food intake per unit of body weight is actually reduced. Obese males are slightly but not significantly heavier than females at all ages. Lean SHR, in contrast, show a marked sex difference in body weight, as is typical for rodent species. Thus, SHROB females weigh about three times as much as SHR females, whereas SHROB males weigh twice as much as SHR males.

Food intake in female and male SHROB and SHR across the life span is illustrated in figure 8.1. Food intake is remarkably constant in lean SHR. Relative to these lean animals, the SHROB overeat by about 50%. There is a tendency for overeating to be greater early in life during the phase of rapid weight gain and then to decline during subsequent weight stability during maturity.

Massive obesity in the presence of relatively mild hyperphagia suggests a reduction in energy expenditure. Consistent with this concept is the reduced expression

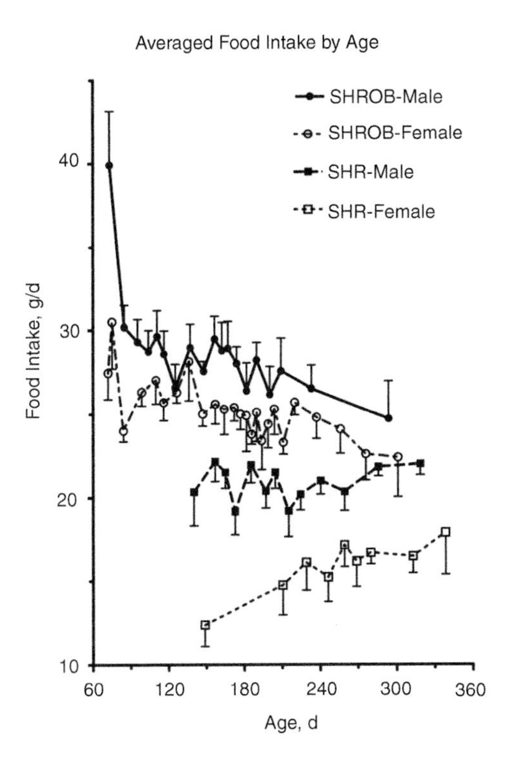

FIG. 8.1 Evolution of food intake across the life span in female and male SHROB/Kol and SHR/Kol rats. Food was weighed at least once a week in the home cage for 60–110 animals in each group of various ages. Standard soy-based diet was given (Ralston-Purina or Teklad).

of messenger RNA for uncoupling protein-1 in brown fat of SHROB relative to SHR (Rhinehart et al. 2004). Relative to Zucker obese rats, SHROB consume less oxygen and produce less carbon dioxide (Strohl et al. 1997). In addition, SHROB show less physical activity in their home cage (unpublished observations).

Enormous deposition of fat occurs throughout the body, most notably in the subscapular depot, which exceeds 30 g in SHROB, but is nearly absent in SHR littermates. Retroperitoneal deposits within the abdomen and the sex-specific epididymal and myometrial depots, as well as the mesenteric fat pads associated with the viscera, are enlarged by up to 19-fold in SHROB relative to SHR controls. Only the head, face, and distal portions of the fore and hind limbs lack massive fat deposits. At least some of the accumulated lipid is synthesized endogenously, since the animals are fed standard chow containing 5% fat by weight. Consistent with this notion, incorporation of labeled glucose into lipids in the heart, diaphragm, skeletal muscle, and adipose tissues and liver in SHROB is threefold that of SHR (O'Dea and Koletsky 1977).

We recently used magnetic resonance imaging (MRI) to map the distribution of abdominal visceral and subcutaneous adipose tissue in SHROB and SHR animals (Wan et al. 2005). In SHROB, the total volume of adipose tissue was 275 mL, of which 113 mL (41%) was visceral and 161 mL (59%) was subcutaneous. In lean

SHR, 22.8 mL of total adipose tissue could be divided into 13.7 mL (60%) of visceral and 9.1 mL (40%) of subcutaneous. Thus, even though the lean SHR have much less adipose tissue than SHROB, the ratio of visceral to subcutaneous fat is higher.

Body length is increased 11% in females with no change in males, implying an acceleration of linear growth in females (Koletsky et al. 2001). Females are shorter in length than males, for both SHROB and SHR. The Lee index, an established measure of rodent obesity, is greater in the obese genotype in both sexes, as is the body mass index (BMI, kilogram/square meter, using length in place of height) (Koletsky et al. 2001).

Hypertension

SHROB and their SHR littermates develop high blood pressure spontaneously before weaning around 30 days of age (fig. 8.2). In both SHROB and SHR, systolic blood pressure reaches hypertensive levels (>150 mmHg) prior to weaning (<28 days of age) and then rises progressively between two and four months. Hypertension is maintained until death. Direct mean arterial pressure (MAP) under urethane anesthesia (Ernsberger, Koletsky, Collins, et al. 1996; Ernsberger et al. 1994a) and radiotelemetry studies (Ernsberger and Rao 2000) confirm the tail cuff data. SHR

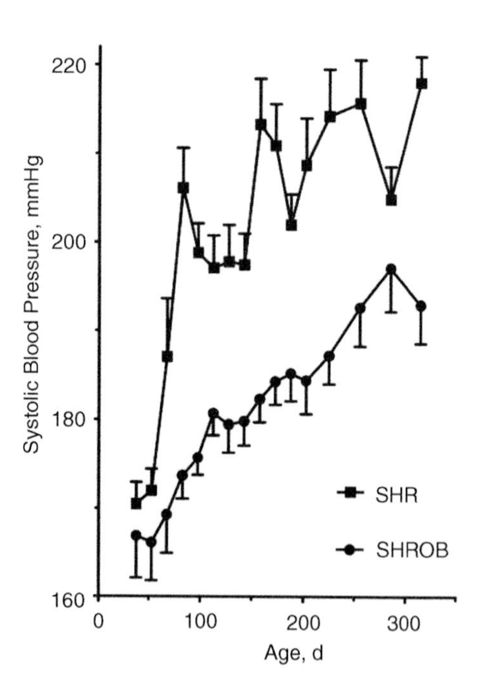

FIG. 8.2 Evolution of blood pressure across the life span in SHROB/Kol and SHR/Kol rats. Rats were placed in restrainers and systolic blood pressure was determined by a tail cuff device as previously described (Ernsberger et al. 1994a). Each point represents the mean ± standard error of 12–72 rats. Standard error bars are not visible on some points because they are smaller than the plot symbols.

males show significantly higher blood pressure than SHR females between two and eight months of age, whereas SHROB show no sex difference in blood pressure at any age (Koletsky et al. 2001). Thus, similar to findings in humans, severe obesity removes the protective effect of female gender on blood pressure.

Under basal conditions while maintained on regular chow *ad libitum*, the SHROB have slightly lower blood pressures than their lean littermates throughout the life span (fig. 8.2). Blood pressure rises more gradually with age in SHROB relative to SHR. This suggests that obesity does not contribute to hypertension in this model. Furthermore, caloric restriction and weight loss do not normalize blood pressure in SHROB and may even increase it (Ernsberger et al. 1994a; Koletsky and Puterman 1976). The reduced blood pressure of SHROB relative to SHR littermates probably reflects loss of the blood pressure elevating effects of the leptin receptor system. The abrogation of leptin signaling in the SHROB model may break the link between hypertension and obesity. Leptin is known to activate sympathetic nervous system activity and thus increase vasoconstriction (Mark et al. 1999). High levels of leptin in leptin-resistant states where signaling pathways are intact may account for the link between obesity and hypertension.

The mechanisms responsible for hypertension in SHROB and their lean littermates are polygenic in origin, as evidenced by multiple quantitative trait loci distributed across many chromosomes (Schork et al. 1995). The SHROB model permits the study of how hypertension-promoting genes are modified by the presence of massive obesity (Koletsky and Ernsberger 1992).

Cardiac Hypertrophy

Left ventricular wall thickness is similar in SHROB and SHR. Heart weight is not significantly affected by the obese genotype in male rats, while obese female rats had heavier hearts than their lean sisters. These changes are likely to reflect cardiac hypertrophy. Atrial natriuretic peptide levels in plasma are higher in SHROB (1595 ± 371 pg/mL [Mukaddam-Daher et al. 2003]) than in comparable lean controls. High levels of atrial natriuretic peptide are a potential marker of myocardial growth and may be a harbinger of cardiac dysfunction.

Hyperlipidemia

SHROB of either sex uniformly develop hyperlipidemia, characterized by markedly elevated plasma triglycerides (TG) and a moderate rise in plasma cholesterol relative to age-matched SHR (Velliquette et al. 2006; Koletsky et al. 2001; Ernsberger, Koletsky, and Friedman 1999; Koletsky and Ernsberger 1996; Koletsky 1975b). Serum lipid levels in SHROB are elevated as early as five weeks of age and continue to rise throughout life until the last few weeks of life, when the animals are in a terminal decline (Koletsky, 1975b). TG and cholesterol values are not elevated in lean littermates of either sex. Lipid profiles are not different between sexes in either group. JCR:LA-*cp* rats descended from SHROB show a similar hypertriglycerdemic syndrome (Brindley and Russell 2002). Possible mechanisms for elevated cholesterol in SHROB included increased hepatic synthesis (Tan et al. 1976; Velliquette et al.

2006). TG levels decline during weight loss on a very low-calorie diet and promptly rise again during regain of lost weight (Ernsberger et al. 2005).

We recently investigated TG synthesis and secretion into the plasma by using tyloxapol, an inhibitor of plasma TG breakdown by lipoprotein lipase (Velliquette et al. 2006). After an overnight fast, TG were secreted into the plasma at a very high rate, with levels rising nearly to 800 mg/dL over the course of three hours. Stable isotope studies with deuterated water confirmed that a significant fraction of plasma TG was derived from *de novo* synthesis of fatty acids in the form of palmitate (Velliquette et al. 2006). These results are consistent with previous reports that tritiated water incorporation into liver and adipose lipid fractions is increased several fold in SHROB relative to SHR (O'Dea and Koletsky 1977). These data suggest that hepatic and/or adipose lipogenesis is markedly increased in SHROB, even in the fasted state, and newly synthesized fatty acids contribute to excess plasma TG, although re-esterification is likely to contribute the bulk of circulating TG.

Free fatty acids (FFA) following an overnight fast are elevated by only about 25% in SHROB relative to SHR (1.81 ± 0.09, $n = 50$) compared with (1.45 ± 0.05, $n = 51$). FFA levels in SHROB and SHR are more than double the levels typically found in normotensive control rat strains (Velliquette et al. 2002). The high FFA levels in SHROB and SHR may reflect a mutation in the CD36 gene, a fatty acid transporter molecule, which has been identified in SHR (Aitman et al. 1999).

Although plasma FFA is only marginally elevated in SHROB relative to SHR, SHROB show an abnormal regulation of FFA levels following a glucose challenge. FFA levels fell at 30 and 60 min after a glucose challenge in SHR, as expected from the switch from a fasted to fed state (Velliquette et al. 2002). In contrast, FFA levels fail to decline following a glucose load in SHROB. This failure to suppress circulating FFA in response to a test meal has been reported in humans with type 2 diabetes (Iannello et al. 1998) and is thought to reflect adipocyte insulin resistance.

Glucose Intolerance

SHROB and SHR show normal levels of fasting glucose. Following a glucose challenge, SHROB show high glucose levels that return toward fasting levels very slowly (fig. 8.3). Oral glucose tolerance testing in SHROB demonstrated a sustained postchallenge elevation in circulating glucose from 60 to 360 min compared to lean SHR littermates. The area under the curve for glucose was far greater for SHROB than for SHR, suggestive of decreased rate of glucose disposal. Similar results have been obtained for the JCR:LA-*cp* model descended from the SHROB (Brindley and Russell 2002).

A very low-calorie diet actually causes an acute further deterioration of glucose tolerance, followed by a sustained improvement of glucose tolerance during regain of lost weight, apparently from increased glucose disposal (Ernsberger et al. 2005).

Glycated hemoglobin reflects fasting and postprandial glucose control over time. Despite normal fasting glucose, glycated hemoglobin was significantly increased in the plasma SHROB at six to eight months of age ($3.91 \pm 0.21\%$; $n = 13$) versus matched SHR controls ($3.62 \pm 0.13\%$; $n = 10$; $P < 0.05$, t-test). Although statistically

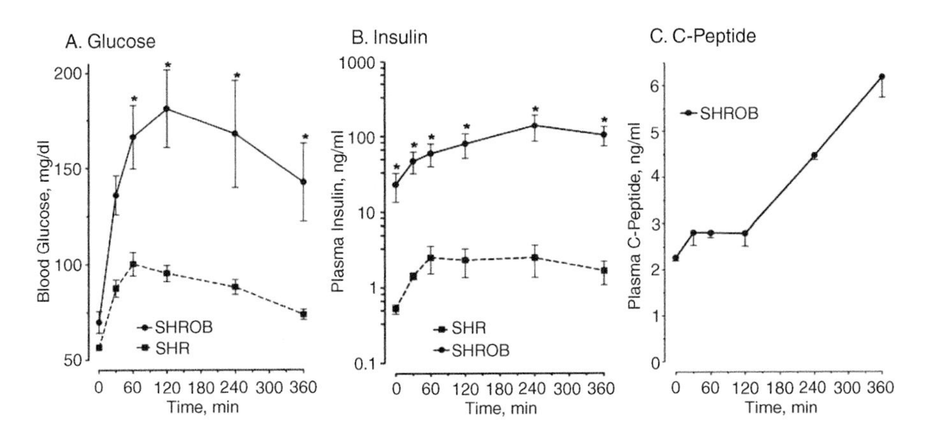

FIG. 8.3 Plasma glucose (A) and insulin (B) during oral glucose tolerance testing in SHROB/Kol and SHR/Kol rats, and C-peptide (C) in SHROB. Rats four to six months of age were fasted for 18 h and given a 6-g/kg glucose load by gavage with a feeding tube at time 0, then blood samples (0.2 mL) were taken at regular intervals. Glucose was assayed plasma enzymatically by glucose oxidase (sigma). Insulin and C-peptide were measured by radioimmunoassay using rat standards and an antibody directed against rat hormone (Linco, St. Charles, Illinois).

significant, a net increase of less than one part in ten probably has no physiological importance.

Kidney Disease

SHROB develop spontaneous glomerulopathy with proteinuria when maintained on a soy-based rat chow (Koletsky et al. 1995; Ernsberger et al. 1993; Abramowsky et al. 1984; Koletsky 1975b). Proteinuria is detected as early as six weeks of age and accelerates exponentially. By six months of age, SHROB exhibit severe proteinuria (Ernsberger, Ishizuka, et al. 1999; Ernsberger, Koletsky, Collins, et al. 1996; Koletsky et al. 1995). SHR do not exhibit significant proteinuria or kidney disease, despite having even more severe hypertension than SHROB. Raising blood pressure via weight cycling or via a high-salt diet (4% NaCl) exacerbates kidney disease in SHROB (Koletsky et al. 1995). Proteinuria increased in SHR fed a high-salt diet, but remained 10-fold lower than SHROB. Morphologically, the kidneys of SHROB show focal segmental glomerulosclerosis and nephrosclerosis, resembling human diabetic and hypertensive glomerular and vascular damage. In spite of the proteinuria, SHROB show increases in creatinine and blood urea nitrogen only in the last few weeks of life.

A possible mediator of renal pathology in the SHROB is angiotensin II because receptor sites for this hormone are down-regulated in SHROB with early renal pathology, consistent with overactivity of the renin-angiotensin system (Ernsberger et al. 1993). Angiotensin II may also contribute to kidney damage caused by diabetes in humans, contributing to the nephroprotective effects of angiotensin II inhibitors (Kasiske et al. 1993).

Vascular Disease

Originally, a proportion of SHROB developed vascular disease characterized by lesions in the mesenteric arteries (Koletsky 1975b). Vessels showed focal or diffuse nodular thickening, beading, and tortuosity with aneurysms and thrombosis. Occasional myocardial necrosis was observed. These pathologic findings were no longer apparent in generations of rats bred after 1980. The genetic defects leading to obesity and hypertension may be separate from the mutation that caused atherosclerosis on the background of these disorders.

The disappearance of the vascular disease may have been caused by a spontaneous mutation or reversion to wild type. Interestingly, one of the many substrains derived from the SHROB breeding stock sent to the NIH in 1972 still appears to develop vascular and myocardial lesions similar to those described in the original SHROB model (Russel and Koeslag 1990). JCR:LA-*cp* rats have the fa^k mutation on a background that includes SHR and LA/N components. In addition, the JCR:LA-*cp* rat shows elevated levels of plasminogen activator inhibitor (PAI-1), a thrombogenic marker linked to human metabolic syndrome (Brindley and Russell 2002). Identification of the genetic differences between JCR:LA-*cp* and SHROB rats may reveal genes contributing to atherosclerosis.

ENDOCRINE FUNCTION

Leptin

Circulating levels of the adipose tissue hormone leptin are elevated 170-fold in the plasma of the SHROB rat relative to SHR littermates (Friedman et al. 1997a). This presumably reflects the profound leptin resistance resulting from the complete absence of functional leptin receptors. Remarkably, the expression of leptin mRNA in adipose tissue is only elevated threefold. This implies that the clearance of leptin from the circulation may be impaired in SHROB rats.

In addition to mediating the hormonal actions of leptin, the leptin receptor may mediate leptin transport from the circulation into the cerebrospinal fluid (Lynn et al. 1996). If so, then SHROB should have negligible levels of leptin in their cerebrospinal fluid. However, the levels of leptin in cerebrospinal fluid samples from SHROB and SHR were similar (Ishizuka et al. 1998). Thus, despite a lack of leptin receptors, leptin was able to penetrate the blood–brain barrier in SHROB (Kastin et al. 1999). This implies a separate transporter for leptin independent of the leptin receptor. However, the ratio of plasma to cerbrospinal fluid leptin was far lower in SHROB, implying impaired leptin transport into the brain.

Hypothalamus

SHROB show changes in hypothalamic neuropeptides that favor the development of obesity. These changes are consistent with the absence of any action of leptin, owing to a knockout of all forms of the leptin receptor. Thus, neuropeptide Y and agouti-related peptide are upregulated, while the expression of CART (cocaine and amphetamine regulated transcript) and POMC (proopiomelanocortin) messages were reduced (Rhinehart et al. 2004). Abnormally low levels of leutenizing hormone releasing hormone (LHRH) were also observed, consistent with the infertility found in SHROB of both sexes (Rhinehart et al. 2004).

Hypothalamic regulation of body temperature is also abnormal in SHROB. Induction of fever by bacterial pyrogen under reduced ambient temperature (22°C) impaired the body temperature response and delayed recovery (Steiner et al. 2004). Furthermore, the hypothermic response to tumor necrosis factor alpha is prolonged. Central regulation of breathing is also altered in SHROB, with differences in the ventillatory responses to reduced oxygen and elevated carbon dioxide (Strohl et al. 1997).

Hypothalamic–Pituitary–Adrenal Axis

SHROB produce twice as much corticosterone as lean littermates while maintaining diurnal rhythm (Koletsky and Ernsberger 1992). Histologically, the adrenal zona fasciculata that produces corticosterone is enlarged in the SHROB (Koletsky 1975b). The production of 18-hydroxy steroids with mineralocorticoid activity is increased (Angelin and Karlmar 1979). A higher dose of the synthetic glucocorticoid dexamethasone is needed in SHROB than in SHR to suppress endogenous corticosterone production (Koletsky and Ernsberger 1992). The SHROB has normal diurnal rhythms in urinary corticosterone, urine volume, food intake, and water consumption, but the amplitude of each rhythm is greater than for their SHR littermates. This partial profile of the hypothalamic–pituitary–adrenal axis suggests a setpoint abnormality at the level of the hypothalamus or pituitary gland (Koletsky et al. 1982). In addition, the response of ACTH and corticosterone to bacterial pyrogen is blunted in SHROB relative to SHR (Steiner et al. 2004), further suggesting abnormal hypothalamic control of pituitary and adrenal function, presumably resulting from an absence of leptin receptors.

Insulin and Insulin Signaling

Pancreatic islets are greatly enlarged in SHROB compared with SHR littermates (Koletsky 1975b). Fasting insulin levels are elevated 44-fold in SHROB in the presence of normal fasting glucose (fig. 8.3). In response to an oral glucose challenge, insulin rose markedly in SHROB (6.0-fold) and SHR (4.8-fold) for up to 360 min, which is apparent even when plotted on a logarithmic scale (fig. 8.3). To assess the rate of endogenous insulin secretion in SHROB, we assayed circulating levels of C-peptide. Levels of C-peptide during fasting are very high in SHROB and rise to a similar extent as insulin in response to an oral glucose load (fig. 8.3). This implies that enhanced insulin secretion, rather than reduced insulin turnover, is the primary contributor to hyperinsulinemia.

The rate of insulin-stimulated 3-O-methylglucose transport was reduced by 68% in isolated epitrochlearis muscles from the SHROB compared to SHR. Insulin-stimulated tyrosine phosphorylation of the insulin receptor β-subunit and IRS-1 in intact skeletal muscle of SHROB were reduced by 36 and 23%, respectively, compared with SHR, due primarily to 32 and 60% decreases in insulin receptor and IRS-1 protein expression, respectively. The levels of p85 regulatory subunit of phosphatidylinositol-3-kinase and the glucose transporter GLUT4 were reduced by 28 and 25% in SHROB muscle compared to SHR. In the liver of SHROB, the ability of insulin to induce tyrosine phosphorylation of IRS-1 was not changed, but insulin receptor phosphorylation was decreased by 41% compared with SHR, due to a 30% reduction in insulin receptor levels (Ernsberger, Ishizuka, et al. 1999; Ishizuka et al. 1998; Friedman et al. 1997a, 1997b, 1998; Ernsberger, Koletsky, Collins, et al. 1996).

Insulin-stimulated phosphorylation of tyrosine residues on the insulin receptor and on the associated docking protein IRS-1 are reduced in skeletal muscle and liver compared with SHR, due mainly to diminished expression of insulin receptor and IRS-1 proteins (Friedman et al. 1997b). Reduced expression of IRS-1 may in particular represent part of the molecular basis for insulin resistance in this model.

Recent data have identified a prominent defect in the insulin signaling cascade downstream of the insulin receptor and IRS-1. Insulin activation of protein kinase B, also known as Akt, is profoundly impaired in adipocytes freshly isolated from SHROB relative to those from SHR littermates (Sun and Ernsberger 2005). The peak activation of protein kinase B after 10 min of treatment with 100 n*M* insulin was fourfold in SHROB compared with ninefold in SHR. Also consistent with reduced insulin sensitivity, the ED_{50} for insulin was higher in SHROB than in SHR (29 ± 3.8 vs. 3.5 ± 0.5 n*M*). Thus, the protein kinase B activation step in the insulin signaling pathway in adipocytes from SHROB shows profound resistance to activation by insulin, suggesting that this step could be an important indicator or target for the treatment of insulin resistance.

Glucagon

Glucagon is a key hormone opposing the actions of insulin by promoting hepatic glycogenolysis and gluconeogenesis (Unger 1971). The molar ratio insulin/glucagon is an indicator of the relative metabolic states of catabolism versus anabolism (Unger 1971). The SHROB have markedly elevated fasting glucagon levels relative to lean SHR littermates (137 ± 13 vs. 80 ± 4 pg/dL, $n = 20$ for both, $P < 0.01$) (Velliquette et al. 2002). The fasting I/G ratio is highly elevated in the SHROB relative to SHR (406 ± 127 vs. 16 ± 4.9, $P < 0.01$).

A glucose load is expected to lower glucagon while elevating the opposing hormone, insulin. In contrast, SHROB respond to an oral glucose challenge with a 560% increase of glucagon at 60 min after the oral glucose challenge, whereas only a modest rise of less than one-third was seen among the lean SHR (fig. 8.4). Glucagon levels only began to decline after 120 min postchallenge. The I/G molar ratio changed in opposite directions in response to an oral glucose challenge in the two genotypes, with a decrease in SHROB at 30 and 60 min, in contrast to the physiologically appropriate increase at 30 and 60 min postchallenge in lean SHR (Velliquette et al. 2002).

The 72% elevation in fasting glucagon in the SHROB raises the question whether this hormonal abnormality has any impact on glucose homeostasis. To test this hypothesis, we injected fasted lean SHR with a low dose of glucagon and followed

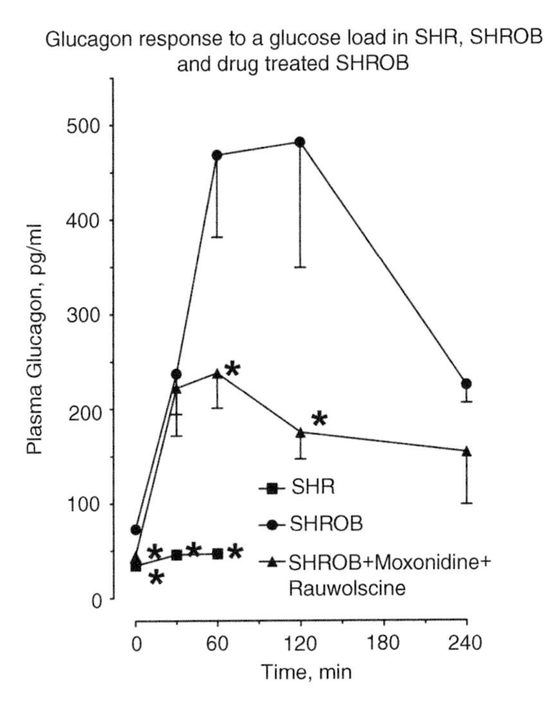

FIG. 8.4 Plasma glucagon responses to an oral glucose challenge in SHROB, SHR, and SHROB treated with moxonidine (0.5 mg/kg) in combination with rauwolscine (7.5 mg/kg) to block α_2-adrenergic receptors. A glucose load (6 g/kg) was given by oral gavage at time 0. Blood samples (0.2 mL) were obtained from the tail at baseline and at the times indicated.

the evolution of glucose and insulin levels. We reasoned that if the higher glucagon levels in SHROB contributed significantly to their insulin resistance, then raising glucagon levels in lean SHR should reproduce a portion of the insulin resistance seen in SHROB. In response to a low dose of glucagon sufficient to raise glucagon levels to those seen in SHROB, SHR showed sharp elevations in glucose and insulin within 30 min and remained elevated throughout the seven-hour observation period (Velliquette et al. 2002). Plasma FFA levels were, however, unchanged after the glucagon injection. These data suggest that at least part of the insulin resistance in SHROB is contributed by excess glucagon. In support of this hypothesis, pharmacological intervention with moxonidine normalized glucagon levels and also normalized glucose intolerance in SHROB (Koletsky et al. 2003; Velliquette and Ernsberger 2003a). The SHROB may be a model for glucagon-induced insulin resistance.

Reproductive Function

Male and female SHROB have consistently failed to produce offspring when mated with each other or with their SHR siblings, even when male SHROB have been treated with daily injections of testosterone (unpublished observations). Microscopic study of gonads demonstrated decreased levels of spermatogenesis or fewer mature ovarian follicles with ova (Koletsky 1975b). The absence of leptin receptors in SHROB probably accounts for their functional sterility, given the importance of leptin in reproduction (Rhinehart et al. 2004; Chehab et al. 1996).

RETINAL ABNORMALITIES

The blood vessels of the retina have been characterized in SHROB and in SHR littermates by fundus photography and fluorescein angiography under ketamine anesthesia (Huang et al. 1995). Retinal whole mounts were photographed by using Ektachrome ASA1600 film and a Leitz fluorescence microscope. The SHROB retina shows signs of neovascularization and progressive capillary dropout (Khosrof et al. 1995). At three months of age, SHROB show arteriolar sclerosis, mild dilatation of larger vessels, increased capillary tortuosity, and formation of primitive vascular tufts (Benetz et al. 1996). By nine months, SHROB demonstrated marked capillary dropout, extreme capillary tortuosity, and vascular abnormalities that leaked fluorescein and extended above the surface of the retina. Microaneurysms and retinal hemorrhages were not observed. Elastase digestion of retinal mounts confirmed the presence of ghost vessels, several vascular compromises, and looping. Pericyte and endothelial abnormalities were also present. This retinopathy occurred in the absence of hyperglycemia, implicating other factors in its pathogenesis, such as hypertension, hyperlipidemia, and hyperinsulinemia (Benetz et al. 1996; Huang et al. 1995; Khosrof et al. 1995).

LIFE SPAN AND CAUSE OF DEATH

SHROB rats live a little more than half as long as their SHR siblings (fig. 8.5). SHROB die predictably within a short time frame between 7 and 13 months of age, while their SHR siblings show <5% mortality at these ages. Median life expectancy (MLE) does not differ by sex in SHROB or SHR. The shortened life span of SHROB is primarily a result of kidney disease, protein wasting, and, ultimately, renal failure. The terminal course extends several weeks and is characterized by marked anorexia and weight loss ending with a failure to consume fluids and dehydration.

Dietary protein restriction is an established treatment for renal failure in humans (Mitch, 1991). In a pilot study, we tested the effects of reducing dietary protein by one-half, from 24% of calories in regular chow to 12% of calories. Sucrose contributed to 50% of calories. This diet is designated low protein, high sucrose (LPHS) in figure 8.5. Median life expectancy was extended by nearly 100 days by the LPHS diet, a highly significant effect ($P < 0.005$ by chi-square analysis). Dietary protein restriction delayed the onset of gross symptoms of renal failure.

USE OF SHROB RAT AS AN EXPERIMENTAL MODEL

Metabolic syndrome consists of insulin resistance as a primary defect associated with compensatory hyperinsulinemia, impaired glucose tolerance, dyslipidemia, and hypertension (Zavaroni et al. 1994). Obesity is often present. A very similar constellation of abnormalities is found in SHROB. The relationships between the diverse phenotypic features of human metabolic syndrome are unknown. The use of SHROB

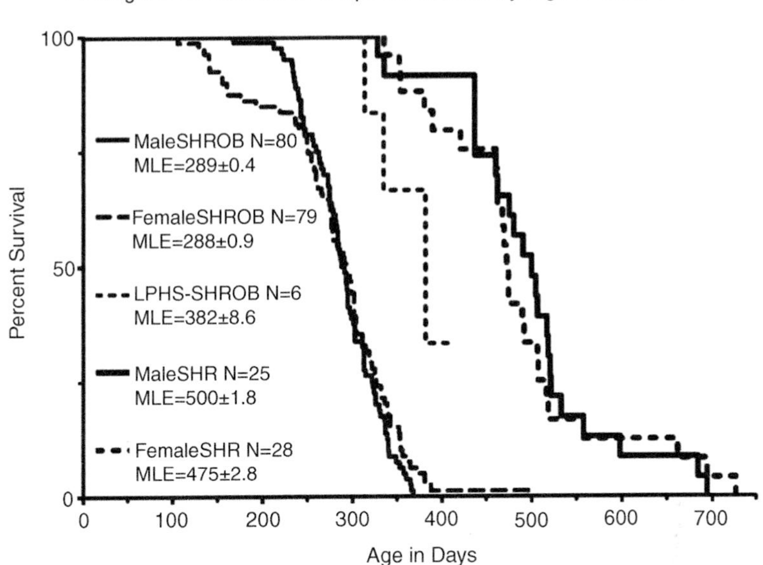

FIG. 8.5 Survival curves and median life expectancy for female and male SHROB. Age at death was determined in the colony over a period of three years. Median life expectancy is estimated by the life table method using Prism 4.0 (GraphPad Software). (Motulsky, HJ and Ransnas, LA. 1987. *FASEB J* 1:365–374.)

as a model might provide information on the interactions between the various components of the syndrome.

Dietary Manipulations

Low-Calorie Diet

Weanling SHROB, restricted to one-third of their usual consumption, lost 30% of their body weight but remained obese (Koletsky and Puterman 1976, 1977). Plasma cholesterol and triglyceride levels nearly normalized, but, surprisingly, blood pressure rose significantly during caloric restriction. Proteinuria and renal disease were significantly reduced. Fasting insulin fell in calorically restricted SHROB but still exceeded levels in SHR (O'Dea and Koletsky 1977). Lipogenesis was reduced but remained elevated in calorically restricted SHROB, particularly in adipose tissue. Glycogen synthesis was increased in most tissues after caloric restriction. Thus, caloric restriction failed to correct all of the metabolic abnormalities in SHROB, suggesting intrinsic deficits, but the life span of SHROB was doubled.

Weight Cycling

Obese humans commonly alternate between dietary restriction and bingeing and show repeated cycles of weight loss and regain (Folsom et al. 1996). The SHROB rat has been used to model this "yo-yo syndrome" by alternating a very low-calorie diet (VLCD) with *ad libitum* refeeding (Ernsberger et al. 1994b; Ernsberger, Koletsky, Baskin, et al. 1996; Ernsberger et al. 1998, 2005; Koletsky et al. 1995; Ernsberger and Koletsky 1993). The fluctuations of body weight have deleterious cardiovascular effects and increase mortality by unknown mechanisms. SHROB were subjected to one or more cycles of VLCD followed by regain of lost weight after restoration of free access to food. SHROB lost weight on the restricted diet, but subsequently regained more than was lost despite similar total overall cumulative food intakes as *ad libitum* fed SHROB. Excess fat accumulated selectively within abdominal depots. Cycled animals lost less weight after each succeeding cycle and metabolic efficiency was enhanced. Blood pressure in weight-cycled obese animals increased by 27 mmHg beyond baseline levels of *ad libitum* fed SHROB (Ernsberger et al. 1994a). In association with the rise in blood pressure, urinary protein excretion rose, and renal damage was enhanced (Koletsky and Ernsberger 1996).

The likely mechanism for the rise in blood pressure during weight regain is increased sympathetic activity. The evidence implicating the sympathetic nervous system in refeeding hypertension includes down-regulation of cardiac β-adrenergic receptors (Ernsberger et al. 1994a) and kidney α_2-adrenergic receptors (Ernsberger, Koletsky, Kline, et al. 1999) in weight-cycled SHROB. Furthermore, ganglionic blockade selectively abolished the elevation in blood pressure (Ernsberger et al. 1994a; Ernsberger, Koletsky, Baskin, et al. 1996). Moreover, urinary catecholamine levels were elevated during the weight-regaining phase of weight cycling and declined during caloric restriction (Ernsberger, Koletsky, Kline, et al. 1999).

We recently examined the impact of a single cycle or three cycles of weight loss and regain on glucose and insulin metabolism. Glucose tolerance was impaired immediately after a 14-day VLCD (Ernsberger et al. 2005). During and following regain of lost weight, glucose tolerance was actually improved and insulin resistance reduced as indicated by lower fasting insulin levels. Areas under the curve during oral glucose tolerance tests were reduced for glucose and insulin in weight-cycled animals relative to weight-stable controls, despite identical body weights. Weight regain may create an anabolic state with increased disposal of glucose. Beneficial effects of weight loss in the SHROB model persist following regain of the lost weight.

High-Salt Diet

Elevated intake of salt can increase blood pressure in some SHR strains and some humans (Jin et al. 1989). After two weeks on 4% NaCl, SHROB showed a 47 mmHg increase in tail cuff blood pressure, while SHR showed a smaller 22 mmHg rise (Ernsberger, Koletsky, and Collins 1994). Blood pressure remained elevated in SHROB at the same level for at least seven weeks on high salt. The hypertensive effect of dietary salt was confirmed by direct mean blood pressure under anesthesia. This difference was abolished by ganglionic blockade, suggesting a role for the sympathoadrenal system in promoting salt sensitivity in SHROB. Other laboratories have implicated the sympathetic nervous system in the response in SHR to excess dietary salt (Wyss et al. 1992).

SHROB fed the high-salt diet showed nearly threefold higher excretion of protein in the urine and greater renal damage relative to SHROB fed regular chow (Ernsberger, Koletsky, Collins, et al. 1996; Koletsky and Ernsberger 1996; Koletsky et al. 1995). At autopsy, the size of the heart and the thickness of the left ventricular wall were increased by dietary salt loading in SHROB, but not in SHR. On the high-salt diet SHROB suffered 57% mortality compared with 4% mortality on regular chow. The cause of death was uncertain, but renal failure, stroke, and cardiac death may all contribute. SHROB are markedly responsive to dietary NaCl and represent a novel model of salt-sensitive obese hypertension.

High-Sucrose Diet

Excess dietary sucrose is thought to exacerbate the metabolic components of metabolic syndrome in humans (Young and Landsberg 1982). In some obese rat models, excess dietary sucrose can elicit or exacerbate diabetes (Velasquez et al. 1995). SHROB placed on a 60% sucrose diet did not gain additional weight or change their fat distribution (Ernsberger, Koletsky, Kline, et al. 1999). Fasting glucose levels were actually slightly lower with sucrose feeding, yet glucose intolerance was markedly exacerbated. Fasting insulin and the insulin response to glucose challenge were further increased from the elevated levels found in SHROB relative to SHR. Thus, a high-sucrose diet is not diabetogenic in the SHROB model, but the abnormalities associated with metabolic syndrome are further exaggerated. As noted earlier, a high-sucrose diet that is also lower in protein may extend life expectancy in SHROB (fig. 8.5).

Feeding a high-sucrose supplement to SHROB in a liquid form also failed to cause weight gain or other metabolic changes (Johnson et al. 2005). Feeding the identical supplement to lean SHR littermates induced weight gains of up to 25%. Thus, SHROB may be resistant to dietary obesity.

RESPONSE TO PHARMACOTHERAPY

Antihypertensive Drugs

Treatment of SHROB with a combination of hydralazine, hydrochlorothiazide and propranolol eliminated hypertension (Koletsky and Snajdar 1979). This treatment reduced incidence of vascular disease and prolonged life span despite a lack of effect on blood lipid levels. When the rats were treated with antihypertensive drugs and simultaneously maintained on a low-calorie diet, which alleviated the hyperlipidemia, the result was complete elimination of vascular disease and increase in longevity to that of a normal rat. This supports the view that high blood pressure and hyperlipidemia both contribute to cardiovascular and renal damage and treatment of either abnormality is beneficial, but treatment of both confers even greater benefits.

The ganglionic blocker chlorisondamine was administered to SHROB and SHR to eliminate the sympathoadrenal contribution to maintenance of blood pressure. The lower blood pressure associated with the obese genotype persisted following ganglionic blockade, suggesting that obesity per se and the sympathetic nervous system are not the causes of hypertension in this model (Koletsky and Ernsberger 1992, 1996).

Treatment of SHROB with the direct vasodilator hydralazine alone lowered blood pressure but exacerbated several aspects of metabolic syndrome, including worsening of glucose tolerance, suppression of insulin secretion in response to a glucose load, and slightly increasing glucagon secretion during glucose loading (Velliquette and Ernsberger 2003a). Treatment of SHROB with the centrally acting sympatholytic agent α-methyldopa controlled hypertension to a similar extent as moxonidine and hydralazine but worsened glucose tolerance and impaired insulin secretion in response to a glucose load. The most dramatic effect of α-methyldopa was a profound exaggeration of the inappropriate glucagon secretion in response to a glucose load (Velliquette and Ernsberger 2003a). Thus, some antihypertensive agents may have unfavorable metabolic effects while others have primarily beneficial actions.

Moxonidine Treatment

Moxonidine is a centrally acting selective I_1-imidazoline receptor agonist that inhibits sympathetic activity. Moxonidine effectively reduced blood pressure to a similar extent in SHROB and SHR (Friedman et al. 1998) consistent with human data showing the effectiveness of this agent in obese humans (Doggrell 2001). Moxonidine treatment reduced fasting insulin levels by 71% in SHROB and lowered plasma FFA by 25%. In SHR, moxonidine treatment did not affect insulin levels but decreased FFA by 17%. During an oral glucose tolerance test, blood glucose levels in moxonidine-treated SHROB were reduced relative to untreated SHROB from 60 min onwards. Insulin secretion was facilitated at 30 (83% greater) and 60 min (67% greater) postchallenge compared with control SHROB. In skeletal muscle, moxonidine treatment increased the expression of the insulin receptor protein by 19% in SHROB, but was without effect in SHR. The level of IRS-1 protein was decreased by 60% in untreated SHROB compared with SHR.

Moxonidine treatment enhanced the expression and insulin-stimulated phosphorylation of IRS-1 protein in skeletal muscle by 74 and 27%, respectively, in SHROB, and by 40 and 56% in SHR. Moxonidine increased the levels of expression of IRS-1 protein in liver by 275% in SHR and 260% in SHROB, respectively. These findings indicate that chronic inhibition of sympathetic activity with moxonidine therapy can lower FFA and significantly improve insulin secretion, glucose disposal, and expression of key insulin signaling intermediates in an animal model of obese hypertension.

We have recently shown that moxonidine treatment lowers TG and cholesterol levels in SHROB, whether given acutely (Velliquette et al. 2006) or chronically (Velliquette and Ernsberger 2003a; Ernsberger, Koletsky, Collins, et al. 1996). This effect appears to be mediated by a direct action on imidazoline receptors in the liver to reduce the assembly of plasma TG and the synthesis of fatty acids.

As shown in figure 8.4, treatment with the imidazoline agonist moxonidine lowered fasting glucagon levels in SHROB and essentially normalized fasting hyperglucagonemia. In this experiment, the α_2-adrenergic blocker rauwolscine was administered to block opposing actions and isolate effects specific to imidazoline receptors (Velliquette and Ernsberger 2003b). Following moxonidine therapy, the paradoxical increase in glucagon elicited by a glucose load was still present, but reduced by one-half (fig. 8.4). This suggests that imidazoline agonists may be useful in countering syndromes of excess glucagon secretion. Lowering glucagon secretion may be useful in correcting metabolic syndrome.

CONCLUSION

The SHROB rat is a unique animal model expressing multiple abnormal phenotypic features, including genetic obesity, spontaneous hypertension, hyperinsulinemia, and hyperlipidemia. These features closely resemble those found in the human metabolic syndrome. The SHROB also has a spontaneous and progressive nephrotic syndrome, which is a potential model for human diabetic and hypertensive nephropathies as well as a retinopathy that resembles diabetes. Overactivity of the SNS may be a common link between hypertension and insulin resistance. We have used dietary modifications and pharmacologic interventions to study the physiologic changes that alter the various phenotypic components of metabolic syndrome. Further studies directed at understanding the mechanisms and interactions between the multiple abnormal phenotypes and their underlying genotypes will lead to better understanding of these commonly intertwined clinical problems.

AVAILABILITY

SHROB/Kol and SHR/Kol littermates are commercially available exclusively from Charles River Laboratories in Wilmington, Massachusetts. Phone: 1-877-CRIVER-1. Web: www.criver.com. SHROB animals may require additional veterinary support because of increased incidence of superficial skin infections, particularly if soft bedding is not provided and not changed frequently. Surgical procedures require special care because of increased risk of infection, poor wound healing, and circulatory disturbances related to hypertension and obesity.

REFERENCES

Abramowsky CR, Aikawa M, Swinehart GL, Snajdar RM. (1984) Spontaneous nephrotic syndrome in a genetic rat model. *Am J Pathol* 117:400–408.

Aitman TJ, Glazier AM, Wallace CA, et al. (1999) Identification of Cd36 (Fat) as an insulin-resistance gene causing defective fatty acid and glucose metabolism in hypertensive rats. *Nat Genet* 21:76–83.

Angelin B, Karlmar KE. (1979) Adrenal hydroxylations in genetically obese and hypertensive rats. *Biochim Biophys Acta* 574:344–350.

Benetz BA, Khosrof SA, Huang SS, et al. (1996) Age of onset of retinal vascular changes in obese SHR. *Invest Ophthalmol Vis Sci* 37:S695.

Brindley DN, Hales P, Al-Sieni AII, Russell JC. (1992) Sustained decreases in weight and serum insulin, glucose, triacylglycerol and cholesterol in JCR:LA-corpulent rats treated with D-fenfluramine. *Br J Pharmacol* 105:679–685.

Brindley DN, Russell JC. (2002) Animal models of insulin resistance and cardiovascular disease: Some therapeutic approaches using JCR:LA-*cp* rat. *Diabetes Obesity Metab* 4:1–10.

Buñag RD, Barringer DL. (1988) Obese Zucker rats, though still normotensive, already have impaired chronotropic baroreflexes. *Clin Exp Hypertension [A]* 10: Suppl 1, 257–262.

Chehab FE, Lim ME, Lu RH. (1996) Correction of the sterility defect in homozygous obese female mice by treatment with the human recombinant leptin. *Nat Genet* 12:318–320.

Doggrell SA. (2001) Moxonidine: Some controversy. *Expert Opin Pharmacother* 2:337–350.

Ernsberger P, Ishizuka T, Liu S, et al. (1999) Mechanisms of antihyperglycemic effects of moxonidine in the obese spontaneously hypertensive Koletsky rat (SHROB). *J Pharmacol Exp Ther* 288:139–147.

Ernsberger P, Koletsky RJ. (1993) Weight cycling and mortality: Support from animal studies. *J Am Med Assoc* 269:1116.

Ernsberger P, Koletsky RJ, Baskin JS, Collins LA. (1996) Consequences of weight cycling in obese spontaneously hypertensive rats. *Am J Physiol* 270: R864–R872.

Ernsberger P, Koletsky RJ, Baskin JS, Foley, M. (1994a) Refeeding hypertension in obese spontaneously hypertensive rats. *Hypertension* 24:699–705.

Ernsberger P, Koletsky RJ, Baskin JZ, Foley M. (1994b) Refeeding hypertension in obese SHR. *Hypertension* 24:699–705.

Ernsberger P, Koletsky RJ, Collins LA. (1994) Lethal consequences of a high-salt diet in obese SHR. *Hypertension* 24:376.

Ernsberger P, Koletsky RJ, Collins LA, Bedol D. (1996) Sympathetic nervous system in salt-sensitive and obese hypertension: Amelioration of multiple abnormalities by a central sympatholytic agent. *Cardiovasc Drugs Ther* 10:Suppl 1, 275–282.

Ernsberger P, Koletsky RJ, Collins LA, Douglas JG. (1993) Renal angiotensin receptor mapping in obese spontaneously hypertensive rats. *Hypertension* 21:1039–1045.

Ernsberger P, Koletsky RJ, Friedman JE. (1999) Molecular pathology in the obese spontaneous hypertensive Koletsky rat: A model of syndrome X. *Ann NY Acad Sci* 892:272–288.

Ernsberger P, Koletsky RJ, Kilani A, et al. (1998) Effects of weight cycling on urinary catecholamines: Sympathoadrenal role in refeeding hypertension. *J Hypertension* 16:2001–2005.

Ernsberger P, Koletsky RJ, Kline DD, et al. (1999) The SHROB model of syndrome X: Effects of excess dietary sucrose. *Ann NY Acad Sci* 892:315–318.

Ernsberger P, Rao S. (2000) Moxonidine's antihypertensive actions do not involve α_2-adrenergic receptors in radiotelemetered SHR. *J Hypertension* 18: S234.

Ernsberger P, Velliquette RA, Johnson JL, Koletsky RJ. (2005) Improvements in glucose tolerance and insulin resistance following weight loss persist after regain of the lost weight. *Obesity Res* 13:A53.

Folsom AR, French SA, Zheng W, et al. (1996) Weight variability and mortality: The Iowa Women's Health Study. *Int J Obesity* 20:704–709.

Friedman JE, de Vente JE, Peterson RG, Dohm GL. (1991) Altered expression of muscle glucose transporter GLUT-4 in diabetic fatty Zucker rats (ZDF/Drt-fa). *Am J Physiol* 261:E782–E788.

Friedman JE, Ishizuka T, Liu S, et al. (1997a) Metabolic consequences of a nonsense mutation in the leptin receptor gene (*fa*^{k}) in the obese spontaneously hypertensive Koletsky rat (SHROB). *Exp Clin Endocrinol Diabetes* 105: (Suppl. 3), 82–84.

Friedman JE, Ishizuka T, Liu S, et al. (1997b) Reduced insulin receptor signaling in the obese spontaneously hypertensive Koletsky rat. *Am J Physiol* 273:E1014–E1023.

Friedman JE, Ishizuka T, Liu S, et al. (1998) Antihyperglycemic activity of moxonidine: Metabolic and molecular effects in obese spontaneously hypertensive rats. *Blood Press Suppl* 3:32–39.

Greenhouse DD, Hansen CT, Michaelis OE. (1990) Development of fatty and corpulent rat strains. *ILAR News* 32:2–4.

Huang SS, Khosrof SA, Koletsky RJ, et al. (1995) Characterization of retinal vascular abnormalities in lean and obese spontaneously hypertensive rats. *Clin Exp Pharmacol Physiol Suppl* 22:S129–S131.

Iannello S, Campione R, Belfiore F. (1998) Response of insulin, glucagon, lactate, and nonesterified fatty acids to glucose in visceral obesity with and without NIDDM: Relationship to hypertension. *Mol Genet Metab* 63:214–223.

Ishizuka T, Ernsberger P, Liu S, et al. (1998) Phenotypic consequences of a nonsense mutation in the leptin receptor gene (fak) in obese spontaneously hypertensive Koletsky rats (SHROB). *J Nutr* 128:2299–2306.

Jin H, Chen Y-F, Yang R-H, Oparil S. (1989) Atrial natriuretic factor in NaCl-sensitive and NaCl-resistant spontaneously hypertensive rats. *Hypertension* 14:404–412.

Johnson JL, Wan DP, Koletsky RJ, Ernsberger P. (2005) A new rat model of dietary obesity and hypertension. *Obesity Res* 13:A113–A114.

Kasiske BL, Cleary MP, O'Donnell MP, Keane WF. (1985) Effects of genetic obesity on renal structure and function in the Zucker rat. *J Lab Clin Med* 106:598–604.

Kasiske BL, Kalil RS, Ma JZ, et al. (1993) Effect of antihypertensive therapy on the kidney in patients with diabetes: A meta-regression analysis. *Ann Intern Med* 118:129–138.

Kastin AJ, Pan W, Maness LM, et al. (1999) Decreased transport of leptin across the blood–brain barrier in rats lacking the short form of the leptin receptor. *Peptides* 20:1449–1453.

Koletsky RJ, Boccia J, Ernsberger P. (1995) Acceleration of renal disease in obese SHR by exacerbation of hypertension. *Clin Exp Pharmacol Physiol* 22: Suppl 1, S254–S256.

Koletsky RJ, Dluhy RG, Crantz FR, Williams GH. (1982) Cortisol suppression test in patients with elevated adrenocorticotropic hormone levels. *Ann Intern Med* 96:277–280.

Koletsky RJ, Ernsberger P. (1992) Obese SHR (Koletsky rat): A model for the interactions between obesity and hypertension. In *Genetic Hypertension*, ed. Sassard, J. Libbey, London, 373–375.

Koletsky RJ, Ernsberger P. (1996) Phenotypic characterization of genetically obese and hypertensive rat strain: SHROB/Kol. *Rat Genome* 2:10–22.

Koletsky RJ, Friedman JE, Ernsberger P. (2001) The obese spontaneously hypertensive rat (SHROB, Koletsky rat): A model of metabolic syndrome X. In *Animal Models of Diabetes: A Primer*, ed. Sima, AAF, Shafrir, E. Harwood Academic Press, Amsterdam, 143–158.

Koletsky RJ, Velliquette RA, Ernsberger P. (2003) The role of I(1)-imidazoline receptors and alpha(2)-adrenergic receptors in the modulation of glucose and lipid metabolism in the SHROB model of metabolic syndrome X. *Ann NY Acad Sci* 1009:251–261.

Koletsky S. (1972) New type of spontaneously hypertensive rats with hyperlipemia and endocrine gland defects. In *Spontaneous Hypertension: Its Pathogenesis and Complications*, ed. Okamoto, K. Igaku Shoin, Tokyo, 194–197.

Koletsky S. (1975a) Animal model: Obese hypertensive rat. *Am J Pathol* 81:463–466.

Koletsky, S. (1975b) Pathologic findings and laboratory data in a new strain of obese hypertensive rats. *Am J Pathol* 80:129–142.

Koletsky S, Puterman DI. (1976) Effect of low calorie diet on the hyperlipidemia, hypertension, and life span of genetically obese rats. *Proc Soc Exp Biol Med* 151:368–371.

Koletsky S, Puterman DI. (1977) Reduction of atherosclerotic disease in genetically obese rats by low calorie diet. *Exp Mol Pathol* 26:415–424.

Koletsky S, Snajdar RM. (1979) Reduction of vascular disease in genetically obese rats treated for hypertension and hyperlipidemia. *Exp Mol Pathol* 30:409–419.

Khosrof SA, Huang SS, Benetz BA, et al. (1995) Characterization of retinal vascular abnormalities in lean and obese SHR: A new model of type 2 diabetic retinopathy. *Invest Ophthalmol Vis Sci* 36:S172.

Kurtz TW, Montano M, Chan L, Kabra P. (1989) Molecular evidence of genetic heterogeneity in Wistar–Kyoto rats: Implications for research with spontaneously hypertensive rats. *Hypertension* 13:188–192.

Lynn RB, Cao GY, Considine RV, et al. (1996) Autoradiographic localization of leptin binding in the choroid plexus of *ob/ob* and *db/db* mice. *Biochem Biophys Res Commun* 219:884–889.

Mark AL, Correia M, Morgan DA, et al. (1999) State-of-the-art lecture: Obesity-induced hypertension: New concepts from the emerging biology of obesity. *Hypertension* 33:537–541.

Motulsky HJ, Ransnas LA. (1987) Fitting curves to data using nonlinear regression: A practical and nonmathematical review. *FASEB J* 1:365–374.

Mukaddam-Daher S, Menaouar A, El Ayoubi R, et al. (2003) Cardiac effects of moxonidine in spontaneously hypertensive obese rats. *Ann NY Acad Sci* 1009:244–250.

O'Dea K, Koletsky S. (1977) Effect of caloric restriction on basal insulin levels and the *in vivo* lipogenesis and glycogen synthesis from glucose in the Koletsky obese rat. *Metabolism* 26:763–772.

Okamoto K, Aoki K. (1963) Development of a strain of spontaneously hypertensive rats. *Jpn Circ J* 27:282–293.

Rhinehart EK, Kalra SP, Kalra PS. (2004) Neuropeptidergic characterization of the leptin receptor mutated obese Koletsky rat. *Regul Pept* 119:3–10.

Russel JC, Koeslag DG. (1990) Jcr:LA-corpulent rat: A strain with spontaneous vascular and myocardial disease. *ILAR News* 32:27–32.

Schork NJ, Krieger JE, Trolliet MR, et al. (1995) A biometrical genome search in rats reveals the multigenic basis of blood pressure variation. *Genome Res* 5:164–172.

Steiner AA, Dogan MD, Ivanov AI, et al. (2004) A new function of the leptin receptor: Mediation of the recovery from lipopolysaccharide-induced hypothermia. *FASEB J* 18:1949–1951.

Strohl KP, Thomas AJ, St Schlenker EH, et al. (1997) Ventilation and metabolism among rat strains. *J Appl Physiol* 82:317–323.

Sun Z, Ernsberger P. (2005) Profound insulin resistance of PKB (Akt) activation in intra-abdominal adipocytes from obese hypertensive SHROB rats. *Obesity Res* 13:A102.

Takaya K, Ogawa Y, Hiraoka J, et al. (1996) Nonsense mutation of leptin receptor in the obese spontaneously hypertensive Koletsky rat. *Nat Genet* 14:130–131.

Tan E, Butkus A, Koletsky S. (1976) Hepatic cholesterol metabolism *in vitro* in the obese spontaneously hypertensive, hyperlipemic and atherosclerotic rat. *Exp Mol Pathol* 25:142–151.

Unger RH. (1971) Glucagon and the insulin: Glucagon ratio in diabetes and other catabolic illnesses. *Diabetes* 20:834–838.

Velasquez MT, Abraham AA, Kimmel PL, et al. (1995) Diabetic glomerulopathy in the SHR/N-corpulent rat: Role of dietary carbohydrate in a model of NIDDM. *Diabetologia* 38:31–38.

Velliquette RA, Ernsberger P. (2003a) Contrasting metabolic effects of antihypertensive agents. *J Pharmacol Exp Ther* 307:1104–1111.

Velliquette RA, Ernsberger P. (2003b) The role of I(1)-imidazoline and alpha(2)-adrenergic receptors in the modulation of glucose metabolism in the spontaneously hypertensive obese rat model of metabolic syndrome X. *J Pharmacol Exp Ther* 306:646–657.

Velliquette RA, Friedman JE, Shao J, et al. (2005) Therapeutic actions of an insulin receptor activator and a novel peroxisome proliferator-activated receptor {gamma} agonist in the spontaneously hypertensive obese rat model of metabolic syndrome X. *J Pharmacol Exp Ther* 314:422–430.

Velliquette RA, Koletsky RJ, Ernsberger P. (2002) Plasma glucagon and free fatty acid responses to a glucose load in the obese spontaneous hypertensive rat (SHROB) model of metabolic syndrome X. *Exp Biol Med (Maywood)* 227:164–170.

Velliquette RA, Kossover R, Previs SF, Ernsberger P. (2006) Lipid-lowering actions of imidazoline antihypertensive agents in metabolic syndrome X. *Naunyn Schmiedebergs Arch Pharmacol* 372:300–312.

Wan DP, Johnson DH, Johnson JL, et al. (2005) Magnetic resonance imaging of visceral and subcutaneous fat distribution in genetic versus dietary obesity. *Obesity Res* 13:A172.

White BD, Corll CB, Porter JR. (1989) The metabolic clearance rate of corticosterone in lean and obese male Zucker rats. *Metabolism* 38:530–536.

Wyss JM, Oparil S, Sripairojthikoon W. (1992) Neuronal control of the kidney: Contribution to hypertension. *Can J Physiol Pharmacol* 70:759–770.

Yamashita T, Murakami T, Iida M, et al. (1997) Leptin receptor of Zucker fatty rat performs reduced signal transduction. *Diabetes* 46:1077–1080.

Yen TT, Shaw WN, Yu PL. (1977) Genetics of obesity in Zucker rats and Koletsky rats. *Heredity* 38:373–377.

Young JB, Landsberg L. (1982) Diet-induced changes in sympathetic nervous system activity: possible implications for obesity and hypertension. *J Chronic Dis* 35:879–886.

Zavaroni I, Bonini L, Fantuzzi M, et al. (1994) Hyperinsulinaemia, obesity, and syndrome X. *J Intern Med* 235:51–56.

9 OLETF Rats: Model for the Metabolic Syndrome and Diabetic Nephropathy in Humans

Kazuya Kawano

CONTENTS

DISCOVERY AND ESTABLISHMENT OF OLETF RATS

We developed three kinds of model animals from rats that had been purchased from Charles River Canada Inc. in 1982. One strain was developed by selective breeding of these rats and designated OLETF (Otsuka Long–Evans Tokushima fatty) (Kawano et al. 1992). From the same Long–Evans colony stock, a line developing type 1-like insulin-dependent diabetes mellitus was isolated and designated Long–Evans Tokushima lean (LETL) (Kawano et al. 1991). A control line, Long–Evans Tokushima Otsuka (LETO), was also established.

In this chapter we describe the diabetic characteristics of OLETF rats and introduce the metabolic syndrome characterized by the visceral fat accumulation. We also describe the diabetic nephropathy in OLETF rats and consider the causes for the appearance of the disease.

The Long–Evans rats were selectively bred based on a hypothesis proposed by Goto et al. (1975). We performed an oral glucose tolerance test (OGTT) on male and female animals through three generations and selectively mated rats with high total plasma glucose. The same procedure was repeated in the offspring, but their plasma glucose levels were not elevated. However, one rat that differed from spontaneous LETL rats was discovered. The rat was obese and demonstrated glucosuria, yet developed only mild subsequent weight loss. An OGTT performed on all its sibling rats revealed that all males and one female were diabetic. The offspring of these brother–sister rats had already been born. We attempted to develop a model of type 2 diabetes (T2D) from these siblings.

Selective brother–sister mating was then initiated using males of >400 g and apparently normal females at 9–10 weeks. At least 20 pairs were mated in each generation. The males used for mating were examined by OGTT at 25 weeks of age. Only rats with diabetes and their offspring that developed diabetes were bred, while rats of the normal OGTT and their offspring were sacrificed. Until June 1989, this process was repeated through 20 generations. This established strain of T2D model animals was named the OLETF strain.

The LETO line was obtained by original mating that is different from that for OLETF rats, but both strains originated from the same colony of Long–Evans rats. The LETO line did not show the diabetic syndrome from the F1 to the F20 generation.

CLINICAL FEATURES

Table 9.1 shows serial data of food intake, body weight, and nonfasting plasma glucose and triglyceride (TG) levels in OLETF and LETO rats. The food intake of the male OLETF rats increased to ~30g/day from 10 to 70 weeks of age. Because food intake was greater in animals demonstrating glucosuria after 30 weeks, however, the standard deviation for food intake was greater after 50 weeks of age. The food intake of male LETO rats was roughly 22 g/day by 70 weeks. From 10 to 70 weeks, food intake increased by ~20 g/day in female OLETF rats and ~16 g/day in female LETO rats.

The average body weight of male OLETF rats at 10 weeks was 381 g, ~80 g more than that of LETO rats at the same age. Thereafter, both strains continued to gain weight. Because OLETF rats manifesting glucosuria after 30 weeks began to lose weight, the standard deviation for body weight was greater after 50 weeks. Average body weight of female OLETF rats at 10 weeks of age was 244 g, ~40 g more than that of LETO rats at the same age. Subsequently, both strains of rats continued to gain weight. By 70 weeks, the average body weight of female OLETF and LETO rats was 532 and 333 g, respectively—a difference of almost 200 g.

TABLE 9.1
Characterization of Food Intake, Body Weight, Fed Plasma Glucose, and Triglyceride Levels in OLETF and LETO Rats

Weeks of Age	Strain	Sex	Food Intake (g/day)	Body Weight (g)	Fed Plasma Glucose (mg/dL)	Fed Plasma Triglyceride (mg/dl)
10	OLETF	Male	29.8 ± 2.0	381 ± 22	168 ± 7	123 ± 28
	LETO	Male	22.5 ± 1.0	301 ± 13	137 ± 5	68 ± 11
	OLETF	Female	21.1 ± 1.0	244 ± 17	147 ± 11	155 ± 35
	LETO	Female	17.6 ± 0.8	203 ± 9	142 ± 7	86 ± 12
30	OLETF	Male	27.2 ± 1.1	659 ± 42	162 ± 22	294 ± 110
	LETO	Male	22.3 ± 0.7	528 ± 22	119 ± 10	89 ± 18
	OLETF	Female	17.4 ± 1.6	418 ± 54	143 ± 13	321 ± 104
	LETO	Female	13.6 ± 0.3	301 ± 14	120 ± 8	83 ± 24
50	OLETF	Male	29.5 ± 3.2	682 ± 68	262 ± 128	299 ± 46
	LETO	Male	20.0 ± 0.6	584 ± 26	130 ± 9	90 ± 22
	OLETF	Female	19.9 ± 2.0	505 ± 66	141 ± 22	520 ± 132
	LETO	Female	14.3 ± 0.2	324 ± 18	107 ± 12	113 ± 35
70	OLETF	Male	29.0 ± 5.9	652 ± 117	326 ± 203	426 ± 181
	LETO	Male	22.6 ± 1.6	620 ± 25	113 ± 7	98 ± 29
	OLETF	Female	20.1 ± 2.0	532 ± 82	126 ± 15	440 ± 175
	LETO	Female	15.8 ± 0.9	333 ± 20	106 ± 16	168 ± 76

Note: Data are represented as mean ± SD.

Fed plasma glucose levels in male OLETF rats began to rise after 30 weeks, with values of 262 and 326 mg/dL at 50 and 70 weeks, respectively. On the other hand, plasma glucose in male LETO rats at 10 weeks was ~130 mg/dL. Plasma glucose in these rats did not increase with age. Plasma glucose in female OLETF rats was 140 and 126 mg/dL at 50 and 70 weeks, respectively. Unlike in the males, plasma glucose did not increase with age in the females. Plasma glucose in female LETO rats at 10 weeks of age was 142 mg/dL. At 70 weeks, this value was roughly 110 mg/dL.

At 10 weeks, plasma TG in male OLETF rats was almost twice that of LETO rats. At 70 weeks, the TG was 426 mg/dL, ~4.3 times that of LETO rats at the same age. TG levels in male LETO rats at 10 weeks and 70 weeks were 68 and 90 mg/dL, respectively. TG increased with age in both strains of female rats. However, the degree of increase was greater in OLETF rats, with an increase from 155 mg/dL at 10 weeks to 520 mg/dL at 50 weeks. TG levels in female LETO rats at 30, 50, and 70 weeks were 83, 113, and 168 mg/dL, respectively.

ORAL GLUCOSE TOLERANCE TEST

Table 9.2 shows changes in plasma glucose and insulin levels in the OGTT in male OLETF and LETO rats at 10, 30, 50, and 70 weeks. Peak and 120-min plasma glucose levels increased with age in male OLETF rats in the OGTT. At 30 weeks, peak and 120-min plasma glucose were greater than 300 and 200 mg/dL, respectively. However, plasma glucose did not increase with age in male LETO rats. Furthermore, peak or 120-min plasma glucose did not exceed 300 or 200 mg/dL, respectively. Plasma insulin levels in OGTT did not differ between the strains at 10 weeks. However, plasma insulin was markedly higher in OLETF rats at 30 and 50 weeks. In contrast, plasma insulin was even lower in 70-week OLETF rats with severe diabetes than in LETO rats. Peak plasma glucose in the OGTT in female OLETF rats did not significantly change from 10 to 70 weeks. Compared with LETO rats, 70-week-old female OLETF rats showed a marked increase in plasma insulin. Insulin levels in LETO rats did not rise with age.

INSULIN RESISTANCE

From 12 weeks of age, male OLETF rats showed insulin resistance evaluated by euglycemic hyperinsulinemic clamp experiment (Sato et al. 1995). At 30 weeks, the total content of insulin-regulated glucose transporter (GLUT4) was significantly decreased in the muscles of diabetic rats compared with the control LETO rats. The plasma membrane content of the GLUT4 protein in muscles of OLETF rats was increased in a basal state. Furthermore, activity and protein content of muscle hexokinase type II in the OLETF rats were significantly decreased. In this way, the distribution of GLUT4 and glucose phosphorylation was impaired. These abnormalities seem to cause the development of insulin resistance in the skeletal muscles of OLETF rats (Sato et al. 1997).

TABLE 9.2
Characterization of Plasma Glucose and Insulin Levels in OGTT in Male OLETF and LETO Rats

Weeks of Age		Plasma Glucose (mg/dL)					Plasma Insulin (pg/mL)		
	Strain	0 min	30 min	60 min	90 min	120 min	0 min	60 min	120 min
10	OLETF	96 ± 10	258 ± 32	211 ± 20	165 ± 18	123 ± 16	302 ± 162	1,533 ± 741	812 ± 445
	LETO	79 ± 8	175 ± 9	159 ± 8	139 ± 11	109 ± 7	198 ± 104	1,643 ± 750	508 ± 139
30	OLETF	146 ± 7	352 ± 46	355 ± 61	354 ± 56	228 ± 40	9086 ± 4963	14,003 ± 6,012	14433 ± 6040
	LETO	114 ± 5	157 ± 6	146 ± 11	165 ± 6	135 ± 11	2,012 ± 633	6,440 ± 2,087	3,238 ± 1,120
50	OLETF	164 ± 19	374 ± 71	428 ± 83	405 ± 62	299 ± 74	7,427 ± 1,799	10,425 ± 2,315	9,657 ± 1,852
	LETO	123 ± 7	142 ± 10	158 ± 10	159 ± 5	146 ± 5	2,302 ± 861	8,184 ± 2,000	6,925 ± 2,226
70	OLETF	136 ± 26	294 ± 69	390 ± 94	395 ± 75	342 ± 71	1,776 ± 878	1,914 ± 1,247	1,858 ±1,130
	LETO	112 ± 8	149 ± 11	146 ± 7	164 ± 16	146 ± 7	1,899 ± 447	3,726 ± 976	3,514 ± 913

Note: Data are represented as mean ± SD.

ABDOMINAL FAT BY MAGNETIC RESONANCE IMAGING (MRI)

We measured abdominal fat mass (intra-abdominal visceral fat, summing retroperitoneal, mesenteric, and epididymal fat, and subcutaneous fat) and analyzed the abdominal fat distribution of OLETF and LETO rats using MRI. Intra-abdominal visceral and subcutaneous fat was highly correlated with body weight in OLETF and in LETO rats. Intra-abdominal visceral and subcutaneous fat of OLETF rats significantly accumulated compared with that of LETO rats. Intra–abdominal visceral fat mass of OLETF rats was positively correlated with subcutaneous fat mass. The accumulation of intra-abdominal visceral fat mass of OLETF rats was about 3.5 times that of subcutaneous fat. Thus, the obesity of OLETF rats was characterized by the marked accumulation of intra-abdominal visceral fat compared with that of subcutaneous fat and OLETF rats are a diabetic model of visceral fat accumulation (Ishikawa and Koga 1998).

TRIGLYCERIDE (TG) STORE IN THE PANCREATIC ISLETS

Hypertriglyceridemia is known to be a feature of obesity-related T2D, but the pathoetiological significance of this association is obscure. The effects of TG on beta cell functions and morphological changes in the pancreas were examined using *in vivo* and *in vitro* approaches in male OLETF rats at 6, 12, and 30 weeks of age, and in LETO rats.

Plasma TG was increased 2.7-fold at 6 weeks, 3.1-fold at 12 weeks, and 5.3-fold at 30 weeks in OLETF rats, compared with age-matched LETO rats. The TG content in the islets from 12-week-old OLETF rats was significantly increased compared with their counterparts. However, the islet TG content was not increased in 6-week-old OLETF rats. For this reason, the islets from 6-week-old rats were cultured with free fatty acid (FFA) or TG for 72 h. Abnormalities in OLETF rats were evident, in contrast to the results in LETO rats:

Glucose-induced insulin secretion was more inhibited by FFA or TG in the presence of 27.7 m*M* glucose, which was associated, at least in part, with reduced glucokinase activity in the islets.

Marked elevation of TG content was found in the islets.

Deposition of fat droplets in the enlarged islets, even in beta cells, was found by Oil Red O insulin double staining at 30 weeks.

In conclusion, hypertriglyceridemia resulted in significant TG store in the islets, which subsequently inhibited glucose-induced insulin secretion, at least in part, via reduced glucokinase activity in the islets. Fat droplets in the islets, therefore, may play an important role in hastening the development of NIDDM in this rat model (Man et al. 1997).

VISCERAL FAT WEIGHT

Absolute and relative visceral fat weight significantly increased with age in male OLETF rats compared with LETO rats. At 6, 12, and 30 weeks of age, the total weight of epididymal, mesenteric, and retroperitoneal fat was significantly increased in male OLETF compared with LETO rats.

INTESTINAL MONOACYLGLYCELOL ACYLTRANSFERASE (MGAT) ACTIVITIES

OLETF rats, characterized by obesity and hypertriglyceridemia, were divided into an *ad libitum* fed group and a food-restricted group. To eliminate the effects of hyperphagia on obesity, the amount of daily food intake was the same as that of their control strain LETO rats. Changes in body weight, body fat, intra-abdominal fat weight, and TG content in the liver were measured. Biochemical blood tests and the activity assay of TG synthase (MGAT) and diacylglycerol acyltransferase (DGAT) were performed:

The body weight of restricted OLETF rats was significantly decreased to 71.7% of that of satiated OLETF rats, which was almost the same as that of LETO rats. However, body fat and intra-abdominal fat weight were significantly increased in restricted OLETF rats and satiated OLETF rats compared with LETO rats.

Plasma TG, insulin, glucose, leptin, and hepatic TG content were significantly higher in OLETF rats than in LETO rats.

MGAT activity in the small intestine was significantly higher in satiated and restricted OLETF rats than in LETO rats.

DGAT activity in OLETF rats was not significantly different from that in LETO rats. Body fat weight and plasma TG were still significantly accelerated in OLETF rats at the same food intake as LETO rats. Therefore, high MGAT activity in the small intestine may play an important role in HTG and obesity, subsequently hastening the development of T2D in OLETF rats (Luan at al. 2002).

BLOOD PRESSURE

Blood pressure, measured by the tail-cuff method, gradually increased with age in male OLETF rats and became >150 mmHg at 50 weeks of age. In male LETO rats, the blood pressure did not increase with age. Therefore, male OLETF rats should be regarded as an animal model of hypertension.

CHOLECYSTOKININ (CCK)-A RECEPTOR GENE

The receptors for CCK in peripheral tissues and the central nervous system have been classified into two subtypes: CCK-A and -B receptors. Comparing the DNA sequences of the OLETF and LETO CCK-A receptor genes and the normal gene, Takiguchi (1997, 1998) found deletion in the OLETF gene, 6,847 bases in length, which was flanked by two 3-base pair direct repeats (5′-TGT-3′) at positions –2407/–2405 and 4441/4443, numbered according to the LETO gene sequence, one of which was lost. The promoter region and the first and second exons were missing in the mutant, although CCK-B receptors were intact. OLETF rats are naturally occurring knockout rats with the homozygously disrupted CCK-A receptor gene.

The genetic background of T2D in OLETF rats is very complicated (Hirashima et al. 1996, 1998). According to the results of the study of Funakoshi et al. (1994), a null mutation in the CCK-A receptor gene seems to elevate plasma glucose levels in OLETF rats in cooperation with other factors (Takiguchi et al. 1998).

We have shown that two recessive genes, ODB1 mapped on the X chromosome and ODB2 mapped on chromosome 14, are involved in the induction of diabetes in OLETF rats (Hirashima et al. 1995, 1996). We focused on the genotype of the CCK-A receptor gene and the ODB1 gene in the regulation of glucose homeostasis in the F2 cross of OLETF rats. Relatively high plasma glucose levels were observed in the F2 offspring with the homozygously disrupted CCK-A receptor gene. The disrupted CCK-A receptor gene and the ODB1 gene synergistically increased plasma glucose levels in F2 rats. The CCK-A receptor gene was found to be mapped very close to ODB2 by a linkage analysis using microsatellite markers (Takiguchi et al. 1998).

DIABETIC NEPHROPATHY

The glomerular lesions found in OLETF rats, similar to those in human diabetic nephropathy, are exudative lesions characterized by fibrin-cap, capsular drop, and aneurismal dilatation of intraglomerular vessels. However, the pathogenesis of diabetic nephropathy in OLETF rats demonstrating these findings remains to be clarified. Comparing the renal histological changes of four different diabetic rat species (Manuel et al. 1990; Mori et al. 1988), only OLETF rats develop exudative lesions and eventually obsolescent glomeruli, although the other three strains exhibit widening of mesangial matrix and thickening of basement membrane.

As reported by Kawano et al. (1996), renal histopathological changes of OLETF rats become visible after 23 weeks of age and thereafter become more severe, showing segmental PAS-positive lesion (29 weeks), exudative lesion with so-called capsular drop lesion, expansion of the mesangial matrix, the aneurysmal dilatation of intraglomerular vessels (55 weeks), and, finally, obsolescent glomeruli at end-stage kidney (after 80 weeks).

EFFECTS OF FOOD RESTRICTION

Using the OLETF rats, we examined the effect of diet control on the prevention or improvement of T2D and accompanying nephritic complications. The rate of onset of T2D was significantly suppressed when the animals were fed a restricted diet. The effect of 30% restricted feeding is clearly seen in those that received this diet during 6–80 weeks of age (percent incidence < 50%). Urinary protein concentration was significantly decreased in the food restriction group. The absolute weights of kidneys were significantly lower than those of the *ad libitum* group. Thickening of the glomerular basement membrane (GBM) and widening of the mesangial matrix were seen diffusely in the kidneys of the *ad libitum* group. Exudative lesion with fibrin-cap and glomerular obsolescence were also frequently observed. Conversely, renal histological changes in food restriction groups were less pronounced than those in the *ad libitum* group, yet mesangial cell proliferation and exudative changes were observed less frequently and less severely (Mori et al. 1996c).

PROTEIN LOADING

We performed 40% protein loading in OLETF rats and examined the involvement of protein in deterioration of diabetic nephropathy. In this experiment, kidney weights were 2.2 g (0.46 g% body weight) in the control LETO group and 3.57 g (0.52 g%) in the control OLETF group. In control OLETF group, kidney weight was markedly increased compared with that in control LETO group. In addition, in the protein-loaded OLETF group, kidney weight (3.88 g, 0.58 g%) was more markedly increased compared with that in the control OLETF group (3.57 g, 0.52 g%).

Incidence of glomerular lesion in control OLETF and protein-loaded OLETF rats at 30 weeks of age is shown in figure 9.1. In both OLETF groups, glomerular hypertrophy, exudative lesions with fibrin-cap, and dilatation of the mesangial area related to proliferation of mesangial cells in the kidney glomerulus were observed. In the control OLETF group, exudative lesions comprised 1.2 ± 1.1% of the entire glomerulus. In the protein-loaded group, the percentage was significantly increased to 4.8 ± 1.7%. Expansion in the mesangial area was 31.1 ± 5.9% in the control group and 26.6 ± 5.1% in the protein-loaded group. There was no significant

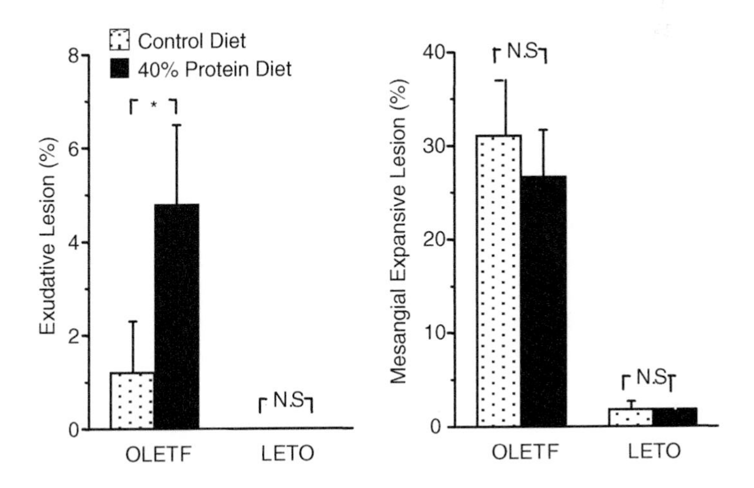

FIG 9.1 Incidence of glomerular lesions when a 40% protein diet was given between 5 and 30 weeks of age. *P < 0.05 versus each control diet (Wilcoxon rank sum test).

difference. Furthermore, glomerular hypertrophy in the protein-loaded OLETF group was more marked than that in the control group.

In the protein-loaded OLETF group, the number of polyethyleneimine (PEI) on GBM was significantly decreased compared with that in the control group. The number of PEI in the control OLETF group was significantly decreased compared with that in LETO rats. However, among LETO rats, there was no significant difference between the protein-loaded group and control group. GBM thickness in the control OLETF group was significantly greater than that in control LETO rats. However, there was no significant difference between the protein-loaded group and the control group. Foot processes in control and protein-loaded LETO groups showed a normal height and interval on the GBM. However, in control and protein-loaded groups of OLETF rats, the height and interval of foot processes were irregular and fused. In the protein-loaded OLETF group, rupture of the GBM was observed, although the incidence was low (Kawano et al. 1999).

The internal and external hyaline layers of the GBM function as a charge barrier via heparan sulfate and inhibit the passage of negatively charged substances as albumin (Kanwar et al. 1984). However, it has been reported that heparan sulfate levels are decreased under diabetic conditions, destroying the charge barrier and enhancing the permeability of GBM (Rosenzweig and Kanwar 1982).

DISCUSSION AND CONCLUDING REMARKS

Metabolic Syndrome

Obesity is considered to be one of the causes of T2D. Obese patients frequently have glucose intolerance, hyperlipidemia, and hypertension. However, the mechanism responsible for obesity has not yet been fully elucidated.

Many recent studies have shown that body fat distribution is more important in the morbidity of obesity than the extent of body fat accumulation (Kanai et al. 1990; Nakajima et al. 1989; Nakamura et al. 1994). Metabolic disorders such as glucose intolerance and hyperlipidemia develop more frequently in visceral fat obesity than in subcutaneous fat obesity (Fujioka et al. 1987; Kawano et al. 1992). In addition to metabolic disorders, visceral fat accumulation has been shown to be closely correlated with cardiovascular complications (Kanai et al. 1990; Nakajima et al. 1989; Nakamura et al. 1994). Thus, visceral fat accumulation provides a pathogenic background in hyperlipidemia. As mentioned above, visceral fat accumulation is more frequently involved in metabolic disorders than subcutaneous fat accumulation.

OLETF rats show obesity, hyperlipidemia, and T2D. They show hyperphagia and obesity a few weeks after birth and gradually gain weight with age. The difference in body weight between OLETF and LETO rats reaches approximately 200 g at 40 weeks of age (Sato et al. 1995). The obesity of OLETF rats was more markedly characterized by MRI, showing the marked accumulation of intra-abdominal visceral fat, than by that of subcutaneous fat (Ishikawa and Koga 1998). Concerning obesity, OLETF rats are characterized by obesity with intra-abdominal visceral fat accumulation and differ from other genetically obese rats, such as Zucker fatty rats, which have prominent subcutaneous fat accumulation (Stolba et al. 1992).

OLETF rats resemble human obese patients with T2D. There have been some reports on lipid production in OLETF rats. For example, hepatic acyl-coenzyme A synthase (ACS) and microsomal TG transfer protein (MTP) messenger RNAs were shown to be expressed at high levels in these animals (Kuriyama et al. 1998). However, it has been reported that OLETF rats show hyperphagia, a major factor responsible for obesity (Kobayashi et al. 1999). Satiated rats were used in the previous study. It is difficult to determine whether these results were affected by hyperphagia.

To eliminate the effects of hyperphagia, OLETF rats were given the same amount of daily food as LETO rats. We found that body fat weight was still significantly increased in restricted OLETF rats, indicating that there may be some other factors causing obesity in tissues besides hyperphagia, such as differences in enzyme activity or genetic factors. Food intake was the same in the restricted OLETF and LETO groups, so it was possible to make comparisons between the two groups to find abnormalities in lipid metabolism in OLETF rats.

To determine whether TG is overproduced in OLETF rats, we focused on the roles of two key enzymes in TG biosynthesis: MGAT and DGAT. MGAT uniquely catalyzes the conversion of extraneous lipid to TG in the intestinal mucosa. DGAT is believed to control the one common step of synthesis of triglycerides in all tissues. Results showed that MGAT activities were significantly higher in OLETF rats than in LETO rats. On the other hand, no significant difference in DGAT activity was noted between OLETF and LETO rats. The small intestine is the main tissue involved in TG synthesis. Although it is possible that increased MGAT activity may cause overproduction of TG, it has not been confirmed *in vivo*.

From our study, even though restricted OLETF rats were given the same amount of daily food as LETO rats, body fat weight, plasma insulin, glucose, and leptin were increased compared with LETO rats, suggesting that the restricted OLETF rats will be a useful model for studying obesity and its relationship with NIDDM. High MGAT activity in the small intestine may play an important role in hypertriglyceridemia and obesity, subsequently hastening the development of T2D in OLETF rats (Kanai et al. 1990). Thus, OLETF rats may be a useful model for studying the pathogenesis of the metabolic syndrome including obesity, hypertriglyceridemia, hypertension, and T2D.

SUPPLY SYSTEM OF OLETF RATS

Basic Principles

Four-week-old rats will be supplied.
No females will be supplied.
Each laboratory that needs older rats is recommended to maintain those until use.

Procedures to Use to Apply for Rats

Submit the application form, memorandum, experimental design, or protocol to the Tokushima Research Institute, Otsuka Pharmaceutical Co., Ltd. We indicate acceptance of the application with a shipment order. Animal cost will not be billed, but users should pay transportation fees.

REFERENCES

Fujioka S, Matsuzawa Y, Tokunaga K, et al. (1987) Contribution of intra-abdominal fat accumulation to the impairment of glucose and lipid metabolism in human obesity. *Metabolism* 36:54–59.

Funakoshi A, Miyasaka K, Jimi A, et al. (1994) Little or no expression of the cholecystokinin-A receptor gene in the pancreas of the diabetic rats (Otsuka Long–Evans Tokushima fatty = OLETF rats). *Biochem Biophys Res Commun* 199:482–488.

Goto Y, Kakizaki M, Masaki N. (1975) Spontaneous diabetes produced by selective breeding of normal Wistar rats. *Proc Jpn Acad* 51:80–85.

Hirashima T, Kawano K, Mori S, et al. (1995) A diabetogenic gene (ODB1) assigned to the X-chromosome in OLETF rats. *Diabetes Res Clin Prac* 27:91–96.

Hirashima T, Kawano K, Mori S, et al. (1996) A diabetogenic gene (ODB2) identified on chromosome 14 of the OLETF rats and its synergistic action with ODB1. *Biochem Biophys Res Commun* 224:420–425.

Ishikawa M, Koga K. (1998) Measurement of abdominal fat by magnetic resonance imaging of OLETF rats, an animal model of NIDDM. *Magn Reson Imag* 16:45–53.

Kanai H, Matsuzawa Y, Kotani K, et al. (1990) Close correlation of intra-abdominal fat accumulation to hypertension in obese women. *Hypertension* 16:484–490.

Kanwar YS, Kanwar YS, Veits A, Kimma JH, Jakubowski ML. (1984) Characterization of heparan sulfate proteoglycan of glomerular basement membranes. *Proc Natl Acad Sci USA* 81:76–81.

Kawano K, Hirashima T, Mori S, et al. (1992) Spontaneous long-term hyperglycemic rat with diabetic complications Otsuka Long–Evans Tokushima fatty (OLETF) strain. *Diabetes* 41:1422–1428.

Kawano K, Hirashima T, Mori S, et. al. (1991) New inbred strain of Long–Evans Tokushima lean rats with IDDM without lymphopenia. *Diabetes* 40:1375–1381.

Kawano K, Hirashima T, Mori S, et al. (1996) Spontaneously diabetic rat "OLETF" as a model of human diabetes. In *Lessons from animal diabetes*, ed. E. Shafrir. Birkhauser, Boston, 6:225–236.

Kawano K, Mori S, Hirashima T, et al. (1999) Examination of the pathogenesis of diabetic nephropathy in OLETF rats. *J Vet Med Sci* 61:1219–1228.

Kobayashi S, Miyasaka K, Funakoshi A. (1999) Behavior abnormalities in the OLETF rat. In *Obesity and NIDDM: Lessons from the OLETF rat*, ed. K. Shima. Elsevier, Amsterdam, 191–198.

Kuriyama H, Yamashita S, Shimomura I, et al. (1998) Enhanced expression of hepatic acyl-coenzyme A synthetase and microsomal triglyceride transfer protein messenger RNAs in the obese and hypertriglyceridemic rat with intraabdominal fat accumulation. *Hepatology* 27:557–561.

Luan Y, Hirashima T, Man Z-W, et al. (2002) Pathogenesis of obesity by food restriction in OLETF rats increased intestinal monoacylglycerol acyltransferase activities may be a crucial factor. *Diabetes Res Clin Prac* 57:75–82.

Man Z-W, Zhu M, Noma Y, et al. (1997) Impaired b-cell function and deposition of fat droplets in the pancreas as a consequence of hypertriglyceridemia in OLETF rat, a model of spontaneous NIDDM. *Diabetes* 46:1718–1724.

Manuel TV, Paul LK, Otho EM. (1990) Animal models of spontaneous diabetic kidney disease. *FASEB J* 4:2850–2859.

Mori Y, Yokoyama J, Nishimura M, et al. (1988) Studies of a new diabetic strain of rat (WBN/kob). Third report (diabetic nephropathy). *J Jpn Diab Soc* 31:909–915.

Mori S, Kawano K, Hirashima T. (1996a) Study on progress of diabetes induced sucrose-load in OLETF rats. In *13th Jpn Assoc Anim Diabetes Res* (in Japanese).

Mori S, Kawano K, Hirashima T. (1996b) Study on progress of diabetes and diabetic nephropathy-induced sucrose load in OLETF rats. *J Jpn Diabetes Soc* 39: (Suppl. 1) (in Japanese).

Mori S, Kawano K, Hirashima T. (1996c) Relationships between diet control and the development of spontaneous type II diabetes and diabetic nephropathy in OLETF rats. *Diabetes Res Clin Prac* 33:145.

Nakajima T, Fujioka S, Tokunaga K, et al. (1989) Correlation of intraabdominal fat accumulation and left ventricular performance in obesity. *Am J Cardiol* 64:369–373.

Nakamura T, Tokunaga K, Shimomura I, et al. (1994) Contribution of visceral fat accumulation to the development of coronary artery disease in non-obese men. *Arteriosclerosis* 107:239–246.

Rosenzweig LJ, Kanwar YS. (1982) Removal of heparan sulfate of nonsulfate (hyaluronic acid) glycosaminoglycans results in increased permeability of the glomerular basement membrane to ^{125}I-BSA. *Lab Invest* 47:177.

Sato T, Asahi Y, Toide K, et al. (1995) Insulin resistance in skeletal muscle of the male Otsuka Long–Evans Tokushima fatty rat, a new model of NIDDM. *Diabetologia* 38:1033–1041.

Sato T, Man Z-W, Toide K, et al. (1997) Plasma membrane content of insulin-regulated glucose transporter in skeletal muscle of the Otsuka Long–Evans Tokushima fatty rat, a model of non-insulin-dependent diabetes mellitus. *FEBS Lett* 407:329–332.

Stolba P, Dobesova Z, Husek P, et al. (1992) The hypertriglyceridemic rat as a genetic model of hypertension and diabetes. *Life Sci* 51:733–740.

Takiguchi S, Takata Y, Funakoshi A, et al. (1997) Disrupted cholecystokinin A receptor (CCKAR) gene in OLETF rats. *Gene* 197:169–175.

Takiguchi S, Takata Y, Takahashi N, et al. (1998) A disrupted cholecystokinin A receptor gene induces diabetes in obese rats synergistically with ODB1 gene. *Am J Physiol* 274:E265–E270.

10 Neonatally Streptozotocin-Induced (n-STZ) Diabetic Rats: A Family of Type 2 Diabetes Models

B. Portha, J. Movassat, C. Cuzin-Tourrel, D. Bailbe, M. H. Giroix, P. Serradas, M. Dolz, and M. Kergoat

CONTENTS

Syndromes resembling human diabetes occur spontaneously in some animal species. Alternatively, they can also be induced by treating animals with drugs or viruses, excising their pancreases, or manipulating their diet. Of course, none of the known animal models can be taken to reproduce human diabetes, but they are believed to illustrate various types of etiological and pathogenic mechanisms that could also operate in humans. Among these models, diabetes induced in rats by neonatal streptozotocin (STZ) administration (so-called "n-STZ models") has been recognized during the last two decades as an adequate tool to study the long-term consequences of a gradually reduced beta-cell mass (Weir et al. 1986; Grill and Östenson 1988; Portha et al. 1990). We and others have found that defects in insulin secretion and action, which in many ways resemble those described in human type 2 diabetes (T2D), develop in these n-STZ models (Leahy et al. 1984; Leahy and Weir 1985). Our review aims to describe the methods necessary to generate the n-STZ rodents, to sum up the information so far collected in this family of diabetes models, and to highlight their potential as well as their limitations for future research.

n-STZ MODELS: HOW TO GENERATE RAT VARIANTS WITH GRADED SEVERITY OF DIABETES IN ADULTHOOD

The diabetic syndrome in this model was described for the first time after injecting rats (Sherman or Wistar strains) on the day of their birth (n0 = birth) i.v. (saphenous vein) or i.p. with 100 mg/kg STZ (Portha et al. 1974). STZ is a 2-deoxymethyl-nitrosoureaglycopyranose molecule that produces a selective toxic effect on beta-cells and induces diabetes in most laboratory adult animals (Rerup 1970). Although the exact mechanism of its toxicity is still a matter of debate, one proposed site for the action of STZ is the nuclear DNA. During decomposition of STZ, highly reactive carbonium ions are formed that alkylate DNA bases (LeDoux et al. 1986). In the following phase of the excision DNA repair, the nuclear enzyme poly (ADP-ribose) synthase becomes activated to such an extent that cellular levels of its substrate NAD become critically depleted, leading to cell death (Okamoto 1981).

The neonatal rats treated with STZ at birth exhibit acute insulin-deficient diabetes three to five days after birth (fig. 10.1). They are really diabetic during this period: Plasma glucose is high (345 ± 37 mg/dL), pancreatic insulin stores show a 93% decrease, plasma insulin is low considering the high glucose level, and plasma glucagon is high despite unchanged pancreatic glucagon content (Portha et al. 1974, 1979). All the surviving pups (mortality < 30%) can be easily kept to adulthood. It is remarkable that the marked hyperglycemia observed in the neonates following STZ is only transient (fig. 10.1) (Portha et al. 1974, 1979). This may explain why some authors (Junod et al. 1969) unduly reported that neonatal rats were resistant to STZ. By the end of the first postnatal week, plasma glucose and insulin values no longer differ significantly from those of controls.

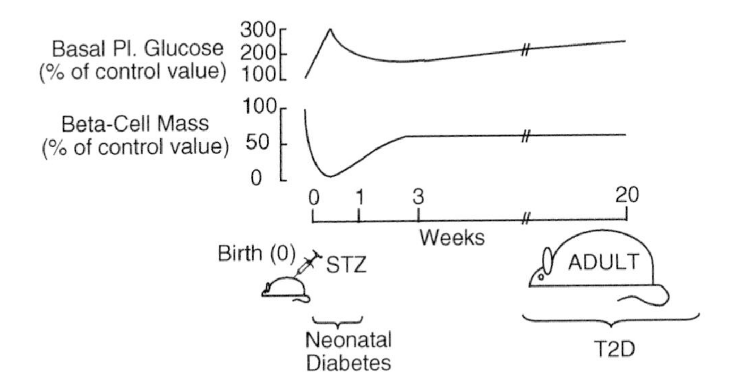

FIG. 10.1 Evolution of streptozotocin (STZ) effect in the Wistar rat neonate after one injection (100 mg/kg) on day of birth (n0-STZ rat model). The values for the streptozotocin-injected animals are expressed as a percentage of those observed in control animals killed on the same day. See text for comments on pathogenic progression toward T2D in the adults.

Such a transitory nature of overt diabetes is a unique characteristic of rat n-STZ diabetes, as compared with STZ diabetes induced in adult rats. At three to four weeks of age (i.e., at weaning), body weight and basal plasma glucose values in the n-STZ rats cannot be distinguished from control values. However, by eight weeks of age and thereafter, n0-STZ rats show mild basal hyperglycemia (150–180 mg/dL), abnormal responses to i.p. or i.v. glucose tolerance tests, and a 50% decrease in pancreatic insulin without change in pancreatic glucagon (fig. 10.1) (Portha et al. 1979).

An interesting but more severe variant of this model has been reported by Bonner-Weir et al. (1981) and Weir et al. (1981). Sprague–Dawley pups are injected i.p. on day 2 after birth with 90 mg/kg STZ (n2-STZ). Only n2-STZ-treated pups with plasma glucose values of 275 mg/dL or higher are selected and raised. By six weeks of age these animals show marked basal hyperglycemia (>200 mg/dL) and abnormal glucose tolerance. Mild hypoinsulinemia is also seen by this time. Since the studies of Bonner-Weir et al. differed from ours not only with respect to different times of STZ administration but also with regard to the strain of rat employed, one cannot discern whether the difference in subsequent severity of diabetes is due to the timing of toxin exposure or to a difference in strain-related capacity for spontaneous remission.

In view of this heterogeneity, we compared Wistar rats made diabetic with STZ injection on day 2 (n2-STZ version) and on day 5 after birth (n5-STZ version). As a consequence of such an approach, three models exhibiting non-insulin-dependent diabetes with graded severity were obtained in the adult rat. The n2-STZ exhibited characteristics (growth, basal plasma glucose and insulin levels, lack of insulin release in response to glucose *in vivo*, glucose intolerance, depletion of pancreatic insulin stores) very similar to those obtained in the n0-STZ version (Portha et al. 1974, 1979). By contrast the n5-STZ version was shown to exhibit a frank basal hyperglycemia with glucose intolerance, raised glycohemoglobin, a strong reduction of the pancreatic insulin stores, a 50% decreased basal plasma insulin level, and a lack of plasma insulin response to glucose *in vivo*.

The development and progression of the hyperglycemia in the Wistar n5-STZ demonstrated many similarities to those described by others (Levy et al. 1984; Schaffer et al. 1985; Grill and Rundfeldt 1986; Grill et al. 1986; Fantus et al. 1987) using the procedure described by Bonner-Weir et al. (1981)—that is, the i.p. administration of 90 mg/kg STZ in female Sprague–Dawley rats aged two days. The onset of hyperglycemia occurred after four weeks of age and a consistent decrease in the nonfasted plasma insulin levels was found from this time; the pancreatic insulin stores were markedly decreased to 11% of control values (10 weeks of age) and the perigonadal and retroperitoneal adipose tissue masses and adipocyte volumes were significantly smaller at 10 weeks of age (Blondel et al. 1989). At adult age, male n2-STZ rats are more severely affected than females (Bonner-Weir et al. 1981). This difference is not related to sex-related dissimilarity in beta-cell susceptibility to STZ during the neonatal period. Rather, the rise in androgens starting with puberty is at least partly responsible for the more severe diabetic state in the males (Ostenson et al. 1989; Iwase et al. 1996).

Note that, in mice also (NMRI strain), neonatal (day 2 and day 3 after birth) STZ (150 mg/kg, i.p.) has been reported to result in permanent diabetes with impaired glucose-induced insulin secretion *in vivo* (Ahren and Skoglund 1989). We and others (Weir et al. 1981; Levy et al. 1984; Schaffer et al. 1985; Grill and Rundfeldt 1986; Blondel et al. 1989) have found that the n-STZ models develop defects in insulin secretion and action that in many ways resemble those described in human T2D.

BETA-CELL FUNCTION IS ALTERED IN n-STZ ADULT RATS

Insulin Secretion in Response to Glucose and Nonglucose Secretagogues

We were first to show that, when they are adults, the n0-STZ rats are characterized by a low insulin release *in vivo* in response to glucose or amino acids (Portha et al. 1979). Because the beta-cell number and the insulin stores in the pancreases of these diabetic rats were low, the defective insulin response observed *in vivo* could be attributed to these quantitative abnormalities of the islets. In fact, additional changes in the beta-cell responsiveness to various stimuli were present. Insulin secretion studies were carried out primarily in 10- to 16-week-old n0-STZ rats using the isolated pancreas perfusion (Giroix et al. 1983). Insulin response to glucose stimulation over the range of 5.5–22 m*M* was lacking, thus indicating loss of beta-cell sensitivity to glucose. In contrast, glyceraldehyde elicited insulin release as high as that obtained in control pancreases. Mannose stimulated insulin secretion less in the diabetics than in the controls. The insulin secretion obtained in response to isoproterenol indicated that the ability of the adenylcyclase to generate cAMP in beta-cells of the diabetic was not decreased. In the absence of glucose, beta-cells of the diabetics were unexpectedly hypersensitive to arginine and leucine, and also the insulinotropic action of acetylcholine was increased (Giroix et al. 1983; Kergoat et al. 1987c).

While some disagreement exists about quantitative results obtained in the different diabetic variants (Wistar n5-STZ, Sprague–Dawley n2-STZ) (Giroix et al. 1983; Levy et al. 1984; Fantus et al. 1987; Okabayashi et al. 1989), there is universal agreement that the beta-cells in n-STZ models are essentially insensitive to glucose but retain their responsiveness to nonglucose secretagogues. Not only the response of beta-cells to glucose alone was abnormal, but also the glucose potentiating effect on nonglucose secretagogues was lost. In control pancreas the response to 19 m*M* arginine in the presence of low glucose concentration was much smaller than that found at higher levels; in marked contrast, the insulin responses in the n0-STZ rats were similar at low and high glucose concentrations (Kergoat et al. 1987). Also, in islets isolated from n0-STZ rats, the beta-cells have been found to exhibit poor sensitivity to glucose (Halban et al. 1983; Portha 1985). However, it is clear that a high glucose concentration was able to trigger a significant release from the diabetic rat islets, in contrast to findings with perfused pancreas (Giroix et al. 1983).

In the islets of the n0-STZ rats, insulin biosynthesis (measured by the incorporation of ^{3}H-phenylalanine into immunoprecipitable proinsulin) also appeared to be less stimulated by glucose than in control islets (Portha 1985).

In the n2-STZ and n5-STZ diabetic rats, it has also been reported that the secretion ratio of amylin to insulin is dramatically increased compared with that of nondiabetics (Inoue et al. 1992). To what extent the higher proportion of amylin in secretory granules might participate in the mechanism of beta-cell dysfunction is not known. The abnormal secretion in the n-STZ models bears a resemblance to the insulin secretory characteristics of human diabetes. It is noteworthy, however, that in the perfused rat pancreas first- and second-phase insulin responses are severely impaired, whereas in humans with T2D, significant preservation of second-phase release is usually found. The reason for this difference is unknown.

Incompetence of Beta-Cells to Glucose in n-STZ Rats: Intracellular Mechanisms

The reasons for the glucose incompetence in the beta-cells of n-STZ rats referred to previously are at present only partially understood. In the normal beta-cell, most evidence suggests that an increased rate of glucose metabolism is necessary for initiation of glucose-stimulated insulin release (Malaisse 1991). Based on the observation that the n-STZ models show a reduction in the amount of GLUT-2 transporter protein in their beta-cells (Thorens et al. 1990), it has been proposed that impaired glucose entry into beta-cells causes the deterioration of glucose-induced insulin secretion. Our measurements of 3-0-methyl-glucose uptake by intact islet cells of n0-STZ diabetic rats support the view that there is indeed a defect of the beta-cell hexose transport system (Giroix, Rasschaert, et al. 1992; Giroix, Baetens, et al. 1992). However, such a conclusion needs to be considered with caution.

First, our results could reflect a change in the relative abundance of beta- and nonbeta-islet cells since the fraction of the islet mass occupied by insulin-producing cells is indeed decreased by 50%. Second, it is not obvious that the anomaly of hexose transport plays any major role in the impaired secretory response of the beta-cell to glucose in the n0-STZ model, since it has been repeatedly emphasized that the efficiency of glucose transport should be decreased by at least one if not two orders of magnitude to become a rate-limiting step in glucose catabolism (Malaisse 1991). The finding that over 90–120 min incubation the overall rate of glucose utilization is affected little or not at all in the islets of diabetic rats (Portha, Giroix et al. 1988) further suggests that the anomaly of hexose transport does not result, under steady-state conditions, in any marked change in the intracellular glucose concentration. The preferential alteration of the beta-cell response to glucose in the n0-STZ model is not attributable to a decrease in hexokinase and glucokinase activities (islet homogenates) or an altered binding of these isoenzymes to mitochondria (Giroix et al. 1990). In the case of the high Km glucokinase, the sole perturbation in the diabetic animals was an apparently decreased affinity for glucose (Giroix et al. 1990). These findings support the view that the transport of glucose across the plasma membrane and its subsequent phosphorylation by glucokinase should not be considered significantly responsible for the glucose response impairment of the diseased beta-cells in the n0-STZ model.

Our results indicate that the rate of total glycolysis, as judged from the production of $^{3}H_{2}0$ from [5-^{3}H]glucose, is not significantly different in islets from n0-STZ and control rats (Portha et al. 1988; Giroix et al. 1990), at least when related to the protein or DNA content of islets. By contrast, the oxidative response to a high concentration of glucose was severely affected, especially in terms of [6-^{14}C]glucose oxidation, an estimation of the oxidation of the glucose-derived acetyl residues in the Krebs cycle. Thus, the ratio [6-^{14}C]glucose oxidation/[5-^{3}H]glucose utilization was much less increased in response to a rise in hexose concentration (Giroix et al. 1990, 1993). This metabolic defect contrasts with a normal or even increased capacity of n0-STZ islets to oxidize [6-^{14}C]glucose, [2-^{14}C]pyruvate, [U-^{14}C]glutamine, and [U-^{14}C]leucine at low, noninsulinotropic concentrations of these substrates, as well as lesser impairment of the oxidation of [U-^{14}C]leucine tested at a high concentration (Giroix et al. 1990).

We have shown that, in mitochondria of n0-STZ islets, there was a severe decrease in the basal (no Ca^{2+}) generation of $^{3}H_{2}0$ from [2-^{3}H]glycerol-3-phosphate and the Ca^{2+}-induced increment in [^{3}H]glycerophosphate detritiation (Giroix et al. 1991; Giroix et al. 1992). This coincided with the fact that a high glucose concentration failed to increase significantly [2-^{3}H]glycerol conversion to $^{3}H_{2}0$ in intact islets from n0-STZ rats, in contrast to the situation in nondiabetic rats (Giroix et al. 1992). The n0-STZ mitochondria were also less efficient than those of control animals in generating $^{14}CO_{2}$ from [1-^{14}C]-2-ketoglutarate. However, the change in the 2-ketoglutarate dehydrogenase was less marked than in the FAD-linked glycerophosphate dehydrogenase and a normal to slightly elevated glutamate dehydrogenase activity was found in n0-STZ islet mitochondria (Giroix et al. 1991, 1992). Therefore, it is suggested that, in the n0-STZ model, the impairment of glucose-induced insulin secretion is, at least in part, due to a deficiency in the activity of mitochondrial

FAD-linked glycerophosphate dehydrogenase, leading to an altered transfer of reducing equivalents into the mitochondria by the glycerol phosphate shuttle.

We have also investigated the possibility that a defect in the islet cAMP production and/or in the islet phosphoinositide metabolism could be involved in the failure of the glucose-induced insulin secretion in the n0-STZ model (Dachicourt et al. 1996; Morin et al. 1997). Concerning the first possibility, we have shown that there is no major alteration of the functionality of the adenylate cyclase/phosphodiesterase/cAMP system in the n0-STZ beta-cells (in terms of its modulation by forskolin, glucagon, pertussis toxin, glucagon, GIP, GLP-1, or IBMX); however, there is a defective glucose-induced cAMP generation (Dachicourt et al. 1996) that could be explained by a block in the step(s) linking glucose metabolism and activation of adenylate cyclase and/or the lack of expression of the G_{olf} protein isotype despite increased expression of the ACII and ACIII adenylate cyclase isoforms (Frayon et al. 1999). Concerning the second possibility, we have reported that the glucose-induced polyphosphoinositides (PPI) hydrolysis was severely diminished on the n0-STZ beta-cells and have proposed that a reduced phosphatidylinositol kinase activity, concomitantly with a decreased Ca^{2+}-stimulated phospholipase C activity, may participate in alteration of the phosphoinositide pathway, limitation of the inositol phosphate production, and, finally, impairment of the glucose-induced insulin release (Morin et al. 1996, 1997).

Is the function of ionic channels defective in the n-STZ beta-cell? The closure of the K^+-ATP channels is a key step in the mechanism of insulin secretion in response to glucose. The resultant depolarization allows Ca^{2+} entry through voltage-dependent Ca^{2+} channels, thereby triggering insulin secretion. Tsuura et al. (1992) have studied the properties of the K^+-ATP channels in single beta-cells of n-STZ diabetic rats using the patch-clamp technique. The unitary conductance of the channel in diabetic beta-cells was virtually identical to that in control beta-cells and there was no difference between the diabetic and control groups in the sensitivity of K^+-ATP channels to ATP and glibenclamide. In response to glucose, the activity of the K^+-ATP channels was diminished in a concentration-dependent manner in control and diabetic-intact cells. However, the inhibition of the K^+-ATP channels in intact beta-cells of n-STZ rats was significantly less than that in control cells. Even in the presence of high glucose, the openings of a few single K^+-ATP channels were consistently observed in cell-attached patch membranes of diabetic but not control beta-cells.

Finally, it appears that the impaired insulinotropic action of glucose in beta-cells of n-STZ rats is associated with a reduced sensitivity of the K^+-ATP channel to extracellular glucose, but not to intracellular ATP. These conclusions fit very well with our recent observation of an impaired reduction of the ^{86}Rb efflux when perfused n0-STZ islets are exposed to glucose (Giroix et al. 1993). Therefore, the glucose insensitivity of the K^+-ATP channels in n-STZ beta-cells may be the result of insufficient ATP production caused by impaired glucose metabolism in diabetic beta-cells, rather than by a defect in the K^+-ATP channel. Within the frame of this hypothesis, we have found in n0-STZ islets that the ATP/ADP ratio at high glucose concentration was lower than in nondiabetic islets (Giroix et al. 1993).

Using a complementary approach aimed at understanding the diminished oxidative capacity of the diabetic beta-cell, data related to the expression of the mitochondial genome in the n0-STZ beta-cells have been gained. The contents of mitochondrial cytochrome-b mRNA and mitochondrial 12S rRNA were lower (50–30%) in n0-STZ islets compared with nondiabetic Wistar rats (Welsh et al. 1991; Serradas et al. 1993). The content of mitochondrial DNA (Southern blotting) was markedly decreased (70%) in islets from n0-STZ rats, but no deletion was detected (Welsh et al. 1991; Serradas et al. 1993). Therefore, since a lower mitochondrial RNA content may result in a diminished oxidative capacity, it is conceivable that this deficiency may contribute to insulin deficiency in the n0-STZ models.

Is there a glucose/glucose-6-phosphate futile cycling in the beta-cell? An additional explanation for the defect of the beta-cell secretory response to glucose in the n0-STZ rats has been proposed by Kahn et al. (1990). It is related to the induction of glucose-6-phosphatase activity in n0-STZ islets resulting in an ATP-wasting futile cycling between glucose and glucose-6-phosphate. In islets from normal rats, such futile cycling is negligible. We were unable to confirm Kahn's group's observations since we found that the activity of islet glucose-6-phosphatase is equally low in diabetic and control rats (Giroix et al. 1992).

Are sympathetic and parasympathetic influences on the beta-cell impaired? The parasympathetic nervous system exerts an important influence on insulin secretion. The receptors mediating the cholinergic effect are muscarinic in nature since acetylcholine-induced insulin secretion is readily blocked by atropine. We (Giroix et al. 1983; Kergoat et al. 1987) and others (Bonner-Weir et al. 1988; Grill and Ostenson 1988) have found a significantly enhanced response to acetylcholine in n-STZ rats when insulin release was tested using islets or perfused pancreas. This was associated with an increase of muscarinic receptors in the pancreatic islets as proposed by Ostenson and Grill (1987). Moreover, insulin treatment of the n-STZ rats for three days lowered blood glucose, diminished binding of [^{3}H]methylscopolamine, and abolished the insulin secretory hyperresponse to carbamylcholine. When normal islets were kept in tissue culture for three days, binding was higher when a high rather than a low concentration of glucose was included in the culture medium (Ostenson and Grill 1985). These authors have suggested that in the n-STZ model, the hyperglycemia *in vivo* is a determinant of the number of muscarinic receptors in the beta-cells and that such regulation is associated with the increase in the cholinergic induced insulin secretion.

We have also tested the notion that increased alpha-adrenergic receptor activity could contribute to suppress the glucose-induced insulin secretion in the perfused pancreas of n0-STZ rats by the addition of the selective alpha-adrenergic blocking agent UK-14304 to the perfusion medium. In the diabetic pancreas, the inhibitory effect of UK-14304 upon maximally stimulated-insulin release (by a glucose/arginine mixture) was significantly higher than that exerted on the nondiabetic pancreas, thus suggesting that the reactivity of beta-cells to sympathetic neural activation is enhanced in the n0-STZ model. Such an assumption is also supported by the data of Östenson et al. (1989) and by the demonstration by Kurose et al. (1992) that electrical splanchnic nerve stimulation decreases insulin secretion (perfused pancreas) more efficiently in n-STZ rats than in normal rats.

Finally, it of interest that alpha- and delta-cells in n-STZ models have been found insensitive to some modalities of glucose regulation (despite the lack of significant changes in glucagon and somatostatin stores), suggesting that incompetence to glucose is qualitatively similar for glucagon, somatostatin, and insulin secretion (Giroix et al. 1984, 1992; Weir et al. 1986; Ostenson et al. 1990).

Loss of Beta-Cell Competence to Glucose in n-STZ Rats: Not Irreversible and Glucotoxicity Related

The available evidence related to the etiology of the inability of beta-cells to respond to glucose (so-called "glucose incompetence") in the n-STZ model is contradictory. The first proposal—that it is irreversibly damaged in n-STZ rats—was based on experiments in which islets of adult Wistar–Furth n0-STZ rats were transplanted under the kidney capsule of syngeneic normal or diabetic recipients and subsequently tested (perfusion of the graft-bearing kidneys) for glucose-induced insulin release (Inoue et al. 1994). Against such a view, we are inclined to exclude the possibility that the pathogenesis of the beta-cell incompetence to glucose in adult n0-STZ rats can be entirely explained by a permanent toxic action of STZ since the beta-cells of adults are mostly regenerated beta-cells resulting from neogenesis taking place at a time when STZ is no longer present in the body fluids.

We have directly evaluated the possibility that the STZ treatment per se causes a chronic impairment of the insulin response to glucose by testing the insulin response at intervals after STZ treatment from day 1 to day 21 using perfusion of pancreatic fragments (Portha, Kergsat 1985). While the glucose-induced insulin release was completely abolished (2% of normal response) on day 1 after STZ, it could be demonstrated after day 3. Moreover, it increased as a function of age (6 and 36% of the normal response on day 5 and day 14, respectively). This restoration of the insulin response to glucose closely paralleled the recovery of pancreatic insulin stores (6 and 51% of normal values on days 5 and 14, respectively). In sharp contrast with the lack of glucose response observed *in vitro* in the adult, glucose-induced insulin release was still detected on day 21. Accordingly, these data support the notion that the pathogenesis of the beta-cell lesion in n0-STZ rats is caused mainly by factors other than a primary toxic effect of STZ.

The sustained glucose load acting on a reduced beta-cell mass (so-called "glucotoxicity") could represent one among these factors leading to the functional abnormalities of the beta-cells in this diabetic model. To test this hypothesis, n0-STZ islets were cultured five days at 5.5 or 11 m*M* glucose to determine whether their beta-cell derangements could be modified by changing the environmental conditions. The insulin release and the (pro)insulin biosynthesis (measured in basal or stimulated states) were then found to be similar in the islets of diabetics and control islets after the 5.5-m*M* glucose culture period. By contrast, after the 11-m*M* glucose culture period, the insulin release and the (pro)insulin biosynthesis in the islets of diabetic rats were found significantly less stimulated by 16.5 m*M* glucose than in control islets (Portha 1985). According to these data and those of Welsh and Hellerström (1990), who used a very similar protocol, it seems likely that chronic hyperglycemia is responsible for these defects.

A different approach to determine whether the glucose defect could be counteracted was designed by normalizing the diabetic state in n0-STZ rats using insulin therapy (Kergoat et al. 1987). Insulin was given over one day or five consecutive days and insulin secretion was studied the morning after the last insulin injection with the isolated perfused pancreas preparation. Basal plasma glucose levels decreased in diabetic rats from 183 ± 8 to 136 ± 10 mg/dL after five-day insulin treatment (vs. 116 ± 3 mg/dL in control rats). Although the one-day insulin treatment did not modify the lack of glucose response in the diabetic rats, the five-day treatment improved the glucose-induced insulin secretion.

Moreover, insulin therapy improved the priming effect of glucose on a second stimulation with glucose. The return of this glucose effect was hardly detectable after the one-day insulin therapy but was clearly present after the five-day treatment. The hyperresponse to arginine characteristic of the untreated diabetic rats was similar to controls after a one-day insulin therapy, and it was again amplified at high glucose levels, although amplification remained lower than that in control rats. This indicates that the potentiating effect of glucose on the response to arginine was regained more precociously than the acute insulin response to glucose after insulin therapy. Nevertheless the improvement of the beta-cell secretory function in the n0-STZ rats remained strictly dependent upon correction of the hyperglycemic–hypoinsulinemic pattern by insulin treatment: Ten days after exogenous insulin, the impaired glucose-stimulated insulin secretion and hyperresponse to arginine-stimulated insulin secretion were back. However, these results do not allow the distinction of whether the abnormality of the insulin secretion is due to high levels of glucose per se or to other factors associated with the diabetic state.

It was therefore investigated whether a long-lasting *in vitro* glucopenia during perfusion of pancreas could restore beta-cell sensitivity to glucose in n0-STZ and n5-STZ rats (Portha et al. 1990). In nondiabetic rats, after 50-min perfusion with a medium containing no glucose, the integrated insulin response to subsequent stimulation with 16 mmol/L glucose was not significantly different from that obtained after a 20-min glucose omission period. (This last condition represents our current procedure in the pancreas perfusion experiments.) Conversely, in the n0-STZ rats, the incremental insulin response to 16 mmol/L glucose after 50-min glucose omission was enhanced fivefold relative to the response obtained 20 min after isolation. Moreover, the normal biphasic pattern of response to glucose was restored (Portha et al. 1990). The effect of glucose omission seems specific for the diabetic pancreas since it failed to exert a significant effect on the subsequent response to glucose in pancreases from nondiabetic rats.

Finally, we have recently shown that the glucose-incompetent n0-STZ beta-cells can be rendered glucose competent by *in vitro* GLP-1 or acetylcholine exposure (Dachicourt et al. 1997; Kergoat 2005 unpublished data; Dolz 2006 unpublished data) and more generally by artificially raising their intracellular cAMP (Dachicourt et al. 1996; Dolz 2006 unpublished data).

Nevertheless, as far as the reversibility of the beta-cell secretory lesion is concerned, different conclusions have been reported in the n2-STZ and n5-STZ models. The defective glucose-induced insulin secretion in the Sprague–Dawley n2-STZ rat was not restored by insulin treatment (Leahy et al. 1985; Grill and Rundfeldt 1986); only the ability of glucose to potentiate arginine-induced insulin release was normalized. We also failed to improve significantly the glucose-induced insulin secretion in the Wistar n5-STZ model by chronic insulin treatment (Serradas and Portha unpublished results).

Moreover, in n5-STZ rats, the incremental insulin release in response to 16 m*M* glucose after a 50-min glucose omission period remained unchanged compared with a 20-min period of glucose omission (Portha et al. 1990; Serradas et al. 1991).

We are left without an explanation as to why the defective glucose-induced insulin secretion in the n5-STZ model cannot be reversed. However, it is pertinent that the basal plasma glucose level was consistently more elevated in the n5-STZ rat (17 m*M*) than in the n0-STZ rat (7 m*M*). In the case of the n5-STZ Wistar and the n2-STZ Sprague–Dawley models, a circumstantial explanation cannot be eliminated because it is not possible to exclude a chronic STZ effect on the remaining islets, in addition to the acute destruction of beta-cells. *In vitro* findings in the adult mouse islets suggest that after a cytotoxic STZ injury there remains a population of partially damaged beta-cells with a severely impaired ability to recognize glucose as a stimulus (Eizirik et al. 1988). Such a possibility in the n5-STZ rats is in fact supported by the lack of significant beta-cell regeneration in their pancreases during two weeks following the beta-cell insult. In contrast, in the n0-STZ Wistar rats, spontaneous beta-cell regeneration after the STZ exposure enables a partial replenishment of beta-cell mass by newly formed beta-cells (Cantenys et al. 1981; Portha et al. 1990; Wang et al. 1994; Movassat et al. 1997).

INSULIN ACTION IS IMPAIRED IN n-STZ ADULT RATS

It is recognized that severe reduction in beta-cells, as obtained in subjects with insulin-dependent-diabetes or animals after alloxan or STZ injection, is associated with resistance of target tissues to the action of insulin (Reaven et al. 1977; De Fronzo et al. 1982; Bevilacqua et al. 1985). However, at the time at which the n-STZ models were introduced, there was no clear answer to the question of whether or not a more modest reduction in beta-cell mass was associated with insulin resistance and, on a more general background, whether or not a primary reduction of beta-cell mass necessarily leads to the development of insulin resistance.

In adult n0-STZ females, we have shown that the hepatic glucose production measured in the basal state was higher in diabetics than in controls, despite similar peripheral insulin levels in both groups. The factors responsible for maintaining this elevated rate of basal hepatic glucose production in the n0-STZ rats remain to be identified. This could be a reflection of eliminated hepatic insulin resistance, since the liver of the n0-STZ diabetic rats was in fact hyperresponsive to submaximal insulin levels (Melin et al. 1991). During the clamp studies, the sensitivity of the liver to insulin's suppressive effect on glucose production was enhanced (Kergoat and Portha 1985). An abnormality in the mechanism of hyperglycemia suppression of hepatic glucogeneogenesis in diabetic rats can be implied and it is also possible that other factors are operative that stimulate hepatic glucose production in diabetics by increasing glycogenolysis and/or gluconeogenesis. The properties of the liver insulin receptor in the n0-STZ model have been studied (Portha, Chamras, et al. 1983; Kergoat et al. 1988; Melin et al. 1991). Liver insulin receptors are not up-regulated and the tyrosine kinase activity remains unaffected, suggesting that the increased insulin effect in the liver of n0-STZ rats is probably distal to the insulin-regulated receptor kinase.

The insulin-mediated glucose uptake by the whole body as estimated during the clamp studies was found to be normal in the n0-STZ rats (Kergoat and Portha 1985). In fact, data related to glucose utilization *in vivo* by individual peripheral tissues show that even if the insulin action is not increased in the whole body, this effect could indeed be greater in some tissues (Kergoat et al. 1991). This is the case at the levels of white adipose tissue (paraovarian and inguinal) and brown adipose tissue. In these tissues, increased sensitivity and increased responsiveness to insulin action are detected when comparing diabetic females with control ones. By contrast, insulin action is normal in skeletal muscles and diaphragm of the same adult diabetic females. This observation of a normal insulin action in muscles together with an enhanced insulin action in the liver and white and brown adipose tissues indicates that glucose is preferentially channeled toward the liver and adipose tissues in the n0-STZ females. Finally, in the n0-STZ model, we were unable to demonstrate any insulin resistance.

In the n2-STZ rats, glucose utilization rate induced by hyperinsulinemia was found to be normal, whereas the hepatic glucose production rate was significantly higher in the basal state, but normally suppressed by hyperinsulinemia (Blondel et al. 1989). The data obtained in the basal state under postabsorptive conditions could be interpreted as suggesting a discrete alteration of hepatic insulin action. However, one cannot presently eliminate the possibility that increased glucagon secretion or some indirect mechanism elicited by beta-cell deficiency could be responsible for such an alteration.

In contrast, in n5-STZ rats, clamp studies indicate that the glucose utilization rate induced by hyperinsulinemia was significantly reduced and the hepatic glucose production rate was less efficiently suppressed by hyperinsulinemia. Thus, insulin resistance was present *in vivo* at the level of peripheral tissues and the liver (Blondel et al. 1989). Our data related to impairment of insulin action *in vivo* in the Wistar n5-STZ model are therefore consistent with data obtained in the Sprague–Dawley n2-STZ model, describing a poor response of the heart or adipocytes to insulin *in vitro* (Trent et al. 1984; Fantus et al. 1987) or a reduced rate of glucose disposal *in vivo* during an insulin suppression test (Levy et al. 1984).

Through the comparison of these models, it is possible to indicate some correlations that delineate the conditions necessary for the emergence of insulin resistance (fig. 10.2). First, peripheral insulin action was kept normal, despite reduction of the beta-cell mass by 50%, loss of the insulin secretion in response to glucose, very mild hypoinsulinemia, and mild hyperglycemia. Second, insulin resistance developed only under the conditions expressed in n5-STZ rats, reflecting a dramatic reduction of beta-cell mass, an overt basal hypoinsulinemia, and a frank basal hyperglycemia. Taken together, these data lend support to the suggestion that a certain degree of insulin deficiency is necessary to induce a clear-cut decrease of insulin action (Portha et al. 1995).

Finally, the n-STZ models have contributed to defining whether some metabolic derangement (hyperglycemia) that occurs secondarily to the insulin deficiency or insulin deficiency per se (hypoinsulinism) is responsible for the development of insulin resistance. The impact of chronic hyperglycemia per se was evaluated in n5-STZ rats receiving a continuous infusion of phlorizin for four weeks (Blondel, Bailbe, et al. 1990). Phlorizin treatment in the diabetic rats efficiently decreased their basal plasma glucose levels from 16 to 7 m*M*. The level of glycohemoglobin

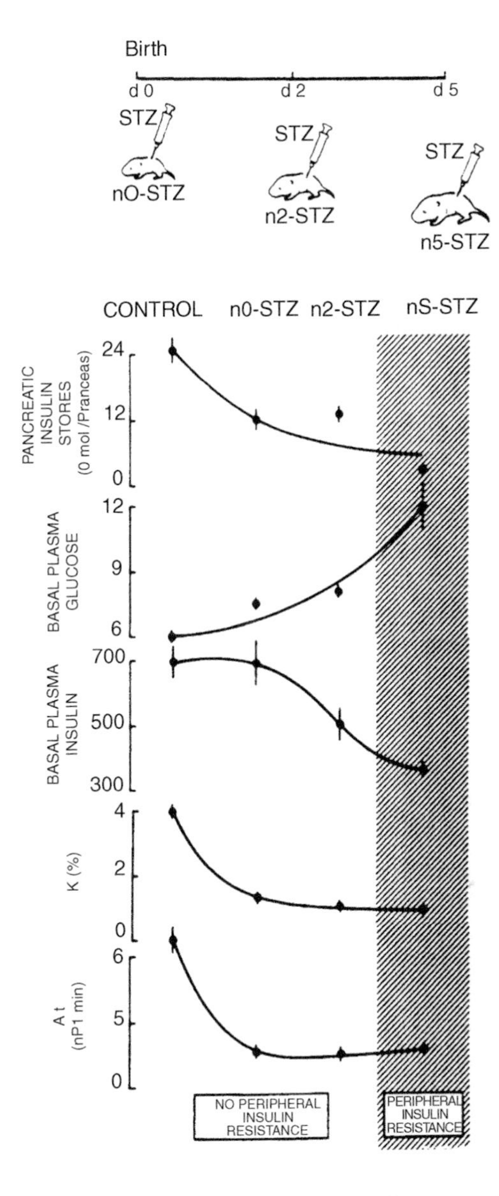

FIG. 10.2 Scheme of how progressive decline in beta-cell mass, as induced by injecting neonatal rats with streptozotocin (n-STZ) may lead to sequential appearance of abnormalities found in T2D. Events (pancreatic insulin stores, basal plasma glucose and insulin levels, glucose disappearance rates [K], and mean incremental insulin areas [ΔI]) are based on comparison of data obtained from three models of neonatally induced diabetes in Wistar rats: n0-STZ, 100 mg/kg STZ at birth; n2-STZ, 80 mg/kg STZ two days after birth; n5-STZ, 80 mg/kg STZ five days after birth. See text for comments on pathogenic progression toward impairment of insulin action in peripheral tissues.

returned to normal. In contrast, the basal plasma insulin levels and the glucose-stimulated insulin secretion remained as low as in the untreated diabetic rats. The basal glucose utilization and the glucose production rates were normalized by the phlorizin treatment and, when measured following submaximal hyperinsulinemia, both returned to normal. Since phlorizin treatment in control rats did not affect any of the preceding parameters, our data demonstrate that the sole restoration of normoglycemia in the n5-STZ rats (in the absence of any improvement in the endogenous insulin release) can completely correct the impairments of the insulin action upon the glucose uptake by the peripheral tissues and glucose production by the liver. Therefore, in the n5-STZ insulin-resistant diabetic model, hyperglycemia can be viewed as a pathogenic factor of its own (glucotoxicity) in the development of insulin resistance.

To evaluate the consequences of chronic hypoinsulinism per se (in the absence of hyperglycemia), clamp experiments were carried out in four-week-old n0-STZ rats at a time when basal plasma glucose is still normal. In these rats, the basal plasma insulin was significantly reduced and the glucose-induced insulin secretion *in vivo* was markedly decreased (Kergoat et al. 1991). Clamp studies revealed that the overall glucose utilization in the n0-STZ rats was significantly higher in the basal state and after submaximal hyperinsulinemia (Kergoat et al. 1991). We also verified that this was correlated with increased stimulation of glucose utilization in soleus muscle, diaphragm, and white and brown adipose tissues.

The *in vivo* observations related to white adipose tissue agree well with our *in vitro* studies of inguinal adipocytes (Kergoat et al. 1991). Specifically, the subcutaneous inguinal adipocytes were more sensitive to insulin in the four-week-old n0-STZ females than in control females, with respect to glucose conversion to total lipids only. The insulin-induced increase in glucose oxidation was the same in adipocytes from n0-STZ females and controls, suggesting that the enhancement of the rate of glucose metabolism in the n0-STZ rat adipocytes reflects a change beyond glucose transport and glucose metabolism along the glycolytic pathway. From these data, it was concluded that, in young normoglycemic n0-STZ rats, the insulin-dependent glucose utilization by white and brown adipose tissues is clearly increased as a consequence of the mild chronic (four weeks' duration) hypoinsulinism. This suggests that glucose is preferentially directed to adipose tissues in young n0-STZ rats, resulting in hypertrophied adipocytes and raising the possibility that the emergence of an increased body fat mass may result from a primary mild insulin deficiency.

n-STZ ADULT RATS: ADEQUATE TOOLS FOR STUDYING INTERACTIONS BETWEEN T2D AND DIET, OBESITY, OR GESTATION

n-STZ Models and Diet

The appropriate amount of carbohydrate and fat to be included in the diet of diabetic patients still remains highly controversial (Mann 1980; Reaven 1980). Experimental data obtained from diabetic animals are conflicting or inappropriate for several reasons: Investigations are always carried out in severely diabetic animals with no insulin response and are based on short periods of observation because of the high death rate of acutely diabetic rats without insulin treatment. Data have also been obtained in spontaneously diabetic rodents (Gutzeit et al. 1979; Leiter et al. 1981); however, in these models, obesity is also present and as a consequence it is difficult to dissociate a specific metabolic consequence due to dietary factors from those linked to obesity per se.

The n-STZ models are appropriate to circumvent these limitations. The effects of chronic high-sucrose feeding for one month on *in vivo* and *in vitro* insulin secretion and on *in vivo* insulin action have been studied in Wistar n0-STZ rats (Kergoat et al. 1987). The data clearly showed significant deterioration of glucose tolerance in these rats. This impairment results from two additive changes in insulin's effect on the target tissues: The insulin-mediated glucose uptake by peripheral tissues is decreased and the liver becomes resistant to insulin action due to diminished ability of insulin to suppress the hepatic glucose output. These results suggest that high-sucrose (instead of complex carbohydrate) feeding in rats with mild diabetes is likely to develop insulin resistance in target tissues. Such a pattern is directly related to the insulin-deficient state in these rats, since the same sucrose diet induces an enhanced insulin-mediated uptake in non-insulin-deficient animals (Kergoat et al. 1987).

A high-lipid diet may also be regarded as an aggravating factor of glucose handling in Wistar n0-STZ rats. We have shown that, in n0-STZ rats on a high-lipid food for one month, the glucose-induced insulin secretion was not significantly enhanced (at variance with the effect observed in control rats) and the glucose tolerance deteriorated, while it remained normal in control rats (Portha et al. 1982).

n-STZ Models and Obesity

In most of the spontaneous animal models of T2D, it is not easy to evaluate the effect of obesity on diabetes since these two pathological states are always interrelated. In experimental diabetes induced in adult rats by pancreatectomy, alloxan, or STZ, the modulatory effect of obesity upon the evolution of diabetes could not hitherto be studied since obesity was never obtained (Young et al. 1965; York and Bray 1972; Friedman 1972).

We have used n0-STZ rats to answer two questions: (1) whether obesity can be obtained in these animals; and, if so, (2) whether obesity modifies the course of diabetes. Two experimental designs were used to associate obesity with experimental diabetes in the adult rat. One was an electrolytic lesion of the ventromedial hypothalamus in adult diabetic Wistar n0-STZ rats. The other used genetically obese rats from the Zucker *fa/fa* strain made diabetic by a neonatal STZ injection (Portha et al. 1983). In diabetic rats, weight gain was similar to that in the nondiabetic rats, whether hyperphagia was due to a ventromedial hypothalamic lesion or to a genetic factor. Glucose-induced insulin release *in vivo* was increased in obese diabetic rats compared with nonobese diabetic rats.

Although glucose-induced insulin secretion was increased in diabetic rats becoming obese (in n0-STZ *fa/fa* Zucker rats and n0-STZ rats with hypothalamic obesity), these animals clearly did not ameliorate their glucose tolerance, suggesting the development of a state of insulin resistance related to obesity. Indeed, some of these animals developed overt diabetes with permanent or transient glucosuria and basal hyperglycemia. This was noted in almost one-third of the diabetic obese rats.

Thus, obesity may be looked upon as a contributing factor in the development of diabetes in insulin-deficient rats. The obese rat with mild n-STZ diabetes appears to be a suitable tool for study of the relationship between obesity and diabetes.

Alternative procedures to obtain an obese diabetic rat model have been proposed by Kawai, Suzuki, et al. (1991) using STZ (90 mg/kg on day 2 after birth) and monosodium glutamate (2g/kg, from day 1 to day 5). Also, the combination of neonatal administration (60 mg/kg on day 2) and postnatal overnutrition produced by breeding rats in small litters (four pups/litter) (Mende et al. 1996) or ICR mice on a cafeteria diet (Hasegawa et al. 1989) has been shown to promote higher body weight as well as increased blood glucose levels.

n-STZ Models and Pregnancy

The effect of pregnancy on the course of diabetes has been studied in Wistar n0-STZ females (Triadou et al. 1982). In pregnant n0-STZ rats (late pregnancy), the glucose-induced insulin secretion was found increased compared with that in the virgin state. Basal plasma glucose was decreased, but plasma glucose levels after glucose load were similar to values found in the virgin state, thus suggesting decreased glucose tolerance. Glucose tolerance remained impaired one and two months postpartum, while insulin secretion returned to the range found in the virgin state. These findings indicate that glucose tolerance is and remains impaired by pregnancy in n0-STZ females, but it is and remains unchanged in normal female rats. Thus, despite increased insulin response to glucose during late gestation in the n0-STZ rats, the diabetogenicity of pregnancy is confirmed with this experimental model. Neonatal STZ treatment (75 mg/kg STZ in spontaneously hypertensive female rat at two days of age) has been used to develop an animal model of high-risk pregnancy complicated by hypertension and diabetes mellitus (Wada et al. 1995).

n-STZ ADULT DIABETIC RATS: COMPLICATIONS OF LONG-STANDING T2D

The n-STZ models are especially attractive in this perspective since the rats can be easily kept for more than 20 months under standard breeding conditions. Their usefulness in studies related to the pathogenesis of nephropathy, hypertension, or osteopenia under conditions of chronic hyperglycemia is illustrated thereafter.

It is well known that hypertension is frequently associated with T2D in humans. However, it is difficult to disclose the mechanism of this combination as well as its long-term effect on the development and progression of nephropathy. It has been shown that n-STZ treatment carried out in spontaneously hypertensive rats (SHRs) offers an appropriate model for studying the interactions between hyperglycemia and hypertension, as well as their influence upon renal damage (Iwase et al. 1987, 1994; Wakisaka et al. 1988). Various degrees of hyperglycemia can be achieved in male adult SHR rats by varying the neonatal STZ i.p. dose from 37.5 to 75 mg/kg. Exogenous insulin is not required for long-term survival and hypertension develops normally (as in nondiabetic SHRs) in male SHRs receiving STZ on day 2 after birth. Using these models, it has been demonstrated that the combination of hypertension and hyperglycemia accelerates not only the progression of the established nephropathy but also the early development of nephropathy. The more severe hyperglycemia was also associated with the more severe nephropathy (Wakisaka et al. 1988). Also in conjunction with the development of nephropathy is the report that kidney basolateral membranes from n2-STZ rats exhibited impaired activity of [Ca^{2+},Mg^{2+}] ATPase and abnormal phospholipid content (Levy et al. 1988).

Clinical studies in humans have revealed the existence of a T2D-induced cardiomyopathy that is independent of atherosclerotic coronary artery disease, hypertension, or vascular disease. This syndrome is characterized by mechanical and metabolic abnormalities of the heart. Studies into the etiology of this myopathy are questionable since most of them were carried out in models more closely resembling insulin-dependent diabetes with overt hyperglycemia and continued waste throughout their lifetime. A more adequate animal model has become available following the report that Wistar n2-STZ rats develop cardiomyopathy (Schaffer et al. 1985; Schaffer et al. 1986). Hearts isolated from 12-month-old diabetic rats exhibited reduced rates of contractility and relaxation. Associated with the abnormality in contractility was redistribution in myosin isozyme content. Defects in myocardial relaxation also occurred concomitantly with impaired handling of calcium (Schaffer et al. 1989).

Alterations in femoral bone histomorphometry and vitamin D metabolism have been described in long-standing neonatally STZ-induced rats (n2-STZ, n5-STZ). The growth and strength of femurs decreased in the diabetic rats. Histomorphometric parameters such as cortical bone thickness, number of metaphysical trabeculae, and percent trabecular volume of metaphysical area all significantly decreased in the n-STZ rats. There were no significant differences in parameters between the n2-STZ and n5-STZ diabetic rats. Histological findings demonstrated no significant change in the number of femoral osteoclasts or change corresponding to osteomalacia. The plasma calcium level did not change in the n-STZ rats, although their plasma phosphate or A1-p levels increased. Circulating 24,25(OH)2D3 was significantly lower in the n-STZ rats. However, 25(OH)D3 or biologically active 1,25(OH)2D3 was not different between the controls and n-STZ rats. Osteopenia is thus present in the femurs of long-standing n-STZ rats, due in part to abnormal vitamin D metabolism (Umeda et al. 1992).

A selective endothelial dysfunction has been identified in the n-STZ adult rat through recording of endothelium-dependent and -independent relaxations to acetylcholine and sodium nitroprusside in thoracic aortic rings. This is probably related to the imbalance in oxidant/antioxidant status of the n-STZ rats since the diminished acetylcholine response, the enhanced aortic malondialdehyde level, and the decreased plasma vitamin levels were all restored after vitamin therapy (food supplemented with 2% vitamin E and 4% vitamin C) (Alper et al. 2005).

n-STZ ADULT DIABETIC RATS: ADEQUATE TOOLS FOR TESTING HYPOGLYCEMIC DRUGS

Selection of animal models for studying new hypoglycemic drugs depends upon the particular features of T2D required. In the n-STZ rat models of T2D, the various lesions contributing to the hypoglycemia represent many potential sites at which new hypoglycemic agents could be directed. These targets include (1) beta-cell lesions (n0-STZ model) such as altered rate of renewal (replication, neogenesis), decreased insulin biosynthesis, and decreased insulin release in response to glucose; (2) insulin target cell lesions (n2- and n5-STZ models); (3) hepatic glucose output lesions (n2- and n5-STZ models); and (4) imbalance of the glucose–fatty acid cycle (n5-STZ model). Accordingly, the n-STZ rats, which are relatively recent additions to the list of animal models of T2D, have so far been useful for testing many hypoglycemic drugs—for example:

sulfonylureas (Serradas et al. 1989; Ohnota et al. 1996)
nateglinide (A4166) or repaglinide (Kergoat, unpublished data)
KAD-1229 (Ohnota et al. 1994)
M-16209 (Nakayama et al. 1995)
JTT-608 (Ohta et al. 1999)
nicorandil (N-2hydroxyethyl-nicotinamide nitrate, an ATP-K^+ channel opener) (Kasono et al. 2004)
the selective alpha2-adrenoceptor antagonist SL840418 (Angel et al. 1996)
benfluorex (Portha et al. 1993)
metformine (Kergoat, unpublished data; Ashokkumar and Pari 2005)
pioglitazone (Kergoat, unpublished data)
N-benzoyl-D-phenylalanine (Pari and Ashokkumar 2005)
GLP-1 (Dachicourt et al. 1997)
combined drugs such as the association insulin/gliclazide (Kawai et al. 1991)
phlorizin (Serradas et al. 1991; Blondel et al. 1990)
vanadate and BMOV (bis-maltolato-oxovanadium-IV) (Blondel et al. 1990; Shinde et al. 2001)
tungstate (Barbera et al. 1997)

Additionally, they are also convenient models for testing the impact of islet grafting in non-insulin-dependent diabetes (Elian et al. 1996; Tormo et al. 1997).

n-STZ NEONATAL DIABETIC RATS: APPROPRIATE TOOLS FOR STUDY OF SPONTANEOUS REMISSION FROM HYPERGLYCEMIA AND OF BETA-CELL REGENERATION

The unique characteristic of STZ diabetes in rat neonates is the transitory nature of the overt beta-cell deficiency as compared with STZ diabetes in the adult rat. In the n0-STZ model, we and others have shown that the recovery of plasma glucose to normal values as soon as one week after birth was related to recovery of the pancreatic insulin content and the beta-cell mass (Portha et al. 1974, 1979; Cantenys et al. 1981; Wang et al. 1994; Ferrand et al. 1995; Movassat et al. 1997). From postnatal day 4 onward, signs of regeneration became apparent in that numerous insulin positive cells were found throughout the acinar parenchyma and within the duct epithelium (Cantenys et al. 1981). The apparent budding of islets from the ducts was a common feature (Dutrillaux et al. 1982). A study of the mitotic rate in colchicine-treated animals suggested that many beta-cells were formed by mitosis of undifferentiated cells (Dutrillaux et al. 1982). However, the long-term impairment of the plasma glucose homeostasis (Portha et al. 1979) and the persistence of beta-cell mass reduced to 50% in adult animals (Portha et al. 1989) are proof that the regeneration process was incomplete.

Notably, we have shown that insulin therapy in the n0-STZ model during the neonatal hyperglycemic phase markedly improves the recovery of the insulin stores in the pancreas (Portha and Picon 1982) and the total beta-cell mass (Movassat et al. 1997), as well as leads to a significantly improved glucose tolerance in adult life (Portha and Picon 1982). Another factor that seems to be important to the efficiency of the regeneration process is the timing of the STZ injection, since it coincides with the normal development of the islet mass in the rat (McEvoy and Madson 1980). Indeed, we found that Wistar rats injected with STZ at seven days of age did not recover from the insult (Dutrillaux et al. 1982). In the n5-STZ rats, the lack of significant reaccumulation of insulin of the pancreas during the two weeks following the beta-cell insult contrasts with the recovery of the insulin stores as found in the pancreases of the n2-STZ and n0-STZ rats.

This confirms that there is some capacity of beta-cell regeneration in the neonatal rat pancreas and that the capacity for beta-cell regeneration in the Wistar strain decreases quickly during the first postnatal week and thereafter is no longer significant. It is noticeable that the Wistar n2-STZ model exhibits less severe diabetes than the Sprague–Dawley n2-STZ model. This difference is probably not related to gender selection because female Wistar rats were used in our studies, whereas other authors have used male Sprague–Dawley rats. We have checked that the hyperglycemia in the adult male Wistar n2-STZ rats was similar to that in the corresponding females (unpublished data). More realistic is the possibility that the difference in diabetes severity is linked to the selection of pups after STZ injection. In our study, all injected pups were investigated at adult age, whereas in other studies only those with blood glucose > 12 m*M* two days after injection were accepted (Bonner-Weir et al. 1981).

An alternative but not exclusive explanation is that regeneration of beta-cells in Sprague–Dawley neonates is less efficient than in the Wistar ones. This assumption that the capacity for beta-cell regeneration is dependent on the strain of rat is supported by the observation that severity of diabetes after neonatally induced STZ is increased in the SHR as compared with Wistar rats (Iwase et al. 1987). Additionally, the extent of beta-cell regeneration as well as its mechanism seems to be influenced by the timing of STZ injection and/or the strain of the rat. In the Sprague–Dawley n2-STZ model, the partial replenishment of the beta-cell mass has been claimed to reflect mainly replication of existing beta-cells (Weir et al. 1986) rather than neogenesis from undifferentiated precursors.

It has been recently proposed that the beta-cell regeneration after neonatally induced STZ was linked to hyperplasia of alpha-cells with an altered phenotype of increased GLP-1 synthesis; the target cells of GLP-1 were putatively immature beta-cells that coexpress proglucagon. These conclusions rely on a variant of the n-STZ model, using Wistar rats administered (70 mg/kg) on postnatal day 4, and the endocrine pancreas being examined between 4 and 40 days later. STZ-treated rats showed an approximate 60% loss of existing beta-cells and a moderate hyperglycemia of <15 m*M* glucose, with levels returning to near-control values after 20 days. Within pre-existing islets, there was increased cell proliferation in insulin- and glucagon-positive cells at eight days, as well as alpha-cell hyperplasia. This was associated with increased content of glucagon in the pancreas and in circulation. Pancreatic levels of glucagon-like polypeptide-1 (GLP-1) were increased eight days after STZ compared with controls and the GLP-1/glucagon ratio changed in favor of GLP-1. Administration of a GLP-1 receptor antagonist, GLP-1 (9-39), resulted in decreased recovery of beta-cells following STZ and worse glucose tolerance. Atypical glucagon-positive cells that colocalized Pdx-1 or Glut2 were found within islets; insulin-positive cells that colocalized glucagon and GLP-1 were found after eight days (Thyssen et al. 2006).

Increased clusterin expression has been described in alpha-cells of n-STZ neonatal rats, in close correlation with islet cell proliferation, higher transcription of insulin mRNA, and MAPKs activation. This led to the proposal that up-regulation of clusterin in alpha-cells might induce beta-cell proliferation and thus restore beta-cell population after islet injury and that, doing so, clusterin could be considered as a growth factor-like molecule stimulating islet-cell proliferation by paracrine action (Kim et al. 2001).

STZ treatment of neonatal rodents was also useful for evaluation of the regenerative capacity of endogenous beta-cells in rodents with programmed reduced beta-cell mass as encountered spontaneously in the GK rat (Movassat and Portha 1999) or in the normal rat after maternal undernutrition (Garofano et al. 2000). In both situations, it was reported that beta-cell neogenesis was severely depressed. This provided an explanation for the beta-cell mass deficit in the GK rat (the primary defect leading to T2D in the adult) (Movassat and Portha 1999) and for the impaired capacity of malnourished rats to adapt their beta-cell mass during aging or pregnancy with a subsequent glucose tolerance deterioration (Garofano et al. 2000).

n-STZ RATS: ADEQUATE TOOLS FOR TESTING BETA-CELL REGENERATIVE THERAPIES

The n-STZ rat models are particularly helpful to assess the effectiveness of potential therapeutics aimed to enhance beta-cell growth and/or survival and to understand the mechanisms of compensation of a reduced beta-cell mass. We have previously shown that neonatal treatment of n0-STZ neonates with exogenous insulin for a few days enhanced their total beta-cell mass (Movassat et al. 1997). More recently, using n0-STZ Wistar neonates, we have investigated the capacity of an early treatment with GLP-1 or exendin-4 to promote beta-cell regeneration and thereby to improve islet functions on a long-term range when animals become adults (Tourrel et al. 2001).

To this end, n0-STZ rats were submitted to GLP-1 or exendin-4 from postnatal day 2 to day 6 only, and their beta-cell mass and pancreatic functions were tested on day 7 and at two months. On day 7, both treatments increased body weight, decreased basal plasma glucose, decreased insulinemia, and increased pancreatic insulin content in n0-STZ rats. At the same age, the beta-cell mass, measured by immunocytochemistry and morphometry methods, was strongly increased in n0-STZ/GLP-1 and n0-STZ/Ex rats compared to n0-STZ rats, representing respectively

51 and 71% of the normal beta-cell mass; n0-STZ beta-cell mass represented only 21% of the control value. Such early improved beta-cell mass was maintained at adult age. These findings in the n0-STZ model indicate for the first time that GLP-1 or exendin-4 applied during the neonatal diabetic period exerts short- and long-term beneficial effects on beta-cell mass recovery and glucose homeostasis.

Activin A and betacellulin (BTC), previously reported to regulate differentiation of pancreatic beta-cells during development and regeneration of beta-cells in adults, have been found to promote regeneration of beta-cells also in n0-STZ neonatal rats. One-day-old STZ-treated rats have been administered for seven days with activin A and/or BTC. The pancreatic insulin content and beta-cell mass in rats treated with activin A and BTC were significantly increased compared with the control group on day 8 and at two months. Treatment with activin A and BTC significantly increased replication of preexisting beta-cells, ductal cells, and delta-cells. The number of islet cell-like clusters and islets was significantly increased by treatment with activin A and BTC. In addition, the number of insulin/somatostatin-positive cells and pancreatic duodenal homeobox-1/somatostatin-positive cells was significantly increased (Li et al. 2004). Similar beta-regenerative activity was exerted by a novel, alternatively spliced mRNA transcript of the BTC gene. This splice isoform, termed BTC-delta4, lacks the C-loop of the epidermal growth factor motif and the transmembrane domain as a result of exon 4 skipping. When recombinant BTC-delta4 was administered to n0-STZ neonatal rats, it significantly increased the insulin content, the beta-cell mass, and the numbers of islet-like cell clusters and PDX-1-positive ductal cells (Ogata et al. 2005).

More recently, ghrelin, a stomach-derived hormone functioning in multiple biological processes, was also shown to regenerate beta-cells in n0-STZ neonatal rats. Following administration of ghrelin from postnatal days 2–8, beta-cell mass, mRNA expression levels of insulin and Pdx 1, and beta-cell replication were significantly increased (Irako et al. 2006).

Oral sodium tungstate treatment of n-STZ rats has also been shown to regenerate a stable, functional pancreatic beta-cell population, which leads to and maintains normoglycemia. Tungstate treatment increased extra-islet beta-cell replication without modifying intra-islet beta-cell replication rates, together with an increase of insulin-positive cells located close to ducts and of PDX-1 positive cells scattered in the exocrine tissue, suggesting active neogenesis (Fernandez-Alvarez et al. 2004).

Treatment of n-STZ neonatal rats with conophylline (a vinca alkaloid extracted from the tropical plant *Ervatamia microphylla*) also activates beta-regeneration through an increased number of islet-like cell clusters and pancreatic duodenal homeobox-1-positive ductal cells. This demonstrates that conophylline induces differentiation of pancreatic precursor cells and increases the formation of beta-cells (Ogata et al. 2004).

SUMMARY

The study of the different variants of n-STZ rat models has provided new insights into the pathogenesis of T2D as the rats develop defects in insulin secretion and action that, in many ways, resemble those described in human T2D. The advantages of these easily produced rodent models are several fold:

They provide interesting models for the study of beta-cell growth and regeneration since new beta-cells partially repopulate the pancreas during the spontaneous remission occurring after overt neonatal diabetes. They are particularly suitable to assess the effectiveness of therapeutics aimed to enhance beta-cell growth and/or survival and to understand the mechanisms for compensation growth of the beta-cell mass.

Since the rats can be easily kept for more than 20 months under standard breeding conditions, they allow study in the adult of the long-term consequences of a primary reduced beta-cell mass. They are suitable to evaluate the effect of the major modulators of diabetes, such as composition of the diet, obesity, gestational diabetes, and aging. They have proved to be useful for studies related to the pathogenesis of nephropathy, vascular complications, hypertension, or osteopenia under conditions of chronic hyperglycemia.

They are valuable for testing hypoglycemic drugs and identifying their mechanisms of action.

REFERENCES

Ahren B, Skoglund G. (1989) Insulin secretion in neonatally streptozotocin-injected mice. *Diabetes Res* 11:185–190.

Alper G, Olukman M, Irer S, et al. (2005) Effect of vitamin E and C supplementation combined with oral antidiabetic therapy on the endothelial dysfunction in the neonatally streptozotocin injected diabetic rat. *Diabetes Metab Res Rev* 22:190–197.

Angel I, Burcelin R, Prouteau M, et al. (1996) Normalization of insulin secretion by a selective alpha2-adrenoceptor antagonist restores Glut-4 glucose transporter in adipose tissues of type II diabetic rats. *Endocrinology* 137:2022–2027.

Ashokkumar N, Pari L. (2005) Effect of N-benzoyl-D-phenylalanine and metformin on carbohydrate metabolic enzymes in neonatal streptozotocin diabetic rats. *Clin Chim Acta* 351:105–113.

Barbera A, Fernandez-Alvarez J, Truc A, et al. (1997) Effects of tungstate in neonatally streptozotocin-induced diabetic rats: Mechanism leading to normalization of glycemia. *Diabetologia* 40:143–149.

Bevilacqua S, Barett EJ, Smith D, et al. (1985) Hepatic and peripheral insulin resistance following streptozotocin-induced insulin deficiency in the dog. *Metabolism* 34:817–825.

Blondel O, Bailbe D, Portha B. (1989) Relation of insulin deficiency to impaired insulin action in NIDDM adult rats given streptozotocin as neonates. *Diabetes* 36:610–617.

Blondel O, Bailbe D, Portha B. (1990) Insulin resistance in rats with non-insulin-dependent diabetes induced by neonatal (5 days) streptozotocin: Evidence for reversal following phlorizin treatment. *Metabolism* 39:787–793.

Blondel O, Simon J, Chevalier B, Portha B. (1990) Impaired insulin action but normal insulin receptor activity in the liver of diabetic rats. Effect of vanadate treatment. *Am J Physiol* 258:E459–E467.

Bonner-Weir S, Orci L. (1982) New perspectives on the microvasculature of the islets of Langerhans in the rat. *Diabetes* 31:883–889.

Bonner-Weir S, Trent DF, Honey RN, Weir GC. (1981) Responses of neonatal islets to streptozotocin: Limited B-cell regeneration and hyperglycemia. *Diabetes* 30:64–69.

Cantenys D, Portha B, Dutrillaux MC, et al. (1981) Histogenesis of the endocrine pancreas in newborn rats after destruction by streptozotocin. An immunocytochemical study. *Virchows Arch Cell Pathol* 35:109–122.

Dachicourt N, Serradas P, Bailb D, et al. (1997) GLP-1(7-36) amide confers glucose sensitivity to previously glucose-incompetent B-cells in diabetic rats. *In vivo* and *in vitro* studies. *J Endocrinol* 155:369–376.

DeFronzo RA, Hendler R, Simonson D. (1982) Insulin resistance is a prominent feature of insulin-dependent diabetes. *Diabetes* 31:795–801.

Dutrillaux MC, Portha B, Roze C, Hollande E. (1982) Ultrastructural study of pancreatic B-cell regeneration in newborn rats after destruction by streptozotocin. *Virchows Arch Cell Pathol* 39:173–185.

Eizirik D, Sandler S, Welsh N, Hellerstrom C. (1988) Preferential reduction of insulin production in mouse pancreatic islet maintained in culture after streptozotocin exposure. *Endocrinology* 122:1242–1249.

Elian N, Bensimon C, Chapa O, et al. (1996) Pancreatic transplantation in experimental non-insulin dependent diabetic rats. *Transplantation* 61:696–700.

Fantus IG, Chayoth R, O'Dea L, et al. (1987) Insulin binding and glucose transport in adipocytes in neonatal streptozotocin-injected rat model of diabetes mellitus. *Diabetes* 36:654–660.

Fernandez-Alvarez J, Barbera A, Nadal B, et al. (2004) Stable and functional regeneration of pancreatic beta-cell population in nSTZ-rats treated with tungstate. *Diabetologia* 47:470–477.

Ferrand N, Astesano A, Phan HH, et al. (1995) Dynamics of pancreatic cell growth and differentiation during diabetes reversion in STZ-treated newborn rats. *Am J Physiol* 269:C1250–1264.

Frayon S, Pessah M, Giroix MH, et al. (1999) Golf identification by RT-PCR in purified normal pancreatic B cells and in islets from rat models of non-insulin dependent diabetes. *Biochem Biophys Res Commun* 254:269–272.

Friedman MI. (1972) Effects of alloxan diabetes on hypothalamic hyperphagia and obesity. *Am J Physiol* 222:174–178.

Garofano A, Czernichow P, Breant B. (2000) Impaired beta-cell regeneration in perinatally malnourished rats: A study with STZ. *FASEB J* 14:2611–2617.

Giroix M-H, Baetens D, Rasschaert J, et al. (1992) Enzymic and metabolic anomalies in islets of diabetic rats: Relationship to B-cells mass. *Endocrinology* 130:2634–2640.

Giroix MH, Portha B, Kergoat M, et al. (1983) Glucose insensitivity and amino-acid hypersensitivity of insulin release in rats with non-insulin-dependent diabetes: A study with the perfused pancreas. *Diabetes* 32:445–451.

Giroix M-H, Portha B, Kergoat M, Picon L. (1984) Glucagon secretion in rats with non-insulin-dependent diabetes: an *in vivo* and *in vitro* study. *Diabètes Metabol* 10:12–17.

Giroix MH, Rasschaert J, Bailbe D, et al. (1991) Impairment of glycerol phosphate shuttle in islets from rats with diabetes induced by neonatal streptozotocin. *Diabetes* 40:227–232.

Giroix MH, Rasschaert J, Sener A, et al. (1992) Study of hexose transport, glycerol phosphate shuttle and Krebs cycle in islets of adult rats injected with streptozotocin during the neonatal period. *Mol Cell Endocrinol* 83:95–104.

Giroix MH, Sener A, Bailbe D, et al. (1990) Impairment of the mitochondrial oxidative response to D-glucose in pancreatic islets from adult rats injected with streptozotocin during the neonatal period. *Diabetologia* 33:654–660.

Giroix MH, Sener A, Portha B, Malaisse WJ. (1993) Preferential alteration of oxidative relative to total glycolysis in pancreatic islets of two rat models of inherited or acquired type 2 (non-insulin-dependent) diabetes mellitus. *Diabetologia* 36:305–309.

Grill V, Ostenson CG. (1988) The influence of a diabetic state on insulin secretion: Studies in animal models of non-insulin dependent diabetes. In *Pathogenesis of Non-Insulin-Dependent Diabetes Mellitus,* ed. Grill V, Efendic S. Raven Press, New York, 93–106.

Grill V, Rundfeldt M. (1986) Abnormalities of insulin responses after ambient and previous exposure to glucose in streptozotocin-diabetic and dexamethasone-treated rats: Role of hyperglycemia and increased B-cell demands. *Diabetes* 35:44–51.

Grill V, Welsberg C, Ostenson CG. (1987) b-Cell sensitivity in a rat model of non-insulin-dependent diabetes. Evidence for a rapidly reversible effect of previous hyperglycemia. *J Clin Invest* 80:664–669.

Gutzeit A, Renold AE, Cerasi E, Shafrir E. (1979) Effect of diet-induced obesity on glucose and insulin tolerance of a rodent with a low insulin response (*Acomys cahirinus*). *Diabetes* 28:777–784.

Halban PA, Bonner-Weir S, Weir GC. (1983) Elevated proinsulin biosynthesis *in vitro* from a rat model of non-insulin-dependent diabetes mellitus. *Diabetes* 32:277–283.

Hasegawa G, Mori H, Sawada M, et al. (1989) Overt diabetes induced by overeating in neonatally STZ-treated impaired glucose tolerant mice: Long-term follow-up study. *Endocrinol Jpn* 36:471–479.

Inoue K, Cetkovic-Cvrlje M, Eizirik D, Grill V. (1994) Irreversible loss of normal beta-cell regulation by glucose in neonatally streptozotocin diabetic rats. *Diabetologia* 37:351–357.

Inoue K, Hisatomi A, Umeda F, Nawata H. (1992) Relative hypersecretion of amylin to insulin from rat pancreas after neonatal STZ treatment. *Diabetes* 41:723–727.

Irako T, Akamizu T, Hosoda H, et al. (2006) Ghrelin prevents development of diabetes at adult age in streptozotocin-treated newborn rats *Diabetologia.* In press [Epub ahead of print].

Iwase M, Kikuchi M, Nunoi K, et al. (1987) Diabetes induced by neonatal streptozotocin treatment in spontaneously hypertensive rats. *Metabolism* 36:654–657.

Iwase M, Nunoi K, Himeno H, et al. (1994) Susceptibility to neonatal streptozotocin-induced diabetes in spontaneously hypertensive rats. *Pancreas* 9:344–348.

Iwase M, Wakisaka M, Yoshinari M, et al. (1996) Effect of gonadectomy on the development of diabetes mellitus, hypertension, and albuminuria in the rat model. *Metabolism* 45:155–161.

Junod A, Lambert A, Stauffacher W, Renold AE. (1969) Diabetogenic action of streptozotocin: Relationship of dose to metabolic response. *J Clin Invest* 48:2129–2139.

Kahn A, Chandramouli V, Ostenson CG, et al. (1990) Glucose cycling in islets from normal and diabetic rats. *Diabetes* 39:456–459.

Kasono K, Yasu T, Kakehashi A, et al. (2004) Nicorandil improves diabetes and rat islet beta-cell damage induced by streptozotocin *in vivo* and *in vitro. Eur J Endocrinol* 151:277–285.

Kawai K, Suzuki S, Murayama Y, et al. (1991) Comparison of beta-cell function after long-term treatment with insulin, insulin plus gliclazide or gliclazide in neonatal streptozotocin-induced non-insulin-dependent diabetic rats. *Diabetes Res Clin Pract* 12:163–172.

Kawai K, Suzuki S, Watanabe Y, et al. (1991) Neonatal streptozotocin monosodium-glutamate injection in rats: A model of obese NIDDM. In *Lessons from Animal Diabetes. The Tokyo Workshop,* ed. Goto Y, Kanazawa Y. Smith–Gordon, London, 45–49.

Kergoat M, Bailbe D, Portha B. (1987a) Insulin treatment improves glucose-induced insulin release in rats with NIDDM induced by streptozotocin. *Diabetes* 36:971–977.

Kergoat M, Bailbe D, Portha B. (1987b) Effect of high-sucrose diet on the insulin secretion and insulin action. A study in the normal rat. *Diabetologia* 30:252–258.

Kergoat M, Bailbe D, Portha B. (1987c) Effect of high sucrose diet on insulin secretion and insulin action. A study in rats with non-insulin dependent diabetes induced by streptozotocin. *Diabetologia* 30:666–673.

Kergoat M, Guerre-Millo M, Lavau M, Portha B. (1991) Increased insulin action in rats with mild insulin deficiency induced by neonatal streptozotocin. *Am J Physiol* 260:E561–567.

Kergoat M, Portha B. (1985) *In vivo* hepatic and peripheral insulin sensitivity in rats with non-insulin-dependent diabetes induced by streptozotocin. *Diabetes* 34:574–579.

Kergoat M, Simon J, Portha B. (1988) Insulin receptor binding and tyrosine kinase activity in the liver of rats with non-insulin-dependent diabetes. *Biochem Biophys Res Commun* 152:1015–1022.

Kim BM, Ha YM, Shin YJ, et al. (2001) Clusterin expression during regeneration of pancreatic islets cells in streptozotocin-induced diabetic rats. *Diabetologia* 44:2192–2202.

Kurose T, Tsuda K, Ishida H, et al. (1992) Glucagon, insulin and somatostatin secretion in response to sympathetic neural activation in streptozotocin-induced diabetic rats. A study with the isolated perfused rat pancreas *in vitro*. *Diabetologia* 35:1035–1041.

Leahy JL, Bonner-Weir S, Weir GC. (1984) Abnormal glucose regulation of insulin secretion in models of reduced B-cell mass. *Diabetes* 33:667–673.

Leahy JL, Bonner-Weir S, Weir GC. (1985) Abnormal insulin secretion in a streptozotocin model of diabetes. Effects of insulin treatment. *Diabetes* 34:660–666.

Leahy JL, Weir GC. (1985) Unresponsiveness to glucose in a streptozotocin model of diabetes; inappropriate insulin and glucagon responses to a reduction of glucose concentration. *Diabetes* 34:653–659.

LeDoux SP, Woodley SE, Patton NJ, Wilson GL. (1986) Mechanisms of nitrosourea-induced b-cell damage. Alterations in DNA. *Diabetes* 35:866–872.

Leiter EH, Coleman DL, Eisenstein AB, Strack I. (1981) Dietary control of pathogenesis in C57 BL/KsJ *db/db* diabetic mice. *Metabolism* 30:554–562.

Levy J, Gavin JR, Fausto A, et al. (1984) Impaired insulin action in rats with non-insulin-dependent diabetes. *Diabetes* 33:901–906.

Levy J, Suzuki Y, Avioli L, et al. (1988) Plasma membrane phospholipid content in non-insulin-dependent streptozotocin-diabetic rats—Effect of insulin. *Diabetologia* 31:315–321.

Li L, Yi Z, Seno M, Kojima I. (2004) Activin A and betacellulin: Effect on regeneration of pancreatic beta-cells in neonatal streptozotocin-treated rats. *Diabetes* 53:608–615.

Malaisse WJ. (1991) Regulation of insulin secretion by nutrients. In *Nutrient Regulation of Insulin Secretion,* ed. Flatt PR. Portland Press, London, 83–100.

Mann JI. (1980) Diet and diabetes. *Diabetologia* 18:89–95.

McEvoy R, Madson KL. (1980) Pancreatic insulin, glucagon and somatostatin positive islet cell populations during perinatal development of the rat. Morphometric quantitation. *Biol Neonate* 37:248–254.

Melin B, Caron M, Cherqui G, et al. (1991) Increased insulin action in cultured hepatocytes from rats with diabetes induced by neonatal streptozotocin. *Endocrinology* 182: 1693–1701.

Mende S, Mühle W, Peters W. (1996) Influence of postnatal overnutrition and pregnancy on non-insulin dependent diabetes induced in Wistar rats by neonatal streptozotocin. *Horm Metab Res* 28:81–85.

Morin L, Giroix M-H, Gangnerau M-N, et al. (1997) Impaired phosphoinositide metabolism in glucose-incompetent islets of neonatally streptozotocin-diabetic rats. *Am J Physiol* 272:E737–E745.

Movassat J, Portha B. (1999) Beta-cell growth in the neonatal Goto–Kakisaki rat and regeneration after treatment with streptozotocin at birth. *Diabetologia* 42:1098–1100.

Movassat J, Saulnier C, Portha B. (1997) Insulin administration enhances growth of the beta-cell mass in streptozotocin-treated newborn rat. *Diabetes* 46:1445–1452.

Nakayama K, Murakami N, Ohta M, et al. (1995) Effects of M16209 on insulin secretion in isolated, perfused pancreases of normal and diabetic rats. *Eur J Pharmacol* 276:85–91.

Ogata T, Dunbar AJ, Yamamoto Y, et al. (2005) Betacellulin-delta4, a novel differentiation factor for pancreatic beta-cells, ameliorates glucose intolerance in streptozotocin-treated rats. *Endocrinology* 146:4673–4681.

Ogata T, Li L, Yamada S, Yamamoto Y, Tanaka Y, Takei I, Umezawa K, Kojima I. (2004) Promotion of beta-cell differentiation by conophylline in fetal and neonatal rat pancreas. *Diabetes* 53:2596–2602.

Ohnota H, Kitamura T, Kinikawa M, et al. (1996) A rapid and short-acting hypoglycemic agent KAD-1229 improves postprandial hyperglycemia and diabetic complications in streptozotocin-induced non-insulin-dependent diabetes mellitus rats. *Jpn J Pharmacol* 71:315–323.

Ohnota H, Koizumi M, Momose Y, Sato F. (1994) Normalization of impaired glucose tolerance by the short-acting hypoglycemic agent calcium(2S)-2-benzyl-3-(cis-hexahydro-2-isoindolinylcarbonyl)propionate dihydrate (KAD-1229) in non-insulin-dependent diabetes mellitus rats. *Can J Pharmacol* 73:1–6.

Ohta T, Furukawa N, Yonemori F, Wakitani K. (1999) JTT-608 controls blood glucose by enhancement of glucose-stimulated insulin secretion in normal and diabetic mellitus rats. *Eur J Pharmacol* 367:91–99.

Okabayashi Y, Otsuki M, Ohki A, et al. (1989) Increased B-cell secretory responsiveness to ceruletide and TPA in streptozotocin-induced mildly diabetic rats. *Diabetes* 38:1042–1047.

Okamoto H. (1981) Regulation of proinsulin synthesis in pancreatic islets and a new aspect to insulin-dependent diabetes. *Mol Cell Biochem* 37:43–61.

Östenson CG, Cattaneo AG, Doxey JC, Efendic S. (1989) a-Adrenoceptors and insulin release from pancreatic islets of normal and diabetic rats. *Am J Physiol* 257:E439–E443.

Östenson CG, Efendic S, Grill V. (1990) Abnormal regulation by glucose and somatostatin secretion in the perfused pancreas of NIDDM rats. *Pancreas* 5:347–353.

Östenson CG, Grill V. (1985) Glucose exerts opposite effects on muscarinic receptor binding to A and B cells of the endocrine pancreas. *Endocrinology* 116:1741–1745.

Östenson CG, Grill V. (1987) Evidence that hyperglycemia increases muscarinic binding in pancreatic islets of the rat. *Endocrinology* 121:1705–1710.

Östenson C, Grill V, Roos M. (1989) Studies on sex dependency of b-cell susceptibility to streptozotocin in a rat model of type II diabetes mellitus. *Exp Clin Endocrinol* 93:241–247.

Pari L, Ashokkumar N. (2005) Effect of N-benzoyl-D-phenylalanine, a new potential oral antidiabetic agent, in neonatal streptozotocin-induced diabetes in rats. *Pharmacol Rep* 57:498–503.

Portha B. (1985) Decreased glucose-induced insulin release and biosynthesis by islets of rats with non-insulin-dependent diabetes: Effects of tissue culture. *Endocrinology* 117:1735–1741.

Portha B, Blondel O, Serradas P, et al. (1989) The rat models of non-insulin dependent diabetes induced by neonatal streptozotocin. *Diabetes Metab* 15:61–75.

Portha B, Chamras H, Broer Y, et al. (1983) Decreased glucagon-stimulated cyclic AMP production by isolated liver cells of rats with type II diabetes. *Mol Cell Endocrinol* 32:13–26.

Portha B, Giroix MH, Picon L. (1982) Effect of diet on glucose tolerance and insulin response in chemically diabetic rats. *Metabolism* 21:1194–1199.

Portha B, Giroix MH, Serradas P, Malaisse WJ, et al. (1988) Insulin production and glucose metabolism in isolated pancreatic islets of rats with NIDDM. *Diabetes* 37:226–233.

Portha B, Goursot R, Giroix MH, et al. (1983) Experimental hypothalamic or genetic obesity in the non-insulin-dependent diabetic rat. *Diabetologia* 25:51–55.

Portha B, Kergoat M. (1985) Dynamics of glucose-induced insulin release during the spontaneous remission of streptozotocin diabetes induced in the newborn rats. *Diabetes* 34:574–579.

Portha B, Kergoat M, Blondel O, et al. (1995) Pathogenesis of impaired insulin action in rat models of insulin deficiency. In *Lessons from Animal Diabetes,* ed. Shafrir E. Smith–Gordon, London, 5:83–91.

Portha B, Levacher C, Picon L, Rosselin G. (1974) Diabetogenic effect of streptozotocin during the perinatal period in the rat. *Diabetes* 23:889–895.

Portha B, Picon L. (1982) Insulin treatment improves the spontaneous remission of neonatal streptozotocin diabetes in the rat. *Diabetes* 31:165–169.

Portha B, Picon L, Rosselin G. (1979) Chemical diabetes in the adult rat as the spontaneous evolution of neonatal diabetes. *Diabetologia* 17:371–377.

Portha B, Serradas P, Bailbe D, et al. (1993) Effect of Benfluorex on insulin secretion and insulin action in streptozotocin diabetic rats. *Diabetes Metab Rev* 9:57–63.

Portha B, Serradas P, Blondel O, et al. (1988) Relation between hyperglycemia and impairment of insulin secretion and action. Information from the n-STZ rat models. In *Lessons from Animal Diabetes*, ed. Shafrir E. Smith–Gordon, London, 3:334–341.

Reaven GM. (1980) How high the carbohydrate? *Diabetologia* 19:409–413.

Reaven GM, Sageman WS, Swenson RS. (1977) Development of insulin resistance in normal dogs following alloxan-induced insulin deficiency. *Diabetologia* 13:459–462.

Rerup C. (1980) Drugs producing diabetes through damage of the insulin secreting cells. *Pharmacol Rev* 22:485–518.

Schaffer S, Tan B, Wilson G. (1985) Development of a cardiomyopathy in a model of non-insulin-dependent diabetes. *Am J Physiol* 248:H179–H185.

Schaffer SW, Mozaffari S, Artman M, Wilson G. (1989) Basis for myocardial mechanical defects associated with non-insulin-dependent diabetes. *Am J Physiol* 256:E25–E30.

Schaffer SW, Seyed-Mozaffari M, Cutcliff CR, Wilson GL. (1986) Postreceptor myocardial metabolic defect in a rat model of non-insulin-dependent diabetes mellitus. *Diabetes* 35:593–597.

Serradas P, Bailbé D, Blondel O, Portha B. (1991) Abnormal B-cell function in rats with non-insulin dependent diabetes induced by neonatal streptozotocin: Effect of *in vivo* insulin, phlorizin or vanadate treatments. *Pancreas* 6:54–62.

Serradas P, Bailbé D, Portha B. (1989) Chronic gliclazide treatment improves the *in vitro* glucose-induced insulin release in rats with non-insulin-dependent diabetes induced by neonatal streptozotocin. *Diabetologia* 32:577–584.

Serradas P, Blondel O, Bailbé D, Portha B. (1993) Benfluorex normalizes basal hyperglycemia and reverses hepatic resistance in streptozotocin-diabetic rats. *Diabetes* 43:564–570.

Shinde UA, Mehta AA, Goyal RK. (2001) Effect of chronic treatment with Bis(maltolato)oxovanadium (IV) in rat model of non-insulin-dependent diabetes. *Indian J Exp Biol* 39:864–870.

Thorens B, Weir GC, Leahy JL, et al. (1990) Reduced expression of the liver/beta cell glucose transporter isoform in glucose-insensitive pancreatic beta cells of diabetic rats. *Proc Natl Acad Sci USA* 87:6492–6496.

Thyssen S, Arany E, Hill DJ. (2006) Ontogeny of regeneration of beta-cell in the neonatal rat following treatment with streptozotocin. *Endocrinology* 147:2346–2356.

Tormo MA, Leon-Quinto T, Saulnier C, et al. (1997) Insulin secretion and glucose tolerance after islet transplantation in rats with non-insulin dependent diabetes induced by neonatal streptozotocin. *Cell Transplant* 6:23–32.

Tourrel C, Bailbè D, Meile MJ, et al. (2001) Glucagon-like peptide-1 and exendin-4 stimulate b-cell neogenesis in streptozotocin treated newborn rats, resulting in persistently improved glucose homeostasis at adult age. *Diabetes* 50:1562–1570.

Trent DF, Fletcher DJ, May JM, et al. (1984) Abnormal islet and adipocyte function in young b-cell deficient rats with near normoglycemia. *Diabetes* 33:170–175.

Triadou N, Portha B, Picon L, Rosselin G. (1982) Experimental chemical diabetes and pregnancy in the rat. Evolution of glucose tolerance and insulin response. *Diabetes* 31:75–79.

Tsuura Y, Ishida H, Okamoto Y, et al. (1992) Impaired glucose sensitivity of ATP-sensitive K^+ channels in pancreatic B-cells in streptozotocin-induced NIDDM rats. *Diabetes* 41:861–865.

Umeda F, Inoue K, Hirano K, et al. (1992) Alterations in femoral bone histomorphometry and vitamin D metabolism in neonatal streptozotocin-induced diabetic rat. *Fukuoka Igaku Zasshi* 83:403–408.

Wada M, Iwase M, Wakisaka M, et al. (1995) A new model of diabetic pregnancy with genetic hypertension: Pregnancy in spontaneously hypertensive rats with neonatal streptozotocin-induced diabetes. *Am J Obstet Gynecol* 72:626–630.

Wakisaka M, Nunoi K, Iwase M, et al. (1988) Early development of nephropathy in a new model of spontaneously hypertensive rat with non-insulin-dependent diabetes mellitus. *Diabetologia* 31:291–296.

Wang RN, Bouwens L, Klppel G. (1994) Beta-cell proliferation in normal and streptozotocin-treated newborn rats: Site, dynamics and capacity. *Diabetologia* 37:1088–1096.

Weir GC, Clore EE, Zmachinsky CJ, Bonner-Weir S. (1981) Islet secretion in a new experimental model for non-insulin-dependent diabetes. *Diabetes* 30:590–595.

Weir GC, Leahy JL, Bonner-Weir S. (1986) Experimental reduction of B-cell mass: Implications for the pathogenesis of diabetes. *Diabetes Metab Rev* 2:125–161.

Welsh N, Hellerström C. (1990) *In vitro* restoration of insulin production in islets from adult rats treated neonatally with streptozotocin. *Endocrinology* 126:1842–1848.

Welsh N, Pääbo S, Welsh M. (1991) Decreased mitochondrial gene expression in isolated islets of rats injected neonatally with streptozotocin. *Diabetologia* 34:626–631.

York DA, Bray GA. (1972) Dependence of hypothalamic obesity on insulin, the pituitary and the adrenal gland. *Endocrinology* 90:885–894.

Young TK, Liu AC. (1965) Hyperphagia, insulin and obesity. *Chin J Physiol* 19:247–253.

11 The Rhesus Monkey (*Macaca mulatta*) Manifests All Features of Human Type 2 Diabetes

Barbara C. Hansen and Xenia T. Tigno

CONTENTS

THE RHESUS MONKEY (*MACACA MULATTA*) AS IDEAL MODEL OF HUMAN TYPE 2 DIABETES

The nonhuman primate, particularly the rhesus monkey (*Macaca mulatta*), provides the most human-like extensively studied model of metabolic disorders. Hamilton and Ciaccia (1978) presented the first findings on diabetes mellitus in the rhesus monkey. Several early studies documented overt diabetes mellitus in individual rhesus monkeys, including those of Kirk and colleagues (1972), Di Giacomo and associates (1971), and Uno and coworkers (1985). A number of comprehensive reviews have been published regarding diabetes mellitus in nonhuman primates (Howard 1983; Howard and Yusada 1990; Hansen, 1992, 1996, 2004e; Hansen et al. 1995; Bodkin 1996). Many primate species that have been identified to develop type 2 diabetes mellitus (T2D), often in the zoological literature, are summarized in table 11.1.

The Obesity, Diabetes, and Aging Research Center (ODARC) situated at the University of South Florida currently houses 150 monkeys ranging in age from 8 to 40 years. The colony includes normal lean adult, obese with or without the metabolic syndrome (MetSynd), and obese with T2D. For convenience in reporting about diabetes in monkeys, we have used the American Diabetes Association (ADA) criteria

TABLE 11.1
Primate Species with Diabetes

Cebus apella	Capuchin
Cercopithecus aethiops	African green monkey, vervet, or grivet
Cercopithecus cephus	Moustached green guenon
Cercopithecus diana	Diana monkeys
Cercopithecus mitis	Blue, Sykes, silver, golden, or Samango monkey
Cercopithecus mona	Mona monkey
Colobus polykomos	King Colobus monkey
Galago crassicaudatus	African bushbaby
Macaca fasicularis	Cynomolgus or crab-eating macaque
Macaca cyclopis	Taiwan or Formosan rock macaque
Macaca mulatta	Rhesus monkey
Macaca nemestrina	Pig-tailed macaque
Macaca radiata	Bonnet macaque
Mandrillus leucophaeus	Mandrill baboon
Papio hamadryas	Sacred baboon
Saguinus fuscicollis	Tamarin
Saquinus oedipus	Tamarin
Saimiri sciureus	Squirrel monkeys
Pan troglodytes	Chimpanzee
Pan paniscus	Bonobo

Source: Hansen, BC. (2004) In *Diabetes Mellitus: A Fundamental and Clinical Text*, ed. LeRoith, D, Olefsky, JM, and Taylor, S. Philadelphia: Lippincott Williams and Wilkins, 1059–1074.

for classification of impaired glucose tolerance and diabetes as defined for humans (ADA-Expert Committee, 2003; ADA, 2006). Thus, impaired fasting glucose is defined as ranging from 100 to 125 mg/dL, and diabetes is defined as a glucose ≥126 mg/dL. For normal glycemic humans, the glucose level averages ~85 mg/dL, while for rhesus monkeys the equivalent fasting level is ~65 mg/dL. In view of this difference in normal fasting glucose, it is alternatively possible to define the respective threshold for monkeys as 20 mg/dL lower than human levels (e.g., impaired fasting glucose is identified for rhesus monkeys as 80–105 mg/dL and a glucose level of ≥105 mg/dL may be used to identify overt diabetes in rhesus monkeys.

By tradition, we and others have used the human definitions in publications to date. In monkeys, we have further defined MetSynd as the presence of impaired fasting glucose, impaired glucose tolerance (glucose disappearance rates during an IVGTT of <2.0%/min), or insulin resistance (insulin-mediated glucose disposal rate <7.5 mg/kg FFM/min), and two or more of the following characteristics: adiposity (body fat > 22%); hyperinsulinemia (fasting insulin values > 70 microunits/mL), hypertension (systolic ≥ 130 mmHg; diastolic ≥ 80 mmHg), and dyslipidemia (fasting triglycerides ≥ 80 mg/dL or HDL-cholesterol < 60 mg/dL).

There are no statistically valid databases from which to derive the incidence or prevalence of T2D in monkeys. Our laboratory has evaluated and followed prospectively several hundred adult middle-aged and older monkeys. The earliest observed age of onset of overt diabetes is 10 years, and the latest onset was at age 29. Onset most commonly occurs between ages 15 and 22. Our experience suggests that if all rhesus monkeys lived out their full natural life span under protected and individual housing with *ad libitum* feeding of chow, then approximately 25% of monkeys would develop overt T2D. An additional 25–40% will manifest adiposity and features of the MetSynd during their lifetimes. An estimated ~35% will remain lean, normoglycemic, normolipidemic, and normotensive for their lifetime.

There have been no cases of spontaneous type 1 diabetes reported to date in any nonhuman primate. The natural history of diabetes in rhesus monkeys has been described in previous publications (Hansen et al. 1986, Hansen and Bodkin 1990; Bodkin et al. 1989; Hansen 2004a). Given the existence of extensive medical records of all monkeys in the colony, frequent assessments of the metabolic status of each individual in the colony, excellent husbandry and careful monitoring of each subject (including daily food intake and random glucose values), we are able to identify which normoglycemic monkeys are actually in an early prediabetic phase. The colony serves as a valuable resource for studies related to aging and the natural development of aging-associated diseases, as well as for therapeutic trials where novel interventions can be initiated prior to or after the conversion to T2D. Our previous publication on age-related changes identifies disease-related perturbations in common clinical variables—those that occur in the course of natural aging (Erwin et al. 2004; Tigno et al. 2004).

Furthermore, observations on these monkeys have shown that all of the diabetes complications of diabetes in humans are also present in monkeys. These include retinopathy (Cornblath et al. 1989; Kim et al. 2004; Johnson et al. 2005), neuropathy (Cornblath et al. 1989), nephropathy (Cusumano et al. 2002), and other microvascular derangements (Tigno et al. 2006). Dyslipidemia, again a recognized predictor of cardiovascular disease and diabetes in humans, also develops in rhesus monkeys—despite a low-fat, low-cholesterol, and high-fiber diet—and this hyperlipidemia demonstrates a profile very similar to that found in humans (Ding et al. 2006). This high degree of resemblance between monkey and human complications makes the rhesus model far superior to any other model, natural or induced, for T2D and the most appropriate one for studies related to diabetic complications. Limitations include the numbers of animals available in existing colonies and the cost of maintenance and study.

OVERVIEW OF DISCOVERIES RELATED TO OBESITY AND DIABETES

Acquisition of Obese Monkeys and Early Characterization of Type 2 Diabetes

The ODARC colony was started in 1969 with a few young adult lean rhesus monkeys. In its first phase, increased adiposity was experimentally induced by forced overfeeding. In 1980 with the assistance of Dr. Eliot Stellar, the provost of the University of Pennsylvania, a large group of obese middle-aged monkeys was obtained from the primate colony originally described and studied by the late Dr. Charles Hamilton (Hamilton and Ciaccia 1978). These unrelated, laboratory-maintained monkeys became the initial group of longitudinally studied primates. This animal resource has grown and developed over the last 30 years, with gradual increase in the steady-state colony size to >150 monkeys.

The characterization in the rhesus monkey of obesity and its spontaneous development, including physiological and morphological characteristics of increased adiposity, have been well established (Jen et al. 1985; Hansen et al. 1988; Bodkin et al. 1989, 1990; Hotta et al. 1996, 1999; Hotta, Gustafson, Ortmeyer, et al. 1998; Hotta, Gustafson, Yoshioka, et al. 1998; Hotta, Matsukawa, et al. 2001; Hansen, 2004b), as well as the association of various metabolic disorders, including hyperinsulinemia, hyperleptinemia (Bodkin et al. 1996), insulin resistance, reduced adiponectinemia (Hotta, Funahashi, et al. 2001), and dyslipidemia, including increased plasma VLDL triglycerides and decreased HDL cholesterol concentrations (Uno et al. 1985; Hannah et al. 1991; Bodkin et al. 1993; Ding et al. 2006). Early studies in the ODARC primates also provided evidence of increased beta-cell tropin concentrations in obese monkeys (Morton et al. 1992) and decreased plasma insulin-like growth factor-I (IGF-I) concentrations in aging monkeys (Bodkin et al. 1991).

Description of Natural History of T2D and the Metabolic Syndrome

Monkeys gradually progress from overweight to obesity and then some naturally and spontaneously progress to classical non-insulin-dependent T2D (Hansen 1985; Hansen and Bodkin 1986). This disease as manifest in rhesus monkeys appears to be identical to the usual obesity-associated T2D of humans, including the same *in vivo* molecular and biochemical pathophysiological disturbances. The prediabetic and diabetic rhesus monkey is therefore an exciting and valuable model providing insight into human diabetes.

Initially, serial observations in young, lean monkeys were obtained to establish the reference values under well-controlled conditions, as well as to provide measurements in older obese but metabolically normal monkeys. All monkeys were systematically followed longitudinally (some for >20 years) with measurements taken at least annually. These studies demonstrated that T2D was a progressive disorder with distinct serial changes in body weight, body fat, and plasma insulin (Hansen and Bodkin 1986) and acute insulin response to glucose and glucose tolerance (Hansen 1985). The metabolic changes were found to be transitional and progressive rather than clustered or stepwise in changes. Hansen and coworkers proposed the term "phases" to identify the gradual metabolic transition and breakpoints in the diabetic progression. Briefly, phase 1 includes all normal lean young adult monkeys; phase 2 comprises aging, nonobese, normal monkeys and obese but otherwise normal monkeys; phases 3–7 are composed of monkeys in a period of progression through obesity, hyperinsulinemia, and the MetSynd; phase 8 includes overt clinical T2D (Hansen and Bodkin 1986; Hansen et al. 1988); and phase 9 is made up of severe insulin-treated diabetes (Bodkin et al. 1989).

These initial primate studies were instrumental in establishing that the earliest detectable abnormality in the development of T2D in primates was not an absolute deficiency of insulin, but rather an enhanced basal insulin secretion and enhanced insulin response to glucose—findings that have now been well supported in other nonhuman primate species and in humans. The late change in fasting plasma glucose (hence its unsuitability for use as a very early criterion) was also an important finding. More recent cross-sectional and longitudinal data from the colony further validate the conclusion that hyperinsulinemia is a reliable predecessor to the future development of T2D. As noted, insulin-resistant monkeys are those with M-rates < 7.5 mg/kg FFM/min as determined from a euglycemic-hyperinsulinemic clamp (see discussion in the next section). The M-rate represents maximal insulin-stimulated glucose uptake primarily into peripheral tissues.

Figure 11.1 demonstrates the characteristics of a cross-sectional sample of our rhesus monkeys. Prediabetic MetSynd subjects weigh the most and have circulating levels of insulin significantly greater than normal (fig. 11.1). At this stage, glucose intolerance and insulin resistance are evident. The panels clearly show that long before diagnosis of T2D, glucose homeostasis and insulin action are already severely impaired. The ability to characterize this phase provides the special advantage that interventions, including weight reduction and therapeutic agents, can be initiated early in the course of the disease so that further progression into overt diabetes may be averted. This in fact has been confirmed by different studies performed in the laboratory where calorie restriction (Hansen et al. 1999) and pharmaceutical agents have been demonstrated to reverse some of the symptoms and disturbances in metabolic variables.

Insulin Resistance and Use of the Euglycemic Clamp

An important step in study of the natural history of diabetes in the rhesus monkey followed from the development of the euglycemic-hyperinsulinemic clamp (De Fronzo et al. 1979) for the assessment of human insulin sensitivity *in vivo.*

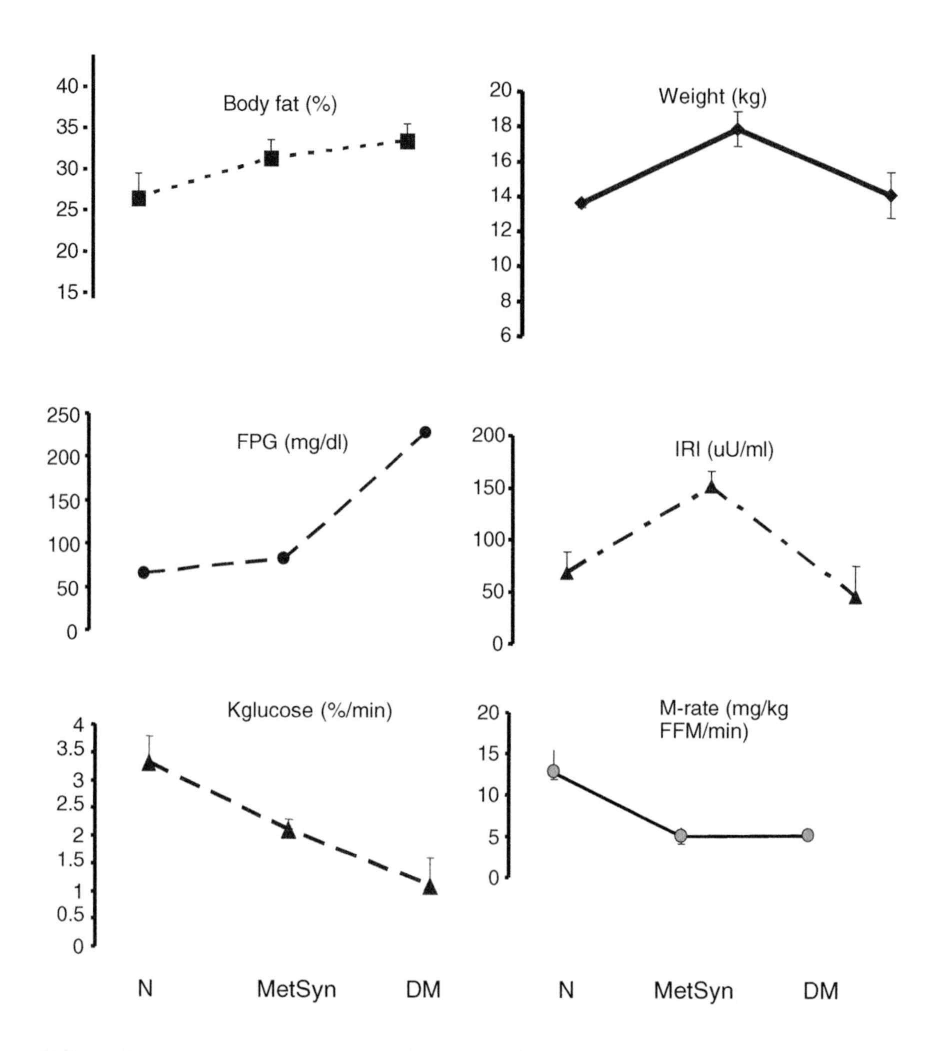

FIG. 11.1 The data show that body weight, body fat, and fasting insulin are all increased in the prediabetic/metabolic syndrome subjects compared with normal subjects, while the fasting glucose values lie between those of the truly normal and overtly diabetic monkeys. In prediabetes, insulin sensitivity (M-rate) and glucose tolerance (K_{gluc}) are already declining, even though fasting plasma glucose values are well below the diabetic threshold.

This method was then adapted for use in rhesus monkeys (Bodkin et al. 1989) and in diabetes research is still considered to be the gold standard for the measurement of *in vivo* insulin-stimulated glucose uptake. In the rhesus monkey, it is carried out under light ketamine hydrochloride sedation (10–15 mg/kg) or under fentanyl citrate anesthesia (1–2 μg/kg/h). Briefly, a priming dose of regular insulin is administered followed by continuous insulin infusion (400 mU/m^2 body surface area/min^{-1} for 120 min) with a variable 20% glucose infusion in order to maintain plasma glucose at a target steady-state glucose level of 85 mg/dL. The insulin dose was selected to assure maximum effect, as previously determined in dose–response curves (Bodkin et al. 1989).

The glucose disposal rate (M-rate) is calculated during the steady-state plasma glucose level, usually between one and two hours after the initiation of the clamp and is adjusted for fat-free mass, as determined by tritiated water or by dual energy x-ray absorptiometry (DXA) (Bodkin et al. 1993).

In the primates (as in humans), the euglycemic-hyperinsulinemic clamp technique allows the estimation of insulin sensitivity, which usually develops prior to defective beta-cell function and is considered by many to be the primary defect in T2D. Since 1985, the study of insulin sensitivity has been the focus of many diabetes research groups, who have actively sought to identify the major early defect(s) in the insulin action pathway. Many groups have contributed to the advances made in this period and to the present-day understanding of the pathology of insulin resistance. Tissue biopsies obtained under basal and insulin-stimulated conditions during the euglycemic-hyperinsulinemic clamp have been key to these studies.

The optimal size of the rhesus monkey allowing concurrent tissue biopsies of key organs involved in the development of T2D (muscle, liver, subcutaneous and omental adipose tissue) during the euglycemic-hyperinsulinemic clamp has provided insight into the possible tissue specific defect(s) in insulin action pathways in monkeys and humans. Findings in the T2D rhesus have included early defective insulin action on skeletal muscle glycogen synthase (Ortmeyer, Bodkin, and Hansen 1993b) and cAMP-dependent protein kinase activity (Ortmeyer 1997) and adipose tissue glycogen synthase (Ortmeyer, Bodkin, and Hansen 1993a), although there is normal insulin action on liver glycogen synthase in the early phases of insulin resistance (Ortmeyer et al. 1997; Ortmeyer and Bodkin, 1998).

In addition, in the insulin-resistant prediabetic phases, urinary excretion of D-chiroinositol (a component of a putative mediator or modulator of insulin action) decreases in diabetic monkeys and humans (Kennington et al. 1990; Ortmeyer, Bodkin, Lilley, et al. 1993). Insulin action on muscle glycogen synthase and phosphorylase activity was enhanced when D-chiroinositol was administered to monkeys (Ortmeyer 1995). Glucose and insulin lowering in insulin-resistant and diabetic monkeys was also enhanced after D-chiroinositol administration (Ortmeyer, Huang, et al. 1993; Ortmeyer et al. 1995).

STUDIES OF PHARMACOLOGICAL AGENTS

Rhesus monkeys have served in this and other laboratories as excellent models for the study of mechanisms of action and efficacy of many pharmaceutical agents. Over the past 10 years we have reported studies involving antiobesity agents (Young et al. 1999; Hansen 2000), lipid-lowering agents (Hannah et al. 1995), antihypertensives (Bodkin and Hansen 1995), insulin sensitizers (Ortmeyer et al. 1995; Ortmeyer et al. 2000, 2005; Oliver et al. 2001; Winegar et al. 2001; Schafer et al. 2004), and glucose-lowering agents (Bodkin et al. 1991). These studies in the primates offer many advantages in the complex steps of pharmaceutical compound development, including the identification of appropriate target groups (e.g., prediabetic vs. diabetic subjects), appropriate dose–response data, the optimal therapeutic window, bioavailability, and pharmacokinetic data, all obtained under well-controlled conditions.

Methods for Obtaining Optimal Results

Well-Controlled Environmental Conditions

As part of the ODARC standard operating procedures, temperature and humidity are consistently maintained in each of the animal areas at 72°F or 22°C and 30–70% relative humidity as noted in the *Guide for the Care and Use of Laboratory Animals* (National Institutes of Health Committee on Care and Use of Laboratory Animals 1996).

Ad Libitum *Food Intake*

The food intake and body weights of each animal are monitored continuously. Food intake measurement is carried out and recorded at a minimum of twice daily on each monkey. Fresh food (Monkey Diet #5038, Purina-Mills, Inc., St. Louis, Missouri) is provided to the monkey and recorded each morning and each afternoon.

Daily food intake evaluation is particularly important to the management of the diabetic monkeys due to the need to have ample fresh food available throughout the day and night in order to minimize the possibility of a hypoglycemic reaction. Any alterations from normal feeding behavior are immediately investigated. A monkey with decreased food intake from its normal level is provided with a fruit supplement and carefully evaluated to determine the cause of the change in appetite. If the monkey is receiving daily insulin, the dose is adjusted with any significant decrease in food intake. All monkeys receive a daily multivitamin.

Body Weight Monitoring

Body weight of all monkeys is determined a minimum of once weekly. The weights are recorded on the individual food intake sheets so that fluctuations or variability can be noted immediately. We have found that decreased food intake and even small decrease in body weight, particularly in the older and/or diabetic monkeys, is almost always indicative of an impending disorder or illness. Thus, this regimen allows for early detection and follow-up of such problems.

Activity

While there are some variations in normal activity levels across monkeys, most adult middle-aged rhesus monkeys are relatively sedentary. Thus, if left alone, adult monkeys do not choose to exercise in any large measure. The genetic abnormalities that underlie the expression of obesity and diabetes in humans are likely to be present in a substantial number of rhesus monkeys, although the identification of such genetic contributors is still in the future. No single gene mutation has been identified to cause obesity in nonhuman primates. The phenotypic expression is enhanced by *ad libitum*-fed conditions. The development of obesity and diabetes in free-ranging monkeys on the island of Cayo Santiago has been also been described (Schwartz et al. 1993). The Cayo Santiago monkeys live in an environment protected from predators, with *ad libitum* provisioning of food by humans. Their activity is unrestricted.

Health Monitoring

The monkeys of the ODARC colony are generally in excellent health, especially given the advanced ages of many of the prediabetic and diabetic monkeys. A practical overview of basic protocols and clinical monitoring has been previously published.

The oldest monkey studied in the ODARC colony was a female that lived to be 40 years of age with T2D and had been insulin treated for 10 years prior to death. As in humans, older monkeys may develop occasional medical problems. We have found such problems to include occasional rashes and mild skin breakdown in the very obese monkeys, particularly in the abdominal and inguinal areas, infrequent but occasional diabetic-related skin ulcers of the heel or knee, and tartar buildup or dental caries. In our experience, these medical conditions have always been quickly resolved by appropriate treatment, such as a short period of skin care with topical antibiotics, appropriate antibiotics, and dressings in the care of the diabetics and dental scaling and/or tooth extraction as required. Baseline chemistry and hematology data are available on each monkey in the colony, and these data are very helpful for evaluating changes in white cell and differential counts, BUN, and creatinine and/or liver function tests.

For the monkey's health and well-being, specific important parameters such as food intake, body weight, blood pressure, and hematocrit are carefully checked and recorded prior to each experiment. Hematocrit is particularly important because it reflects the overall general health of the monkey and its potential to tolerate blood sampling during an experimental procedure. The general health of the animals is monitored, including routine tuberculosis testing and physical examinations and all veterinary, medical, surgical, and/or emergency care, including but not limited to radiological examinations and cardiopulmonary evaluations as needed.

Special Care Considerations

Special care is required for many animals in the colony due to aging, obesity, or diabetes-associated complications. The longitudinal and acute metabolic status of each monkey is continuously monitored; therefore, all prediabetic (fasting plasma glucose from 80 to 125 mg/dL) are carefully monitored in regard to stability of body weight, food and water intake, degree of glucosuria, and degree of hyperglycemia. Further, diabetic monkeys are carefully maintained on appropriate insulin therapy so as to assure adequate and appropriate food intake, mild glucosuria (to minimize any possibility of a hypoglycemic reaction), and stable body weight.

Blood and urine glucose levels are monitored and insulin therapy is carefully adjusted for each individual diabetic monkey. All diabetic monkeys (fasting plasma glucose > 126 mg/dL) are carefully maintained so as to assure adequate and appropriate food intake and a stable body weight. The insulin doses are adjusted based on twice daily urine glucose and ketone measurements and on blood glucose determinations by glucometer.

For over 15 years, we have had very good success with insulin dosing adjusted on the basis of glucose and ketone monitoring, probably due to the combination of carefully monitored food intake and activity levels of the diabetic monkeys, the regularity and structure of the insulin treatment/monitoring regimen, and close vigilance. Long-acting insulin plus regular or NPH insulin are given each morning. Humulin insulin is used since monkey insulin has been shown to be structurally identical to human insulin (Naithani et al. 1984) and therefore the development of insulin antibodies in the diabetic monkeys is minimized. Only the free insulin assay is used to determine insulin concentrations of plasma samples from the insulin-treated diabetic monkeys.

The daily insulin doses of the diabetic monkeys range from 5 to 250 units/day. The insulin (U-100) is given subcutaneously in a standard U-100 syringe with a 28-gauge microfine needle. All of the monkeys are accustomed to their daily dose and tolerate it well, and most will readily present an arm or leg when shown the syringe. A small food reward is given.

Primate Environmental Enrichment

The objectives of the ODARC primate enrichment program are to utilize unique enrichment resources and techniques to provide a consistently nonthreatening and interactive environment for the primates. Toys and other objects are available for manipulation.

As all people who work closely with nonhuman primates know, environmental enrichment is a very complex subject and an integral component of primate husbandry and care. The primate environment can be the richest place in the world and yet it can also be the most lethal. In fact "free roaming" monkeys, such as those on the island of Cayo Santiago, are in a very rich environment, but rarely live past the age of 15 years, unless they are placed in an individually protected environment.

Another environmentally enriched setting is the corral-maintained colony. However, it is clear from reports of other colonies that corral maintenance produces significant problems of aggression that are, in fact, lethal for some primates. The dominance hierarchy means that some animals will be injured or deprived, while others will exist at the opposite end of the spectrum. Unfortunately, corral maintenance also makes difficult the early identification of a sick monkey or an impending illness. In fact, most diabetic animals identified in the corral situation are identified only after major weight loss occurs and at a point where their disease has become life threatening. This contrasts with the very early detection and treatment of laboratory-maintained diabetic primates at a point when health can be maintained by optimal treatment.

Within the care program of the ODARC are several key contributors to environmental enrichment. Social grouping of cages and facilitation of visual and auditory interactions between monkeys and between monkeys and caretakers are carefully considered. This colony consists of normal, prediabetic, diabetic, obese, and aged monkeys. Many of these animals have special requirements, such as daily insulin treatment, specific feeding protocols, and close monitoring of behavior and feeding. The research of the ODARC requires that all primates have individual access to *ad libitum* food intake, which is documented daily; therefore, all macaques in this research program are required to be individually housed.

Although these macaques are individually housed, each primate is able to see and hear other monkeys in the room. Generally, cages are arranged so that primates can readily see one another. Very importantly, human contact with the monkeys occurs throughout the day in all areas. Due to the close monitoring of the diabetic and older monkeys, such contact is likely to be much more frequent than in standard primate facilities.

Specialized Primate Methods

Sedation and Anesthesia

The monkeys are lightly sedated (ketamine hydrochloride, 10–15 mg/kg) prior to handling, thereby eliminating any distress that might occur with direct handling and assuring staff safety with these large animals. Ketamine has been shown to be safe and useful as a tranquilizing agent in large species and is the usual drug of choice for handling primates. It has a wide therapeutic index and is administered i.m. (Green 1982). None of our research procedures involves pain; however, for the safety of the staff and the monkeys, the monkeys are maintained under light ketamine sedation during experiments. In addition, our observations have confirmed the reports of others (Kemnitz 1982; Brady and Koritnik 1985) that ketamine has no or minimal effect on glucose and insulin levels.

Blood Sampling Techniques

Blood sampling is carried out by using an Abbocath T-20 or T-22 gauge × 1π-in. radiopaque Teflon catheter inserted into the saphenous or cephalic vein. The femoral vein is infrequently used. The catheter is connected by a stopcock to a syringe filled with 0.9% NaCl or 0.9% NaCl and heparin (1000 U/mL) at a dilution of 7.5 units heparin/milliliter of saline.

Basal fasting samples for substrates and hormones include four 3-mL samples, which are drawn in 3-min intervals and inverted gently for thorough mixing. These four samples are centrifuged, plasma aliquoted, pooled, stored at –20°C, and then assayed for insulin, C-peptide, glucagons, and other hormones and substrates. Glucose levels are assayed using a glucose analyzer. Extra plasma (1–2 mL) is placed in long-term storage tubes in 0.5-mL aliquots and stored at –80°C. Other experiments requiring plasma, such as the i.v. glucose tolerance test, oral glucose tolerance test, and the euglycemic clamp, are sampled in a similar manner.

In addition, under appropriate anesthesia and with sterile surgical technique, basal and insulin-stimulated skeletal muscle (vastus lateralis), subcutaneous and omental adipose tissue, and liver have been obtained for the determination of enyme activities (glycogen synthase and phosphorylase, protein kinase, and protein phosphatase) and substrate concentrations (glucose-6-phosphate and glycogen). Muscle is frozen *in situ* in liquid nitrogen with freeze clamps and stored in liquid nitrogen until processing. Adipose tissue and liver are frozen *ex situ* with freeze clamps and stored in liquid nitrogen until processing. The activities of the enzymes and concentrations of the substrates are then compared; basal versus insulin-stimulated values lend important insight into the pathology involved in development of T2D.

Suppliers and Sources for This Model

The Regional Primate Center of the University of Washington maintains a regular emailed publication, *The Primate Resource Referrals Service* (formerly called *The Primate Clearinghouse*), that announces the availability of nonhuman primates and of tissue or other nonhuman primates' specimens. Other sources include commercial breeding companies, the names of which may also be obtained from the Primate Newsource database.

The cost of a young (five to eight years of age) male rhesus monkey generally ranges from $3,000 to $6,000. These costs can only be estimated initially—that is, for a young research-naïve primate—because obesity and diabetes develop over two to ten years, and the costs are cumulative beyond the initial purchase price. During these developmental years, the longitudinal studies that span the monkey's lifetime are costly, requiring a specialized facility with a well trained and knowledgeable husbandry and research staff.

Most Suitable Areas of Research Applying Particularly to This Model

Metabolic Syndrome and Type 2 Diabetes

The spontaneously obese rhesus monkey is an excellent model for studying the metabolic components of the MetSynd, including its prospective development, sequence, and contribution to the eventual development of T2D. The ability to carry out well-controlled primate studies with subjects of known metabolic status is an advantage not present in human studies. A summary of the characteristics of insulin resistance, prediabetes, and T2D diabetes in the rhesus monkeys, compared with humans, is shown in table 11.2.

In addition, our previous studies have documented the compressed time frame (3–6 years in monkeys, in contrast to 10–25 years in humans) during which the MetSynd may develop and/or the progression of a subject from normal metabolic status to prediabetic MetSynd (obesity, hyperinsulinemia, insulin resistance, and dyslipidemia) and, finally, to overt T2D. At the present time in the ODARC, ongoing studies of the mechanisms underlying these disorders are under way, with complex *in vivo* and *in vitro* methods not possible in human subjects, such as serial measurements of insulin resistance with the euglycemic-hyperinsulinemic clamp, and concurrent tissue biopsies.

Studies of Islet Transplant and Function

The insulin-requiring T2D rhesus monkey is an excellent model for studies of islet transplantation and islet function. Such studies would have exciting potential to provide a more normal insulin replacement therapy for the insulin-requiring type 1 or type 2 diabetic subject. Among the distinct advantages over many other species are the size of the rhesus monkey, the potential to undergo intra-abdominal surgical procedures successfully, the identical structure of rhesus monkey insulin and human insulin (Naithani et al. 1984), and the similarity of the pancreatic islet cell morphology in normal monkeys and in the insulin-requiring T2D monkey to that of humans (Clark et al. 1990, 1995, 2001; Charge et al. 1996).

Complications Associated with Diabetes

The naturally occurring diabetes in the rhesus monkey is associated with all of the complications of T2D known in man, including hypertension (Bodkin 1995), neuropathy (Cornblath et al. 1989), nephropathy (Cusumano et al. 2002), retinopathy and cataracts (Robison 1991; Bernstein et al. 2000; Otsuji et al. 2002; Kim et al.

TABLE 11.2
Comparison of Type 2 Diabetes and Metabolic Syndrome in Monkeys and Humans

	Monkeys	Humans
Insulin resistance (early prediabetes)		
Central obesity	Yes	Yes
Normoglycemia	Yes	Yes
Fasting hyperinsulinemia (greater in monkeys than in humans)	Yes	Yes
Increased insulin response to glucose	Yes	Yes
Insulin molecule structure identical	Yes	Yes
Slightly reduced glucose tolerance	Yes	Yes
Reduced hepatic clearance of insulin	Yes	Yes
Declining whole-body insulin-mediated glucose uptake	Yes	Yes
Reduced insulin activation of muscle glycogen synthase	Yes	Yes
Reduced basal and insulin-stimulated total muscle glycogen synthase	Yes	?
Reduced insulin activation of adipose tissue glycogen synthase	Yes	?
Higher insulin-stimulated muscle glucose-6-phosphate content	Yes	?
Beginning hypertriglyceridemia (sometimes)	Yes	Yes
Hypertension (sometimes)	Yes	Yes
Impaired glucose tolerance (late prediabetes)		
Postprandial glucosuria	Yes	Yes
Slight hyperglycemia	Yes	Yes
Impaired glucose tolerance	Yes	Yes
Decreasing insulin response to glucose	Yes	Yes
Reduced whole-body insulin-stimulated glucose uptake	Yes	Yes
Reduced insulin activation of muscle glycogen synthase	Yes	Yes
Reduced insulin activation of adipose tissue glycogen synthase	Yes	?
Reduced basal and insulin-stimulated total muscle glycogen synthase	Yes	?
Higher insulin-stimulated muscle glucose-6-phosphate content	Yes	?
Hypertriglyceridemia (sometimes)	Yes	Yes
Hypertension (sometimes)	Yes	Yes
Non-insulin-dependent diabetes mellitus		
Middle-age onset most common (in years)	12–25	30–60
Polyphagia	Yes	Yes
Polydipsia	Yes	Yes
Glucosuria	Yes	Yes
Hyperglycemia	Yes	Yes
Hyperfructosemia	Yes	Yes
Normo- or hyperinsulinemia (insulin levels higher in monkeys than in humans)	Yes	Yes
Impaired glucose tolerance	Yes	Yes
Decreased insulin response to glucose	Yes	Yes
Low whole-body insulin-stimulated glucose uptake	Yes	Yes
Lower basal and insulin-stimulated total muscle glycogen synthase	Yes	?
Higher insulin-stimulated muscle glucose-6-phosphate content	Yes	?

(*Continued*)

TABLE 11.2
Comparison of Type 2 Diabetes and Metabolic Syndrome in Monkeys and Humans (Continued)

	Monkeys	Humans
Reduced insulin action on:		
Muscle glycogen synthase	Yes	Yes
Adipose tissue glycogen synthase	Yes	?
Adipocyte glucose oxidation	Yes	Yes
Adipocyte lipid synthesis	Yes	Yes
Increased hepatic glucose production	Yes	Yes
Hypertriglyceridemia (common)	Yes	Yes
Reduced HDL-cholesterol	Yes	Yes
Hypertension (common)	Yes	Yes
Amyloid deposition in pancreatic islets (significant)	Yes	Yes
Complications (frequent):		
Neuropathy	Yes	Yes
Nephropathy	Yes	Yes
Retinopathy	Yes	Yes

Source: Hansen BC. (1994) Primate animal models of type 2 diabetes. In *Diabetes Mellitus: A Fundamental and Clinical Text*, ed. LeRoith D, Olefsky JM, Taylor S, Philadelphia: Lippincott Williams and Wilkins. Reprinted with permission.

2004; Johnson et al. 2005), and infiltration of the pancreatic islets by amyloid (Clark et al. 1990, 1991; de Koning, 1993, 1995, 1998; Clark et al. 1995; Kim et al. 2005). Thus, the rhesus monkey offers exciting opportunities of insight into these often occurring conditions in the T2D human. Table 11.2 provides a comparison of the rhesus monkey and humans, summarizing several characteristics associated with insulin resistance, prediabetes, and T2D.

Tips on Possible Pitfalls

Importance of Longitudinal Characterization

A number of measurements are regularly and periodically carried out to assess longitudinally some of the pathophysiological processes that develop spontaneously and naturally in these monkeys as they undergo the aging process and progress from normal, lean, young animals to obese animals with or without various physiological disturbances. Usually these disturbances develop at varying rates in different monkeys and include hypertension, hyperlipidemia, hyperinsulinemia, hyperglycemia, impaired glucose tolerance, insulin resistance, peripheral neuropathy, impaired renal clearance, and cardiovascular disease—in short, the entire syndrome that becomes manifest in aged and diabetic humans. The longitudinal studies depend specifically on regular assessment of each animal and determination of the rate at which that animal is or is not progressing in the development of pathophysiological disturbances. This allows one to estimate where a given monkey is in regard to the progression to overt diabetes.

Environmental Risks

The procedures, plasma samples, and tissue samples in the rhesus monkey involve situations or materials that may be hazardous to personnel. Therefore, special precautions must be used. Research with nonhuman primates carries the risk of B-virus infection, which is known to be fatal to most humans who contract it. Most cases reported to date have involved the handling of primates without anesthetizing them and/or without proper protective apparel.

Because of the prevalence of B-virus in macaques, all personnel must be well trained regarding the risk of B-virus and how to avoid potential exposure. At the ODARC, in the event of an employee undergoing a potential exposure (bite, scratch, or needle-stick), a strict B-virus protocol is carried out under the direction of the University Employee Health Services and Veterinary Resources. Briefly, the protocol includes immediate and thorough cleansing of the wound and reporting of the incident for prompt medical attention. Baseline serology is drawn on the exposed person for examination as well as baseline serology and virology (buccal and conjunctival samples) on the monkey.

The employee is instructed regarding evidence of infection and closely observed over the next two weeks (symptoms include blisters at the sites of the wound, itching pain or numbness, any neurological illness, or other evidence of the B-virus syndrome). The employee has a follow-up serology drawn 14–28 days after the incident. It should be noted that only 2–3% of macaques shed B-virus at any one time, and in spite of the large number of primates and primate handlers involved in research, less than 40 cases of B-virus have been reported since 1920.

To reduce the potential hazardous exposure, it is recommended that needles not be recapped. Staff training concerning protective laboratory apparel, blood-borne pathogens, and the handling of radioactive substances and wastes is required. In assays and protocols involving radioactive materials, appropriate shielding devices and/or dosimeter badges are used as needed and all environmental health and safety guidelines are followed.

CONCLUDING REMARKS

The rhesus monkey model of spontaneous T2D is an excellent model for studying the natural history of diabetes and its variegated complications. The close phylogenetic proximity of monkeys to humans, development of practically all signs and symptoms of the disease as it occurs in humans, and applicability of using the same biomarkers as in humans to precisely characterize the various stages of the disease process demonstrate that the nonhuman primate is extremely valuable for studying T2D and normal aging. Our past and present studies using this model have established that the rhesus monkey offers a distinct advantage over other animal species, including its ability to predict efficacy of pharmacological and other therapeutic interventions for mitigation of the disease.

ACKNOWLEDGMENTS

The author gratefully acknowledges the scientific contribution of Dr. Noni Bodkin to the prior edition of this chapter and the special scientific contribution of Drs. Heidi Ortmeyer and Kikuko Hotta, as well as many others, to these longitudinal studies of obesity and diabetes in rhesus monkey.

Tribute is particularly due to the late Georgiella Gerzanich who developed and implemented the extensive data management system used to characterize and follow these monkeys and to describe the conditions presented in this chapter.

In addition, many faithful and skilled technical staff members have contributed to the challenging husbandry, research, and managerial responsibilities associated with the primate colony. These include Theresa Alexander, Wallace Evans, Maryne Glowacki, Karen Brocklehurst, Dosu Doherty, Susan Fluck, Dennis Harman, Charla Sweeley, Terry Russell, Dr. Sadaf Aslam, Dr. Joanna Sanford, Allison Gibbs, Jason Markle, Anita Parker, Tawania St. John, and Juliane Krueger.

Finally, we gratefully acknowledge the skilled veterinary expertise provided by our prior veterinarians as well as by clinical primate veterinarian, Dr. Sylvia Gografe.

REFERENCES

ADA-Expert Committee. (2003) The Expert Committee on the Diagnosis and Classification of Diabetes Mellitus: Follow-up report on the diagnosis of diabetes mellitus. *Diabetes Care* 26:3160–3167.

American Diabetes Association. (2006) Diagnosis and classification of diabetes mellitus. *Diabetes Care* 29:S43–S48.

Bernstein SL, Liu AM, Hansen BC, et al. (2000) Heat shock cognate-70 gene expression declines during normal aging of the primate retina. *IOVS* 41:2857–2862.

Bodkin NL. (1996) The rhesus monkey: providing insight into obesity and diabetes. *Lab Anim* 25:33–36.

Bodkin NL, Hannah JS, Ortmeyer HK, et al. (1990) Obesity and hyperlipidemia in the rhesus monkey: Interactions with the development of type 2 diabetes. In *Lessons from Animal Diabetes*, ed. Shafrir E, Renold AE. London: Smith–Gordon, 3:644–649.

Bodkin NL, Hannah JS, Ortmeyer HK, et al. (1993) Central obesity in rhesus monkeys: Association with hyperinsulinemia, insulin resistance and hypertriglyceridemia? *Int J Obesity* 17:53–61.

Bodkin NL, Hansen BC. (1995) Antihypertensive effects of captopril without adverse effects on glucose tolerance in hyperinsulinemic rhesus monkeys. *J Med Primatol* 24:1–6.

Bodkin NL, Metzger BL, Hansen BC. (1989) Hepatic glucose production and insulin sensitivity preceding diabetes in monkeys. *Am J Physiol* 256:E676–681.

Bodkin NL, Nicolson M, Ortmeyer HK, et al. (1996) Hyperleptinemia: Relationship to adiposity and insulin resistance in the spontaneously obese rhesus monkey. *Horm Metab Res* 28:674–678.

Bodkin NL, Sportsman R, DiMarchi RD, et al. (1991) Insulin-like growth factor-I in non-insulin-dependent diabetic monkeys: Basal plasma concentrations and metabolic effects of exogenously administered biosynthetic hormone. *Metabolism* 40:1131–1137.

Brady AG, Koritnik DR. (1985) The effects of ketamine anesthesia on glucose clearance in African green monkeys. *J Med Primatol* 14:99–107.

Charge SB, Esiri MM, Bethune CA, et al. (1996) Apolipoprotein E is associated with islet amyloid and other amyloidoses: Implications for Alzheimer's disease. *J Pathol* 179:443–447.

Clark A, de Koning E, Hansen BC, et al. (1990) Islet amyloid in glucose intolerant and spontaneous diabetic *Macaca mulatta* monkeys. In *Lessons from Animal Diabetes*, ed. Shafrir E, Renold AE. London: Smith–Gordon, 3:502–506.

Clark A, de Koning EJ, Hattersley AT, et al. (1995) Pancreatic pathology in non-insulin dependent diabetes (NIDDM). *Diabetes Res Clin Pract* 28:S39–47.

Clark A, Jones LC, Koning E, et al. (2001) Decreased insulin secretion in type 2 diabetes: A problem of cellular mass or function? *Diabetes* 50 Suppl 1:S169–S171.

Clark A, Morris JF, Scott LA, et al. (1991) Intracellular formation of amyloid fibrils in beta-cells of human insulinoma and prediabetic monkey islets. In *Amyloid and Amyloidosis*, ed. Natvis JB. Amsterdam: Kluwer, 453–456.

Cornblath DR, Hillman MA, Striffler JS, et al. (1989) Peripheral neuropathy in diabetic monkeys. *Diabetes* 38:1365–1370.

Cusumano AM, Bodkin NL, Hansen BC, et al. (2002) Glomerular hypertrophy is associated with hyperinsulinemia and precedes overt diabetes in aging rhesus monkeys. *Am J Kidney Dis* 40:1075–1085.

De Fronzo RA, Tobin JD, Andres R. (1979) Glucose clamp technique: a method for quantifying insulin secretion and resistance. *Am J Physiol* 237:E214–223.

de Koning EJ, Charge S, Morris J, et al. (1995) Macrophages in pancreatic islet amyloidosis. In *Amyloid and Amyloidosis*, ed. Natvis JB. Amsterdam: Kluwer, 405–407.

de Koning EJ, van der Brand JJ, Mott VL, et al. (1998) Macrophages and pancreatic islet amyloidosis. *Amyloid* 1:247–245.

de Koning EJP, Hansen BC, Clark A. (1993) Diabetes mellitus in *Macaca mulatta* monkeys is characterized by severe islet amyloidosis and reduction in beta-cell population. *Diabetologia* 36:378–384.

Di Giacomo RF, Myers RE, Baez LR. (1971) Diabetes mellitus in a rhesus monkey (*Macaca mulatta*). *Lab Anim Sci* 21:572–574.

Ding SY, Tigno, XT, Hansen, BC. (2006) Nuclear magnetic resonance-determined lipoprotein abnormalities in nonhuman primates with the metabolic syndrome and type 2 diabetes mellitus. *Metabolism*. In press.

Erwin JM, Tigno XT, Gerzanich G, et al. (2004) Age-related changes in fasting plasma cortisol in rhesus monkeys: Implications of individual differences for pathological consequences. *J Gerontol A* 59:424–432.

Green C. (1982) Pharmocology of drugs acting on the central nervous system. In *Animal Aneasthesia*, ed. Green C. London: Laboratory Animals Ltd., 29–43.

Hamilton CL, Ciaccia P. (1978) The course of development of glucose intolerance in the monkey (*Macaca mulatta*). *J Med Primatol* 7:165–173.

Hannah JS, Bodkin NL, Paidi MS, et al. (1995) Effects of Acipimox on the metabolism of free fatty acids and very low lipoprotein triglyceride. *Acta Diabetol* 32:279–283.

Hannah JS, Verdery RB, Bodkin NL, et al. (1991) Changes in lipoprotein concentrations during the development of noninsulin-dependent diabetes mellitus in obese rhesus monkeys (*Macaca mulatta*). *J Clin Endocrinol Metab* 72:1067–1072.

Hansen BC. (1992) Obesity and diabetes in monkeys. In *Obesity*, ed. Brodoff PBaBN. New York: J.B. Lippincott Co., 256–265.

Hansen BC. (1996) Primate animal models in non-insulin dependent diabetes mellitus. In *Diabetes Mellitus: A Fundamental and Clinical Text*, ed. LeRoith D, Olesfky JM. Philadelphia: Lippincott–Raven Publishers, 595–603.

Hansen BC. (2000) Primates in the experimental pharmacology of obesity. In *Handbook of Experimental Pharmacology: Obesity, Pathology and Therapy*, ed. Lockwood D, Heffner TG. Heidelberg: Springer–Verlag, 461–489.

Hansen BC. (2004a) Primate animal models of type 2 diabetes. In *Diabetes Mellitus: A Fundamental and Clinical Text*, ed. LeRoith D, Olefsky JM, Taylor S. Philadelphia: Lippincott Williams and Wilkins, 1059–1074.

Hansen BC. (2004b) Primates in the study of aging-associated obesity. In *Handbook of Obesity*, ed. Bray CB. New York: Marcel Dekker, Inc., 283–299.

Hansen BC, Bodkin NL. (1986) Heterogeneity of insulin responses: Phases leading to type 2 (non-insulin-dependent) diabetes mellitus in the rhesus monkey. *Diabetologia* 29:713–719.

Hansen BC, Bodkin NL. (1990) Beta-cell hyperresponsiveness: Earliest event in development of diabetes in monkeys. *Am J Physiol* 259:R612–617.

Hansen BC, Bodkin NL, Ortmeyer HK. (1999) Calorie restriction in nonhuman primates: Mechanisms of reduced morbidity and mortality. *Toxicological Sci* 52S:56–60.

Hansen BC, Jen KL, Schwartz J. (1988) Changes in insulin responses and binding in adipocytes from monkeys with obesity progressing to diabetes. *Int J Obesity* 12:433–443.

Hansen BC, Ortmeyer, HK, Bodkin NL. (1995) *Obesity, Insulin-Resistance and Non-Insulin-Dependent Diabetes in Aging Monkeys: Implications for NIDDM in Humans*. London: Smith–Gordon.

Hotta K, Funahashi T, Bodkin NL, et al. (2001) Circulating concentrations of the adipocyte protein adiponectin are decreased in parallel with reduced insulin sensitivity during the progression to type 2 diabetes in rhesus monkeys. *Diabetes* 50:1126–1133.

Hotta K, Gustafson TA, Ortmeyer HK, et al. (1996) Regulation of obese (ob) mRNA and plasma leptin levels in rhesus monkeys: Effects of insulin, body weight and diabetes. *J Biol Chem* 271:25327–25331.

Hotta K, Gustafson TA, Ortmeyer HK, et al. (1998) Monkey leptin receptor mRNA: Sequence, tissue distribution, and mRNA expression in the adipose tissue of normal, hyperinsulinemic and type 2 diabetic rhesus monkeys. *Obesity Res* 6:353–360.

Hotta K, Gustafson TA, Yoshioka S, et al. (1998) Relationships of PPARγ and PPARγ2 mRNA levels to obesity, diabetes and hyperinsulinemia in rhesus monkeys. *Int J Obesity* 22:1000–1010.

Hotta K, Gustafson TA, Yoshioka S, et al. (1999) Age-related adipose tissue mRNA expression ADD1, PPARγ, lipoprotein lipase and GLUT4 glucose transporter in rhesus monkeys. *J Gerontol: Biol Sci* 54A:B183–B188.

Hotta K, Matsukawa M, Masahiko T, et al. (2001) Galectin-12, an adipose-expressed galectin-like molecule possessing an activity to induce apoptosis. *J Biol Chem* 276:34089–34097.

Howard C, Yasuda M. (1990) Diabetes mellitus in nonhuman primates: Recent research advances and current husbandry practices. *J Med Primatol* 19:609–625.

Howard JCF. (1983) Diabetes and carbohydrate impairment in nonhuman primates. In *Nonhuman Primate Models for Human Diseases*, ed. Dukelow EWR. Boca Raton, FL: CRC Press, 1–36.

Jen KL, Hansen BC, Metzger BL. (1985) Adiposity, anthropometric measures, and plasma insulin levels of rhesus monkeys. *Int J Obesity* 9:213–224.

Johnson MA, Lutty GA, McLeod DS, et al. (2005) Ocular structure and function in an aged monkey with spontaneous diabetes mellitus. *Exp Eye Res* 80:37–42.

Kemnitz J, Kraemer G. (1982) Assessment of glucoregulation in rhesus monkeys sedated with ketamine. *Am J Med Primatol* 3:201–210.

Kennington AS, Hill CR, Craig J, et al. (1990) Low urinary chiro-inositol excretion in non-insulin-dependent diabetes mellitus. *N Engl J Med* 323:373–378.

Kim SY, Johnson MA, McLeod DS, et al. (2004) Retinopathy in monkeys with spontaneous type 2 diabetes. *Invest Ophthalmol Vis Sci* 45:4543–4553.

Kim SY, Johnson MA, McLeod DS, et al. (2005) Neutrophils are associated with capillary closure in spontaneously diabetic monkey retinas. *Diabetes* 54:1534–1542.

Kirk JH, Casey HW, Harwell JF, Jr. (1972) Diabetes mellitus in two rhesus monkeys. *Lab Anim Sci* 22:245–248.

Morton JL, Davenport M, Beloff-Chain A, et al. (1992) Correlation between plasma beta-cell tropin concentrations and body weight in obese rhesus monkeys. *Am J Physiol* 262:E963–967.

Naithani VK, Steffens GJ, Tager HS, et al. (1984) Isolation and amino-acid sequence determination of monkey insulin and proinsulin. *Hoppe Seylers Z Physiol Chem* 365:571–575.

National Institutes of Health Committee on Care and Use of Laboratory Animals. (1996) *Guide to the Care and Use of Laboratory Animals*. Washington, D.C.: U.S. Government Printing Office.

Oliver WR, Shenk JL, Snaith MR, et al. (2001) A selective peroxisome proliferator-activated receptor and agonist promotes reverse cholesterol transport. *PNAS* 98:5306–5311.

Ortmeyer HK. (1997) Insulin decreases skeletal muscle cAMP-dependent protein kinase (PKA) activity in normal monkeys and increases PKA activity in insulin-resistant rhesus monkeys. *J Basic Clin Physiol Pharmacol* 8:223–235.

Ortmeyer HK, Adall Y, Marciani KR, et al. (2005) Skeletal muscle glycogen synthase subcellular localization: Effects of insulin and PPAR-alpha agonist (K-111) administration in rhesus monkeys. *Am J Physiol Regul Integr Comp Physiol* 288:R1509–1517.

Ortmeyer HK, Bodkin NL. (1998) Lack of defect in insulin action on hepatic glycogen synthase and phosphorylase in insulin-resistant monkeys. *Am J Physiol* 274:G1005–1010.

Ortmeyer HK, Bodkin NL, Hansen BC. (1993a) Adipose tissue glycogen synthase activation by *in vivo* insulin in spontaneously insulin-resistant and type 2 (non-insulin-dependent) diabetic rhesus monkeys. *Diabetologia* 36:200–206.

Ortmeyer HK, Bodkin NL, Hansen BC. (1993b) Insulin-mediated glycogen synthase activity in muscle of spontaneously insulin-resistant and diabetic rhesus monkeys. *Am J Physiol* 265:R552–558.

Ortmeyer HK, Bodkin NL, Hansen BC. (1997) Insulin regulates liver glycogen synthase and glycogen phosphorylase activity reciprocally in rhesus monkeys. *Am J Physiol* 272:E133–138.

Ortmeyer HK, Bodkin NL, Haney J, et al. (2000) A thiazolidinedione improves *in vivo* insulin action on skeletal muscle glycogen synthase in insulin-resistant monkeys. *Int J Exp Diabetes Res* 1:195–202.

Ortmeyer HK, Bodkin NL, Hansen, BC, et al. (1995) *In vivo* D-chiroinositol activates skeletal muscle glycogen synthase and inactivates glycogen phosphorylase in rhesus monkeys. *J Nutr Biochem* 6:499–503.

Ortmeyer HK, Bodkin NL, Lilley K, et al. (1993) Chiroinositol deficiency and insulin resistance. I. Urinary excretion rate of chiroinositol is directly associated with insulin resistance in spontaneously diabetic rhesus monkeys. *Endocrinology* 132:640–645.

Ortmeyer HK, Bodkin, NL, Yoshioka S, et al. (1999) A thiazolidinedione improves insulin action on skeletal muscle glycogen synthase in insulin resistant rhesus monkeys. *Diabetes* 48:A231.

Ortmeyer HK, Huang LC, Zhang L, et al. (1993) Chiroinositol deficiency and insulin resistance. II. Acute effects of D-chiroinositol administration in streptozotocin-diabetic rats, normal rats given a glucose load, and spontaneously insulin-resistant rhesus monkeys. *Endocrinology* 132:646–651.

Ortmeyer HK, Larner J, Hansen BC. (1995) Effects of D-chiroinositol added to a meal on plasma glucose and insulin in hyperinsulinemic rhesus monkeys. *Obesity Res* 3 (Suppl 4):605S–608S.

Otsuji T, McLeod DS, Hansen BC, et al. (2002) Immunohistochemical staining and morphometric analysis of the monkey choroidal vasculature. *Exp Eye Res* 75:201–208.

Robison W, Laver N. (1991) Diabetic retinal microangiopathies in humans and animal models: Similar histological progression. *Inv Ophthal Vis Sci* 32:915.

Schafer SA, Hansen BC, Volkl A, et al. (2004) Biochemical and morphological effects of K-111, a peroxisome proliferator-activated receptor (PPAR) alpha activator, in nonhuman primates. *Biochem Pharmacol* 68:239–251.

Schwartz SM, Kemnitz JW, Howard CF, Jr. (1993) Obesity in free-ranging rhesus macaques. *Int J Obesity Related Metab Disorders* 17:1–9.

Tigno XT, Ding SY, Hansen BC. (2006) Paradoxical increase in dermal microvascular flow in prediabetes associated with elevated levels of CRP. *Clin Hemorheol Microcirc* 34:273–282.

Tigno XT, Gerzanich G, Hansen BC. (2004) Age-related changes in metabolic parameters of nonhuman primates. *J Gerontol A Biol Sci Med Sci* 59:1081–1088.

Uno H, Kemnitz J, Warner J, et al. (1985) Spontaneous diabetes mellitus in aged captive rhesus monkeys. *Lab Invest* 52:572–574.

Winegar DA, Brown PJ, Wilkison WO, et al. (2001) Effects of fenofibrate on lipid parameters in obese rhesus monkeys. *J Lipid Res* 42:1543–1551.

Young AA, Gedulin BR, Bhavsar S, et al. (1999) Glucose-lowering and insulin-sensitizing actions of exendin-4: Studies in obese diabetic (*ob*/*ob*, *db*/*db*) mice, diabetic fatty Zucker rats and diabetic monkeys (*Macaca mulatta*). *Diabetes* 48:1026–103.

12 Pigs in Diabetes Research, with Special Focus on Type 2 Diabetes Research

Marianne O. Larsen and Bidda Rolin

CONTENTS

INTRODUCTION

Animal models are very important tools for investigation of the etiology and pathogenesis of human diabetes and for the development of new treatments for the disease. In general, all animals that are hyperglycemic and glucose intolerant due to insulin resistance and/or pancreatic beta-cell malfunction can be useful models for type 2 diabetes (T2D) (McIntosh and Pederson 1999), but not all aspects of the human disease are likely to be present in any one single animal model. Since T2D is a multifactorial disease and the mechanisms underlying the disorders in insulin action and secretion involved in the development of T2D are multiple (WHO 1999), it seems that a single optimal animal model reflecting the full spectrum of the heterologous nature of the disease as well as displaying the progression of the disease will be difficult to define. Therefore, the use of animal models must be limited to specific and well defined characteristics similar to and predictive of the disease in humans. Furthermore, the choice of animal model must in each case depend on the aspect of the human disease in focus.

Therefore, it is of major importance to use well characterized animal models where similarities and differences to the disease in humans displayed in each model are known in detail. A range of well characterized and widely used models based on rodents is available (McIntosh and Pederson 1999; Shafrir 1992). However, large animal models of diabetes are a valuable complement to rodent models, in many ways, for practical and physiological reasons.

This chapter focuses mainly on the use of pigs in research related to T2D and the metabolic syndrome; aspects specifically related to type 1 diabetes are not discussed in detail. For more details on the use of pigs in type 1 diabetes research, see Larsen's (2004) recent review.

GENERAL CHARACTERISTICS OF THE PIG IN RELATION TO T2D RESEARCH

The pig is very useful in many aspects as a model for human physiology and pathophysiology since several organ systems of the pig, as well as physiological and pathophysiological responses, resemble those of the human (Brown and Terris 1996; Douglas 1972; Reeds and Odle 1996; Swindle and Smith 1998). Of special interest in the field of T2D are the many similarities found in nutrition requirements and the gastrointestinal tract, plasma lipids, pancreas development, and morphology as well as metabolism and glucose tolerance between the two species. These subjects are discussed in more detail later. Furthermore, the cardiovascular system of the pig resembles that of humans anatomically (in size and structure), functionally, and pathologically (Brown and Terris 1996; Swindle and Smith 1998). Swine have been known to develop spontaneous atherosclerosis resembling what is seen in man (Brown and Terris 1996) with lipoprotein patterns and predilection sites for formation of atherosclerotic lesions sharing many similarities with the situation in humans (Brown and Terris 1996; Thomas et al. 1986). Therefore, the pig is of particular interest as a model for the study of cardiovascular physiology and pathophysiology in general and in relation to diabetes.

Due to the size of the pig, it is possible to obtain a larger volume of blood in these animals compared to rodents, thereby enabling studies similar to those performed in humans. Furthermore, the pig can be trained to allow performance of experiments in conscious, unstressed animals; this is a big advantage since it allows studies in relation to ingestion of meals. The lack of easily accessible superficial veins in the pig makes blood sampling a potentially stressful procedure. This can be circumvented by implantation of permanent central venous catheters, which can be kept patent for long periods of time. However, implantation of catheters poses a considerable risk for catheter-related infections (Barth et al. 1990), but this risk can, to a large extent, be reduced by application of hygiene precautions (O'Grady et al. 2002).

STRAINS OF PIGS

Several different strains of pigs have been used in diabetes research. The common domestic pig (especially Yorkshire [Gerrity et al. 2001; Natarajan et al. 2002]) and Landrace pigs (Jacobsson 1986) have been used within diabetes research and several strains of minipigs are available. Minipigs are particularly useful for long-term studies due to their small size and ease of handling even at full maturity. Some of the most commonly used strains of minipigs are the Göttingen minipig (Kjems et al. 2001; Larsen, Wilken, et al. 2002; Larsen 2004), Yucatan minipigs (Boullion et al. 2003; Panepinto et al. 1982; Phillips et al. 1982; Sebert et al. 2005), and Sinclair minipigs (Dixon et al. 1999). Recently, the Chinese Guizhou minipig has also been reported to be a relevant strain within the field of diabetes (Xi et al. 2004) and considerable interest has been taken in following the Ossabaw pig (Cote et al. 1982; Etherton and Kris-Etherton 1980; Wangsness et al. 1980).

NUTRITION AND GASTROINTESTINAL TRACT

Pigs share several similarities with humans when nutrition requirements and the function of the gastrointestinal tract are considered, even though there are some anatomical differences between the two species (Bentouimou et al. 1997; Brown and Terris 1996; Dixon et al. 1999; Huge et al. 1995; Miller and Ullrey 1987; Phillips et al. 1979; Swindle and Smith 1998). Pigs and humans are highly dependent on dietary quality, since symbiotic microorganisms in the gut play a relatively minor role in modifying ingested nutrients (Miller and Ullrey 1987). Although ingesta transit times are slightly longer in pigs (Barth et al. 1990), the digestive effectiveness is comparable (Miller and Ullrey 1987). Furthermore, the possibility of feeding well defined meals to pigs is a great advantage in the study of glucose tolerance compared to the more continuous feeding pattern seen in rodents.

PLASMA LIPIDS

Plasma cholesterol is lower in most pigs compared with humans, whereas triglyceride levels are in the same range (Ellegaard et al. 1995; Wilson et al. 1986). Changes in cholesterol levels induced by dietary manipulation have been shown to be qualitatively and quantitatively very similar in the Göttingen minipig and man (Barth et al. 1990). Up to 10-fold increases in cholesterol levels can be obtained in domestic and minipigs fed a high-cholesterol diet (around 1–2% cholesterol) (Berlin et al. 1991; Boullion et al. 2003; Dixon et al. 1999; Gerrity et al. 2001; Jacobsson 1986, 1989; Korte et al. 2005), whereas a high-fat diet without increased cholesterol content gives rise to smaller increases in cholesterol levels (Johansen et al. 2001; Larsen, Rolin, et al. 2002). Severe hypercholesterolemia associated with coronary atherosclerosis has been described in a single strain of pigs during low-fat feeding and is related to a mutation in the LDL receptor (Hasler-Rapacz et al. 1998; Prescott et al. 1995; Rapacz et al. 1986).

The distribution between LDL and HDL fractions is similar to that found in humans (Jacobsson 1986), although this was only seen in the Göttingen minipig fed a high-fat diet (Barth et al. 1990). A comparison of domestic swine and Göttingen minipigs showed that the cholesterol levels and distribution in LDL, VLDL, and HDL fractions were similar in these two breeds, whereas basal triglyceride levels were higher in the Göttingen minipig (Jacobsson 1986). When they were fed an atherogenic diet, plasma cholesterol rose faster and reached higher levels in the Göttingen minipig than in the domestic swine. Minipigs had higher levels of HDL cholesterol compared to domestic swine, but in both strains most of the elevated plasma cholesterol was found in the LDL fraction and the levels of triglycerides did not change consistently (Jacobsson 1986). Sinclair minipigs fed an atherogenic hyperlipidemic diet for eight weeks have shown comparable changes in LDL, HDL, and total plasma cholesterol levels and, similarly to what was seen in Göttingen minipigs, total plasma triglyceride concentration did not change (Dixon et al. 1999).

In pigs, dietary fatty acids are not significantly metabolized or oxidized but are deposited in adipose tissues mainly as they are fed (Bee et al. 2002; Leskanich et al. 1997). In humans, *de novo* lipogenesis has been reported to occur in adipose tissue and the liver, with adipose tissue being the main site according to some (Minehira et al. 2003; Aarsland et al. 1997) but not all (Diraison et al. 2003; Large et al. 2004; Shrago and Spennetta 1976) studies. In pigs, *de novo* lipogenesis occurs mainly in the adipose tissues, with very little lipogenesis in the liver (Ding et al. 2000; Gondret et al. 2001; Pullen et al. 1990). Since the consumption of dietary fat is relatively high in humans, *de novo* lipogenesis accounts for a relatively small proportion of the fatty acids secreted by the liver under normal conditions. However, this proportion is increased in response to hyperinsulinemia and low-fat diets as well as excess energy intake (Lammert et al. 2000; Large et al. 2004; Minehira et al. 2003; Parks 2002; Schwarz et al. 2003; Aarsland et al. 1997). Similarly, an increased energy intake in pigs leads to increased *de novo* lipogenesis, mainly in adipose tissues (Bee et al. 1999).

With regard to dynamics of lipoproteins, pigs differ from humans by having very low activity of cholesteryl ester transfer protein (CETP) whereas phospholipid transfer protein (PLTP) activity is similar to what is seen in humans. This could possibly explain the susceptibility of humans and pigs to atherosclerosis (Cheung et al. 1996; Guyard-Dangremont et al. 1998; Speijer et al. 1991).

PANCREAS

The porcine pancreas resembles the human pancreas in size, shape, and position. Porcine and human insulin differ in only one amino acid at the 30 position of the B-chain (Ganong 1991). Endocrine cells are mainly found in the islets of Langerhans, although single cells or small clusters of alpha- and/or beta-cells are also seen (Jay et al. 1999; Wieczorek et al. 1998). As in humans, islet amyloid polypeptide (IAPP) is expressed mostly in beta-cells but also in some alpha- and delta-cells (Lukinius et al. 1996). The sequence of IAPP in the amyloidogenic domain is not similar in pigs and humans, and pigs are not prone to formation of pancreatic amyloid whereas humans are (Betsholtz et al. 1989; Klöppel et al. 1985).

The beta-cell content of the endocrine tissue in the normal minipig is in the same range as the 60–80% reported in humans (Clark et al. 1988; Gepts and

Lecompte 1981; Kobayashi et al. 1997; Larsen, Rolin, et al. 2003a). For a more detailed discussion of the pancreas, see Larsen (2004).

METABOLISM AND GLUCOSE TOLERANCE

Metabolic changes induced by fasting are quite similar in man and pig, showing the rise of plasma FFA, glycerol, and ketone bodies and decline in glucose and insulin. Pigs have lower fasting FFA and ketone bodies, maybe because of an approximately doubled transit time for food compared with man (Barth et al. 1990). This difference in transit time most probably reflects the difference in the length of the intestines of the two species. During fasting, hepatic glucose production is almost twice as high in conscious pigs compared to what is seen in humans (4 vs. 2 mg/kg/min) (DeFronzo et al. 1983; Krssak et al. 2004; Müller et al. 1983). It seems that hepatic gluconeogenesis plays a major role in the maintenance of normoglycemia in the anesthetized pig in contrast to what is seen in humans, where extrahepatic gluconeogenesis can maintain normoglycemia during anesthesia (Lauritsen et al. 2002). Differences between strains of pigs with respect to hepatic glucose production have also been reported; it has been shown in the obesity-prone Ossabaw pig that glucose turnover is higher than in Yorkshire pigs, indicating a higher rate of hepatic gluconeogenesis in the Ossabaw pig (Cote et al. 1982).

Several observations have indicated that pigs, especially Göttingen minipigs, are more glucose tolerant than humans, probably reflecting higher insulin sensitivity. These observations include lower fasting plasma levels in the minipig compared to humans for insulin (Eriksson et al. 1989; Faber et al. 1978; Fritsche et al. 2000; Larsen et al. 2001; Larsen, Wilken, et al. 2002; Matthews et al. 1983; Reaven et al. 1976) and glucose (3.1–3.6 vs. 4.2–5.1 m*M*) (Canavan et al. 1997; Ellegaard et al. 1995; Eriksson et al. 1989; Fritsche et al. 2000; Larsen et al. 2001; Larsen, Wilken, et al. 2002; Matthews et al. 1983; Reaven et al. 1976; Weyer et al. 1999), whereas domestic pigs have fasting glucose in the range of humans (Barb et al. 1992; Ramsay and White 2000; Wilson et al. 1986). As has also been reported in humans (Rosenthal et al. 1982), plasma levels of glucose and insulin increase with age in minipigs (Larsen et al. 2001), thereby indicating reduced insulin sensitivity and/or beta-cell function (Matthews et al. 1985).

We (Larsen, Rolin, et al. 2002; Larsen, Wilken, et al. 2002) have confirmed the observation from several investigators that pigs show better glucose tolerance (i.e., less increase in plasma glucose) after an oral glucose load (Ferrannini et al. 1985; Hanawalt et al. 1947; Kruszynska et al. 1993; Reed and Kidder 1971; Weyer et al. 1999) and dispose an intravenous glucose load more efficiently than humans (Ahren and Pacini 1998; Anderson 1973; Anderson and Elsley 1970; Cerasi and Luft 1967; Deacon et al. 1998; Hanawalt et al. 1947; Ritzel et al. 1995). These characteristics of the pig question the relevance of using the human criteria for abnormal glucose tolerance in the evaluation of glucose tolerance of a diabetes model in minipigs and evaluation of glucose tolerance should therefore take these differences into consideration.

Due to the limited permeability to glucose (Higgins et al. 1982), porcine erythrocytes contain relatively little intracellular glucose in the normal animal compared with humans and many other species. Therefore, porcine hemoglobin contains very few glycated components (HbA_{1C}). Evaluation of long-term glycemia in the pig should therefore not include measurement of HbA_{1C}, which is used routinely in humans (Gabbay et al. 1977; Koenig et al. 1976). Fructosamine should be preferred as an intermediate term (one to three weeks) marker of glucose control as it is also used in humans (Baker et al. 1983, 1985; Cockram et al. 1990; Lloyd et al. 1985).

With regard to stimulated insulin secretion, we have shown that normal Göttingen minipigs have an extensive capacity for insulin secretion after stimulation with glucose and/or arginine *in vivo* (Larsen, Rolin, et al. 2003a). Furthermore, the pig is a model of particular interest in the area of dynamics of insulin secretion, since we and others have shown that peripheral insulin concentrations have very rapid dynamics that facilitate insulin pulse detection (Kjems et al. 2001; Larsen, Elander, et al. 2002).

DIABETES IN PIGS

A single case of spontaneous diabetes with islet atrophy and degeneration has been reported in the domestic pig and the etiology of this case was probably infectious (Biester 1925). Thus, spontaneous diabetes in the pig is extremely rare and type 1-like diabetes must be surgically or chemically induced, as discussed next.

Before inducing a disease experimentally in any animal, the ethical aspects of this procedure should always be carefully considered and the potential benefits to society of the project must outweigh the costs to the animals. The induction as well as the long-term diabetic state can give rise to adverse effects, but suffering and lasting harm can be avoided by defining humane endpoints and by keeping the pigs under competent veterinary supervision to ensure monitoring and correction of health status of each individual animal. With such veterinary care, diabetic pigs can be kept for the long term with good regulation of the diabetic state and only minor adverse effects.

Several of the hallmarks of diabetic late complications seen in the blood vessels, kidneys, and eyes have been described in diabetic pigs. For a more detailed discussion of this, see Larsen (2004).

CHEMICAL INDUCTION OF DIABETES

Several stable models of severe diabetes with markedly increased fasting plasma glucose and decreased glucose-stimulated insulin secretion have been established in domestic pigs and minipigs by the use of pharmacological induction of beta-cell damage with streptozotocin (STZ) (Barb et al. 1992; Gerrity et al. 2001; Grussner et al. 1993; Marshall et al. 1975, 1980; Marshall 1979; Mesangeau et al. 2000; Wilson et al. 1986) or alloxan (ALX) (Boullion et al. 2003; Dixon et al. 1999; Korte et al. 2005; Mokelke et al. 2003; Otis et al. 2003; Phillips et al. 1980). STZ is a 1 methyl-1-nitrosurea linked to position C2 of D-glucose (Herr et al. 1967). The compound enters the beta-cells by the GLUT2 transporter (Elsner et al. 2000; Ledoux and Wilson 1984). STZ induces DNA strand breaks and thereby activates repair mechanisms that result in a reduction in cellular NAD and ATP below physiological levels leading to cell death (Yamamoto et al. 1981). The mechanism of action of ALX is similar to that of STZ (Yamamoto et al. 1981) and their diabetogenic actions share many similarities (Junod et al. 1967; Rerup and Tarding 1969).

Streptozotocin

Low doses of STZ (35–40 mg/kg) have no major effects on glucose metabolism in domestic pigs or minipigs (Gabel et al. 1985; Marshall et al. 1975), whereas a dose of 85 mg/kg has been found to induce diabetes in pigs that was reversible within two weeks (Gabel et al. 1985). Doses of 100–150 mg/kg STZ have been reported to induce insulin-dependent diabetes in domestic and minipigs in several studies (Barb et al. 1992; Gabel et al. 1985; Grussner et al. 1993; Larsen, Wilken, et al. 2002; Wilson et al. 1986). A dosing scheme using two low doses of STZ (60 mg/kg followed by 30 mg/kg eight days apart) has been used for induction of diabetes in Hanford minipigs. However, these animals did not have residual insulin secretory capacity (Marshall 1979). The diabetic state induced in the study by Marshall was stable in four of seven animals whereas metabolic control was improved over 12 months in three of seven animals.

The same approach has recently been used in Yucatan minipigs with two doses of 55 and 50 mg/kg eight days apart. These animals were severely diabetic and most probably even in a catabolic state, because their weight was reduced dramatically (Mesangeau et al. 2000). Similarly, in Yorkshire swine, 50 mg/kg STZ on three consecutive days induced stable, severe diabetes (Gerrity et al. 2001). We have shown that the dose–response relationship between STZ and level of hyperglycemia in the minipig is very steep (Larsen, Wilken, et al. 2002). It is therefore difficult to obtain moderate states of diabetes after STZ. Nicotinamide (NIA) can prevent the acute fall in NAD and ATP levels in beta-cells by inhibiting the DNA repair mechanisms associated with STZ (Masiello et al. 1990; Yamamoto et al. 1981). Similarly to what has previously been published in rats (Masiello et al. 1998; Novelli 2001, 2004), we have shown that a combination of NIA and STZ in minipigs can provide more moderate degrees of fasting and postprandial hyperglycemia with retained residual insulin secretory capacity that is stable for at least two months (Larsen, Wilken, et al. 2002), thereby adding mild STZ-induced diabetes to the range of models available in minipigs.

Alloxan

ALX has also been used for induction of diabetes in Yucatan and Göttingen minipigs with doses between 100 and 200 mg/kg resulting in severe diabetes (Boullion et al. 2003; Dixon et al. 1999; Korte et al. 2005; Mokelke et al. 2003; Otis et al. 2003; Phillips et al. 1980). ALX in a dose of 80 mg/kg has been reported to induce diabetes with moderate hyperglycemia and partial loss of beta-cell mass in Göttingen minipigs (Kjems et al. 2001).

Models of diabetes based on chemical damage of beta-cells cover a wide spectrum of hyperglycemia (Barb et al. 1992; Boullion et al. 2003; Canavan et al. 1997; Gabel et al. 1985; Grussner et al. 1993; Kjems et al. 2001; Larsen, Wilken, et al.

2002; Otis et al. 2003; Wilson et al. 1986), but all are characterized by reduced insulin secretion and beta-cell mass, two of the major characteristics of human diabetes (Butler et al. 2003; Cerasi 1995; Gerich 2000; Sakuraba et al. 2002; Taylor et al. 1994). Another important feature of human diabetes is insulin resistance (DeFronzo et al. 1982; Kolterman et al. 1981). However, although insulin resistance has been reported in severely hyperglycemic pigs (Otis et al. 2003) and dogs (Reaven et al. 1977), this is not characteristic in models with more moderate levels of hyperglycemia (Larsen, unpublished observations; Reaven et al. 1977).

Furthermore, in contrast to observations in humans (Clark et al. 1988; Klöppel et al. 1985), amyloidosis has not been reported in diabetic pigs and would not be expected due to the different sequence of IAPP in pigs compared with humans (Betsholtz et al. 1989).

OBESITY AND DIETARY MANIPULATIONS TO INDUCE DIABETES

Obesity can be induced in several strains of pigs by means of high-energy and/or high-fat diet feeding. Some strains, such as the Ossabaw pig (Cote et al. 1982; Zafar et al. 2004), are more prone to obesity than others, and female Göttingen minipigs grow much more obese than the males when fed an *ad libitum* diet (Bollen et al. 2005). Furthermore, some of the genes associated with obesity and diabetes in humans (such as 11 beta hydroxysteroid dehydrogenase isoform 1) have been shown to be associated with obesity traits in pigs (Otieno et al. 2005).

A common aspect of obesity in pigs is a lower GH level than in lean animals (McCusker et al. 1985; Mersmann 1991; Wangsness et al. 1977), an observation similar to that seen in humans (Kjeldsen et al. 1975). However, this has not been shown in minipigs, where levels were similar in lean and obese animals (Johansen et al. 2001). As it has been reported in humans (Klöppel et al. 1985), we have shown that in minipigs obesity is associated with increased pancreatic beta-cell volume (Larsen et al. 2004).

High-fat feeding has been reported to induce slight hyperglycemia in domestic pigs after isocaloric or 30% increased energy intake (30% fat) (Ponter et al. 1991) and Sinclair minipigs after 35% increased energy intake (21% fat) (Dixon et al. 1999), as well as development of insulin resistance, which has been reported in Göttingen minipigs after feeding them a high-fat, high-energy diet (Johansen et al. 2001; Larsen, Rolin, et al. 2002). Furthermore, we have shown that high-fat feeding and obesity result in slightly disturbed pulsatile insulin secretion (Larsen et al. 2004). However, others have reported that isocaloric high-fat feeding did not change postprandial glucose and insulin levels in domestic pigs (35% fat) (Berschauer et al. 1983).

Several strains of pigs that are genetically predisposed to obesity, including the Ossabaw pig, have been described (Mersmann 1991). These animals show increased food intake in combination with reduced basal metabolic rate (Mersmann 1991). Furthermore, plasma insulin levels are higher in Ossabaw pigs than in lean animals (McCusker et al. 1985) and the clearance of a glucose load is reduced in combination with an increased insulin response, thereby indicating insulin resistance (Wangsness et al. 1977). Other obese pigs show normal glucose tolerance and insulin levels (Mersmann 1991). Ossabaw pigs have been shown to have increased plasma levels of cholesterol, triglycerides, and VLDL and HDL lipoproteins, whereas LDL levels were similar to those in lean animals (Etherton and Kris-Etherton 1980). However, obese animals have not been shown to be more susceptible to high-fat diet-induced atherogenesis (Mersmann 1991).

Feeding genetically predisposed Yucatan minipigs a 20% fat, 29% high-sucrose diet has been reported to induce marked glucose intolerance and insulin resistance resembling T2D (Panepinto et al. 1982), although this could not be reproduced by other investigators (Hand et al. 1987). Of special interest are two recent publications describing diet (10% fat, 37%)-induced sucrose diabetes in Chinese Guizhou minipigs that, after only a few months of diet feeding, displayed mild hyperglycemia, hypertriglyceridemia, insulin resistance, atherosclerosis and reduced beta-cell mass (Xi et al. 2004; Yin 2004).

Recently, STZ-diabetic Yorkshire pigs fed a high-fat, high-cholesterol diet have been reported to develop hypertriglyceridemia and accelerated atherosclerosis, whereas this was not seen in STZ-diabetic animals on low-fat diet or in normoglycemic animals fed a high-fat, high-cholesterol diet (Gerrity et al. 2001). In contrast to this, a high-cholesterol diet increased atherosclerosis in normoglycemic and ALX-diabetic Yucatan miniature pigs, whereas a slight increase in triglyceride levels was seen in ALX-diabetic pigs only, with no difference between control and high-cholesterol diets (Mokelke et al. 2003). Thus, the combination of chemically induced beta-cell damage and high-fat, high-cholesterol diet could be of considerable interest as a method of establishing models of T2D in pigs, although marked differences seem to exist between strains of pigs.

USE OF PIG MODELS TO TEST PHARMACOLOGICAL TREATMENT OF DIABETES

Insulin and insulin analogues have been tested extensively in pigs (Kurtzhals and Ribel 1995; Markussen et al. 1996; Ribel et al. 1990) because this species is particularly suited to model absorption of insulin from the human subcutaneous tissue. Another example of a pharmacological agent that has been tested in pigs is the long-acting GLP-1 analogue, liraglutide. This compound can improve glucose tolerance in diabetic Göttingen minipigs (Ribel et al. 2002) and reduce food intake and body weight in obese Göttingen minipigs (Raun et al. 2003) similarly to what has been seen with native GLP-1 in humans (Zander et al. 2002). Yet another example is the evaluation of the DPPIV inhibitor valine pyrrolidide, which improves glucose tolerance in diabetic Göttingen minipigs acutely (Larsen, Rolin, et al. 2003). This effect, together with a reduction in HbA1c, was also reported with a DPPIV inhibitor in humans (Ahren et al. 2004).

Another agent that has been tested in pigs is the lipoprotein lipase activator NO-1886, which is proposed to have the potential to reduce free fatty acids in humans. In minipigs, this compound has been shown to protect pancreatic beta-cells (Yin 2004), but whether this is predictive of the effect in humans remains to be determined.

FUTURE PERSPECTIVES

Several models of dyslipidemia and obesity in different strains of pigs are available based on dietary manipulations, but most of these models do not have hyperglycemia. Similarly, several models of reduced insulin secretion after chemical reduction of beta-cell mass are available, but these models do not display insulin resistance and dyslipidemia. Therefore, a single model incorporating all of the main characteristics of human T2D is still to be developed in the pig. Such a model might be possible to establish by combining dietary manipulations with chemical reduction of beta-cell mass. Another possibility would be to select animals with a high predisposition for dyslipidemia and hyperglycemia such as the Ossabaw or Guizhou pigs and some of the inbred strains of pigs used in intensive pig farming (Brambilla and Cantafora 2004). However, most of these strains of pigs are not currently available in a microbiologically standardized way globally, and this could make comparisons between these and other strains of pigs potentially difficult.

AVAILABILITY OF PIGS AND HUSBANDRY

Some of the pig strains mentioned in this chapter are commercially available—for example, the Göttingen and Yucatan minipigs as well as most of the domestic pig strains. However, some of the strains, such as the Ossabaw pig, are not commercially available and, to our knowledge, the Chinese strains, such as the Guizhou pig, are only available in China.

Pigs are social animals living in families consisting of females (very often siblings) and young males. Sexually mature boars are normally solitary because they will often fight. The omnivorous pig uses up to 80% of its time in the search for feed, and even animals that are not hungry search for feed. Since animals do not exert unnecessary behavior, the satisfied animal must still have a pronounced need for performing snatching seeking for feed. Bedding and enrichment feed could be introduced in order to activate the pigs. Another very important activity is rubbing behavior, which to some extent may conflict with the equipment used during the experiment, so this calls for some compromises. For more details on housing of pigs, see Grandin (2002).

CONCLUDING REMARKS

Well characterized animal models of T2D are a valuable tool to increase our understanding of the disease in humans. Due to the high degree of physiological and pathophysiological similarity to humans, the pig is of particular interest as a large animal model to complement the range of models available in rodents. Since T2D does not develop spontaneously in pigs, it must be induced chemically or by means of dietary manipulations. Several groups have reported promising models of diabetes in pigs after different degrees of chemical damage to the beta-cells, and especially during the last few years there has been increased interest in developing models of moderate diabetes in pigs. These models are very useful for studying different aspects of diabetes in humans, such as beta-cell function and mass as well as effect of pharmacological interventions, but since insulin resistance is not a characteristic included in these models, not all aspects of T2D are incorporated. Although some studies regarding the effects of high-fat feeding and obesity on glucose tolerance have been performed in various strains of pigs using various diet regimens, results are so far inconclusive and further study is warranted.

REFERENCES

Aarsland A, Chinkes D, Wolfe RR. (1997) Hepatic and whole-body fat synthesis in humans during carbohydrate overfeeding. *Am J Clin Nutr* 65:1774–1782.

Ahren B, Pacini G. (1998) Age-related reduction in glucose elimination is accompanied by reduced glucose effectiveness and increased hepatic insulin extraction in man. *J Clin Endocrinol Metab* 83:3350–3356.

Ahren BM, Landin-Olsson M, Jansson PA, et al. (2004) Inhibition of dipeptidyl peptidase-4 reduces glycemia, sustains insulin levels, and reduces glucagon levels in type 2 diabetes. *J Clin Endocrinol Metab* 89:2078–2084.

Anderson DM. (1973) The effect of fasting and glucose load on insulin secretion and the Staub–Ttraugott phenomenon in pigs. *J Endocrinol* 58:613–625.

Anderson DM, Elsley FWH. (1970) The intravenous glucose tolerance test in the pig. *Q J Exp Physiol* 55:104–111.

Baker JR, Metcalf PA, Holdaway IM, Johnson RN. (1985) Serum fructosamine concentration as measure of blood glucose control in type I (insulin dependent) diabetes mellitus. *Br Med J* 290:352–355.

Baker JR, O'Connor JP, Metcalf PA, et al. (1983) Clinical usefulness of estimation of serum fructosamine concentration as a screening test for diabetes mellitus. *Br Med J* 287:863–867.

Barb CR, Cox NM, Carlton CA, et al. (1992) Growth hormone secretion, serum, and cerebral spinal fluid insulin and insulin-like growth factor-I concentrations in pigs with streptozotocin-induced diabetes mellitus. *Proc Soc Exp Biol Med* 201:223–228.

Barth CA, Pfeuffer M, Scholtissek J. (1990) Animal models for the study of lipid metabolism, with particular reference to the Göttingen minipig. *Adv Anim Physiol Anim Nutr* S20:39–49.

Bee G, Gebert S, Messikommer R. (2002) Effect of dietary energy supply and fat source on the fatty acid pattern of adipose and lean tissues and lipogenesis in the pig. *J Anim Sci* 80:1564–1574.

Bee G, Messikommer R, Gebert S. (1999) Dietary fats and energy levels differently affect tissue lipogenic enzyme activity in finishing pigs. *Fett* 101:336–342.

Bentouimou N, Vaugelade P, Bernard F, et al. (1997) Compared metabolic effects of seaweed fibers in pigs and humans. *EAAP Publication* 88:57–60.

Berlin E, Khan MA, Henderson GR, Kliman PG. (1991) Dietary fat and cholesterol induced modification of minipig lipoprotein fluidity and composition. *Comp Biochem Physiol* 98:151–157.

Berschauer F, Ehrensvard U, Gaus G. (1983) Nutritive physiological effect of dietary fats in rations for growing swine. 2. Effect of an isocaloric exchange of carbohydrate energy versus fat energy in piglets on growth and various metabolic parameters in the subsequent fattening period. *Arch Tierernahr* 33:761–780.

Betsholtz C, Svensson V, Rorsman F, et al. (1989) Islet amyloid polypeptide IAPP complementary DNA cloning and identification of an amyloidogenic region associated with the species specific occurrence of age related diabetes mellitus. *Exp Cell Res* 183:484–493.

Biester HE. (1925) Diabetes in a pig showing pancreatic lesions. *J Am Vet Med Assn* 67: 99–109.

Bollen PJ, Madsen LW, Meyer O, Ritskes-Hoitinga J. (2005) Growth differences of male and female Göttingen minipigs during *ad libitum* feeding: A pilot study. *Lab Anim* 39:80–93.

Boullion RD, Mokelke EA, Wamhoff BR, et al. (2003) Porcine model of diabetic dyslipidemia: Insulin and feed algorithms for mimicking diabetes mellitus in humans. *Comp Med* 53:42–52.

Brambilla G, Cantafora A. (2004) Metabolic and cardiovascular disorders in highly inbred lines for intensive pig farming: How animal welfare evaluation could improve the basic knowledge of human obesity. *Ann Ist Super Sanita* 40:241–244.

Brown D, Terris J. (1996) Swine. In *Physiological and Pathophysiological Research*, ed. Tumbleson ME, Schook LB. Plenum Press: New York, 5–6.

Butler AE, Janson J, Bonner-Weir S, et al. (2003) Beta-cell deficit and increased beta-cell apoptosis in humans with type 2 diabetes. *Diabetes* 52:102–110.

Canavan JP, Flecknell PA, New JP, et al. (1997) The effect of portal and peripheral insulin delivery on carbohydrate and lipid metabolism in a miniature pig model of human IDDM. *Diabetologia* 40:1125–1134.

Cerasi E. (1995) Insulin deficiency and insulin resistance in the pathogenesis of NIDDM: Is a divorce possible? *Diabetologia* 38:992–997.

Cerasi E, Luft R. (1967) The plasma insulin response to glucose infusion in healthy subjects and in diabetes mellitus. *Acta Endocrinol* 55:278–304.

Cheung MC, Wolfbauer G, Albers JJ. (1996) Plasma phospholipid mass transfer rate: Relationship to plasma phospholipid and cholesteryl ester transfer activities and lipid parameters. *Biochim Biophys Acta* 1303:103–110.

Clark A, Wells CA, Buley ID, et al. (1988) Islet amyloid, increased A-cells, reduced B-cells and exocrine fibrosis: Quantitative changes in the pancreas in type 2 diabetes. *Diabetes Res* 9:151–159.

Cockram CS, Pui PC, Keung CC, et al. (1990) A comparison of fructosamine and glycosylated hemoglobin measurements at a diabetic clinic. *Diabetes Res Clin Pract* 9:43–48.

Cote PJ, Wangsness PJ, Varela-Alvarez H, et al. (1982) Glucose turnover in fast-growing, lean and in slow-growing, obese swine. *J Anim Sci* 54:89–94.

Deacon CF, Hughes TE, Holst JJ. (1998) Dipeptidyl peptidase IV inhibition potentiates the insulinotrophic effect of glucagon-like peptide 1 in the anesthetized pig. *Diabetes* 47:764–769.

DeFronzo RA, Ferrannini E, Hendler R, et al. (1983) Regulation of splanchnic and peripheral glucose uptake by insulin and hyperglycemia in man. *Diabetes* 32:35–45.

DeFronzo RA, Simonson D, Ferrannini E. (1982) Hepatic and peripheral insulin resistance: A common feature of type 2 (non-insulin-dependent) and type 1 (insulin-dependent) diabetes mellitus. *Diabetologia* 23:313–319.

Ding S, Schinckel AP, Weber TE, Mersmann HJ. (2000) Expression of porcine transcription factors and genes related to fatty acid metabolism in different tissues and genetic populations. *J Anim Sci* 78:2127–2134.

Diraison F, Yankah V, Letexier D, et al. (2003) Differences in the regulation of adipose tissue and liver lipogenesis by carbohydrates in humans. *J Lipid Res* 44:846–853.

Dixon JL, Stoops JD, Parker JL, et al. (1999) Dyslipidemia and vascular dysfunction in diabetic pigs fed an atherogenic diet. *Arterioscler Thromb Vasc Biol* 19:2981–2992.

Douglas WR. (1972) Of pigs and men and research: A review of applications and analogies of the pig, *Sus scrofa*, in human medical research. *Space Life Sci* 3:226–234.

Ellegaard L, Jørgensen KD, Klastrup S, et al. (1995) Hematologic and clinical chemical values in 3- and 6-months-old Göttingen minipigs. *Scand J Lab Anim Sci* 22:239–248.

Elsner M, Guldbakke B, Tiedge M, et al. (2000) Relative importance of transport and alkylation for pancreatic beta-cell toxicity of streptozotocin. *Diabetologia* 43:1528–1533.

Eriksson J, Franssila-Kallunki A, Ekstrand A, et al. (1989) Early metabolic defects in persons at increased risk for non-insulin-dependent diabetes mellitus. *N Engl J Med* 321:337–343.

Etherton TD, Kris-Etherton PM. (1980) Characterization of plasma lipoproteins in swine with different propensities for obesity. *Lipids* 15:823–829.

Faber OK, Hagen C, Binder C, et al. (1978) Kinetics of human connecting peptide in normal and diabetic subjects. *J Clin Invest* 62:197–203.

Ferrannini E, Bjorkman O, Reichard GA, et al. (1985) The disposal of an oral glucose load in healthy subjects. A quantitative study. *Diabetes* 34:580–588.

Fritsche A, Stefan N, Hardt E, et al. (2000) Characterization of beta-cell dysfunction of impaired glucose tolerance: Evidence for impairment of incretin-induced insulin secretion. *Diabetologia* 43:852–858.

Gabbay KH, Hasty K, Breslow JL, et al. (1977) Glycosylated hemoglobins and long-term blood glucose control in diabetes mellitus. *J Clin Endocrinol Metab* 44:859–864.

Gabel H, Bitter-Suermann H, Henriksson C, et al. (1985) Streptozotocin diabetes in juvenile pigs. Evaluation of an experimental model. *Horm Metab Res* 17:275–280.

Ganong WF. (1991) Endocrine functions of the pancreas. In *Review of Medical Physiology*, Appleton & Lange: New York, 312–333.

Gepts W, Lecompte PM. (1981) The pancreatic islets in diabetes. *Am J Med* 70:105–115.

Gerich JE. (2000) Insulin resistance is not necessarily an essential component of type 2 diabetes. *J Clin Endocrinol Metab* 85:2113–2115.

Gerrity RG, Natarajan R, Nadler JL, Kimsey T. (2001) Diabetes-induced accelerated atherosclerosis in swine. *Diabetes* 50:1654–1665.

Gondret F, Ferre P, Dugail I. (2001) ADD-1/SREBP-1 is a major determinant of tissue differential lipogenic capacity in mammalian and avian species. *J Lipid Res* 42:106–113.

Grandin T. (2002) Comfortable quarters for pigs in research institutions. In *Comfortable Quarters for Laboratory Animals*, ed. Reinhardt V, Reinhardt A. Animal Welfare Institute: Washington, D.C., 78–82.

Grussner R, Nakhleh R, Grussner A, et al. (1993) Streptozotocin-induced diabetes mellitus in pigs. *Horm Metab Res* 25:199–203.

Guyard-Dangremont V, Desrumaux C, Gambert P, et al. (1998) Phospholipid and cholesteryl ester transfer activities in plasma from 14 vertebrate species. Relation to atherogenesis susceptibility. *Comp Biochem Physiol* 120:517–525.

Hanawalt VM, Link RP, Sampson J. (1947) Intravenous glucose tolerance tests on swine. *Proc Soc Exp Biol Med* 65:41–44.

Hand MS, Surwit RS, Rodin J, et al. (1987) Failure of genetically selected miniature swine to model NIDDM. *Diabetes* 36:284–287.

Hasler-Rapacz J, Ellegren H, Fridolfsson AK, et al. (1998) Identification of a mutation in the low-density-lipoprotein receptor gene associated with recessive familial hypercholesterolemia in swine. *Am J Med* 76:379–386.

Herr RR, Jahnke JK, Argoudelis AD. (1967) The structure of streptozotocin. *J Am Chem Soc* 89:4808–4809.

Higgins PJ, Garlick RL, Bunn HF. (1982) Glycosylated hemoglobin in human and animal red cells. Role of glucose permeability. *Diabetes* 31:743–748.

Huge A, Weber E, Ehrlein HJ. (1995) Effects of enteral feedback inhibition on motility, luminal flow, and absorption of nutrients in proximal gut of minipigs. *Dig Dis Sci* 40:1024–1034.

Jacobsson L. (1986) Comparison of experimental hypercholesterolemia and atherosclerosis in Göttingen mini-pigs and Swedish domestic swine. *Atherosclerosis* 59:205–213.

Jacobsson, L. 1989 Comparison of experimental hypercholesterolemia and atherosclerosis in male and female minipigs of the Gottingen strain. *Artery* 16:105–117.

Jay TR, Heald KA, Carless NJ, et al. (1999) The distribution of porcine pancreatic beta-cells at ages 5, 12 and 24 weeks. *Xenotransplantation* 6:131–140.

Johansen T, Hansen HS, Richelsen B, Malmlöf K. (2001) The obese Göttingen minipig as a model of the metabolic syndrome: Dietary effects on obesity, insulin sensitivity and growth hormone profile. *Comp Med* 51:150–155.

Junod A, Lambert AE, Orci L, et al. (1967) Studies of the diabetogenic action of streptozotocin. *Proc Soc Exp Biol Med* 126:201–205.

Kjeldsen H, Hansen AP, Lundbaek K. (1975) Twenty-four-hour serum growth hormone levels in maturity-onset diabetics. *Diabetes* 24:977–982.

Kjems LL, Kirby BM, Welsh EM, et al. (2001) Decrease in beta-cell mass leads to impaired pulsatile insulin secretion, reduced postprandial hepatic insulin clearance, and relative hyperglucagonemia in the minipig. *Diabetes* 50:2001–2012.

Klöppel G, Öhr M, Habich K, et al. (1985) Islet pathology and the pathogenesis of type 1 and type 2 diabetes mellitus revisited. *Surv Synth Pathol Res* 4:110–125.

Kobayashi T, Nakanishi K, Nakase H, et al. (1997) *In situ* characterization of islets in diabetes with a mitochondrial DNA mutation at nucleotide position 3243. *Diabetes* 46:1567–1571.

Koenig RJ, Peterson CM, Jones RL, et al. (1976) Correlation of glucose regulation and hemoglobin AIc in diabetes mellitus. *N Engl J Med* 295:417–420.

Kolterman OG, Gray RS, Griffin J, et al. (1981) Receptor and postreceptor defects contribute to the insulin resistance in noninsulin-dependent diabetes mellitus. *J Clin Invest* 68:957–969.

Korte FS, Mokelke EA, Sturek M, McDonald KS. (2005) Exercise improves impaired ventricular function and alterations of cardiac myofibrillar proteins in diabetic dyslipidemic pigs. *J Appl Physiol* 98:461–467.

Krssak M, Brehm A, Bernroider E, et al. (2004) Alterations in postprandial hepatic glycogen metabolism in type 2 diabetes. *Diabetes* 53:3048–3056.

Kruszynska YT, Meyer-Alber A, Darakhshan F, et al. (1993) Metabolic handling of orally administered glucose in cirrhosis. *J Clin Invest* 91:1057–1066.

Kurtzhals P, Ribel U. (1995) Action profile of cobalt(III)-insulin. A novel principle of protraction of potential use for basal insulin delivery. *Diabetes* 44:1381–1385.

Lammert O, Grunnet N, Faber P, et al. (2000) Effects of isoenergetic overfeeding of either carbohydrate or fat in young men. *Brit J Nutr* 84:233–245.

Large V, Peroni O, Letexier D, et al. (2004) Metabolism of lipids in human white adipocyte. *Diabetes Metab* 30:294–309.

Larsen MO, Elander M, Sturis J, et al. (2002) The conscious Göttingen minipig as a model for studying rapid pulsatile insulin secretion *in vivo*. *Diabetologia* 45:1389–1396.

Larsen MO, Juhl CB, Porksen N, et al. (2004) Beta-cell function and islet morphology in normal, obese and obese beta-cell mass reduced Göttingen minipigs. *Am J Physiol* 288:E412–E421.

Larsen MO, Rolin B. (2004) Use of the Göttingen minipig as a model of diabetes, with special focus on type 1 diabetes research. *ILAR J* 45:303–313.

Larsen MO, Rolin B, Wilken M, et al. (2003a) Measurements of insulin secretory capacity and glucose tolerance to predict pancreatic beta-cell mass *in vivo* in the nicotinamide/streptozotocin Göttlinger minipig, a model of moderate insulin deficiency and diabetes. *Diabetes* 52:118–123.

Larsen MO, Rolin B, Ribel U, et al. (2003) Valine pyrrolidide preserves intact glucose-dependent insulinotropic peptide and improves abnormal glucose tolerance in minipigs with reduced beta-cell mass. *Exp Diabesity Res* 4:93–105.

Larsen MO, Rolin B, Wilken M, et al. (2001) Parameters of glucose and lipid metabolism in the male Göttingen minipig: Influence of age, body weight, and breeding family. *Comp Med* 51:436–442.

Larsen MO, Rolin B, Wilken M, Svendsen O. (2002) High fat high energy feeding impairs fasting glucose and increases fasting insulin levels in the Göttingen minipig. *Ann NY Acad Sci* 967:414–423.

Larsen MO, Wilken M, Gotfredsen CF, et al. (2002) Mild streptozotocin diabetes in the Göttingen minipig. A novel model of moderate insulin deficiency and diabetes. *Am J Physiol* 282:E1342–E1351.

Lauritsen TLB, Grunnet N, Rasmussen A, et al. (2002) The effect of hepatectomy on glucose homeostasis in pig and in man. *J Hepatol* 36:99–104.

Ledoux SP, Wilson GL. (1984) Effects of streptozotocin on a clonal isolate of rat insulinoma cells. *Biochim Biophys Acta* 804:387–392.

Leskanich CO, Matthews KR, Warkup CC, et al. (1997) The effect of dietary oil containing (n-3) fatty-acids on the fatty-acid, physicochemical, and organoleptic characteristics of pig meat and fat. *J Anim Sci* 75:673–683.

Lloyd DR, Nott M, Marples J. (1985) Comparison of serum fructosamine with glycosylated serum protein (determined by affinity chromatography) for the assessment of diabetic control. *Diabetic Med* 2:474–478.

Lukinius A, Korsgren O, Grimelius L, Wilander E. (1996) Expression of islet amyloid polypeptide in fetal and adult porcine and human pancreatic islet cells. *Endocrinology* 137:5319–5325.

Markussen J, Havelund S, Kurtzhals P, et al. (1996) Soluble, fatty acid acylated insulins bind to albumin and show protracted action in pigs. *Diabetologia* 39:281–288.

Marshall M. (1979) Induction of chronic diabetes by streptozotocin in the miniature pig. *Res Exp Med* 175:187–196.

Marshall M, Oberhofer H. Staubesand J. (1980) Early micro- and macro-angiopathy in the streptozotocin diabetic minipig. *Res Exp Med* 177:145–158.

Marshall M, Sprandel U, Zollner N. 1975. Streptozotocin diabetes in a miniature pig. *Res Exp Med* 165:61–65.

Masiello P, Broca C, Gross R, et al. (1998) Experimental NIDDM: Development of a new model in adult rats administered streptozotocin and nicotinamide. *Diabetes* 47:224–229.

Masiello P, Novelli M, Fierabracci V, Bergamini E. (1990) Protection by 3-aminobenzamide and nicotinamide against streptozotocin-induced beta-cell toxicity *in vivo* and *in vitro*. *Res Commun Chem Path Pharm* 69:17–32.

Matthews DR, Hosker JP, Rudenski AS, et al. (1985) Homeostasis model assessment: Insulin resistance and β-cell function from fasting plasma glucose and insulin concentrations in man. *Diabetologia* 28:412–419.

Matthews DR, Lang DA, Burnett MA, Turner RC. (1983) Control of pulsatile insulin secretion in man. *Diabetologia* 24:231–237.

McCusker RH, Wangsness PJ, Griel LC, Kavanaugh JF. (1985) Effects of feeding, fasting and refeeding on growth hormone and insulin in obese pigs. *Physiol Behav* 35:383–388.

McIntosh CHS, Pederson RA. (1999) Noninsulin-dependent animal models of diabetes mellitus. In *Experimental Models of Diabetes Mellitus*, ed. McNeill JH. CRC Press: Boca Raton, FL, 337–386.

Mersmann HJ. (1991) Characteristics of obese and lean swine. In *Swine Nutrition*, ed. Miller ER, Ullrey DE, Lewis, AJ. Butterworth–Heinemann: Boston, 75–89.

Mesangeau D, Laude D, Elghozi JL. (2000) Early detection of cardiovascular autonomic neuropathy in diabetic pigs using blood pressure and heart rate variability. *Cardiovasc Res* 45:889–899.

Miller ER, Ullrey DE. (1987) The pig as a model for human nutrition. *Annu Rev Nutr* 7:361–382.

Minehira K, Bettschart V, Vidal H, et al. (2003) Effect of carbohydrate overfeeding on whole body and adipose tissue metabolism in humans. *Obesity Res* 11:1096–1103.

Mokelke EA, Hu Q, Song M, et al. (2003) Altered functional coupling of coronary K^+ channels in diabetic dyslipidemic pigs is prevented by exercise. *J Appl Physiol* 95:1179–1193.

Müller MJ, Paschen U, Seitz HJ. (1983) Glucose production measured by tracer and balance data in conscious miniature pig. *Am J Physiol* 244:E236–E244.

Natarajan R, Gerrity RG, Gu JL, et al. (2002) Role of 12-lipoxygenase and oxidant stress in hyperglycemia-induced acceleration of atherosclerosis in a diabetic pig model. *Diabetologia* 45:125–133.

Novelli M. (2001) Metabolic and functional studies on isolated islets in a new rat model of type 2 diabetes. *Mol Cell Endocrinol* 175:57–66.

Novelli M. (2004) Alteration of beta-cell constitutive NO synthase activity is involved in the abnormal insulin response to arginine in a new rat model of type 2 diabetes. *Mol Cell Endocrinol* 219:77–82.

O'Grady NP, Alexander M, Dellinger EP, et al. (2002) Healthcare infection. Guidelines for the prevention of intravascular catheter-related infections. *Am J Inf Contr* 30:476–489.

Otieno CJ, Bastiaansen J, Ramos AM, Rothschild MF. (2005) Mapping and association studies of diabetes related genes in the pig. *Anim Genet* 36:36–42.

Otis CR, Wamhoff BR, Sturek M. (2003) Hyperglycemia-induced insulin resistance in diabetic dyslipidemic Yucatan swine. *Comp Med* 53:53–64.

Panepinto LM, Phillips RW, Westmoreland NW, Cleek JL. (1982) Influence of genetics and diet on the development of diabetes in Yucatan miniature swine. *J Nutr* 112:2307–2313.

Parks EJ. (2002) Changes in fat synthesis influenced by dietary macronutrient content. *Proc Nutr Soc* 61:281–286.

Phillips RW, Panepinto LM, Spangler R, Westmoreland NW. (1982) Yucatan miniature swine as a model for the study of human diabetes mellitus. *Diabetes* 31:30–36.

Phillips RW, Panepinto LM, Will DH. (1979) Genetic selection for diabetogenic traits in Yucatan miniature swine. *Diabetes* 28:1102–1107.

Phillips RW, Panepinto LM, Will DH, Case GL. (1980) The effects of alloxan diabetes on Yucatan miniature swine and their progeny. *Metabolism* 29:40–45.

Ponter AA, Salter L, Morgan LM, Flatt PR. (1991) The effect of energy source and feeding level on the hormones of the entero-insular axis and plasma glucose in the growing pig. *Br J Nutr* 66:187–197.

Prescott MF, Hasler-Rapacz J, Linden-Reed J, Rapacz J. (1995) Familial hypercholesterolemia associated with coronary atherosclerosis in swine bearing different alleles for apolipoprotein B. *Ann NY Acad Sci* 748:283–292.

Pullen DL, Liesman JS, Emery RS. (1990) A species comparison of liver slice synthesis and secretion of triacylglycerol from nonesterified fatty acids in media. *J Anim Sci* 68:1395–1399.

Ramsay TG, White ME. (2000) Insulin regulation of leptin expression in streptozotocin diabetic pigs. *J Anim Sci* 78:1497–1503.

Rapacz JJ, Hasler-Rapacz J, Taylor KM, et al. (1986) Lipoprotein mutations in pigs are associated with elevated plasma cholesterol and atherosclerosis. *Science* 234:1573–1577.

Raun K, Von Voss P, Ankersen T, et al. (2003) The GLP-1 derivative NN2211 normalizes food intake and lowers body weight in a hyperphagic minipig model. *Diabetes* 52:A325–2003.

Reaven GM, Bernstein R, Davis D, Olefsky JM. (1976) Nonketotic diabetes mellitus: Insulin deficiency or insulin resistance? *Am J Med* 60:80–88.

Reaven GM, Sageman WS, Swenson RS. (1977) Development of insulin resistance in normal dogs following alloxan-induced insulin deficiency. *Diabetologia* 13:459–462.

Reed JH, Kidder DE. (1971) The oral glucose tolerance test in the young pig. *Br Vet J* 127:318–326.

Reeds P, Odle J. (1996) Pigs as models for nutrient functional interaction. In *Advances in Swine in Biomedical Research*, 2nd ed., ed. Tumbleson ME, Schook LB. Plenum Publishers: New York, 709–711.

Rerup CC, Tarding F. (1969) Streptozotocin- and alloxan-diabetes in mice. *Eur J Pharmacol* 7:89–96.

Ribel U, Hougaard P, Drejer K, Sørensen AR. (1990) Equivalent *in vivo* biological activity of insulin analogues and human insulin despite different *in vitro* potencies. *Diabetes* 39:1033–1039.

Ribel U, Larsen MO, Rolin B, et al. (2002) NN2211: A long-acting glucagon-like peptide-1 derivative with antidiabetic effects in glucose-intolerant pigs. *Eur J Pharmacol* 451:217–225.

Ritzel R, Orskov C, Holst JJ, Nauck MA. (1995) Pharmacokinetic, insulinotropic, and glucagonostatic properties of GLP-1 7-36 amide after subcutaneous injection in healthy volunteers. Dose–response relationships. *Diabetologia* 38:720–725.

Rosenthal M, Doberne L, Greenfield M, et al. (1982) Effect of age on glucose tolerance, insulin secretion, and *in vivo* insulin action. *J Am Geriatr Soc* 30:562–567.

Sakuraba H, Mizukami H, Yagihashi N, et al. (2002) Reduced beta-cell mass and expression of oxidative stress-related DNA damage in the islet of Japanese type II diabetic patients. *Diabetologia* 45:85–96.

Schwarz J, Linfoot P, Dare D, Aghajanian K. (2003) Hepatic *de novo* lipogenesis in normoinsulinemic and hyperinsulinemic subjects consuming high-fat, low-carbohydrate and low-fat, high-carbohydrate isoenergetic diets. *Am J Clin Nutr* 77:43–50.

Sebert SP, Lecannu G, Kozlowski F, et al. (2005) Childhood obesity and insulin resistance in a Yucatan minipiglet model: putative roles of IGF-1 and muscle PPARs in adipose tissue activity and development. *Int J Obesity* 29:324–333.

Shafrir E. (1992) Animal models of non-insulin-dependent diabetes. *Diabetes Metab Rev* 8:179–208.

Shrago E, Spennetta T. (1976) The carbon pathway for lipogenesis in isolated adipocytes from rat, guinea pig, and human adipose tissue. *Am J Clin Nutr* 29:540–545.

Speijer H, Groener JEM, Van Ramshorst E, Van Tol A. (1991) Different locations of cholesteryl ester transfer protein and phospholipid transfer protein activities in plasma. *Atherosclerosis* 90:159–168.

Swindle MM, Smith AC. (1998) Comparative anatomy and physiology of the pig. *Scand J Lab Anim Sci* 25:11–21.

Taylor SI, Accili D, Imai Y. (1994) Insulin resistance or insulin deficiency. Which is the primary cause of NIDDM? *Diabetes* 43:735–740.

Thomas WA, Lee KT, Kim DN. (1986) Pathogenesis of atherosclerosis in the abdominal aorta and coronary arteries of swine in the first 90 days on a hyperlipidemic diet. In *Swine in Biomedical Research*, ed. Tumbleson ME. Plenum Press: New York, 1511–1525.

Wangsness PJ, Martin RJ, Gahagan JH. (1977) Insulin and growth hormone in lean and obese pigs. *Am J Physiol* 233:E104–E108.

Wangsness PJ, Martin RJ, Gatchel BB. (1980) Insulin induced growth hormone response in fast-growing, lean and in slow-growing, obese pigs. *Growth* 44:318–326.

Weyer C, Bogardus C, Pratley RE. (1999) Metabolic characteristics of individuals with impaired fasting glucose and/or impaired glucose tolerance. *Diabetes* 48:2197–2203.

WHO. (1999) Definition, *Diagnosis and Classification of Diabetes Mellitus and its Complications. Part 1: Diagnosis and Classification of Diabetes Mellitus*, ed. Alberti KG, Zimmet P. Geneva: World Health Organization, Department of Noncommunicable Disease Surveillance.

Wieczorek G, Pospischil A, Perentes E. (1998) A comparative immunohistochemical study of pancreatic islets in laboratory animals (rats, dogs, minipigs, nonhuman primates). *Exp Toxicol Pathol* 50:151–172.

Wilson JD, Dhall DP, Simeonovic CJ, Lafferty KJ. (1986) Induction and management of diabetes mellitus in the pig. *Aust J Exp Biol Med Sci* 64:489–500.

Xi S, Yin W, Wang Z, et al. (2004) A minipig model of high-fat/high-sucrose diet-induced diabetes and atherosclerosis. *Int J Exp Pathol* 85:223–231.

Yamamoto H, Uchigata Y, Okamoto H. (1981) Streptozotocin and alloxan induce DNA strand breaks and poly(ADP-ribose) synthetase in pancreatic islets. *Nature* 294:284–286.

Yin W. (2004) NO-1886 decreases ectopic lipid deposition and protects pancreatic beta-cells in diet-induced diabetic swine. *J Endocrinol* 180:399–408.

Zafar MN, Kaser SG, Alloosh M, et al. (2004) Ossabaw swine having the metabolic syndrome exhibit greater neointimal hyperplasia after coronary stent placement than lean Yucatan swine. *FASEB J* 18:2004.

Zander M, Madsbad S, Madsen JL, Holst JJ. (2002) Effect of 6-week course of glucagon-like peptide 1 on glycemic control, insulin sensitivity, and beta-cell function in type 2 diabetes: A parallel-group study. *Lancet* 359:824–830.

13 *Psammomys Obesus:* Nutritionally Induced Insulin Resistance, Diabetes, and Beta Cell Loss

Ehud Ziv, Rony Kalman, and Eleazar Shafrir

CONTENTS

ABSTRACT

The gerbil *Psammomys obesus* (sand rat) is a model of nutritionally induced type 2 diabetes. The progression of diabetes in *Psammomys* resembles in many respects the development of insulin resistance and diabetes in certain human populations.

This animal is well adapted to its environment, where there is a constant supply of low-energy (LE) diet. The high metabolic efficiency of the animals is part of the thrifty metabolism, and it represents a natural adaptation to life in areas where only low-energy diet is available. *Psammomys obesus* are prone to develop hyperglycemia, hyperinsulinemia, and obesity when fed a high-energy (HE) diet. *Psammomys* express four different phenotypic stages: stage A—normoinsulinemia and normoglycemia; stage B—hyperinsulinemia but normoglycemia; stage C—hyperinsulinemia and hyperglycemia; and stage D—hypoinsulinemia and hyperglycemia as a result of loss of insulin secretion capacity. The animals in the Jerusalem colony were separated into two outbred distinct lines differing phenotypically and genotypically: diabetes-prone (DP) and diabetes-resistant (DR) animals.

Psammomys from the DP line become diabetic at a very young age when transferred to HE diet. Within seven days from weaning, 81% of the animals reach blood glucose levels over 200 mg/dL. Percentage of diabetic animals increases to over 90% within 14 days from weaning (when weaned on HE diet). In animals older than eight months, the potential to develop diabetes and obesity decreases. These changes are in correlation with the decrease in fertility in *Psammomys* from DP as well as DR lines.

There is no hyperphagia in DP and DR lines when the animals are fed on HE or LE diets. Metabolic efficiency in DP line *Psammomys* fed on all diets was 6.0–6.6 kcal/g of weight increase while in the DR line metabolic efficiency was 9.0–9.6 kcal/g of weight increase.

Psammomys fed on an LE diet keep normoglycemia in the fed state and do not lapse into hypoglycemia, despite high levels of exogenous insulin administered to them. Results clearly indicate that hyperglycemia in HE-fed *Psammomys* is not due to lack of insulin and that the primary cause for the development of diabetes in *Psammomys* is a primary, inherited insulin resistance expressed mainly in the fed state.

Primary insulin resistance as a species characteristic of *Psammomys* was confirmed by the hyperinsulinemic-euglycemic clamp studies. Hepatic glucose production (HGP) was only partially reduced by insulin infusion (from 10.0 ± 0.6 to 3.8 ± 0.4 mg/min/kg). Also, the limited elevation of total glucose transport (TGT) in the clamped *Psammomys* attests to the fact that the peripheral glucose utilization is low enough to be compensated by gluconeogenesis, thus avoiding lapse into hypoglycemia during exogenous insulin treatment. It is well emphasized that the sum of the reactions of the two different organs determines the diabetic potential of each *Psammomys*.

The adaptation of *Psammomys* to desert conditions is determined by the low expression of certain cellular proteins (low receptor tyrosine kinase and low PTPase activity, low amount of GLUT4 mRNA and protein), but overexpression of protein kinase C epsilon.

Presence of primary insulin resistance and absence of gluconeogenesis restraint in animal and human populations, together with rich nutrient intake, leads to enhancement of insulin secretion, excessive lipogenesis followed by hyperglycemia. This path of events with eventual obesity with potentiation of insulin resistance is well exemplified in *Psammomys*.

INTRODUCTION

Diabetes in *Psammomys* was discovered in the Nile Valley in desert rodents collected by the U.S. Naval Medical Research Unit in Egypt in the 1960s. The animals were trivially nicknamed "sand rats," which is a misnomer since they are not murines but gerbils belonging to the family Gerbillinae and should be referred to as *Psammomys obesus* according to the classification of Thomas (1908). The sand rats were sent to Duke University in Durham, North Carolina, where the first report originated that diabetes occurs in the majority of *Psammomys* maintained on *ad libitum* regular laboratory chow but not on vegetable diet (Schmidt-Nielsen et al. 1964; Hackel et al. 1965). The diabetes ranged from mild hyperglycemia with hyperinsulinemia, to hypoinsulinemia with terminal ketoacidosis. The early investigations were performed mostly on the first generation of Egyptian *Psammomys*, since attempts to establish a multigeneration colony were not successful due to low reproductive capacity on the regular chow.

PSAMMOMYS OBESUS OF THE JERUSALEM COLONY

The Jerusalem colony was started with *Psammomys* collected in 1969 from the arid area north of the Dead Sea in Israel (Adler et al. 1988). The animals were maintained on a "free choice" diet consisting of succulent leaves and branches of the salt bush *Atriplex halimus* collected from the Dead Sea region, fortified with a few pellets of regular rodent chow. During the more than 35 years of the colony's existence, four main stages of progression to diabetes were identified: (A) normoglycemic–normoinsulinemic; (B) normoglycemic–hyperinsulinemic; (C) hyperglycemic–hyperinsulinemic; and (D) markedly hyperglycemic–hypoinsulinemic (Kalderon et al. 1986) (fig. 13.1).

The high insulin secretion in groups B and C of *Psammomys* failed to promote peripheral glucose uptake, as determined by 2-deoxyglucose uptake. It also failed to restrain hepatic gluconeogenesis, as indicated by increased alanine conversion to glucose by isolated hepatocytes and the elevated activity of phosphoenolpyruvate carboxykinase (PEPCK) (Kalderon et al. 1986; Shafrir and Gutman 1993; Shafrir and Ziv 1998). In these groups, there was still a considerable deposition of adipose tissue with a characteristic superscapular hump and hypertriglyceridemia, demonstrating active hepatic lipogenesis and transport of lipoprotein-borne triglycerides to adipose tissue and overriding the insulin resistance of this tissue.

It should be emphasized that the *Psammomys* liver is rich in lipogenic enzyme activity and is the main site of fat synthesis, whereas adipose tissue is poor in lipogenic enzymes and its growth depends mainly on the uptake of preformed fat

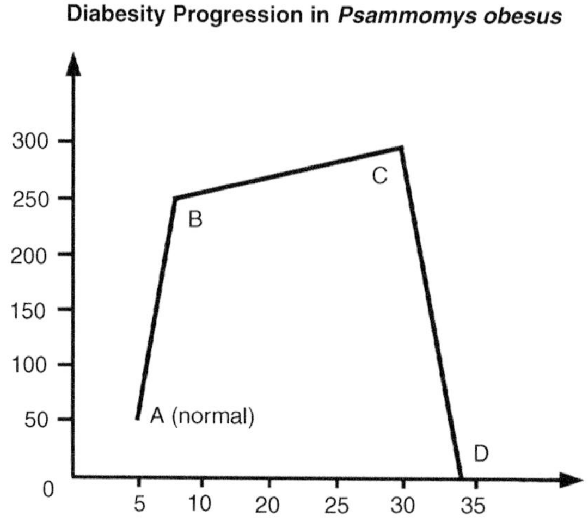

Diabesity Progression in *Psammomys obesus*

FIG. 13.1 "Inverted U" pattern illustrating the progression of *Psammomys* from normalcy (stage A) to hyperinsulinemia (stage B), hyperglycemia (stage C), and hypoinsulinemia with the marked hyperglycemia (stage D). Animals were selected at random from a large colony and kept on a regular rodent chow for two to four weeks. With time, most of the animals progress to stages C and D.

(Kalderon et al. 1983; Gutman et al. 1991). Adipose tissue is rich in lipoprotein lipase and in the capacity to assimilate the preformed lipids (Chajek-Shaul et al. 1988). A disproportionate increase of fatty acid binding proteins was found in the liver of obese, diabetic *Psammomys* (Brandes et al. 1986; Lewandowski et al. 1997). The last stage (D) of insulinopenic and highly hyperglycemic *Psammomys* comprised ~6% of the colony sample. They were all lean and their low plasma insulin levels indicated an exhaustion of insulin secretion (fig. 13.1). This group is on the verge of complete islet necrosis and ketoacidosis, to which the *Psammomys* is particularly sensitive.

DIABETES RESEARCH IN OTHER *PSAMMOMYS* COLONIES

Incidence of overt diabetes has been observed in *Psammomys* originating from Algeria by Marquie et al. (1991). Approximately 40% of the animals developed the diabetes syndrome, with a few dying in ketosis. Aouichat Bouguerra et al. (2001) investigated the role of glucose and insulin in the synthesis of collagen in aortic smooth muscle cells of Algerian *Psammomys*. Gernigon et al. (1994) investigated the seasonal variations in the ultrastructure and production of testosterone dependent proteins in the seminal vesicles of Algerian *Psammomys*. An additional colony of *Psammomys* is maintained by the U.S. Food and Drug Administration, Washington,

D.C., on which beta cell studies and assessment of diabetic lenses were performed (Katzman et al. 2004; Chennault et al. 2002).

A branch of the Israeli *Psammomys* colony is also bred in Australia (Barnett et al. 1994a, b, 1995). It was observed that insulin resistance and hyperinsulinemia appear before weight gain, followed by adipose tissue accretion and diabesity. Tissue TG deposition was driven by the ample hepatic lipogenesis, which continues unabated despite insulin resistance. With regard to the genetic characteristics, it is of interest that the product of a hypothalamic gene discovered by Collier et al. (2000), termed "BEACON," was found to increase the food intake and body weight after intracerebroventricular injection (Walder et al. 2002b). It also induced a twofold increase in hypothalamic neuropeptide Y expression. The Australian investigators have discovered a TANIS gene assumed to be responsible for a link between type 2 diabetes (T2D) and inflammation (Walder et al. 2002c). They also have found a mitochondrial intramembranal protease, known as presenilin-associated rhomboid-like protein (PSARL), to be associated with insulin resistance and possibly a new candidate gene for T2D (Walder et al. 2005). *Psammomys* in Australia was found to be hyperleptinemic and leptin resistant (Walder et al. 1999). Development of obesity was attributed, at least in part, to leptin resistance (Collier et al. 1997, 2000). The Australian investigators also found, along with our experience (Kanety et al. 1994), that the progress of *Psammomys* to diabesity may be reversed by reducing the nutrition in stage C, before beta cell degranulation sets in (Barnett et al. 1994b).

LIPOPROTEIN SYNTHESIS IN *PSAMMOMYS*

Dyslipidemia characterized Israeli *Psammomys* of groups B and C. Both plasma triglyceride (TG) and cholesterol levels were increased, mainly VLDL and LDL. A relation among insulin resistance, diabetes, and hepatic TG-rich lipoprotein overproduction was observed. Insulin resistance and diabetes led to a significant increase in monoacylglycerol acyltransferase (MGAT) and diacylglycerol acyltransferase (DGAT) activity and increased liver secretion of Apo-B-100 (Zoltowska et al. 2004). Free fatty acid (FFA) mobilization from adipose tissue exacerbated insulin resistance and upregulated the expression of microsomal TG transfer protein (MTP) and of the insulin response element in the promoter region of the MTP gene. In addition, FFA influx into the liver increased intracellular TG content and provided lipid substrate for MTP activity and VLDL assembly. Raised availability of lipids as well as augmented MTP activity resulted in lipidation of apo-B and accelerated production of VLDL in the liver of *Psammomys* (Zoltowska et al. 2004).

Like in the liver, insulin resistance and diabetes in *Psammomys* triggers the whole intraenterocyte machinery, leading to lipoprotein assembly and favoring the intestinal oversecretion of apoB-48 lipoprotein, which may contribute to the hypertriglyceridemia. A number of intracellular factors in the intestine of *Psammomys* in stages B and C have been found to be associated with the deregulatory mechanisms, including *de novo* TG synthesis, the monoacylglycerol pathway, liver fatty acid binding protein (L-FABS) mass, reduced proteasomal degradation of apoB-48, and lipoprotein assembly (Zoltowska et al. 2003).

BETA CELL LESION FOLLOWING PROLONGED INSULIN RESISTANCE WITH HYPERINSULINEMIA

Hyperinsulinemia in *Psammomys* does not last long enough to compensate for insulin resistance and the associated hyperglycemia in stage C. The insulin content of beta cells in *Psammomys* is low compared with that of other animal models of diabetes like the *ob/ob* mice or ZDF rats in which the hyperglycemia is effectively compensated (fig. 13.2). The gene expression in *Psammomys* associated with PDX-1 transcription factor is very low or absent and limits the beta cell secretory function (Leibowitz et al. 2001a, b). The secretion pressure leads to apoptosis and beta cell demise (stage D) (Bar-On et al. 1999; Shafrir et al. 1999a).

As reported in numerous previous communications, the "spontaneous" diabetes in *Psammomys* occurs as a result of maintenance on a standard rodent chow, which we label as high-energy diet. *Psammomys* maintained on prolonged HE diet undergo massive beta cell degranulation, loss of insulin immunostaining, apoptosis, and necrosis. Miki et al. (1966) and Like and Miki (1967) performed early microscopic studies of beta cell degranulation associated with glycogen deposition. Excessive glycogen in the islets most probably stems from lack of glucose metabolism because of degeneration of beta cell glycolysis, though islet protein synthesis was not reduced until the late stages.

Jörns et al. (2002) have followed in detail the consecutive beta cell changes during the progression from stages A through D on HE diet, involving a gradual loss of beta cell insulin, glucokinase, and GLUT2 transporter immunoreactivities. After three weeks on HE diet, the reduction was 70–95% of the initial value, in

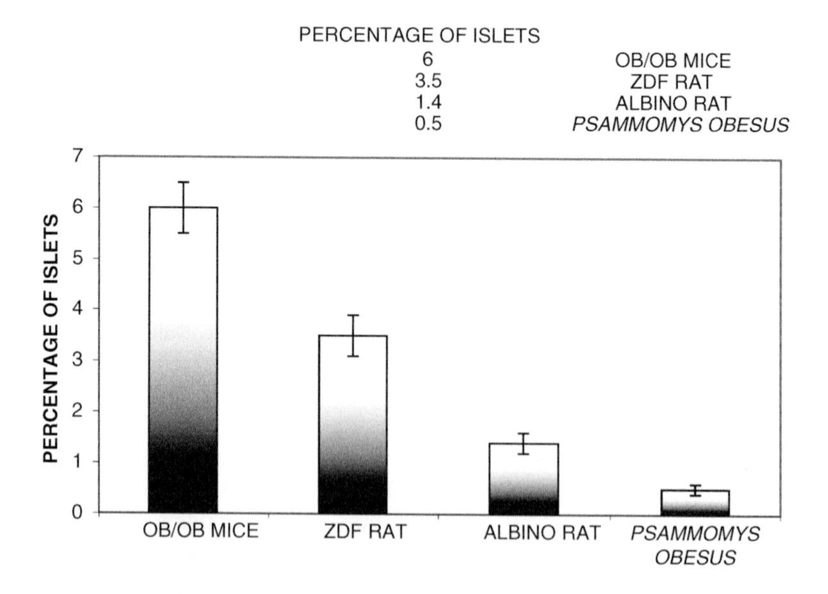

FIG. 13.2 Percentage of islets in the pancreas of *Psammomys* compared with albino rat and with other hyperinsulinemic, insulin-resistant rodents. (E. Shafrir, unpublished observations.)

correlation with the rising blood glucose level. Ultrastructurally, different signs of necrotic destruction of pancreatic beta cells such as the pyknosis of nuclei and a massive vacuolization in the cytoplasm were seen, accompanied by swollen mitochondria and dilated cisternae of the Golgi complex and of the rough endoplasmic reticulum. When the pancreas was removed from animals at stage D, beta cells exhibited apoptosis and DNA fragmentation (Bar-On et al. 1999; Shafrir et al. 1999; Donath et al. 1999; Nesher et al. 1999).

There is no direct evidence for the involvement of gluco- or lipotoxicity in the necrosis of beta cells in *Psammomys* and cytokine causation is also not indicated (Jörns et al. 2006). An attempt to prevent the possible effect of advanced glycation end products or of nitrous oxide by including the glycation inhibitor aminoguanidine in the hyperglycemic incubation medium was not effective in protecting beta cells of *Psammomys*. However, Kaneto et al. (1996) reported that beta cells of streptozotocin diabetic rats were protected by the antioxidants N-acetyl-L-cysteine and aminoguanidine. There may be species difference in reaction to hyperglycemia. In *Psammomys*, the prompt damage of beta cell architecture is probably the result of exhaustion due to the insulin hypersecretion prior to the eventual glucotoxic effect.

Psammomys in stage C shows increased plasma proinsulin levels (Wilke et al. 1979), up to one-half of the circulating immunoassayable total insulin (Gadoth et al. 1994, 1995). The inordinate secretion pressure may cause a swift exocytosis of immature insulin granules escaping prior to the C peptide cleavage. This indicates that the compensation of the delayed glucose removal and suppression of PEPCK by insulin were not effective because proinsulin has only a minute fraction of insulin activity. On the other hand, the high level of circulating proinsulin does not mean that its secretion is similar to that of insulin since the half-life of proinsulin is much longer than that of insulin (Glauber et al. 1986). Deficient insulin secretion, with increased proportions of insulin precursor molecules, is a common feature of T2D and could result from inappropriate beta cell function and reduced beta cell mass.

Bendayan et al. (1995) reported that beta cells of hyperglycemic *Psammomys* are in a chronic high secretory mode with loss of secretory granules and impaired processing of proinsulin by the convertases PC1 and PC2 as compared with normoglycemic animal. Leibowitz et al. (2002) maintained that inadequate regulation of proinsulin gene expression by glucose contributes to the failure of *Psammomys obesus* to cope with the increased demand for insulin associated with caloric excess.

The possible involvement of islet amyloid polypeptide (IAPP) in insulin depletion of diabetic *Psammomys* was also investigated. The immunoreactivity of pancreatic IAPP and insulin were considerably weaker in stage C *Psammomys* compared with normoinsulinemic group A. Plasma IAPP and insulin were significantly elevated, suggesting coexcretion as a result of pancreatic beta cell overstimulation (Leckstrom et al. 1997).

Psammomys is a good model to address questions related to the role of insulin resistance and beta cell failure in nutritionally induced diabetes. Continued HE diet, imposing a gradually rising insulin resistance, results in depletion of pancreatic insulin stores, with increased proportion of insulin precursor molecules in the pancreas and the blood. Inadequate response of the preproinsulin gene to the increased insulin needs is an important cause of diabetes progression. Changes in beta cell mass do not correlate with pancreatic insulin stores and are unlikely to play a role in disease initiation and progression. The major culprit is the inappropriate insulin production with depletion of insulin stores as a consequence (Kaiser at al. 2005a, 2005b). Similar mechanisms could operate during the evolution of T2D in humans.

SELECTION OF DEFINED LINES OF *PSAMMOMYS*

The consumption of HE diet is required for diabetes to be expressed with hyperglycemia developing within 7–21 days in over 80% of the population of the diabetes-prone line. In the original colony, the reaction of randomly chosen individual animals to the same HE diet may differ and the development of diabetes may be slower, and a higher percentage of animals remains normoglycemic even on HE diet.

We used the HE diet to examine the diabetic potential of each individual and to identify the diabetic and nondiabetic margins of the population. By using an assortative mating system based on a minimal inbreeding method (Baker 1979), it was possible to separate the animals in the colony into two distinct lines: diabetes prone (DP) and diabetes resistant (DR), differing phenotypically and genotypically (Kalman et al. 1993). Animals to be mated were chosen according to phenotypic parameters (postprandial blood glucose and plasma insulin levels). The composition and digestibility of the HE and specially prepared low-energy diet are recorded in table 13.1 and fig. 13.3.

METABOLIC AND REPRODUCTIVE EFFICIENCY

The essence of thrifty metabolism is high metabolic efficiency that enables existence in an environment characterized by constant supply of low-energy diets. The artificial laboratory condition of an *ad libitum* accessible HE diet that creates a continuous input of energy leads to hyperinsulinemia and hyperglycemia. The different sensitivity

TABLE 13.1
Composition, Energy, and Digestibility of *Psammomys* Diets[a]

	HE Diet		LE Diet		Salt Bush	
Components	**%**	**Digestible Portion**	**%**	**Digestible Portion**	**%**	**Digestible Portion**
Protein	6	0.73	16.7	0.68	23.3	0.80
Fat	2.4	0.60	3.1	0.81	2.9	0.65
Carbohydrates	68.0	0.79	70.0	0.58	53.3	0.65
Ash	6.0	0.26	10.2	0.43	20.5	0.42
Total digestibility		0.76		0.62		0.69
Total energy (kcal/g)	3.88		3.74		3.32	
Digestible energy (kcal/g)		2.95		2.32		2.29

[a]HE = high energy; LE = low energy; salt bush is native *Psammomys* food staple collected from desert.

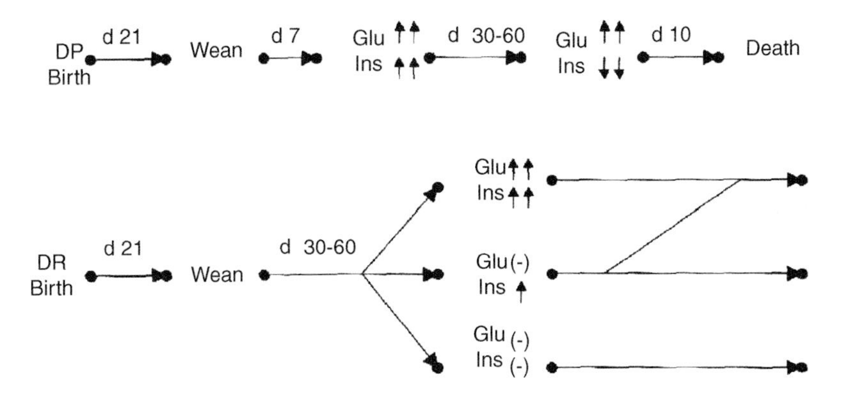

FIG. 13.3 Development of diabetes in diabetes-prone (DP) and diabetes-resistant (DR) *Psammomys* on HE diet from weaning. Animals of the DP line become diabetic at 7–12 days and, if continued on HE diet, they die at 40–70 days. Animals of the DR line may become diabetic in a more attenuated manner than those of the DP line. They may become hyperinsulinemic but normoglycemic or do not develop diabetes at all.

to the development of diabetes between the DP and the DR lines can be caused by one or more factors that include differences in food intake, hepatic and peripheral resistance, pancreas activity, or metabolic efficiency (fig. 13.3).

To measure the metabolic efficiency, we followed animals during the period of their most rapid growth after weaning (2.5–3.0 g/day in males and 2.4–2.9 g/day in females) and calculated metabolic efficiency as the relation between digestible energy intake to weight increment (Kalman et al. 1993). There is no hyperphagia in DP and DR lines when animals are fed HE or LE diets; hyperglycemia is related to energy availability in the diet. Quantity of feces was significantly higher in animals fed LE diet compared with animals fed HE diet. The metabolic efficiency in DP line *Psammomys* fed all diets was 6.0–6.6 kcal/g of weight increase, while in the DR line metabolic efficiency was 9.0–9.6 kcal/g of weight increase. In other studies (Barnett et al. 1995) performed on a branch of the Israeli colony in Australia, hyperphagia was reported in the diabetic state of *Psammomys* when they were fed standard rat maintenance diet, but the weight gain did not exceed 15% of body weight.

Differences between DP and DR lines are not limited to their dietary induced diabetes but also to other zootechnical and reproductive characteristics (Kalman et al. 1996). Reproductive efficiency in *Psammomys* is low compared with outbred rat strains (Baker 1979). Average number of weaned per female per week is 0.28 versus 1.0–1.5 in outbred rats (Weihe 1987). Reproductive efficiency is higher in the DP line compared with DR line females, due to the difference in nonreproductive females (22% in the DP line vs. 41% in the DR line) and to the difference in the average number of births per female (3.3 in the DP line vs. 1.3 in the DR line). These two parameters create a difference in the total number of newborn per female during its reproductive life (11.5 newborn per female in the DP line vs. 5.4 in the DR line). No difference was observed in the average number of newborn per birth (2.8 vs. 2.7, respectively).

The period of most rapid growth of DP *Psammomys* on LE diet is up to 65 days (2.5–3.0 g/day in males and 2.4–2.9 g/day in females). Growth continues up to 180 days but at a slower rate (1.4 and 0.8 g/day, respectively). Average weight of males is 264 ± 5 g and of females is 223.8 ± 7 g. The relative weight of most organs remains unchanged (Kalman et al. 1996). Adrenal glands are the only organ with relative weight significantly different between males and females in all age groups ($p < 0.001$), and it is always higher compared with albino rats (Kalman et al. 1996).

AGE-RELATED DIABETOGENICITY

Psammomys at ages of 1–12 months fed LE diet from weaning and transferred at different ages to HE diet (Ziv et al. 1999) demonstrated that sensitivity to the development of diabetes increases from weaning to be highest at 5 months of age and decreases thereafter. At five months of age, the obesity factor measured as the proportion of epididymal fat weight to total body weight is highest, while the proportion of other organs to total body weight remains unchanged. In animals older than seven or eight months, the potential to develop diabetes and obesity decreases. These changes are in correlation with the decrease in fertility in *Psammomys* from DP and DR lines.

PREGNANCY IN *PSAMMOMYS*

T2D during pregnancy is generally associated with an increased risk for poor reproduction and a high rate of congenital malformations. Pregnancy on the HE diet is longer than on the LE diet (27 vs. 26 days) and litter average lower (2.7 vs. 3.0). At birth, the offspring of the HE diet dams weighed 5.2 vs 7.2 g, had smaller crown–rump length, and presented a 1 to 3 day delay in neurodevelopmental parameters (first turn over, hair appearance, eye-opening, and response to noise). From the fourth week of life they became diabetic and weighed more than the LE diet offspring (Patlas et al., 2006).

INBORN INSULIN RESISTANCE

Insulin resistance in *Psammomys* is an inherent characteristic even in the normoglycemic–normoinsulinemic stage. In experiments where plasma insulin was elevated by intraperitoneal administration of exogenous insulin to normoglycemic–normoinsulinemic animals (stage A), only mild blood glucose reduction was observed compared with severe hypoglycemia in albino rats that received a similar dosage of insulin (Ziv et al. 1996; Ziv and Kalman 2000).

External insulin is effective in *Psammomys* as demonstrated in its hypoglycemic and hypotriglyceridemic effect in insulin-deficient *Psammomys* in stage D. Exogenous bovine insulin in the form of subcutaneous implants releasing 2 U/24 h insulin for 10 days was implanted in stage D *Psammomys* that became metabolically similar to the insulin-resistant stage C *Psammomys*, characterized by endogenous hyperinsulinemia (Ziv et al. 1996). Despite the strong hypoglycemic and hypertriglyceridemic effect in fasted *Psammomys* stage D, the superimposed exogenous insulin was not capable of lowering blood glucose levels in the nonfasting, HE diet-fed stage C *Psammomys*.

Our premise of liver and muscle primary insulin resistance was confirmed by the hyperinsulinemic-euglycemic clamp studies (Ziv et al. 1996). Insulin infusion did decrease the hepatic glucose production and increase the total glucose transport. This demonstrates the effectiveness of exogenous insulin. However, hepatic glucose production was only partially reduced in *Psammomys* whereas, in albino rats under the same conditions, the hepatic glucose production was completely abolished. The limited elevation of total glucose transport in the hyperinsulinemic clamped *Psammomys* attests to the fact that the peripheral glucose utilization is low enough to be compensated by gluconeogenesis, avoiding lapse into hypoglycemia during exogenous insulin treatment. Insulin also failed to suppress in normoglycemic *Psammomys* the activity of hepatic PEPCK, the rate-limiting enzyme of gluconeogenesis (Shafrir 1988) as well as the hepatic glucose output.

The physiologic importance of the innate resistance in a desert animal can be assumed to direct the scarce glucose to the glucose obligatory tissues rather than to the muscle, which can utilize other sources of energy. In this respect, the administration of HE diet failed to induce an increase in muscle malonyl CoA in contrast to albino rats. This fact indicates that fatty acid oxidation is not prevented by the rise of this intracellular inhibitor of fatty acid entry into the mitochondria. This may be relevant to the continuation of muscle fatty acid oxidation even during the availability of glucose (Shafrir et al. 2002).

MECHANISM OF INSULIN RESISTANCE AND ATTENUATION OF RECEPTOR TYROSINE KINASE IN *PSAMMOMYS*

To investigate the cause of insulin resistance, the activity of tyrosine kinase (TK), the receptor enzyme initiating insulin signaling, was studied in the liver and muscle of *Psammomys*. With Kanety et al. (1994), we have found that the density of insulin receptors (IR) is low in *Psammomys* muscle and liver, about one-fifth that of the laboratory albino rat. However, insulin binding and TK activity per wheat germ agglutinin isolated receptor were normal *in vitro* and *in vivo*. It is interesting that Mandarino (1983) was erroneously led to assume that the defect in insulin binding by *Psammomys* hepatocytes was due to "almost complete" absence of insulin receptors (compared to albino rats) and was thus responsible for insulin resistance.

The extent of TK activation by insulin was pronouncedly lower in stages B and C in liver and muscle as compared to stage A (Kanety et al. 1994) (fig. 13.4). The reduced activation by insulin was accompanied by a marked decrease in muscle GLUT4 transporter protein and mRNA (Shafrir and Ziv 1998) to one-half of daily food intake for a few days. The recovery of TK activity was not complete when hyperglycemia was corrected by nutrition restriction, but full after the return to normoinsulinemia. These findings indicate that hyperinsulinemia is the basic event responsible for deficient IR function causing insulin resistance as a result of multisite phosphorylation of the receptor and involving serine sites, inhibitory to tyrosine phosphorylation, on the insulin signal transduction proteins (Shafrir et al. 1999b).

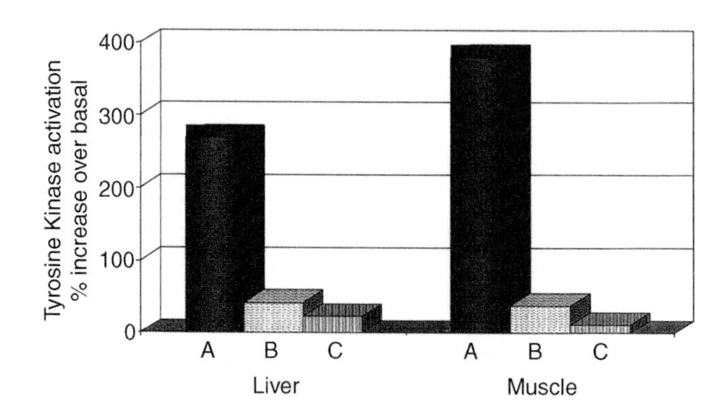

FIG. 13.4 Insulin stimulation of hepatic (left side) and muscle (right side) insulin receptor tyrosine kinase (TK) activity in *Psammomys* of groups A, B, and C, the latter before and after recovery from diabesity. A bolus of 100 μg was administered i.p. to animals 30–40 seconds prior to liver removal. Receptors were purified on wheat germ agglutinin columns, and the phosphorylation of poly(glu:tyr)4:1 substrate was determined using purified receptor. The extent of TK stimulation by insulin is expressed as a percent of change in the basal activity of paired animals. (Adapted from Kanety, H. et al. 1994. *Proc Natl Acad Sci USA* 91:1853–1857.)

OVEREXPRESSION OF PKCε: A NEGATIVE FEEDBACK OF INSULIN SIGNAL TRANSDUCTION

Protein kinase C (PKC) in the gastrocnemius muscle of hyperinsulinemic *Psammomys* was found to be pronouncedly overexpressed and translocated to the cell membrane (Ikeda et al. 2001; Shafrir et al. 1999b) (fig. 13.5). This enzyme group is now widely studied because of its preferential phosphorylation of serine residues on signaling pathway proteins, resulting in the negative feedback of this pathway. The PKC group includes at least 11 isoenzymes, of which PKCε was most pronouncedly overexpressed in *Psammomys* muscle. Translocation to the membrane in addition to overexpression indicates an increased activity (Ikeda et al. 2001). There was a tendency of overexpression of other PKC isoenzymes as well, particularly PKCα and PKCθ. It is of interest that PKCθ rather than PKCε was found to be increased in humans administered a lipid emulsion (Griffin et al. 1999), which generally causes impaired glucose tolerance.

PKCε showed the highest overexpression in the skeletal muscle of *Psammomys* in the hyperglycemic–hyperinsulinemic stage C compared with the DR line. Significant overexpression of PKCε was also seen in the normoglycemic stage A of DP *Psammomys* compared with the DR line and also with the albino rat (Mack et al. to be published), which indicates that PKCε overexpression precedes the onset of overt hyperglycemia. This finding conforms to the innate insulin resistance referred to previously. Thus, PKCε overexpression in stage A may be considered as a marker of “prediabetic” or “preinsulinemic” stages and of propensity of a given individual to progress to overt diabetes on affluent nutrition. It is, however, without untoward consequences as long as the diet is LE.

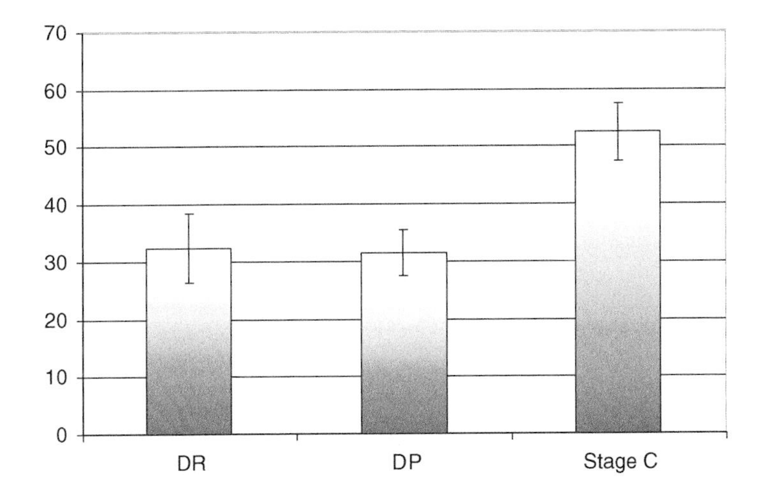

FIG. 13.5 Membrane-associated PKCε in skeletal muscle of diabetes-resistant (DR), diabetes-prone (DP), and diabetic *Psammomys obesus* at stage C. The vertical axis represents optical density units, means ± SE of six animals (P = 0.02). Note the significant increase in the translocation of PKCε from cytosol to the membrane at stage C.

Since PKCε overexpression was associated with impaired TK activation by insulin and reduced GLUT4 mRNA and protein, this indicates that phosphorylation of serine sites on the receptor, IRS, and other insulin signal proteins was detrimental and associated with an impaired PI-3K (phosphoinositol-3-kinase) activation. It was therefore of interest to investigate whether PKCε overexpression induces a further negative downstream effect on insulin signaling. The activity of PKB/Akt, an enzyme regarded as responsible for the activation of pleiotropic metabolic systems, was determined. The transfection of HEK 293 cells with IR and/or PKCε plasmids, followed by stimulation with insulin or phorbol ester, respectively, showed that the activation of PKCε reduced PKB expression and inhibited PKB activation (Ikeda et al. 2001). This may be assumed to be the result of PKCε-effected serine phosphorylation on IRS, on which the PKB and PI-3K functions are dependent.

The increased activity of PKC isoenzymes in muscle membrane, in IR proximity, suggested the involvement of PKCε in the attenuation of IR/TK activation, as described by Kanety et al. (1994). Several PKC isoenzymes were shown to reduce the TK catalyzed phosphorylation of the IR and IRS-1 (see Shafrir 2001). It was found that PKCε overexpression was associated with reduced binding of insulin by muscle IR due to the reduction in the number of IR per cell. The downregulation of IR was demonstrated in HEK 293 cells, which were transfected with human IR and PKCε plasmids. Activation of the PKCε by phorbolester (TPA) reduced the amount of IR to ~40% of the original number (Ikeda et al. 2001). This finding is in accord with observations of degradation of IR induced by PKCε and possibly by other PKC isoforms. It is therefore likely that serine/threonine phosphorylation of IR and/or IRS-1 inhibits the TK activity via a feedback loop and is responsible for the deficient TK activation by insulin and IR degradation, accentuating the insulin resistance in stages B and C in *Psammomys* on HE diet (Kanety et al. 1994).

PKCε OVEREXPRESSION: RELATION TO MUSCLE LIPID AND DIACYLGLYCEROL (DAG) CONTENT

The PKCε overexpression in *Psammomys* was found to be correlated with the increased muscle content of TG and DAG (Ikeda et al. 2001; Shafrir 2001). DAG is an intermediate of fatty acid esterification to TG and TG breakdown to fatty acids and glycerol. The raised muscle concentration of TG and DAG in *Psammomys* occurs in the situation of hyperinsulinemia and hyperglycemia, characteristic of stages B and C.

EFFECT OF EXERCISE

Exercise training of *Psammomys* on HE diet involving a twice daily 90-min walk on a treadmill for four weeks was found to prevent the development of hyperglycemia in the majority of animals (Heled et al. 2002). The animals became normoglycemic and adipose tissue was found to be somewhat reduced but they remained hyperinsulinemic, indicating that insulin resistance still persisted and was not compensated and thus exhibiting the characteristics of stage B, described here, by insulin secretion. PKCδ activity and serine phosphorylation were higher in the exercise group and were suggested to be involved in the adaptive effects of exercise. Nevertheless, physical exercise enhanced the TK activity in stage A *Psammomys* and the activity of receptor TK and insulin receptor substrate (IRS), as well as PI3 kinase, and attenuated the activity of hepatic PEPCK (Heled et al. 2004). The function of insulin receptor was higher in the exercise group and comparable to animals maintained on the LE diet. Physical exercise was also found to increase the expression of TNFα and GLUT1 transporter in muscle of *Psammomys*. Although TNFα is a mediator of insulin resistance, it also increases the activity of GLUT1 transporter and, according to the authors, the elevation in GLUT1 may be responsible for facilitation of muscle glucose uptake (Heled et al. 2005).

EVIDENCE FOR A MAJOR GENE AFFECTING TRANSITION FROM NORMOGLYCEMIA TO HYPERGLYCEMIA IN *PSAMMOMYS OBESUS*

The mode of inheritance of nutritionally induced diabetes in *Psammomys* was investigated following transfer from LE to HE diet, which induces hyperglycemia. *Psammomys* selected for high or low blood glucose level were used as two parental lines. At weeks 1–9 after weaning, a clear bimodal distribution statistically different from unimodal distribution of blood glucose was observed, normoglycemic and hyperglycemic at a 1:1 ratio. This ratio is expected at the first backcross generation for traits controlled by a single dominant gene. From week 0 (prior to the transfer to HE diet) till week 8, the hyperglycemic individuals were significantly heavier (4–17%) than the normoglycemic ones. The bimodal blood glucose distribution in BC generation, with about equal frequencies in each mode, strongly suggests that a single major gene affects the transition from normo- to hyperglycemia. The wide range of blood glucose values among the hyperglycemic individuals (180–500 mg/dL) indicates that several genes and environmental factors influence the extent of hyperglycemia. The diabetes-resistant allele appears to be dominant; the estimate for dominance ratio is 0.97 (Hillel et al. 2005).

PSAMMOMYS AS A MODEL FOR TESTING ANTIHYPERGLYCEMIC DRUGS

Psammomys represents an excellent model to treat diabetes in stages B or C since they are reversible. Stage D, in which the beta cells are already compromised, is not suitable for treatment. Vanadyl sulfate was effectively used to prevent the hyperglycemia and hyperinsulinemia of *Psammomys* maintained on the HE diet (Shafrir et al. 2001). Administration of 5 mg/kg of vanadyl sulfate for five days resulted in a prolonged restoration of normoglycemia and normoinsulinemia and muscle GLUT4 transporter as well. Pretreatment with vanadyl sulfate significantly delayed the onset of hyperglycemia.

Rosiglitazone administered to *Psammomys* on HE diet at 20 mg/kg for two weeks also normalized the hyperglycemia and markedly reduced the hyeperinsulinemia. The glucose metabolism in peripheral tissues was not the primary target of the beneficial effect of rosiglitazone but this thiazolidinedione prevented damage to pancreatic beta cells and loss of insulin, thus enabling insulin secretion to compensate for the peripheral insulin resistance (Hefetz et al. 2005).

Among other beneficial treatments of *Psammomys*, administration of nicotine was shown to decrease the food intake and body weight in *Psammomys* via a non-leptin-dependent pathway (Sanigorski et al. 2002). A sustained hypoglycemic effect was obtained by electroacupuncture (Shapira et al. 2000). Improvement was also obtained following treatment with an analogue of glucagon-like peptide (GLP-1) promoting the beta cell function and insulin availability (Uckaya et al. 2005). Novel peptides modulating a G-protein coupled kinase had an antidiabetic effect in *Psammomys*, *db/db* mice, and ZDF rats (Anis et al. 2004). Marquie et al. (1997) studied an oral agent S15261 (3-[2-[2-[4-[2-[α-fluorenyl acetyl amino ethyl] benzoyloxy] ethyl amino] 1-methoxy ethyl] trifluoromethylbenzene), which also prevented the beta cell lesion in the HE-maintained *Psammomys*. Regranulation of beta cells and restored integrity of cytoarchitecture were observed. Metabolic and antiatherogenic effects were obtained by long-term administration of benfluorex in dyslipidemic insulin-resistant *Psammomys* (Marquie et al. 1998).

DIABETES COMPLICATIONS IN *PSAMMOMYS OBESUS*

Hyperglycemia in *Psammomys* is associated with cataracts (Kohler and Knospe 1980), which appear after two to four months on the HE chow. This was also observed in other *Psammomys* colonies (Hackel et al. 1965; Kuwabara and Okisaka 1976). Cataracts do not occur in the group A and in the normoglycemic–hyperinsulinemic group B of *Psammomys*. There is also evidence of retinal damage by the finding of degeneration in dopamine neurons in hyperglycemic *Psammomys* (Larabi et al. 1991). Also, the Harderian gland, which has lacrimal activity in animals, became necrotic in *Psammomys* maintained on synthetic HE diet (Djeridane 2002).

With respect to the cataract appearance, it was observed that galactokinase activity was very low in *Psammomys* red cells and in ocular tissues, with or without cataracts. This suggests that the formation of cataracts is precipitated by overeating and may involve conversion of glucose and galactose to sorbitol or galacticol, the accumulation of which is favored by hyperglycemia and by galactose content in the diet (Gutman et al. 1975).

Cohen-Melamed et al. (1995) treated the *Psammomys* with acarbose with the profile of glucose and the activity of lens aldose reductase reduced; this was associated with a significant preventive effect of cataract development. Borenshtein et al. (2001) observed that lipoic acid treatment was beneficial with regard to cataract development due to glucose lowering as well as to increase in lens glutathione level.

Microangiopathy, expressed by thickening of the intima and deposition of glycosaminoglycans, was observed in the Algerian *Psammomys* colony (Marquie et al. 1991). Degeneration of the intervertebral discs and spondylosis were noted in the Jerusalem colony (Silberberg et al. 1979; Silberberg 1988; Moskowitz et al. 1990; Amir et al. 1991; Gruber et al. 2002). The hyperglycemic Jerusalem animals also showed evidence of neuropathy manifested as anelgesis, inferred from the high pain threshold and from reduction in nerve conduction velocity (Wuarin-Bierman et al. 1987). Old *Psammomys* exhibited a tendency of hepatic malignancy (Ungar and Adler 1978), uterine neoplasms (Czernobilsky et al. 1982), and thymic tumors (Rosenmann et al. 1982).

Functional otological impairments were seen in the *Psammomys* (Perez et al. 2001, 2002; Sohmer et al. 2004). Regarding the kidney, *Psammomys* is known for a capacity to excrete highly saline urine. The urinary concentrating mechanism was investigated by Jamison et al. (1979) and Barrett et al. (1978). Kidney complications expressed changes in glomerular filtration rate, total protein excretion, and protein creatinine ratio in diabetic *Psammomys*. Histology of diabetic kidneys revealed changes in glomeruli and interstitium with the progress of diabetic nephropathy (Sherzer et al. unpublished).The effects of HE diet on renal sodium pump activity and Na-K-ATPase hyperactivity in diabetic *Psammomys* were also described (Scherzer et al. 2000a, 2000b). The *Psammomys* that expressed these stages of diabetic nephropathy can serve as a model for human diabetic nephropathy and possible interventions to moderate its progression.

CONCLUDING REMARKS

Psammomys in its native habitat is a healthy gerbil with a metabolic–endocrine system adjusted to desert life on a low caloric density food, which enables reasonable survival. *Psammomys* is not hyperphagic, but when high caloric density food becomes available, it is predisposed to weight gain, diabetes, and beta cell overtaxation. This is due to a weak spot in the insulin signaling pathway—particularly increased serine phosphorylation and stoppage of tissue glucose transport but not lipogenesis. This ill adaptation to nutrient excess is an outstanding example of the "thrifty gene" effect (Neil 1962; Wendorf and Goldfine 1991), which represents a suitable model for the study of the mechanism of predisposition to insulin resistance and metabolic syndrome in human populations evolving from scarcity to abundance in nutritional intake.

AVAILABILITY AND SHIPPING

At the time of writing, *Psammomys* is available from the original breeding colony through one of the following addresses:

Rony Kalman, DVM, Animal Facility, Medical School, Hebrew University, Jerusalem 91120, Israel. Fax: 972-2-6424645; e-mail: ronyk@md2.huji.ac.il.

Harlan Laboratories, Ein Karem, Jerusalem 91120, POB 12085, Israel. Tel: 972-2-6439398, fax: 972-2-6439403.

Cost of *Psammomys* ranges from 60 to 80 U.S. dollars, excluding freight costs.

REFERENCES

Adler JH, Lazarovici G, Marton M, et al. 1988. Patterns of hyperglycemia hyper-insulinemnia and pancreatic insufficiency in sand rats (*Psammomys obesus*). In *Lessons from Animal Diabetes*, ed. E. Shafrir, AE Renold. London, J Libbey, 2:275–279.

Amir G, Adler JH, Menczel J. 1991. Histomorphometric analysis of weight bearing bones of diabetic and non-diabetic sand rats (*Psammomys obesus*). *Diabetes Res* 17:135–137.

Anis Y, Leshem O, Reuveni H, et al. 2004. Antidiabetic effect of novel modulating peptides of G-protein-coupled kinase in experimental models of diabetes. *Diabetologia* 47:1232–1244.

Aouichat Bouguerra S, Bourdillon MC, Dahmani Y, Bekkhoucha F. 2001. Non insulin dependent diabetes in sand rat (*Psammomys obesus*) and production of collagen in cultured aortic smooth muscle cells. Influence of insulin. *Exp Diabetes Res* 2:37–46.

Baker DEJ. 1979. Reproduction and breeding. In *The Laboratory Rat*, ed. Baker HJ, Lindsey JR, Weisbroth SH. Academic Press, Orlando, FL, 154–166.

Barnett M, Collier GR, Collier FMcL, et al. 1994a. A cross-sectional and short-term longitudinal characterization of NIDDM in *Psammomys obesus*. *Diabetologia* 37:671–676.

Barnett M, Collier GR, Zimmet P, O'Dea K. 1995. Energy intake with respect to the development of diabetes mellitus in *Psammomys obesus*. *Diabetes Nutr Metab* 8:1–6.

Barnett M, Collier GR, Zimmet P, O'Dea K. 1994b. The effect of restricting energy intake on diabetes in *Psammomys obesus*. *Int J Obesity* 18:789–794.

Bar-On H, Ben-Sasson R, Ziv E, et al. 1999. Irreversibility of nutritionally induced NIDDM in *Psammomys obesus* is related to β-cell apoptosis. *Pancreas* 18:259–265.

Barrett JM, Kriz W, Kaissling B, de Rouffignac C. 1978. The ultrastructure of the nephrons of the desert rodent (*Psammomys obesus*) kidney. II. Thin limbs of Henle of long-looped nephrons. *Am J Anat* 151:499–514.

Bendayan M, Malide D, Ziv E, et al. 1995. Immunocytochemical investigation of insulin secretion by pancreatic beta-cells in control and diabetic *Psammomys obesus*. *J Histochem Cytochem* 43:771–784.

Borenshtein D, Ofri R, Werman M, et al. 2001. Cataract development in diabetic sand rats treated with alpha-lipoic acid and its gamma-linolenic acid conjugate. *Diabetes Metab Res Rev* 17:44–50.

Brandes R, Tsur R, Arad R, Adler JH. 1986. Liver cytosolic fatty acids binding proteins in rats and *Psammomys obesus*: modulation in diabetes. *Comp Biochem Physiol B* 83:837–839.

Chajek-Shaul T, Ziv E, Friedman G, et al. 1988. Regulation of lipoprotein lipase activity in the sand rat: Effect of nutritional stage and cAMP modulation. *Metabolism* 37:1152–1158.

Chennault VM, Ediger MN, Ansari RR. 2002. *In vivo* assessment of diabetic lenses using dynamic light scattering. *Diabetes Technol Therap* 4:651–659.

Cohen-Melamed E, Nyska A, Pollack A, Madar Z. 1995. Aldose reductase (EC 1.1.1.21) activity and reduced-glutathione content in lenses of diabetic sand rats (*Psammomys obesus*) fed with acarbose. *Br J Nutr* 74:607–615.

Collier GR, De Silva A, Sanigorski A et al. 1997. Development of obesity and insulin resistance in the Israeli sand rat (*Psammomys obesus*). Does leptin play a role? *Ann NY Acad Sci* 827:50–63.

Collier GR, McMillan JS, Windmill K, et al. 2000. Beacon: A novel gene involved in regulation of energy balance. *Diabetes* 49:1766–1771.

Czernobilsky B, Ungar H, Adler JH. 1982. Spontaneous uterine neoplasms in the fat sand rat (*Psammomys obesus*). *Lab Animals* 16:285–289.

Djeridane Y. 2002. The Harderian gland in diabetic sand rats (*Psammomys obesus*) a light microscopic study. *Exp Eye Res* 75:753–759.

Donath MY, Gross D, Cerasi E, Kaiser N. 1999. Hyperglycemia-induced β-cell apoptosis in pancreatic islets of *Psammomys obesus* during development of diabetes. *Diabetes* 48:738–744.

Gadoth MG, Leibowitz G, Shafrir E, et al. 1994. Hyperproinsulinemia and insulin deficiency in the diabetic *Psammomys obesus*. *Endocrinology* 135:610–616.

Gadoth M, Ariav Y, Cerasi E, et al. 1995. Hyperproinsulinemia in the diabetic *Psammomys obesus* is a result of increased secretory demand on beta-cells. *Endocrinology* 36:4218–4223.

Gernigon Th, Berger M, Lecher P. 1994. Seasonal variations in the ultrastructure and production of androgen-dependent proteins in the seminal vesicles of a Saharian rodent (*Psammomys obesus*). *J Endocrinol* 142:37–46.

Glauber HS, Revers RR, Henry R, et al. 1986. *In vivo* deactivation of proinsulin action on glucose disposal and hepatic glucose production in normal man. *Diabetes* 35:311–317.

Gruber HE, Johnson T, Norton HJ, Hanley EN Jr. 2002. The sand rat model for disc degeneration: radiologic characterization of age-related changes: cross-sectional and prospective analyses. *Spine* 27:230–234.

Gutman A, Andreus A, Adler JH. 1975. Hyperinsulinemia, insulin resistance and cataract formation in sand rats. *Isr J Med Sci* 11:714–722.

Gutman A, Kalderon B, Levy E, Shafrir E. 1991. Pattern of very low density lipoprotein disposal in the sand rat (*Psammomys obesus*). *Lessons from Animal Diabetes*, ed. Shafrir E, Smith-Gordon, London 3:699–671.

Hackel DB, Schmidt-Nielsen K, Haines HB, Mikat E. 1965. Diabetes mellitus in the sand rat (*Psammomys obesus*). Pathological studies. *Lab Invest* 14:200–207.

Hefetz S, Ziv E, Jorns A, Lenzen S, Shafrir E. 2005b. Prevention of nutritionally induced diabetes by rosiglitazone in the gerbil *Psammomys obesus*. *Diabetes/Metab Rev Res* 22:139–145.

Heled Y, Shapiro Y, Shani Y, et al. 2002. Physical exercise prevents the development of type 2 diabetes mellitus in *Psammomys obesus*. *Am J Physiol Endocrinol Metab* 282:E370–E375.

Heled Y, Shapiro Y, Shani Y, et al. 2004. Physical exercise enhances hepatic insulin signaling and inhibits phosphoenolpyruvate carboxykinase activity in diabetes-prone *Psammomys obesus*. *Metabolism* 53:836–841.

Heled Y, Dror Y, Moran DS, et al. 2005. Physical exercise increases the expression of TNFalpha and GLUT1 in muscle tissue of diabetes-prone *Psammomys obesus. Life Sci* 21:2977–2985.

Hillel J, Gefel D, Kalman R, et al. 2005. Evidence for a major gene effecting the transition from normoglycemia to hyperglycaemia in *Psammomys obesus. Heredity* 95:158–165.

Ikeda Y, Olsen GS, Ziv E, et al. 2001. Cellular mechanism of nutritionally induced insulin resistance in *Psammomys obesus.* Overexpression of protein kinase Cε in skeletal muscle precedes the onset of hyperinsulinemia and hyperglycemia. *Diabetes* 50:584–592.

Jamison RI, Roinel N, de Rouffignac C. 1979. Urinary concentrating mechanism in the desert rodent *Psammomys obesus. Am J Physiol* 236:F448–F453.

Jörns A, Tiedge M, Ziv E, Shafrir E, Lenzen S. 2002. Gradual loss of pancreatic beta-cell insulin, glucokinase and GLUT2 glucose transporter immunoreactivities during the time course of nutritionally induced type-2 diabetes in *Psammomys obesus* (sand rat). *Virchows Arch* 44:63–69.

Jörns A, Rath KJ, Bock O, Lenzen S. 2006. Beta cell death in hyperglycemic *Psammomys obesus* is not cytokine-mediated. *Diabetologia* 49:2704–2712.

Kaiser N, Nesher R, Donath MY, et al. 2005a. Psammomys obesus, a model for environment-gene interactions in type 2 diabetes. *Diabetes* 54:(Suppl. 2) S137–S144.

Kaiser N, Yuli M, Uckaya G, et al. 2005b. Dynamic changes in beta-cell mass and pancreatic insulin during the evolution of nutrition-dependent diabetes in *Psammomys obesus*: Impact of glycemic control. *Diabetes* 54:138–145.

Kalderon B, Adler JH, Levy E, Gutman A. 1983. Lipogenesis in the sand rat (*Psammomys obesus*). *Am J Physiol* 244:E480–486.

Kalderon B, Gutman A, Levy E, Shafrir E, Adler JH. 1986. Characterization of stages in the development of obesity-diabetes syndrome in the sand rat (*Psammomys obesus*). *Diabetes* 35:717–724.

Kalman R, Adler J, Lazarovici G, et al. 1993. The efficiency of sand rat metabolism is responsible for development of obesity and diabetes. *J Basic Clin Physiol Pharmacol* 4:57–68.

Kalman R, Lazarovici G, Bar-On H, Ziv E. 1996. *Psammomys obesus* (sand rat)—Morphological, physiological and biochemical characteristics of a model for type II diabetes. *Lab Anim Sci* 35:567–570.

Kalman R, Ziv E, Shafrir E, et al. 2001. *Psammomys obesus* and the albino rat—Two different models of nutritional insulin resistance, representing two different types of human populations. *Lab Anim* 35:346–352.

Kaneto H, Fujii J, Myint TM, et al. 1996. Reducing sugars trigger oxidative modification and apoptosis in pancreatic beta-cells by provoking oxidative stress through the glycation reaction. *Biochem J* 320:855–863.

Kanety H, Moshe S, Shafrir E, et al. 1994. Hyperinsulinemia induces a reversible impairment in insulin receptor function leading to diabetes in the sand rat model of non-insulin-dependent diabetes mellitus. *Proc Natl Acad Sci USA* 91:1853–1857.

Katzman SM, Messerli MA, Barry DT, et al. 2004. Mitochondiral metabolism reveals a functional architecture in intact islets of Langerhans from normal and diabetic *Psammomys obesus. Am J Physiol* 287:1090–1099.

Kuwabara T, Okisakaa S. 1976. Electron microscopic study of cataractous lenses of diabetic sand rats (*Psammomys obesus*). *Prog Lens Biochem Res* 7–15.

Kohler E, Knospe S. 1980. Glucose metabolism and responsiveness of muscle to insulin during the development of diabetes in sand rats. *Endokrinologie* 75:225–234.

Larabi Y, Dahmani Y, Gernigon T, Nguyen-Legros J. 1991. Tyrosine hydroxylase immunoreactivity in the retina of the diabetic sand rat *Psammomys obesus. J Hirnforsch* 32:525–531.

Leckstrom A, Ziv E, Shafrir E, Westermark P. 1997. Islet amyloid peptide in *Psammomys obesus* (sand rat): effects of nutritionally induced diabetes and recovery on low energy diet or vanadyl sulfate treatment. *Pancreas* 15:358–366.

Leibowitz G, Ferber S, Apelqvist A, et al. 2001a. IPF1 PDX1 deficiency and beta-cell dysfunction in *Psammomy obesus*, an animal with type 2 diabetes. *Diabetes* 50:1799–1806.

Leibowitz G, Melloul D, Yuli M, et al. 2001b. Defective glucose-regulated insulin gene expression associated with PDX-1 deficiency in the *Psammomys obesus* model of type 2 diabetes. *Diabetes* 50:S138–139.

Leibovitz G, Uckaya G, Oprescu et al. 2002. Glucose-regulated proinsulin gene expression is required for adequate insulin production during chronic glucose exposure. *Endocrinology* 143:3214–3220.

Lewandowski P, Cameron-Smith D, Moulton K, et al. 1997. Disproportionate increase of fatty acid binding proteins in the livers of obese diabetic *Psammomys obesus*. *Ann NY Acad Sci* 827:536–540.

Like AA, Miki E. 1967. Diabetic syndrome in sand rats. IV. Morphologic changes in islet tissue. *Diabetologia* 3:143–166.

Mandarino L. 1984. Insulin and glucagons binding to isolated hepatocytes of Egyptian sand rats (*Psammomys obesus*): evidence for an insulin receptor defect. *Biochem Physiol* 78A:519–523.

Marquie G, Duhault J, Espinal J, et al. 1997. S15261, a novel agent for the treatment of insulin resistance. Studies on *Psammomys obesus*. Effect on pancreatic islets of insulin resistant animals. *Cell Molec Biol* 43:243–251.

Marquie G, El Madani T, Solera ML, et al. 1998. Metabolic and anti-atherogenic effects of long-term benfluorex in dyslipidemic insulin-resistant sand rats (*Psammoys obesus*). *Life Sci* 63:65–76.

Marquie G, Duhault J, Hadjiisky P, et al. 1991. Diabetes mellitus in sand rats (*Psammomys obesus*): Microangiopathy during development of the diabetic syndrome. *Cell Mol Biol* 37:651–667.

Marquie G, Hadjiiski P, Arnaud O, Duhault J. 1991. Development of macroangiopathy in sand rats (*Psammomys obesus*), an animal model of non-insulin-dependent diabetes mellitus: Effect of gliclazide. *Am J Med* 90(Suppl 6A): 55–61.

Miki E, Like AA, Steinke J, Soeldner JS. 1967. Diabetic syndrome in sand rats. II. Variability and association with diet. *Diabetologia* 3:135–139.

Moskowitz RW, Ziv I, Denko CW et al. 1990. Spondylosis in sand rats. A model of intervertebral disc degeneration and hyperostosis. *J Orthopaedic Res* 8:401–411.

Neil IV. 1962. Diabetes mellitus, a thrifty genotype rendered detrimental to progress. *Am J Hum Genet* 14:353–362.

Nesher R, Gross D, Donath MY, et al. 1999. Interaction between genetic and dietary factors determines β-cell function in *Psammomys obesus*, an animal model of type 2 diabetes. *Diabetes* 48:731–737.

Patlas N, Avgil M, Ziv E, et al. 2006. Pregnancy outcome in the *Psammomys obesus* gerbil, on low and high energy diets. *Biol Neonate* 90:58–68.

Perez R, Ziv E, Freeman S, Sichel JY, Sohmer H. 2001. Vestibular end-organ impairment in an animal model of type 2 diabetes mellitus. *Laryngoscope* 111:110–113.

Perez R, Freeman S, Cohen D, Sohmer H. 2002. Functional impairment of the vestibular end organ resulting form impulse noise exposure. *Laryngoscope* 112:1110–1114.

Rosenmann E, Adler JH, Ungar H. 1982. Spontaneous thymic tumours in the fat sand rat (*Psammomys obesus*). *J Comp Pathol* 92:349–356.

Sanigorski A, Fahey R, Cameron-Smith D, Collier GR. 2002. Nicotine treatment decreases food intake and body weight via a leptin-independent pathway in *Psammomys obesus*. *Diabetes Obes Metab* 4:345–350.

Scherzer P, Nachliel I, Ziv E, et al. 2000a. Effects of variations in food intake on renal sodium pump activity and its gene expression in *Psammomys* kidney. *Am J Physiol* 279:F1124–F1131.

Scherzer P, Nachliel I, Bar-On H, et al. 2000b. Renal Na-K-ATPase hyperactivity in diabetic *Psammomys obesus* is related to glomerular hyperfiltration but is insulin-independent. *J Endocrinol*167:347–354.

Shafrir E. 2001. Albert Renold memorial lecture: Molecular background of nutritionally induced resistance leading to type 2 diabetes—From animal models to humans. *Int J Exp Diabetes Res* 2:299–319.

Shafrir E, Ben-Sasson R, Ziv E, Bar-On H. 1999a. Insulin resistance, β-cell survival, and apoptosis in type 2 diabetes: Animal models and human implications. *Diabetes Rev* 7:114–123.

Shafrir E, Gutman A. 1993. *Psammomys obesus* of the Jerusalem colony: A model for nutritionally induced, non-insulin dependent diabetes. *J Basic Clin Physiol Pharmacol* 4:83–99.

Shafrir E, Spielman S, Nachliel I, et al. 2001. Treatment of diabetes with vanadium salts: General overview and amelioration of nutritionally induced diabetes in the *Psammomys obesus* gerbil. *Diabetes Metab Res Rev* 17:55–66.

Shafrir E, Ziv E. 1998. Cellular mechanism of nutritionally induced insulin resistance: The desert rodent *Psammomys obesus* and other animals in which insulin resistance leads to detrimental outcome. *J Basic Clin Physiol Pharmacol* 9:347–385.

Shafrir E, Ziv E, Mosthaf L. 1999b. Nutritionally induced insulin resistance and receptor defect leading to β-cell failure in animal models. *Ann NY Acad Sci* 892:223–246.

Shafrir E, Ziv E, Saha AK, Ruderman NB. 2002. Regulation of muscle malonyl-CoA levels in the nutritionally insulin-resistant gerbil, *Psammomys obesus*. *Diabetes Metab Res Rev* 18:217–223.

Shapira MY, Appelbaum EY, Hirshberg B, et al. 2000. A sustained non-insulin related, hypoglycemic effect of electroacupuncture in diabetic *Psammomys obesus*. *Diabetologia* 43:809–813.

Silberberg R, Aufdermaur M, Adler JH. 1979. Degeneration of the intervertebral disks and spondylosis in aging sand rats. *Arch Pathol Lab Med* 103:231–235.

Silberberg R. 1988. The vertebral column of diabetic sand rats *(Psammomys obesus)*. *Exp Cell Biol* 56:217–220.

Sohmer H, Freeman S, Perez R. 2004. Semicircular canal fenestration — improvement of bone — but not air-conducted auditory thresholds. *Hear Res* 187:105–110.

Thomas O. 1908. The *Psammomys* of the alluvial soil of Nile Delta. *Ann Nat Hist* 9:91–92.

Ungar H, Adler JH. 1979. The histogenesis of hepatoma occurring spontaneously in a strain of sand rats (*Psammomys obesus*). *Am J Pathol* 90:399–410.

Walder K, Lewandowski P, Morton G, et al. 1999. Leptin resistance in a polygenic, hyperleptinemic animal model of obesity and NIDDM: *Psammomys obesus*. *Int J Obes* 23:83–89.

Walder K, Ziv E, Kalman R, et al. 2002b. Elevated hypothalamic gene expression in *Pammomys obesus* prone to develop obesity and type 2 diabetes. *Int J Obesity* 26:605–609.

Walder K, Kantham L, McMillan JS, et al. 2002c. Tanis: a link between type 2 diabetes and inflammation? *Diabetes* 51:1859–1866.

Walder K, Kerr-Bayles L, Civitarese A, et al. 2005. The mitochondrial rhomboid protease PSARL is a new candidate gene for type 2 diabetes. *Diabetologia* 48:459–468.

Weihe WH. 1997. The laboratory rat. In *The UFAW Handbook on the Care and Management of Laboratory Animals*, ed. Poole T. Longman Scientific & Tech., England, 309–330.

Wendorf MF, Goldfine ID. 1991. Archeology of diabetes. Excavation of the "thrifty" genotype. *Diabetes* 40:161–165.

Wilke B, Schmidt S, Kloting I, et al. 1979. Studies on (pro) insulin biosynthesis and secretion of pancreatic islets of sand rats (*Psammomys obesus*) after different feeding conditions. *Endokrinologie* 74:73–80.

Wuarin-Bierman L, Zahnd GR, Kaufmann F, et al. 1987. Hyperalgesia in spontaneous and experimental animal models of diabetic neuropathy. *Diabetologia* 30:653–658.

Ziv E, Kalman R. 2000. Primary insulin resistance leading to nutritionally induced type 2 diabetes. In *Animal Models of Diabetes. A Primer*, ed. Sima AAF, Shafrir E. Harwood Academic Publishers, 327–342.

Ziv E, Kalman R, Hershkop K, et al. 1996. Insulin resistance in the NIDDM model *Psammomys obesus* in the normoglycemic, normoinsulinemic state. *Diabetologia* 39:1269–1275.

Ziv E, Shafrir E, Kalman R, et al. 1999. Changing pattern of prevalence of insulin resistance in *Psammomys obesus*, a model of nutritionally induced type 2 diabetes. *Metabolism* 48:1549–1554.

Zoltowska M, Ziv E, Delvin E, et al. 2003. Cellular aspects of intestinal lipoprotein assembly in *Psammomys obesus*. A model of insulin resistance and type 2 diabetes. *Diabetes* 52:2539–2545.

Zoltowska M, Ziv E, Delvin E, et al. 2004. Both insulin resistance and diabetes in *Psammomys obesus* upregulate the hepatic machinery involved in intracellular VLDL assembly. *Atherioscler Thromb Vasc Biol* 24:118–123.

14 The Spontaneously Diabetic Torii (SDT) Rat with Retinopathy Lesions Resembling Those of Humans

Masami Shinohara, Taku Masuyama, and Akihiro Kakehashi

CONTENTS

INTRODUCTION

The number of diabetic patients is growing throughout the world, including developing countries; this has created worldwide insurance issues (King and Zimmet 1988; King and Rewers 1991, 1993). The future world population of diabetic patients is estimated to amount to 221 million by the year 2010 (Amos et al. 1997). At present, various analyses on complicated interaction between hereditary and environmental factors are being undertaken regarding the onset of diabetes.

However, there is a natural limit to information directly obtained from humans, partly because it is extremely difficult to demonstrate the relationship between hereditary and environmental factors in humans and partly because study materials derived from diabetic patients are affected by large ethical restrictions. It is thought that use of animal models is essential for solving this problem and that results of basic studies obtained from animal models of diabetes are useful for the elucidation of the pathogenesis of human diabetes, the study of the causes of diabetic complications, and the development of diabetic medicine. Thus, substantial results contributing to the understanding of clinical diabetes have been obtained from animal models.

The development of diabetic complication models similar to human diabetes calls for urgent attention because it is extremely important to develop animal models with a variety of pathologic conditions similar to those of human diabetes. Also, in recent years type 2 diabetes (T2D) has rapidly and globally increased along with economic development and social modernization. Thus, the development of diabetic complications has become a major concern regarding the prognosis of diabetic patients. Because of this situation, we have established a nonobese T2D model that presents diabetic ocular complications.

Among diabetic ocular complications found in humans, the most frequent are cataracts, diabetic retinopathy, and neovascular glaucoma. Most cataracts can be safely treated surgically. However, despite the development of panretinal photocoagulation and vitreous surgery, diabetic retinopathy has remained a leading cause of vision loss and blindness in adults in most developed countries. Neovascular glaucoma usually follows the development of severe proliferative diabetic retinopathy and is assumed to be the result of advanced retinal hypoxia. This condition is the most severe of ocular complications because it is associated with proliferative diabetic retinopathy and is difficult to treat even with panretinal photocoagulation, vitreous surgery, glaucoma surgery, or all of these.

Diabetic ocular complications do not develop if glycemic control is achieved at the onset of the disease (Diabetes Control and Complications Trial research group 1993; U.K. Prospective Diabetes group 1988). However, many patients pay little attention to poor glycemic control until their vision becomes compromised. Therefore, to prevent diabetic ocular complications, prophylactic medical treatments other than glycemic control are needed. To accomplish this, development of a spontaneously diabetic animal model with severe diabetic complications is needed.

However, no such animal model has been developed. Although numerous spontaneous diabetic animal models have been reported, few models have presented ocular complications. Among them, some experimental diabetic dogs and galactose-fed dogs developed diabetic retinopathy (Engerman et al. 1965; Kador et al. 1975). Although these diabetic dog models have features characteristic of diabetic retinopathy, they have not been widely used in experimental studies of diabetic retinopathy because of the low prevalence of diabetic retinopathy and the prolonged time needed for the complication to develop. Moreover, the dog model is inconvenient for large-scale experiments. In contrast, rat and mouse models are popular for many types of studies, but most show only subclinical diabetic retinopathy.

The Spontaneously Diabetic Torii (SDT) rat is the first animal model with spontaneously occurring advanced diabetic retinopathy (Shinohara et al. 2000; Kakehashi et al. 2006). SDT rats are useful for studying the pathogenesis and treatment of diabetic retinopathy, although further investigation will be needed to clarify the differences in the development of retinopathy between this model and diabetic patients.

ESTABLISHMENT OF THE SDT STRAIN

In 1988, Shinohara found 12-month-old male rats that exhibited polydipsia, polyphagia, polyuria, and glucosuria in an outbred colony of Sprague–Dawley rats purchased from Charles River Japan. To maintain the characteristic traits of these male rats genetically, they were mated with normal female rats of the same strain. In 1991, it was found that several rats among their progeny exhibited glucosuria at a younger age (4 to 5 months). Sister–brother mating was then carried out repeatedly. Finally, in 1997, an inbred strain was established in specific pathogen-free (SPF) facilities in the Research Laboratories of Torii Pharmaceutical Co. This strain was named the Spontaneously Diabetic Torii (SDT) rat (Shinohara et al. 2000). During the successive generations of this inbred strain, diabetes was observed in males from the F1 to the F20 generation; however, in females, it was only noted after the F7 generation. SDT rats have a moderate reproductive performance and, due to their gentle nature, they are easy to handle in breeding and experiments.

DIABETES AND GLUCOSE INTOLERANCE

The onset of diabetes, as indicated by polyuria and glucosuria, is observed by 15–20 weeks of age in male SDT rats. The cumulative incidence of diabetes reaches 100% up to 40 weeks of age. In contrast to males, only 33% of female SDT rats developed diabetes up to 65 weeks of age. The reason for these sexual differences in the development of diabetes in SDT rats remains unclear. Prior to the onset of diabetes, SDT rats are comparable to age-matched normal Sprague–Dawley rats in terms of body weight and body mass. After the onset of diabetes, they show a decrease of body weight and become emaciated with soiled fur. Interestingly, they can survive without insulin treatment for a long time (approximately one year) after the onset of diabetes.

Hyperglycemia in SDT rats is considered to depend on impaired insulin secretion rather than insulin resistance (Masuyama et al. 2004). Prior to the onset of hyperglycemia, the plasma insulin concentration of SDT rats tended to be lower compared to normal Sprague–Dawley rats, and the values decreased significantly after the onset of diabetes. After the animals became diabetic, they showed clinical signs similar to those observed in insulin-dependent diabetes mellitus. It is well known that glucose tolerance is impaired prior to the onset of T2D in humans. Similarly, SDT rats exhibited impaired glucose tolerance by 14–16 weeks of age (at approximately 2 months prior to the onset of diabetes) (Shinohara et al. 2000; Masuyama et al. 2004).

The blood glucose curves on the oral glucose tolerance test (OGTT) elevated progressively with age. This impaired glucose tolerance is associated with decreased insulin secretion, as identified by the decrease of the area under the curve (AUC) of the plasma insulin concentrations during the OGTT (Masuyama et al. 2004). The age at the onset of diabetes in male SDT rats strongly correlates with the AUC of blood glucose concentrations during the OGTT at 20 weeks of age (when they are glucose intolerant but not diabetic) (Masuyama et al. 2003). When SDT rats become diabetic, they exhibit significant fasting hypoinsulinemia and hardly any increase of plasma insulin level is observed after the oral glucose load (Masuyama et al. 2003). In contrast, female SDT rats develop impaired glucose tolerance at a later age (25 weeks) than males (Shinohara et al. 2004).

PATHOLOGICAL CHANGES IN PANCREATIC ISLETS

In male SDT rats, hyperglycemia is considered to be predominantly attributable to an insulin secretory defect resulting from pathological damage of the pancreatic islets. The pancreatic islets of SDT rats are histopathologically normal up to at least 4 weeks of age; however, their beta-cell mass is significantly smaller than that of age-matched Sprague–Dawley rats (Masuyama et al. 2004). Primary pathological changes in the pancreatic islets begin by approximately 8–10 weeks of age when the rats exhibit normal glucose tolerance. At 8–10 weeks of age, congestion and hemorrhage are sporadically observed in the pancreatic islets. These changes are regarded as consecutive events in intraislet microcirculation. After intraislet hemorrhage, inflammatory cells and fibroblasts infiltrate the pancreatic islets, which are then invaded by connective tissue and eventually display fibrosis up to 16 weeks of age.

These inflammatory changes observed in the SDT rat islets are qualitatively different from autoimmune-mediated inflammation such as the typical lymphocyte infiltration consistently observed in autoimmune diabetes. When the animals become mature diabetics, almost all the beta cells disappear from the SDT rat islets (Masuyama et al. 2004). Although the reason for beta-cell disappearance in SDT rat islets remains unclear, exhaustion due to overwork by the surviving beta cells and impaired proliferation (regeneration) of the beta cells may be contributory factors. In this regard, it was recently reported that pancreas transplantation (PTx) significantly delayed the onset of diabetes in SDT rats (Miao et al. 2005). Moreover, insulin-secreting cells with a positive pancreatic duodenal homeobox-1 (PDX-1) were observed and located close to ductal structures in the recipient naive pancreas, indicating that beta-cell regeneration was achieved by PTx (Miao et al. 2005).

OCULAR COMPLICATIONS

There are numerous ocular complications in patients with diabetes, such as cataract, diabetic retinopathy, neovascular glaucoma, optic neuropathy, keratopathy, pupillomotor disturbance, external ophthalmoplegia, accommodation insufficiency, and uveitis. Among these, the most frequent are cataracts, diabetic retinopathy, and neovascular glaucoma (Kakehashi et al. 2006).

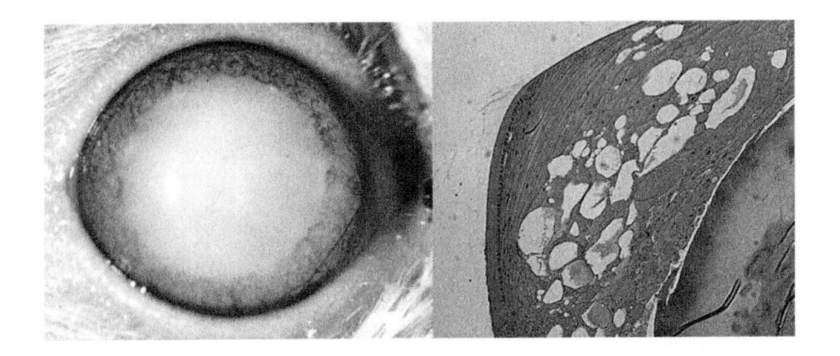

FIG. 14.1 (See color insert following page 338) Cataract development in a 52-week-old male SDT rat. Left: The cataract has developed to maturity; right: pathology of the mature cataract. The sclerotic nucleus floats in a liquefied lens cortex. Vacuolation, disintegration of the lens fibers, and Morgani's globules are observed in the lens cortex.

Cataract

Cataract is the most frequently occurring ocular complication in patients with diabetes. Several types of cataracts (i.e., cortical, nuclear, and subcapsular) are observed in SDT rats after a diabetic crisis. Eventually, the cataracts become hypermature—in most cases, by the time the animals are 40 weeks of age (fig. 14.1, left). The histopathological examination shows that the sclerotic nucleus of the lens floats in a liquefied lens cortex (fig. 14.1, right) in the hypermature stage. Vacuolation, disintegration of the lens fibers, and Morgani's globules are observed in the lens cortex. Some advanced cases of hypermature cataracts include rupture of the posterior capsule. All these features make the SDT rat a good animal model of diabetic cataracts.

Diabetic Retinopathy

In humans, diabetic retinopathy progresses gradually from early to advanced stages. The early stage of diabetic retinopathy in humans is frequently characterized by microaneurysms and retinal hemorrhages resulting from vascular dysfunction. Cotton-wool spots appear as early evidence of capillary occlusion. Eventually, retinal neovascularization occurs as a result of severe retinal ischemia.

The diabetic retinopathy found in SDT rats differs from that in humans. Retinal hemorrhages can be observed in some cases, and microaneurysms are rarely seen. Electron microscopic study of the retinal capillaries in the SDT rats shows that thickening of the basement membrane is an early diabetic change of the retinal vessels. The retinal vasculature evaluated in trypsin-digested retina shows pericyte loss and occluded capillaries as in diabetic retinopathy in humans. Eventually, tractional retinal detachment around the optic disc is observed after the animals reach 50 weeks of age.

Large retinal folds of thickened retina with vitreous traction are the most typical advanced retinopathy in SDT rats (fig. 14.2). At this stage, fluorescein microangiography shows extensive leakage of the dye around the optic disc with vascular tortuosity (fig. 14.3). Capillary occlusion is observed in some cases, but it is not a

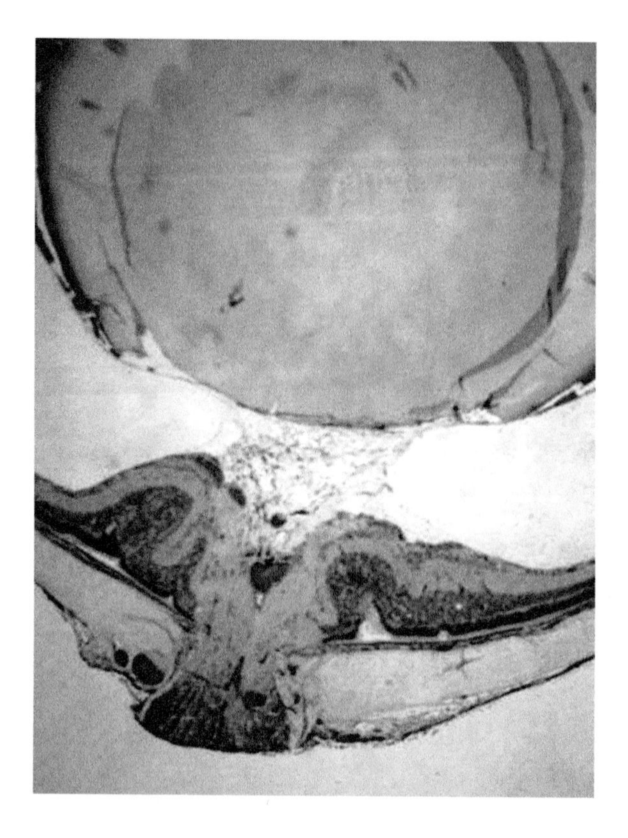

FIG. 14.2 (See color insert) Characteristics of diabetic retinopathy in an 82-week-old female SDT rat. Left: Large retinal folds associated with vitreous traction are seen around the optic disc in an old SDT rat. Some vessels are observed in the vitreous cavity in front of the optic disc. One of the vessels may be a persistent hyaloid artery and the other a new vessel from the optic disc.

frequent finding. The early diabetic retinal changes observed in diabetic retinopathy in humans, such as microaneurysms, small retinal hemorrhages, and capillary non-perfusion, seem to be bypassed in the SDT rats. The findings of proliferative diabetic retinopathy without remarkable retinal ischemia may result from increased expression of vascular endothelial growth factor and pigment epithelium-derived factor (Yamada et al. 2005; Matsuoka et al. 2006). The prevalence of advanced retinopathy in SDT rats is about 80% by the age of 60 weeks, but it is rarely observed under 50 weeks of age.

Neovascular Glaucoma

Neovascular glaucoma iris neovascularization is frequently observed in advanced cases of diabetic retinopathy in humans. The neovascular glaucoma is difficult to treat even with glaucoma surgery. In some advanced cases in SDT rats, iris neovascularization and hemorrhages of the anterior chamber also develop (fig. 14.4).

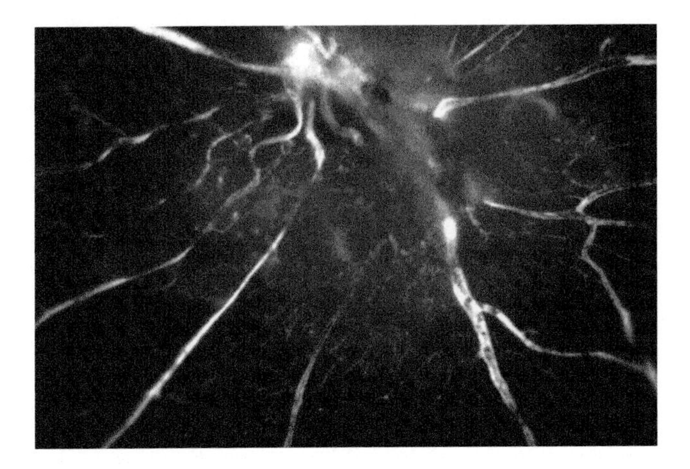

FIG. 14.3 (See color insert) Extensive hyperfluorescence is observed around the optic disc with vascular tortuosity in a 62-week-old male SDT rat.

Fibrovascular tissue around the pupil may cause posterior synechiae, and this condition may induce papillary block and angle-closure glaucoma. The SDT rat is the first animal model of diabetes presenting iris neovascularization; it has not been reported previously in diabetic animals.

Effect of Glycemic Control on Ocular Complications

It was reported that diabetic cataracts and retinopathy were prevented and improved in diabetic SDT rats after attaining glycemic normalization through PTx performed at 5 weeks after the onset of diabetes (Miao et al. 2004). Careful glycemic control with

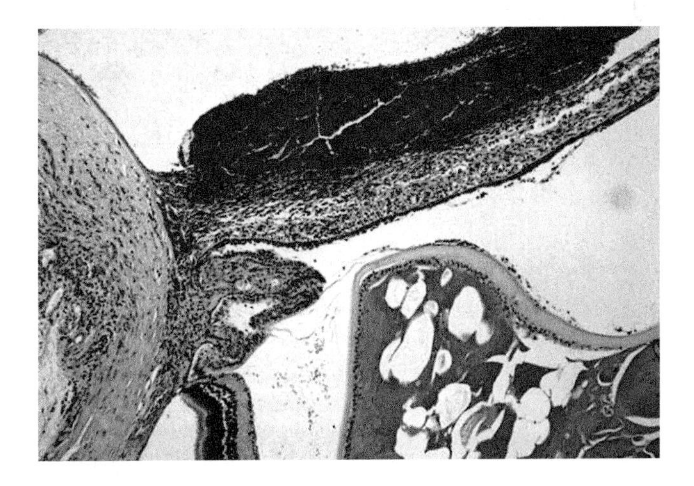

FIG. 14.4 (See color insert) Massive hemorrhage can be seen in the anterior chamber associated with proliferation around the iris in a 77-week-old male SDT rat.

long-term insulin treatment also prevented the development of diabetic cataracts and retinopathy in SDT rats (Sasase et al. 2006). Taken together, the ocular complications observed in diabetic SDT rats are considered to be attributed to chronic hyperglycemia.

GENETICS

The genetic characteristics of diabetes in SDT rats were determined by the quantitative trait locus (QTL) analysis using 319 male backcrosses (Brown Norway [BN] × SDT) × SDT (Masuyama et al. 2003). Three highly significant QTLs (*Gisdt1*, *Gisdt2*, and *Gisdt3*) for glucose intolerance were successfully mapped on rat chromosomes 1, 2, and X, respectively (Masuyama et al. 2003). The SDT allele for these QTLs significantly exacerbated glucose intolerance. It was also observed that each SDT allele of each QTL has a strong effect on the increase of the blood glucose level at 60 min after the glucose load. Moreover, of the three QTLs for glucose intolerance, *Gisdt1* on rat chromosome 1 is involved in regulating the increase of body weight and is transmitted according to a recessive pattern of inheritance. This implies that *Gisdt1* is a locus responsible for increased insulin demand in peripheral tissues rather than decreased insulin secretion.

In contrast, the SDT alleles of *Gisdt2* and *Gisdt3* are not involved in the control of body weight. Thus, *Gisdt2* and *Gisdt3* may be associated with the regulation of glucose homeostasis by pancreatic islets. Furthermore, glucose intolerance exacerbates when the three loci accumulate homozygous SDT allele deletions in the backcross progeny; this demonstrates that glucose intolerance is clearly polygenic in SDT rats and is induced by synergistic interaction of the three QTLs. When the chromosomal positions of the three QTLs (*Gisdt1* ~ 3) are compared with those of other diabetogenic loci in other diabetic rat models such as the Goto–Kagizaki (GK) rat and the Otsuka Long–Evans Tokushima fatty (OLETF) rat reported previously, several of the loci are localized or overlap partly on chromosomal regions containing *Gisdt* QTLs. At present, the causative genes for diabetes in the SDT rat remain unclear. Construction of congenic strains for the three QTLs based on the congenic strategy for positional cloning would be required to apply the positional candidate approach for identifying genes responsible for diabetes in SDT rats.

SDT.Cg-*Lepr^{fa}* CONGENIC RATS

To produce a useful model for studying the metabolic response to insulin resistance in SDT rats, SDT.Cg-*Lepr^{fa}* congenic rats were established by introgressing an *fa* mutation allele of the leptin receptor gene (*Lepr^{fa}*) in the Zucker fatty rat into the genetic background of the SDT rat (Masuyama et al. 2005). The *Lepr^{fa}* gene is well known to produce insulin resistance or obesity due to the loss of leptin action (Chua et al. 1996). This congenic strain of the SDT rat with the *fa* gene was created through the "speed congenic method" using a PCR technique with DNA markers (Markel et al. 1997). The SDT-*fa*/*fa* rats show clinical features that are related to obesity and diabetes in contrast to those of wild-type SDT rats.

SDT-*fa*/*fa* rats of both sexes exhibited obesity, adiposity, and insulin resistance associated with hyperphagia due to the loss of leptin action. Interestingly, they developed diabetes from 5 weeks of age in males and 8 weeks of age in females. The incidences reached 100% at 16 weeks of age in males and 73% at 32 weeks of age in females. These results indicate that the onset of diabetes in SDT rats is accelerated by developing adiposity and/or insulin resistance associated with the introduced *fa* gene. In addition, the SDT-*fa/fa* rat might be a useful model of obesity-related diabetes yet. Diabetic complications including retinopathy in SDT-*fa/fa* rats remain unknown. Furthermore, a long-term observation of the SDT-*fa/fa* rats should be done to discuss the effect of *fa* gene-induced obesity on retinopathy in the SDT rats.

REARING AND HANDLING SDT RATS

Diet, Drinking Water, and Care

In SDT rats, glucosuria appears on ingestion of standard diet for mice and rats. Their diet intake increases (about 1.5–2 times that of SD strain rats) and water intake also significantly increases (200–400 mL/day and more) with the onset of diabetes. It is, therefore, important always to have sufficient amounts of feed and drinking water.

Frequent change of bedding and cleaning of cages are recommended. SDT rats urinate frequently and their bedding easily becomes dirty after the onset of diabetes. Therefore, they can easily develop urinary tract infections. In order to avoid such infections, each rat should be reared alone and requires a large amount of bedding, in a plastic cage. Changing the bedding frequently is also recommended in the case of bracket cage rearing.

Proper Microorganism Control

All rats should be kept and used under SPF conditions of the following microorganisms:

Sendai virus
sialodacryoadenitis virus
pneumonia virus of mice
mouse encephalomyelitis virus
Kilham rat virus
H-1 virus
minute virus of mice
hantavirus
mouse adenovirus
Mycoplasma pulmonis
Bacillus piliformis
CAR bacillus
Bordetella bronchiseptica
Corynebacterium kutscheri
Pasteurella pneumotropica
Pseudomonas aeruginosa
Salmonella spp.
Salmonella typhimurium

Streptococcus pneumoniae
Dermatophytes
Giardia muris
Spironucleus muris
Syphacia spp.

DISTRIBUTOR INFORMATION ON SDT RATS

CLEA Japan, Inc.
1-2-7, Higashiyama, Meguro-ku, Tokyo 153-8533, Japan
TEL: +81-(0)-3-5704-7272, FAX: +81-(0)-3-3791-2859
E-mail: export@clea-japan.com
URL: http://www.CLEA-Japan.com

REFERENCES

Amos AF, McCarty DJ, Zimmet P. (1997) The rising global burden of diabetes and its complications: Estimates and projections to the year 2010. *Diabetes Med* 14:S7–S85.

Chua SC Jr., White DW, Wu-Peng XS, et al. (1996) Phenotype of fatty due to Gln269Pro mutation in the leptin receptor (*Lepr*). *Diabetes* 45:1141–1143.

Diabetes Control and Complications Trial Research Group. (1993) The effect of intensive treatment of diabetes on the development and progression of long-term complications in insulin-dependent diabetes mellitus. *N Engl J Med* 329:977–986.

Engerman RL, Bloodworth JM Jr. (1965) Experimental diabetic retinopathy in dogs. *Arch Ophthalmol* 73:205–210.

Kador PF, Takahashi Y, Wyman M, et al. (1975) Diabetes-like proliferative retinal changes in galactose-fed dogs. *Arch Ophthalmol* 113:352–354.

Kakehashi A, Saito Y, Mori K, et al. (2006) Characteristics of diabetic retinopathy. *Diabetes Metab Res Rev* 22: 455–461.

King H, Rewers M. (1991) Diabetes in adults is now a Third World problem. The WHO Ad Hoc Diabetes Reporting Group. *Bull World Health Org* 69:643–648.

King H, Rewers M. (1993) WHO Ad Hoc Diabetes Reporting Group. Global estimates for prevalence of diabetes mellitus and impaired glucose tolerance in adults. *Diabetes Care* 16:157–177.

King H, Zimmet P. (1988) Trends in the prevalence and incidence of diabetes: Non-insulin-dependent diabetes mellitus. *World Health Stat Q* 41:190–196.

Markel P, Shu P, Ebeling C, et al. (1997) Theoretical and empirical issues for marker-assisted breeding of congenic mouse strains. *Nat Genet* 17:280–284.

Masuyama T, Fuse M, Yokoi N, et al. (2003) Genetic analysis for diabetes in a new rat model of non-obese type 2 diabetes, spontaneously diabetic Torii rats. *Biochem Biophys Res Commun* 304:196–206.

Masuyama T, Katsuda Y, Shinohara M. (2005) A novel model of obesity-related diabetes: Introgression of the $Lepr^{fa}$ allele of the Zucker fatty rat into nonobese spontaneously diabetic Torii (SDT) rats. *Exp Anim* 54:13–20.

Masuyama T, Komeda K, Hara A, et al. (2004) Chronological characterization of diabetes development in male spontaneously diabetic Torii rats. *Biochem Biophys Res Commun* 314:870–877.

Matsuoka M, Ogata N, Minamino K, et al. (2006) High levels of pigment epithelium-derived factor in the retina of a rat model of type 2 diabetes. *Exp Eye Res* 82:172–178.

Miao G, Ito T, Uchikoshi F, et al. (2004) Stage-dependent effect of pancreatic transplantation on diabetic ocular complications in the spontaneously diabetic Torii rat. *Transplantation* 77:658–663.

Miao G, Ito T, Uchikoshi F, et al. (2005) Beneficial effects of pancreas transplantation: Regeneration of pancreatic islets in the spontaneously diabetic Torii rat. *Transplant Proc* 37:226–228.

Sasase T, Ohta N, Ogawa K, et al. (2006) Preventive effects of glycemic control on ocular complications of spontaneously diabetic Torii rat. *Diabetes Obesity Metab* 8:501–507.

Shinohara M, Masuyama T, Shoda T, et al. (2000) A new spontaneously diabetic nonobese Torii rat strain with severe ocular complications. *Int J Exp Diabetes Res* 1:89–100.

Shinohara M, Oikawa T, Sato K, et al. (2004) Glucose intolerance and hyperlipidemia prior to diabetes onset in female spontaneously diabetic Torii (SDT) rats. *Exp Diabesity Res* 5:253–256.

U.K. Prospective Diabetes Study Group. (1988) Intensive blood-glucose control with sulphonylureas or insulin compared with conventional treatment and risk of complications in patients with type 2 diabetes (UKPDS33). *Lancet* 352:837–853.

Yamada H, Yamada E, Higuchi A, et al. (2005) Retinal neovascularization without ischemia in the spontaneously diabetic Torii rat. *Diabetologia* 48:1663–1668.

15 Cohen Diabetic Rat

Sarah Weksler-Zangen, Esther Orlanski, and David H. Zangen

CONTENTS

INTRODUCTION

The Cohen diabetic (CD) rat is an experimental model of type 2 diabetes (T2D) originally developed by A. M. Cohen (1990) at the Hadassah University Hospital in Jerusalem, Israel. Two contrasting strains were derived by selective inbreeding: a sensitive (CDs) rat, which develops T2D only when fed a diabetogenic high-sucrose, copper-poor diet (HSD), and a resistant (CDr) rat that does not develop overt diabetes irrespective of diet. In the CDs rat, the development of T2D is directly attributed to beta-cell dysfunction and reduced insulin secretion (Weksler-Zangen et al. 2001, 2003a).

Similarly to humans, in this rat model hyperglycemia is inducible and reversible by diet adjustment in the early stages of the disease (Cohen 1986). It also expresses many diabetes-related end-organ complications observed in humans, such as nephropathy and retinopathy. As in humans, the diabetic retinopathy and nephropathy in the Cohen rats exhibit pathological lesions in the early and end stages that closely resemble the development of these complications in humans (Cohen 1990; Cohen et al. 1993; Yagil et al. 2005).

ORIGIN AND BREEDING OF THE COHEN DIABETIC RAT

The strain was initiated by Cohen (1990) in the 1950s to test the hypothesis regarding the interaction between the constitutional (genetic) factors and environmental (dietary) influences in the development of T2D. The development of the CD rat strain was prompted by an observation made in Israel on increased incidence of diabetes in new immigrants from Yemen and Kurdistan (Cohen 1961, 1990). Hebrew University strain (HUS) albino rats were fed a high-sucrose diet (HSD) containing salts, vitamins, fats, and proteins and low copper content—1.2 ppm (table 15.1) instead of 6.7 ppm in the regular rat diet. The rats were provided distilled water *ad libitum*. The males and females with the highest blood glucose values 1 and 2 h after the glucose load were selected and mated. The males and females with the lowest blood glucose 1 and 2 h after HSD feeding were also selected and mated.

In each line, brother–sister mating and the method of "two-way selection" were applied (Falconer 1953). This process was repeated until the animals developed overt diabetes characterized by postprandial hyperglycemia and glucosuria (Cohen 1990). The newly inbred CD rat is the result of secondary inbreeding initiated on-site eight years ago. Cohen's original selection criteria were based on an oral glucose tolerance test with blood glucose levels at 2 h of >180 mg/dL for CDs rats and <180 mg/dL for CDr rats (Cohen 1990). In the secondary breeding, more stringent criteria were instituted (table 15.2), wherein blood glucose levels >230 mg/dL qualified as upward line and the rats were renamed CDs (Cohen diabetic sensitive) and <140 mg/dL were considered as downward line CDr (Cohen diabetic resistant) (Weksler-Zangen et al. 2001). Brother–sister mating and reselection by the high/low glucose response was continued for over 10 additional generations, bringing overall inbreeding to >70 generations (Weksler-Zangen et al. 2001). In the newly developed CDs–HSD rat, sustained hyperglycemia is already evident after one month on the HSD. This stands in contrast with Cohen's original "upward" colony, in which sustained hyperglycemia was only fully expressed after 2–2.5 months on the HSD.

TABLE 15.1
Nutrient Composition of HSD

High-Sucrose Copper-Poor Diet (HSD)	
Sucrose	72%
Fat	5%
Protein	18%
Salt mixture	5%
Copper	0.9%
Vitamin mixture	

TABLE 15.2
Reselection of CDs and CDr Rats Using a More Stringent Protocol

High-sucrose (72%) diet, low-copper (0.9 ppm) feeding for 2 months
Subsequent oral GTT
Selective brother–sister mating for 10 additional generations
Resistant (CDr) BG levels < 130 mg/dL
Sensitive (CDs) BG levels > 230 mg/dL

MAINTENANCE AND BREEDING OF THE COHEN DIABETIC RAT: PRACTICAL CONSIDERATIONS

For breeding purposes, CDs and CDr rats should be fed "regular" laboratory chow (RD), which consists of a mixture of ground whole wheat, ground alfalfa, bran, skimmed milk powder, and salts, resulting in 21% protein, 60% carbohydrates, 5% fat, and 0.45% NaCl content (Koffolk Petach-Tikva, Israel). When the rats feed on this diet and are kept in 12 h diurnal light–dark cycles and room temperature between 22 and 25°C, they breed well. However, as will be detailed in the section on embryogenesis, the CDs rats have a significantly reduced pregnancy rate and a low number of live embryos per pregnancy due to increased embryonic death and decreased litter number (Weksler-Zangen et al. 2002, 2003b; Zangen et al. 2006). Therefore, the number of breeding pairs of the CDs should be doubled in order to keep a reasonable breeding rate.

The CDr rats have a slightly reduced breeding rate when compared with outbreed strains but maintain a fairly reasonable number of pregnancies per female and number of live embryos per pregnancy (Weksler-Zangen et al. 2002, 2003; Zangen et al. 2006). During pregnancy and after litter delivery, females are housed in individual cages and pups are weaned at five weeks due to decreased birth weight of the CDs pups as well as increased birth death when weaned earlier (Zusman and Ornoy 1990; Weksler-Zangen et al. 2002). When fed RD, CDs and CDr rats maintain normal blood glucose levels. Full postprandial hyperglycemia develops only in the CDs rats after one month of HSD feeding. HSD consists of 18% casein, 72% sucrose, 4.5% butter, 0.5% corn oil, 5% salt No II USP, water, and fat-soluble vitamins. The HSD is copper poor (0.9 ppm), a prerequisite for CDs rats to develop the full diabetic phenotype (table 15.1) (Weksler-Zangen et al. 2001). Food and distilled water for drinking are provided *ad libitum*.

RAT BODY WEIGHT

Body growth curves served as a reflection of the general state of health of the animal. At weaning, there is no difference in body weight between CDs and CDr rats of either sex. A sex difference developed thereafter. Male CDs and CDr rats fed RD

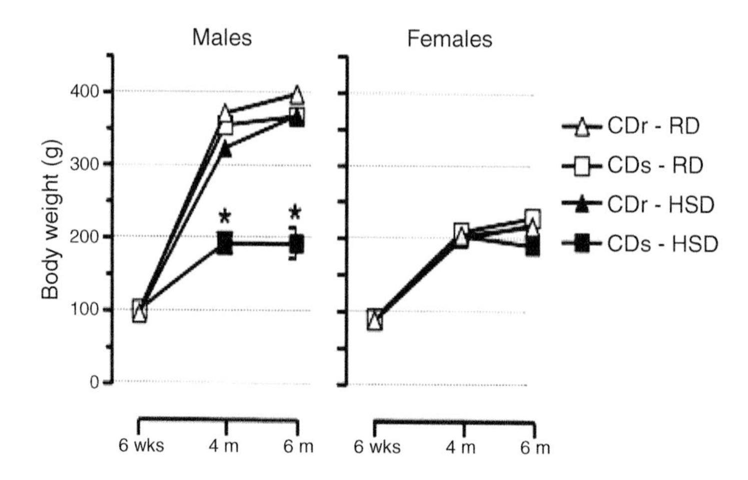

FIG. 15.1 Body weight of CDs and CDr rats that were fed RD or HSD at ages six weeks and four and six months. Data are presented as mean ± SE; *P < 0.01 compared with all other groups. (Weksler-Zangen, S. et al. 2001. *Diabetes* 50:2521–2529.)

continued to grow after weaning at comparable rates, increasing their weight by more than threefold at four and six months (fig. 15.1). Male CDr rats provided HSD chow grew at a rate comparable to those fed RD. In contrast, male CDs rats fed HSD chow failed to thrive and at four months only doubled their weight and displayed no additional growth during the subsequent two months (fig. 15.1). Thus, female CDs rats expressed a different pattern of response to the HSD in terms of growth curve than males. Female CDr and CDs rats fed RD or HSD showed similar growth curves, irrespective of strain or diet (fig. 15.1) (Weksler-Zangen et al. 2001).

Genetic Considerations

The CDs and CDr rats have been selected from the Hebrew University strain (Sabra rats) according to the two-way selection and by inbreeding. In most generations, selection of individuals for breeding was carried out in two litters of the second birth order. The breeding selection resulted in overt diabetes in the CDs rats as of the fifth generation of inbreeding, suggesting that one gene or only very few genes cause diabetes in this model. The hereditary factors responsible for diabetes in this model have been studied with full sibling mating. The results suggest that the inheritance of the phenotype is controlled by few genes, one or more of which are X-linked. Sex influence is also present in this model; the mean blood glucose levels of the males are higher than those of females (Vardi 1990; Weksler-Zangen et al. 2001).

Within-strain homogeneity. The CDs and CDr rats are fully inbred and exhibit a high rate of homozygosity, a measure of genetic homogeneity within strains (table 15.3). The rate of homogeneity within strains was 99% in CDs and 98% in

TABLE 15.3
Genome Screen of the CDs and CDr Strains That Comprise the Cohen Diabetic Rat Model

Chromosome No.	No. of Markers Tested	No. of Polymorphic Markers	Percent Polymorphism
1	117	52	44
2	23	13	57
3	25	11	44
4	24	15	63
5	21	13	62
6	21	9	43
7	18	8	44
8	22	16	73
9	16	4	25
10	28	22	79
11	14	3	21
12	13	10	77
13	6	5	83
14	17	7	41
15	26	7	27
16	8	5	63
17	31	7	23
18	14	5	36
19	11	2	18
20	10	5	50
X	21	1	5
Total	486	220	45

CDr rats. The residual heterozygosity is due in most cases to nonsharing allelic mutations rather than to cross-contamination between the strains (Weksler-Zangen et al. 2001).

Between-strain polymorphism. The rate of genetic polymorphism between CDs and CDr rats, based on 472 microsatellite markers that amplified in both strains, is 43%. This initial genetic screening thus identified 203 microsatellite markers that are informative for linkage analysis in future cross-breeding experiments (table 15.3).

Peripheral Insulin Resistance

Peripheral insulin resistance does not appear to play a major role in initiating T2D in the CDs rats fed HSD (fig. 15.2). Exogenous insulin produces a comparable decrease in blood glucose levels in CDs and CDr rats on either diet, suggesting similar peripheral sensitivities to insulin in the two strains (Weksler-Zangen et al. 2001, 2003a).

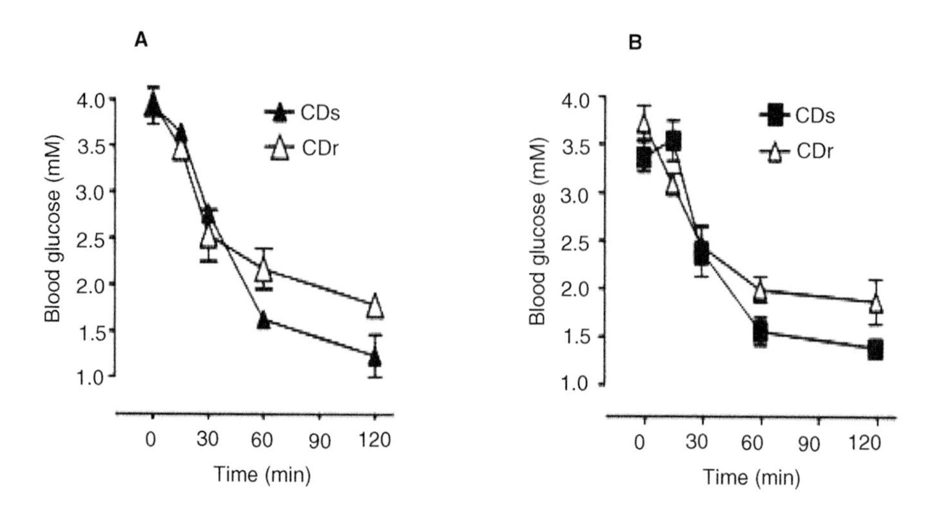

FIG. 15.2 Insulin tolerance test in CDr and CDs rats aged 13 weeks on RD (A) or 1 month of HSD (B). Glucose levels were measured after an overnight fast (time 0) and at 5, 30, 60, and 120 min after the intraperitoneal administration of insulin (1 U/kg body weight). Data are means ± SE for five rats per group.

Islet Morphology and Insulin Content

Immunohistochemistry of islets. CDr and CDs rats fed RD or HSD showed dense insulin immunostaining in pancreatic beta-cell cytoplasm as shown for single islets from the different experimental groups. The alpha-cells of CDr and CDs rats fed RD or HSD were located as a rim in the islet periphery with no significant alterations in staining or distribution (Weksler-Zangen et al. 2003a).

Beta-Cell Function

Postprandial and fasting blood glucose levels. Following one month of HSD feeding, 2-h postprandial blood glucose levels of the CDs rats were >200 mg/dL. Postprandial blood glucose levels increased with time on the HSD and were >300 mg/dL following two months on HSD. These high blood glucose levels were maintained for six months (fig. 15.3). Such levels are maintained for more than a year (not shown). Fasting blood glucose levels were maintained in the normal range for up to five months. Abnormal blood glucose levels of 126–135 mg/dL were observed in males CDs rats after six months on HSD (fig. 15.3) (Weksler-Zangen et al. 2001). The hyperglycemic CDs rats managed to thrive without necessitating administration of exogenous insulin. In marked contrast, CDr rats maintained normal postprandial and fasting blood glucose levels for more than six months on HSD (fig. 15.4).

Intraperitoneal glucose tolerance test (IPGTT). The CDr rats had a normal response to IPGTT irrespective of diet or age. The response of CDs rats to the IPGTT in animals that were fed RD showed a normal pattern, which remained comparable at four and six months. In marked contrast, the response of CDs rats that were

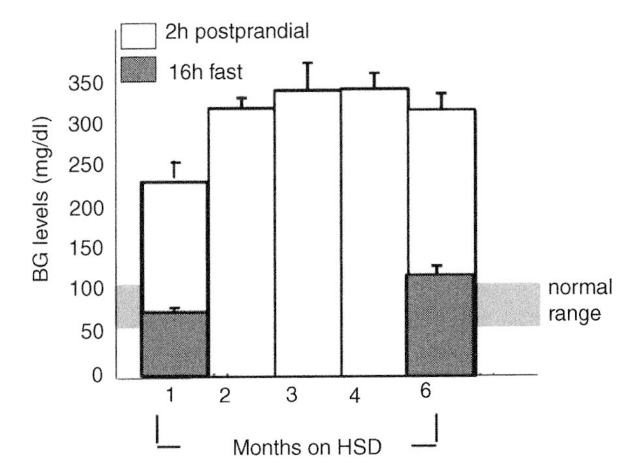

FIG. 15.3 Postprandial and fasting blood glucose levels of male CDs rats, following one to six months on HSD.

provided the HSD was markedly abnormal in all animals studied, with a distinctly more pronounced pattern at six months than at four months and more so in males than in females (fig. 15.5) (Weksler-Zangen et al. 2001).

Exocrine Lesions and Beta-Cell Dysfunction in the Cohen Diabetic Rat Model of T2D

Our recent studies suggested that T2D in the hyperglycemic CDs is mainly due to beta-cell dysfunction. In addition, our data suggest that the exocrine pancreas may

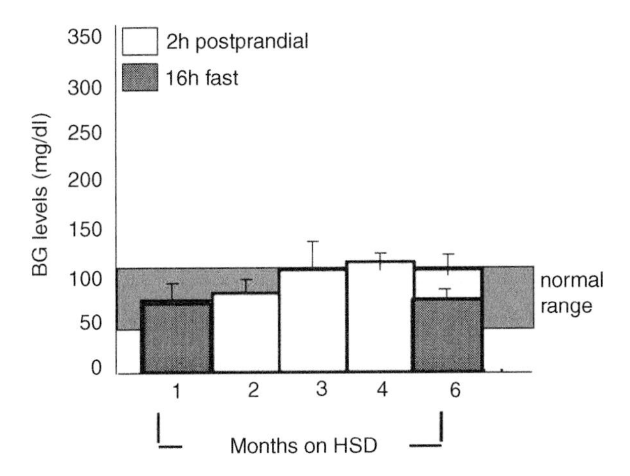

FIG. 15.4 Postprandial and fasting blood glucose levels of male CDr rats, following one to six months on HSD.

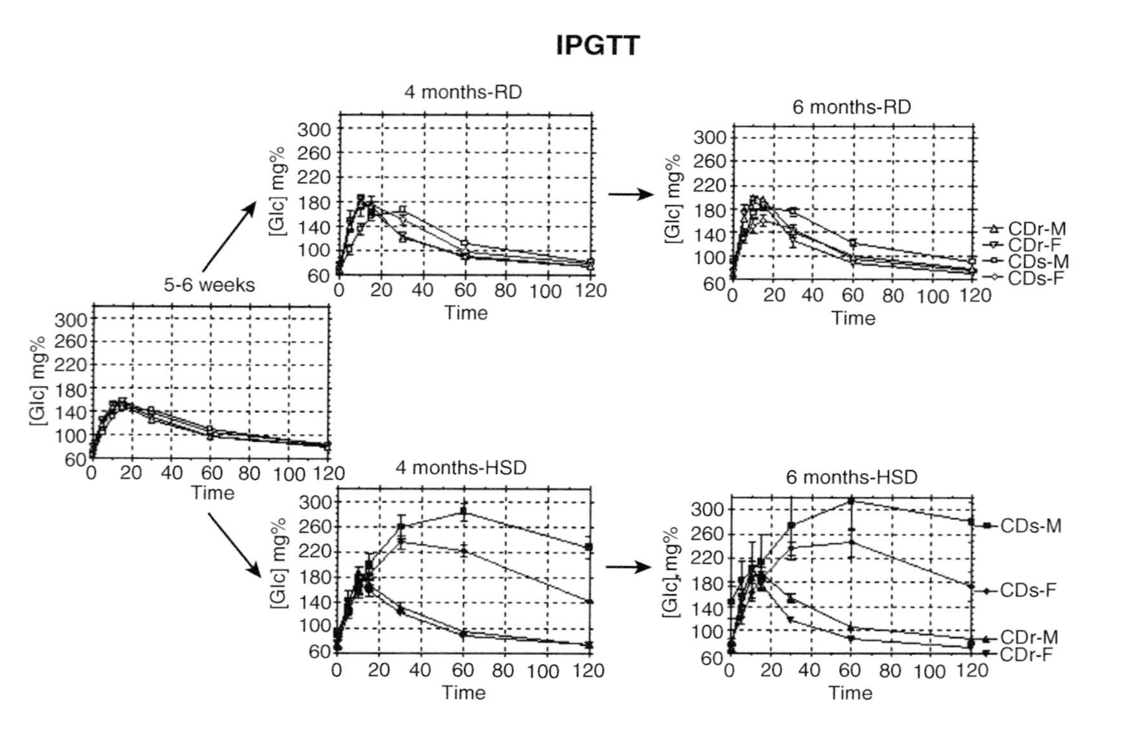

FIG. 15.5 Results of the IPGTT in male and female CDr and CDs rats aged five to six weeks, and four and six months. Animals were fed RD or HSD. Blood glucose concentration was measured after overnight fasting and immediately before i.p. glucose injection (time 0) and at 5, 10, 15, 30, 60, and 120 min thereafter. (Weksler-Zangen, S. et al. 2001, *Diabetes*)

have a role in the inability of beta-cells to secrete sufficient insulin to counteract postprandial hyperglycemia. Histological evaluation of the pancreas of CDs rats fed HSD for 10 weeks showed a significant atrophy of the exocrine acinar tissue while islet morphology and insulin content were preserved (Weksler-Zangen et al. 2003a, 2005, 2006). Our data suggest that exocrine dysmorphogenesis and beta-cell dysfunction are progressive processes accompanied by fat and macrophage infiltration (Weksler-Zangen et al. 2003a, 2005, 2006). These data therefore suggest that exocrine–endocrine interrelationships are important for glucose responsiveness of the beta-cells.

Diabetic Complications

Prolonged hyperglycemia in the CDs leads to nephropathy, retinopathy, testicular atrophy, skeletal pathology, embryopathy, and gastrointestinal disorders resembling human diabetic complications (Cohen and Rosenmann 1990). These findings were presented in Cohen and Rosenmann's book (1990). In this chapter we describe only the recently studied diabetic complications. For further details regarding the complications, refer to the book and additional publications (Ornoy and Zusmann 1990; Rosenmann and Cohen 1990; Silberberg 1990; Yanko et al. 1990).

Nephropathy. In the (upward) original Cohen colony, 90% showed renal lesions and >15% displayed prominent proteinuria (Rosenmann et al. 1971). Recent studies indicated that in the newly bred CDs strain fed HSD, an unusual type of nephropathy evolves, expressed by a reduction in glomerular filtration rate without significant proteinuria (Yagil et al. 2005). The diabetic nephropathy in the CDs rats provided the diabetogenic diet consists of mesangial expansion and widening of the glomerular basement membrane. After three months on the HSD, the kidneys of CDr (as viewed by thin kidney sections stained with hematoxylin and eosin and PAS) are normal. In contrast, the glomeruli in CDs are abnormal with mesangial matrix expansion narrowing the lumen of the capillary loops. The interstitium and the blood vessels are normal. Electron microscopy revealed that the glomeruli of CDr appeared normal while the CDs exhibited increased mesangial matrix protruding into and narrowing the capillary lumen, thickening of the basement membrane, and flattening of the foot processes. Immunocytochemical staining for type IV collagen in CDs revealed increased stain in the glomeruli and, to a lesser degree, in the interstititum.

Retinopathy. The retinal blood vessels present a normal architecture. There are no signs of new blood vessel formation consistent with proliferative diabetic retinopathy. The retinas of CDr rats fed HSD are entirely normal. Severe retinal atrophy thus developed in the CDs rats between 6 and 12 months after initiation of the HSD and the appearance of diabetes (Yagil et al. 2005).

Embryopathy. Ornoy and colleagues have described reduced fertility associated with testicular degeneration in the CDs rat (Ornoy and Cohen 1980; Zusman and Ornoy 1980). When CDs rats were fed HSD after pregnancy was proven, increased incidence of congenital malformation and growth retardation between 9 and 13 days of gestation was demonstrated. In recent studies, it was found that CDs rats fed RD had a decreased rate of pregnancy and increased embryonic resorption (Ornoy and Zusmann 1990; Zangen et al. 2002). CDs rat embryos are smaller than CDr rat

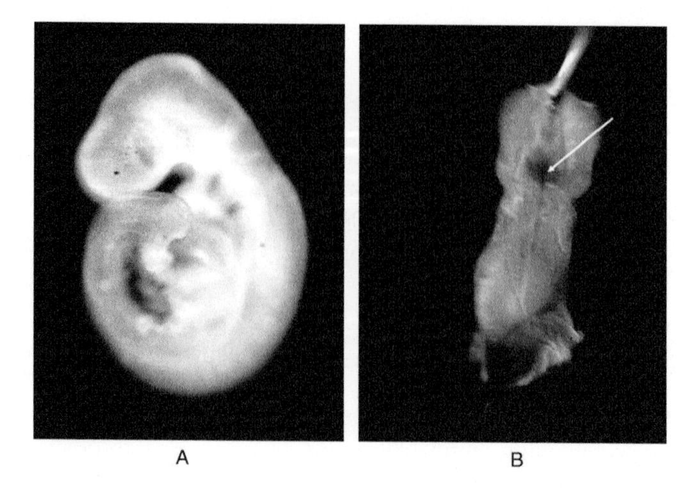

FIG. 15.6 Pictures of normal embryo and embryo with neural tube defect (NTD): (A) 11.5-day-old normal embryo of CDr HSD; (B) 11.5-day-old embryo of CDs HSD with neural tube defect. The arrow indicates cranial neural tube defect.

embryos; 46% are maldeveloped and 7% exhibit neural tube defects (NTD) (fig. 15.6). When CDs rats are fed HSD, the rate of pregnancy is reduced, while the resorption rate is greatly increased (56%; $P < 0.001$); 47.6% of the embryos were retrieved without heartbeats, and 27% exhibited NTD (fig. 15.6). In contrast, all the CDr rat embryos are normal when fed RD or HSD (fig. 15.6). Reduced fertility of the CDs rats, growth retardation, and NTD seem to be genetically determined. Maternal hyperglycemia seems to result in environmentally induced embryonic oxidative stress, resulting in further embryonic damage (Zangen et al. 2002, 2006; Weksler-Zangen et al. 2003b).

THE COHEN–ROSENTHAL DIABETIC HYPERTENSIVE RAT

The Cohen–Rosenthal diabetic hypertensive rat (CDHM) was developed by cross-breeding the CDs rat with SHRs by collaboration with T. Rosenthal (Cohen et al. 1993). In the succeeding generations, sibling pairs with the highest spontaneous blood glucose and blood pressure were selected and mated. In the sixth selected generation, non-insulin-dependent overt diabetes and hypertension were evident. Blood pressure was 171 ± 1 mmHg, spontaneous blood glucose was 340 ± 22 mg/dL, and plasma insulin at 60 min following intragastric glucose load was 83 ± 10 μU/mL. In the CDHM, severe diffuse diabetic glomerulosclerosis and severe hypertensive changes occurred in arteries and arterioles, characterized by fibrinoid necrosis and/or "onion skin" lesions, as well as by smooth muscle cell hyperplasia. Such vascular changes were not observed in the CDs rats or in the SHRs. Myocardial changes were prominent, with foci of ischemic necrosis and hyperplastic vascular changes. This model may be useful in probing the mechanisms potentiating cardiovascular and renal morbid events in the setting of spontaneous hypertension and diabetes.

CONCLUDING REMARKS

The Cohen diabetic rat model is highly homozygous within strains and has a 45% polymorphism between the CDs and CDr strains. This model is thus potentially suitable to elucidate the genetic basis of susceptibility to nutrition-induced T2D. The CD rat is also a useful model for study of the cross-talk between the endocrine and exocrine pancreas and the cellular and molecular factors inducing beta-cell dysfunction. In addition, the Cohen diabetic rat model provides the unique opportunity to study the development of diabetic complications such as nephropathy and retinopathy. As a result, this strain provides a venue for examination of new drugs for treatment of this disorder and its related complications.

LOCATION OF THE COHEN DIABETIC RAT COLONY

The colony of the Cohen diabetic rat is maintained in the animal facility of the Hebrew University–Hadassah School of Medicine in Jerusalem by Dr. Weksler-Zangen (sarahz@hadassah.org.il).

REFERENCES

Cohen AM. (1961) Prevalence of diabetes among different ethnic Jewish groups in Israel. *Metabolism* 10:50–58.

Cohen AM. (1986) Metabolic responses to dietary carbohydrates: interactions of dietary and hereditary factors. *Prog Biochem Pharmacol* 21:74–103.

Cohen AM. (1990) Development of the model. In *The Cohen Diabetic Rat*, ed. Cohen AM, Rosenmann E. Karger, Basel, 1–9.

Cohen AM, Rosenmann E. (1990) *The Cohen Diabetic Rat*. Karger, Basel, 206 pp.

Cohen AM, Rosenmann E, Rosenthal T. (1993) The Cohen diabetic (non-insulin-dependent) hypertensive rat model. Description of the model and pathologic findings. *Hypertension* 6:989–995.

Falconer D. (1953) Selection for large and small size in mice. *J Genet* 51: 470–501.

Ornoy A, Cohen AM. (1980) Teratogenic effects of sucrose diets in diabetic and non-diabetic rats. *Isr J Med Sci* 16:789–791.

Ornoy A, Zusmann I. (1990). Embryonic development in the Cohen diabetic rats: A comparison with human and animal studies. In *The Cohen Diabetic Rat*, ed. Cohen, AM, Rosenmann E. Karger, Basel, 101–116.

Rosenmann E, Cohen AM. (1990) The pathology of the Cohen diabetic rat. In *The Cohen Diabetic Rat*, Cohen AM, Rosenmann E. Karger, Basel, 54–75.

Rosenmann E, Teitelbaum A, Cohen AM. (1971) Nephropathy in sucrose-fed rat. Electron and light microscopic studies. *Diabetes* 20:803–810.

Silberberg R. (1990) Osteopathy and arthropathy. In *The Cohen Diabetic Rat*, ed. Cohen AM, Rosenmann E. Karger, Basel, 87–97.

Vardi R. (1990) Genetic aspects of the Cohen rat model. In *The Cohen Diabetic Rat*, ed. Cohen AM, Rosenmann E. Karger, Basel, 11–23.

Weksler-Zangen S, Chen L, Levy T, Lenzen S, et al. (2005) Reduced glucose stimulated insulin secretion is associated with progressive exocrine pancreatic lesions in the Cohen diabetic sensitive rat. Annual Meeting of EASD, Athens.

Weksler-Zangen S, Yagil C, Zangen DH, Ornoy A, et al. (2001) The newly inbred Cohen diabetic rat: A nonobese normolipidemic genetic model of diet-induced type 2 diabetes expressing sex differences. *Diabetes* 50:2521–2529.

Weksler-Zangen S, Yaffe P, Shechtman S, et al. (2002) The role of reactive oxygen species in diabetes-induced anomalies in embryos of Cohen diabetic rats. *Int J Exp Diab Res* 3:247–255.

Weksler-Zangen S, Oprescu A, Levy C, Zangen DH, et al. (2003a). Nutritionally related pancreatic damage is associated with beta cell dysfunction in the diabetic Cohen rat. International Diabetes Federation Congress (IDF), Paris, France.

Weksler-Zangen S, Yaffe P, Ornoy A. (2003b) Reduced SOD activity and increased neural tube defects in embryos of the sensitive but not of the resistant Cohen diabetic rats cultured under diabetic conditions. *Birth Defects Res* 67:429–437.

Weksler-Zangen S, Lenzen S, Jörns A, et al. (2006) Reduced insulin secretion is coupled with exocrine fat and macrophages infiltration in the Cohen diabetic sensitive rat. American Diabetes Association 66th Scientific Sessions, Washington, D.C.

Yagil C, Barak A, Ben Dor D, et al. (2005) Nonproteinuric diabetes-associated nephropathy in the Cohen rat model of type 2 diabetes. *Diabetes* 54:1487–1496.

Yanko L, Rosenmann E, Cohen AM. (1990) Pathology of the retina in the Cohen diabetic rat. In *The Cohen Diabetic Rat*, ed. Cohen AM, Rosenmann E. Karger, Basel, 76–84.

Zangen S, Yaffe P, Shechtman S, Zangen DH, et al. (2002) The role of reactive oxygen species in diabetes-induced anomalies in embryos of Cohen diabetic rats. *Int J Exp Diab Res* 3:247–255.

Zangen S, Ryu S, Ornoy A. (2006) Alterations in the expression of antioxidant genes and the levels of transcription factor NF-Kappa B in relation to diabetic embryopathy in the Cohen diabetic rat model. *Birth Defects Res A* 76A:107–114.

Zusman I, Ornoy A. (1980) The effect of maternal diabetes and high-sucrose diets on the intrauterine development of rat fetus. *Diabetes Res* 3:153–159.

Zusman I, Ornoy A. (1990) Embryonic development in Cohen diabetic rats: A comparison with human and animal studies. In *The Cohen Diabetic Rat*, ed. Cohen AM, Rosenmann E. Karger, Basel, 101–121.

16 KK and KKA^{y} Mice: Models of Type 2 Diabetes with Obesity

Shigehisa Taketomi

CONTENTS

INTRODUCTION

Type 2 diabetes is a heterogenous disease with multiple etiologies. The development of diabetes characterized by insulin resistance in peripheral tissues and/or impaired insulin secretion is caused by genetic and environmental factors. The basis is thought to be multigenetic rather than monogenic. A series of studies using KK mice has clearly pointed out that the close interaction of genetic and environmental factors such as diet and obesity plays an important role in the development of diabetes.

HISTORY OF DERIVATION OF KK STRAIN

Kondo et al. (1957) selected out and established many mouse strains from Japanese native mice. Among these inbred mouse strains, Nakamura (1962) found that the KK mouse strain, which was named "KK" for its habitat (Kasukabe in Saitama prefecture) (Kondo et al. 1957), is spontaneously diabetic. Thereafter, several investigators (Nakamura 1962; Nakamura et al. 1963, 1967; Iwatsuka, Matsuo, et al. 1970; Dulin et al. 1983) reported that the diabetic state was associated with moderate obesity, sluggishness, polyphagia, polyuria, persistent glucosuria, glucose intolerance, moderate hyperglycemia, hyperlipidemia, insulin resistance in peripheral tissues, hyperinsulinemia, histological changes in the pancreas, and renal glomerular changes. These characteristics suggested several similarities between the diabetic state in the KK mice and human type 2 diabetes (T2D) associated with obesity.

In KK mice, blood glucose levels were higher in males than in females, but there was no sex bias in glucose intolerance (Nakamura 1962; Nakamura et al. 1963, 1967). A genetic study of KK mice indicated that diabetic traits were inherited by polygenes (Nakamura et al. 1963). Because diabetes and obesity in KK mice were relatively moderate, Nishimura (1969) transferred the yellow obese gene (A^y) into KK mice by repeated crossing of yellow obese mice and KK mice. The A^y allele (dominant allele at the mouse agouti locus) is associated phenotypically with yellow fur, hyperphagia, and obesity (Danforth 1927; Bultman et al. 1992; Michaud et al. 1994). This congenic strain of KK mice has been named yellow KK or KKA^y mice. KKA^y mice (yellow) can be easily distinguished by the color of the fur from KK mice (black) at weaning.

The diabetic characteristics of KKA^y mice, such as hyperglycemia, hyperinsulinemia, and obesity, were observed at young ages (six to eight weeks), but reverted apparently to normal after 40 weeks of age (Iwatsuka, Shino, et al. 1970; Iwatsuka, Ishikawa, et al. 1974). This may be due to reduction in food intake because disappearance of diabetic signs in the aged yellow KK mice is not associated with amelioration of insulin resistance or glucose dysmetabolism expected from reduction of adiposity. Therefore, the intrinsic abnormalities such as insulin resistance and glucose dysmetabolism are thought to be primarily responsible for the development of diabetes in the KK strain.

DIABETOGENIC ACTION OF OBESITY

Since the initial studies reported by Nakamura (1962), KK mice have been used as the models for T2D. However, KK mice in our stock derived from the original KK colony at Nagoya University (Japan) showed glucose intolerance and insulin resistance without hyperglycemia and glucosuria, as long as they were kept on a laboratory chow (Iwatsuka, Matsuo, et al. 1970). Accordingly, KK mice could be defined as a strain with diabetic genes. When obesity was produced by transfer of the yellow obese gene (A^y), KK mice exhibited many diabetic symptoms, such as hyperglycemia, glucosuria, hypertrophy of pancreatic islets, and degranulation of beta cells (Iwatsuka, Shino, et al. 1970). Thus, the yellow KK (KKA^y) mice developed more severe diabetes than the original KK mice (table 16.1). These phenomena were also observed in obese KK mice induced by feeding high-calorie diets or hyperphagia due to hypothalamic lesion with goldthioglucose treatment (Matsuo et al. 1970, 1972; Matsuo, Furuno, et al. 1971; Matsuo, Iwatsuka, et al. 1971).

When compared with C57BL or ICR mice, dietary obesity induced hyperglycemia only in KK mice, while goldthioglucose-induced obesity induced hyperglycemia in KK and ICR mice, but not in C57BL mice (fig. 16.1). The degree of glucose intolerance was in the order: KK > ICR > C57BL. KK mice were also more susceptible to diabetogenic effects of obesity than hybrid mice of KK and C57BL, which should carry half of the gene of the KK strain (Iwatsuka and Shino 1970;

TABLE 16.1
Physiological Characteristics in KK and KKA^y Mice

	KK	KKA^y
Body weight	↑	↑↑
Plasma glucose	↑	↑↑
Plasma insulin	↑	↑↑
Plasma triglyceride	—	↑↑
Glucose intolerance	++	++
Insulin resistance	+	++

Notes: ↑ = increase; — = normal; + = mild; ++ = severe.

Iwatsuka, Ishikawa, Shimakawa, et al. 1974). These observations suggest that the difference observed among these mouse strains can be due to genetic factors and that diabetic characteristics in KK mice may be determined by concentration of polygenes, as reported by Nakamura (1962). Furthermore, glucose intolerance and insulin resistance are thought to be good markers for predisposition to diabetes.

INSULIN RESISTANCE

T2D is the most common of metabolic disorders characterized by impaired response of peripheral tissues to insulin. The insulin resistance in liver, muscle, and adipose tissue is thought to lead to perturbation of glucose homeostasis, as well as to impaired insulin secretion in response to glucose. Several investigators have reported insulin

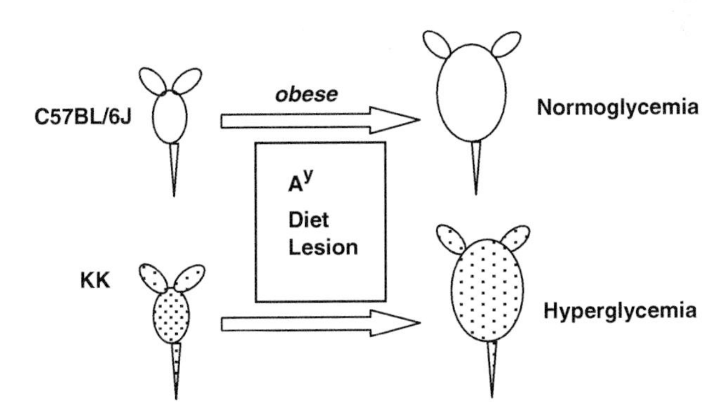

FIG. 16.1 Induction of diabetic alterations by obesity in KK and C57BL mice. Obesity was induced by transfer of the yellow obese gene (A^y), feeding high-calorie diet, or hyperphagia due to hypothalamic lesion caused by hyperglycemia in KK mice, but not in C57BL/6J mice. (Iwatsuka, H, Shino, A, et al. 1970. *Endocr Jpn* 17:23–35; Matsuo, T and Shino, A. 1972. *Diabetologia* 8:391–397.)

resistance and glucose intolerance of the KK strain (Nakamura 1962; Iwatsuka, Matsuo, et al. 1970).

Epididymal adipose tissues of KK mice were less sensitive to the stimulatory action of insulin on glucose oxidation (Iwatsuka, Matsuo, et al. 1970; Iwatsuka, Taketomi, et al. 1974). This phenomenon was confirmed by experiments using adipocytes (Taketomi et al. 1988). Glucose oxidation in KK adipocytes showed decreased insulin sensitivity and responsiveness; the dose–response curve was shifted to the right and the maximum activity was relatively low when compared to adipocytes from C57BL/6J mice. On the other hand, lipid synthesis from glucose or acetate was a normal response to insulin in adipose tissues or adipocytes from KK mice. Such heterogeneity of hormone sensitivity may result in development of diabetes associated with obesity in this strain. Furthermore, in adipose tissue or adipocytes, lipolysis was less sensitive to epinephrine and insulin (Iwatsuka, Matsuo, et al. 1970; Iwatsuka, Taketomi, et al. 1974). Thus, the alterations in lipolysis may also be a risk factor leading to obesity.

These metabolic profiles are very similar to those caused by thrifty genotype in human diabetes proposed by Neel (1962) (Iwatsuka, Taketomi, et al. 1974). Generally, a negative correlation between the cell size and hormone sensitivity was reported (Herberg et al. 1970). However, this phenomenon was not valid in KK mice because KK adipocytes were less sensitive to the hormone than C57BL/6J adipocytes of similar cell size (Iwatsuka, Taketomi, et al. 1974). Therefore, impaired hormonal sensitivity is more likely determined by genetic factors than simply by the changes in cellularity. The F_1-hybrid mice of KK and C57BL mice showed intermediate responses of adipose tissue to insulin (Iwatsuka and Shino 1970) (fig. 16.2). Introduction of the A^y gene into KK and C57BL mice decreased insulin sensitivity of

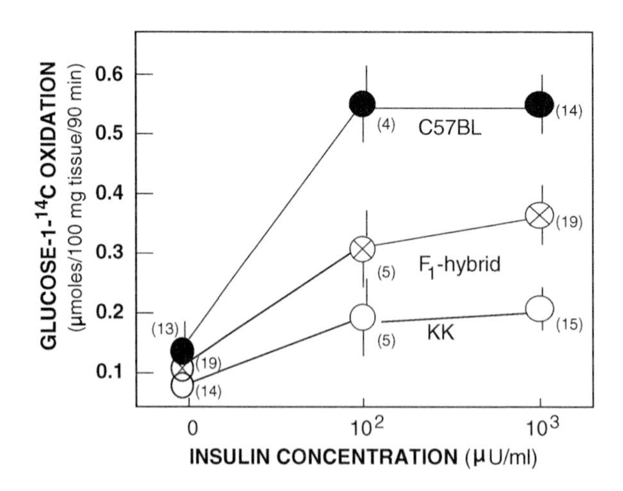

FIG. 16.2 Insulin sensitivity in glucose oxidation of adipose tissue from C57BL, KK, and their F_1 hybrid. The hybrid shows intermediate sensitivity to insulin between the parental strains (Iwatsuka, H and Shino, A. 1970. *Endocr Jpn* 17:535–540.) Numbers in parentheses indicate the number of mice. Mean ± SE.

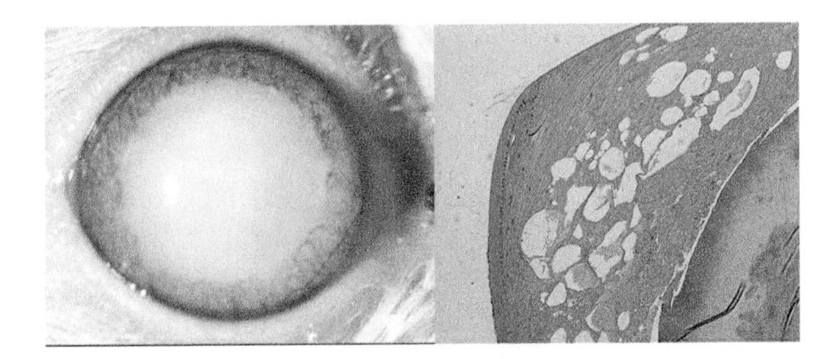

COLOR FIG. 14.1 Cataract development in a 52-week-old male SDT rat. Left: The cataract has developed to maturity; right: pathology of the mature cataract. The sclerotic nucleus floats in a liquefied lens cortex. Vacuolation, disintegration of the lens fibers, and Morgani's globules are observed in the lens cortex.

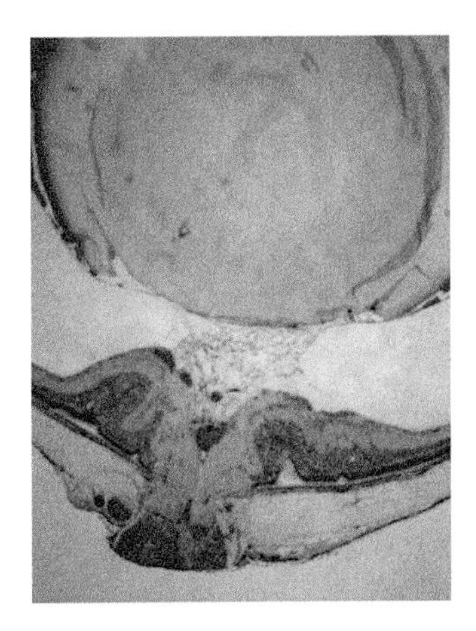

COLOR FIG. 14.2 Characteristics of diabetic retinopathy in an 82-week-old female SDT rat. Left: Large retinal folds associated with vitreous traction are seen around the optic disc in an old SDT rat. Some vessels are observed in the vitreous cavity in front of the optic disc. One of the vessels may be a persistent hyaloid artery and the other a new vessel from the optic disc.

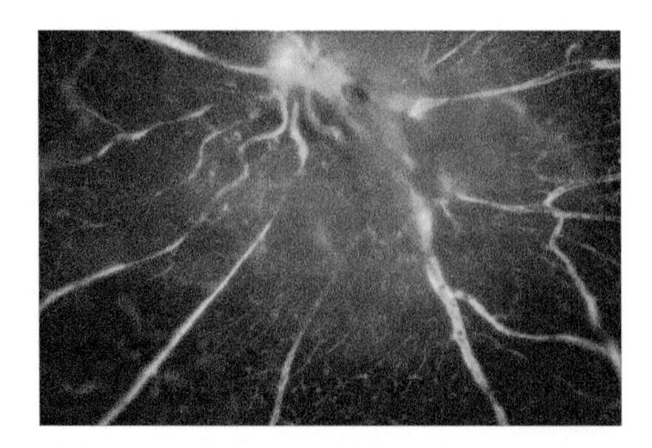

COLOR FIG. 14.3 Extensive hyperfluorescence is observed around the optic disc with vascular tortuosity in a 62-week-old male SDT rat.

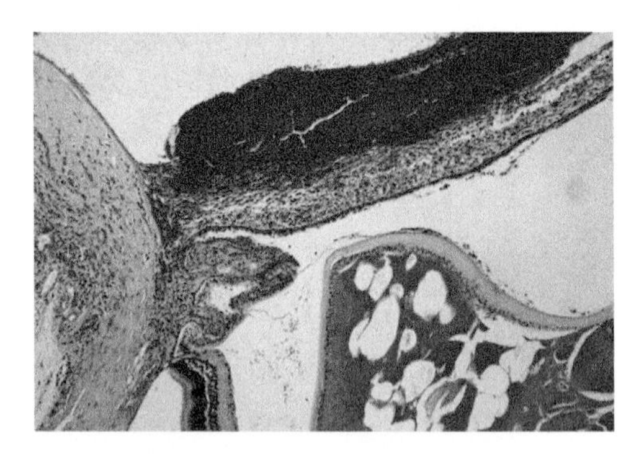

COLOR FIG. 14.4 Massive hemorrhage can be seen in the anterior chamber associated with proliferation around the iris in a 77-week-old male SDT rat.

adipose tissue due to increased cell size (Iwatsuka, Taketomi, et al. 1974). Therefore, this phenomenon may be caused by changes in cellularity. These findings clearly indicate that the insulin resistance of KK mice is inherent and susceptible to the diabetogenic influence of obesity.

Taketomi et al. (1988) found that adipocytes from KK mice were less sensitive and less responsive to insulin with respect to glucose uptake and [1-^{14}C]-glucose oxidation, but normosensitive to insulin with [6-^{14}C]-glucose oxidation or glyceride-glycerol and fatty acid synthesis from both [1-^{14}C]- and [6-^{14}C]-glucose. These findings indicate that, in the adipocytes from KK mice, hormone sensitivity is normal with regard to glycolysis and Krebs cycle and the hormone resistance is restricted to the pentose shunt.

The same type of hormone resistance has been reported in the large adipocytes from spontaneously obese and old rats (Olefsky 1977; Richardson et al. 1978). However, the mechanism of the hormone resistance was quite different between KK mice and the old rats. In the former, [1-^{14}C]-glucose oxidation was insensitive to vitamin K_5, which shows insulin-like actions without binding to insulin receptors, but fatty acid synthesis from glucose responded to the agent in a usual manner. In the latter, [1-^{14}C]-glucose oxidation was normally stimulated by vitamin K_5, but fatty acid from glucose was decreased. In the large cells from the old rats, an accumulation of NADPH induced by decreased fatty acid synthesis seems to result in a feedback inhibition of the pentose shunt. Accordingly, cells from old rats showed a normal response to vitamin K_5, a powerful NADPH oxidant.

On the other hand, the abnormalities in the adipocytes from KK mice were accounted for by the selective defect in the postinsulin binding system. [1-^{14}C]-glucose oxidation in adipocytes of KK mice was also insensitive to insulin-mimicking agents such as H_2O_2. The downregulation of the insulin receptor caused by insulin or insulin imitators in cultured adipose tissue of KK mice was also impaired. In conclusion, the adipocytes of KK mice are insulin resistant with respect to glucose uptake and pentose pathway due to defects in the postinsulin binding signaling pathway.

In KKA^y mice with elevated plasma insulin levels, insulin resistance was associated with decreased expression of GLUT4 transporter protein (GLUT4) in adipocytes and muscles (Hofmann et al. 1991). Furthermore, Yamauchi et al. (2001) reported that adiponectin, an adipocyte-secreted protein, participates in insulin resistance in KKA^y mice. Serum adiponectin levels were decreased in KKA^y mice on a high-fat diet compared with those on a high-carbonate diet. Lower serum adiponectin levels were partially restored by replenishment of recombinant adiponectin and insulin resistance and hypertriglyceridemia. These data suggest that adiponectin is decreased in obesity and that reduction in adiponectin plays causal roles in the development of insulin resistance in KKA^y mice (Yamauchi et al. 2001).

Liver also exhibits insulin resistance. However, hepatic glucose production based upon fasting blood glucose levels is likely to be normally regulated by insulin in prediabetes or mild diabetes because fasting hyperglycemia is not observed (DeFronzo 1997). In moderate/severe diabetes, fasting blood glucose levels are increased, indicating that insulin is unable to prevent excessive hepatic glucose production. The plasma glucose levels after an overnight fast were normal in KK mice, but mildly increased in KKA^y mice (Iwatsuka, Shino, et al. 1970). Wyse et al. (1974) reported that glucose production from pyruvate in the liver of obese and hyperinsulinemic KK mice was not suppressed during the fed state. KKA^y mice with hyperinsulinemia showed higher activities of insulin-suppressive enzymes of gluconeogenesis (glucose-6-phosphatase and fructose-1,6-diphosphatase) compared with KK mice (Taketomi et al. 1973).

From these findings, the response to insulin in the liver of KKA^y mice is impaired. However, KKA^y mice also showed high activities of insulin-inducible enzymes involving glycolysis (glucokinase and pyruvate kinase), pentose phosphate cycle (6-phosphogluconate dehydrogenase), and lipogenesis (ATP citrate lyase, malic enzyme, and acetyl-CoA carboxylase), indicating that insulin normally regulates enzyme activities of glycolysis, pentose phosphate cycle, and lipogenesis. Thus, there was a heterogeneity in the enzyme activity alterations in response to insulin. KKA^y and KK mice showed a correlation between hepatic lipogenic activity and plasma insulin levels (Matsuo, Iwatsuka, et al. 1971; Taketomi et al. 1973). Predisposition to diabetes in KK strain may arise from a normal response to insulin in lipogenesis of liver and adipose tissue concomitant with insulin resistance in the glucose uptake and oxidation of adipose tissue.

Albert et al. (2002, 2003) assessed the effect of selective inhibition of 11β-hydroxysteroid dehydrogenase type I on blood glucose level in KKA^y mice. The enzyme is widely distributed and yields increased local tissue concentration of active glucocorticoid by converting cortisone into cortisol in humans and 11-dehydrocorticosterone into corticosterone in rodents (Seckl et al. 1997; Stewart et al. 1999). The high expression of the enzyme relates with obesity (Rask et al. 2001; Bray and York 1979). The results showed the inhibition of the enzyme improved hepatic insulin sensitivity in KKA^y mice, as observed in lower blood glucose and insulin levels, and reduction of hepatic PEPCK and G6Pase mRNA (Albert et al. 2002, 2003).

Many reports suggest that tumor necrosis factor (TNF-α) is involved in the development of insulin resistance in animal and human T2D by inhibiting tyrosine kinase activity on the β-subunit of the insulin receptor (Hotamisligil and Spiegelman 1994; Hotamisligil et al. 1994). In muscle of KKA^y mice, the RNA transcript encoding the p55 TNF-α receptor is elevated (Hofmann et al. 1994). These findings suggest that the elevated production of TNF-α or other cytokines is responsible for insulin resistance in KKA^y mice.

A convenient method for assessment of the overall insulin sensitivity in small animals (mice) was established using subcutaneous injection of epinephrine, propranolol, and glucose with or without insulin (Taketomi et al. 1982). KK mice showed a reduction of steady-state blood glucose levels in response to insulin compared with C57BL/6J mice.

OTHER CHARACTERISTICS

Leptin regulates appetite and body weight by reducing food intake and increasing energy expenditure. Most obese animals have been linked to defects in the leptin system: deficiency of leptin in *ob*/*ob* mice, which have a mutation in the leptin gene, and defects of leptin receptor in *db*/*db* mice and of all leptin receptor isoforms in Zucker fatty rats, which show overexpression of leptin mRNA (Zhang et al. 1994; Lee et al. 1996; Chen et al. 1996; Phillips et al. 1996; Iida et al. 1996). These animals are also leptin resistant (Frederich et al. 1995). KKA^y mice showed higher leptin mRNA levels in the white adipose tissue (WAT) compared with KK mice (Hayase et al. 1996). In both mice, downregulation of leptin gene expression during fasting was recognized in mesenteric and subcutaneous WAT, but not in epididymal WAT, indicating the presence of regional differences in the regulation of leptin gene expression in adipose tissue.

Masuzaki et al. (1999) obtained genetically obese KKA^y mice with hyperleptinemia by crossing KKA^y mice and transgenic skinny mice overexpressing leptin. This was developed under the control of the liver-specific promoter (Ogawa et al. 1994) with the aim to explore the pathophysiological role of leptin in obesity-associated diabetes. The persistent hyperleptinemia can delay the onset of impaired glucose metabolism and accelerate the recovery from diabetes in KKA^y mice during the course of caloric restriction (Matsuzaki et al. 1999). In addition, in KKA^y mice, leptin is oversecreted from the adipose tissue, which causes blood pressure elevation with increased catecholamine production (Aizawa-Abe et al. 2000). Therefore, KKA^y mice are thought to show different sensitivity to leptin: resistance to the satiety effect of leptin and sensitivity to the hypertensive effect of the hormone. Thus, KKA^y mice are good models for investigating the pathophysiological role of leptin in obesity-related hypertension (Aizawa-Abe et al. 2000).

HISTOLOGICAL CHANGES AND COMPLICATIONS

In KK and KKA^y mice, plasma insulin levels were elevated after a meal, but not after glucose loading, indicating that beta cells of the KK strain are selectively insensitive to glucose (Iwatsuka, Matsuo, et al. 1970; Iwatsuka, Shino, et al. 1970). As for morphological changes in KK mice, beta cell degranulation, glycogen deposition, and hypertrophy were observed at the age of 16 weeks (Iwatsuka, Shino, et al. 1970; Shino and Iwatsuka 1970). KKA^y mice with severe hyperinsulinemia showed more prominent changes in pancreatic islets than KK mice (Shino and Iwatsuka 1970). Degranulation, glycogen deposition, and hypertrophy of beta cells were observed at the ages of 5–10 weeks. These changes became prominent at 16 weeks of age and diminished with increasing age. As shown by electron microscopy, degranulated beta cells contained well-developed Golgi apparatus, abundant granular endoplasmic reticulum, and fine granules of glycogen. Beta cell granules were occasionally facing the cell membrane, indicating emiocytosis. These changes suggest increased synthesis and release of insulin, which corresponds to high levels of plasma insulin.

KK and KKA^y mice showed renal lesions similar to human diabetic nephropathy (Nakamura 1962; Treser et al. 1968; Camerini-Davalos et al. 1970; Iwatsuka, Shino, et al. 1970; Wehner et al. 1972; Reddi et al. 1978; Emoto et al. 1982). Diffuse glomerulosclerosis, nodular changes, and peripheral glomerular basement membrane (GBM) thickening were recognized in KKA^y mice. Diani et al. (1987) found that GBM thickening occurred at an early age and developed rapidly in KKA^y mice in comparison with other diabetic animals. Mesangial enlargement and hypercellularity were also observed in KK mice.

Proteinuria and microalbuminuria were observed in KK and KKA^y mice (Treser et al. 1968; Wehner et al. 1972; Reddi et al. 1978). However, it remains unclear whether the glomerulosclerosis is due to diabetes in the KK strain because a high incidence of the same changes was observed in the prediabetic stage in KK mice.

Additional factors, such as renal amyloidosis, may be involved in the pathogenesis of glomerulosclerosis (Soret et al. 1977). Recently, losartan, an angiotensin II receptor antagonist, was reported to ameliorate progression of glomerular mesangial expansion and glomerulosclerosis with exudative lesion (Sasaki et al. 2004).

CHROMOSOME MAPPING

To identify the genetic factors underlying T2D, quantitative trait loci (QTL) analysis was performed on KK and KKA^y mice by several groups (Suto et al. 1998a, 1998b, 2002; Taylor et al. 1999; Shike et al. 2001). QTLs were identified on some chromosomes for diabetes-related phenotypes such as fasting hyperglycemia, glucose intolerance, hyperinsulinemia, and hyperlipidemia. However, the results of these analyses were inconsistent. The discrepancies might be related to a number of differences, including ages at the time of phenotype evaluation, sample size, methods for measurement of phenotypes, crossing partners (C57BL/6J vs. BALB/c), and potential KK substrain differences, as reported by Shike et al. (2001). Recently, Shike et al. (2005) reported susceptibility loci for the development of albuminuria in diabetic KK mice.

USEFUL FOR EVALUATION OF ANTIDIABETIC DRUGS

Based on the fact that insulin resistance is one of the causes for development of T2D, compounds that reduce insulin resistance or increase insulin sensitivity have been sought using KK mice or KKA^y mice. The first member of the thiazolidinediones group (TZD), ciglitazone (ADD-3878), was discovered in the *in vivo* screening system using KKA^y mice (Fujita et al. 1983). This compound decreased hyperglycemia, hyperlipidemia, and hyperinsulinemia, accompanied by reduced insulin resistance in peripheral tissues in KKA^y mice. Subsequently, pioglitazone (AD-4833) was selected (Ikeda et al. 1990).

TZD was also reported to induce preadipocytes to differentiate into mature adipocytes, acting as a high-affinity ligand for transcription factor PPARγ (peroxisome-proliferation activated receptor) (Tontonoz et al. 1994). The affinity of TZD for PPARγ has been known to correlate with antidiabetic effects (Berger et al. 1996; Wilson et al. 1996). Therefore, TZDs are thought to exert their insulin sensitizing action by binding to PPARγ (Wilson et al. 1996). The effectiveness of pioglitazone in the treatment of T2D patients is generally accepted. Treatment with a vanadium complex was also successful in producing an amelioration in diabetes, obesity, and hypertension in KKA^y mice (Adachi et al. 2006).

CONCLUDING REMARKS

The KK strain of mice exhibits two characteristics in the development of T2D. One is the fact that their diabetic traits are inherited by polygenes. This genetic feature is quite different from some other types of diabetic animals, such as *ob/ob* mice, *db/db* mice, or Zucker diabetic fatty rats, whose diabetic states are induced by mutations of the gene. The other is that the KK strain has a latent diabetic state before the development of hyperglycemia and glucosuria. At this stage, glucose intolerance and insulin resistance are already observed. They become overtly diabetic with aging and/or with increasing body weight.

KK mice are predisposed to diabetes with the onset of obesity induced by feeding of high energy diets, injection of goldthioglucose, or induction of the yellow obese gene A^y. This predisposition is considered to be due to genetically determined resistance to insulin. The genetically obese KK, KKA^y (also known as yellow KK mice), are characterized by rapid onset of diabetes and several renal glomerular lesions. Therefore, the KK and KKA^y mice are suitable for studying the causal genes of diabetes and effects of environmental factors, as well as evaluation of antidiabetic and antiobesity agents designated to ameliorate insulin resistance and obesity-related metabolic disturbances.

HUSBANDRY

KK mice fed a powdered laboratory chow show significantly higher blood glucose levels than those fed the pelleted one, indicating that rigidity or size of the pellet was unsuitable to feed mice (Matsuo, Furuno, et al. 1971). Caging population also affected blood glucose levels. Mice kept together with five heads per cage occasionally fought with each other, which might have resulted in decreased food intake, mental irritation, or increased physical activity. Blood glucose levels and body weight gains were significantly lower in mice kept together than in those kept individually. The severity of their diabetic state was dependent on body weight (fig. 16.3). KKA^y mice were maintained in individual cages.

KK and KKA^y mice can be obtained from Clea Japan (Tokyo, Japan).

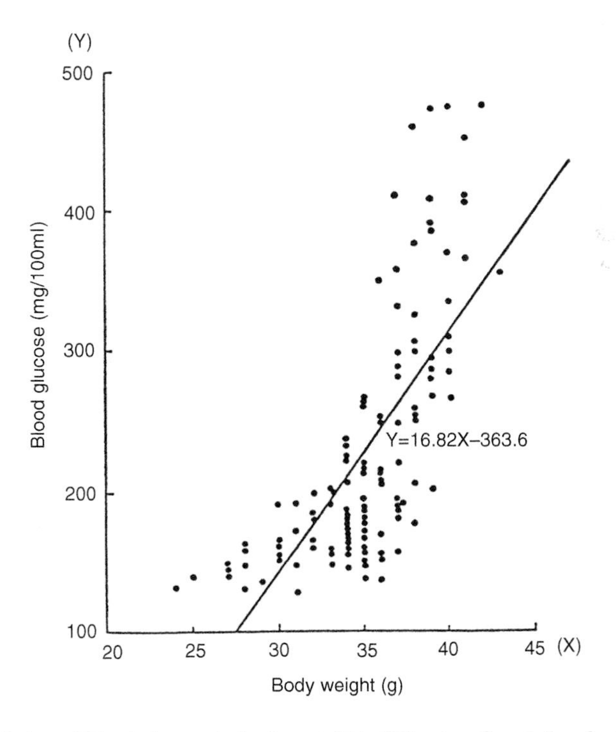

FIG. 16.3 Relation of blood glucose to body weight in KK mice. Correlation for blood glucose versus body weight was 0.715 ($p < 0.01$). (Matsuo, T, Furuno K, 1971. *J Takeda Res Lab* 30:307.)

REFERENCES

Adachi Y, Yoshikawa Y, Yoshida J, et al. (2006) Improvement in diabetes, obesity and hypertension in type 2 diabetic KKA(y) mice by bis(allixunato)oxovanadium(IV) complex. *Biochem Biophys Res Commun* (published in PubMed ahead of print).

Aizawa-Abe M, Ogawa Y, Masuzaki H, et al. (2000) Pathophysiological role of leptin in obesity-related hypertension. *J Clin Invest* 105:1243–1252.

Albert P, Engblom L, Edling N, et al. (2002) Selective inhibition of 11β-hydroxysteroid dehydrogenase type 1 decreases blood glucose concentrations in hyperglycemic mice. *Diabetologia* 45:1528–1532.

Albert P, Nilsson C, Selen G, et al. (2003) Selective inhibition of 11β-hydroxysteroid dehydrogenase type 1 improves hepatic insulin sensitivity in hyperglycemic mice strains. *Endocrinology* 144:4755–4762.

Berger J, Bailey P, Biswas C, et al. (1966) Thiazolidinediones produce a conformational change in peroxisomal proliferation-activated receptor-γ: Binding and activation correlate with antidiabetic actions in *db/db* mice. *Endocrinology* 137:4189–4195.

Bray GA, York DA. (1979) Hypothalamic and genetic obesity in experimental animals: an autonomic and endocrine hypothesis. *Physiol Rev* 59:719–809.

Bultman S, Michaud EJ, Woychik RP. (1992) Molecular characterization of the mouse agouti locus. *Cell* 71:1195–1204.

Camerini-Davalos RA, Oppermann W, Mittl R, et al. (1970) Studies of vascular and other lesions in KK mice. *Diabetologia* 6:324–329.

Chen HO, Charlat LA, Tartaglia EA, et al. (1966) Evidence that the diabetes gene encodes the leptin receptor: Identification of a mutation in the leptin receptor gene in *db/db* mice. *Cell* 84:491–495.

Danforth CH. (1927) Hereditary adiposity in mice. *J Heredity* 18:153–162.

DeFronzo RA. (1997) Pathogenesis of type II diabetes: metabolic and molecular implications for identifying diabetes genes. *Diabetes Rev* 5:177–269.

Diani AR, Sawada GA, Zhang NY, et al. (1987) The KKA^y mouse: A model for the rapid development of glomerular capillary basement membrane thickening. *Blood Vessels* 24:297–303.

Dulin WE, Gerritsen GC, Chang AY. (1983) Experimental and spontaneous diabetes in animals. In *Diabetes Mellitus: Theory and Practice*, ed. M. Ellenberg, H. Rifkin. Medical Examination Publishing Co., New York, 361–408.

Emoto M, Matsutani H, Kimura S, et al. (1982) Nephropathy in KK mice treated with a limited diet. *J Jpn Diabetes Soc* 25:1211–1219.

Frederich R, Hamann A, Anderson S, et al. (1995) Leptin levels reflect body lipid content in mice: evidence for diet-induced resistance to leptin action. *Nat Med* 1:1311–1314.

Fujita T, Sugiyama Y, Taketomi S, et al. (1983) Reduction of insulin resistance in obese and/or diabetic animals by 5-{4-(1-methylcyclohexylmethoxy)benzyl}-thiazolidine-2,4-dione (ADD-3878, U-63,287, ciglitazone), a new antidiabetic agent. *Diabetes* 32:804–810.

Hayase M, Ogawa Y, Katsuura, G, et al. (1996) Regulation of obese gene expression in KK mice and congenic lethal yellow obese KKAy mice. *Am J Physiol* 271:E333–E339.

Herberg L, Gries FA, Hesse-Wortmann C. (1970) Effect of weight and cell size on hormone-induced lipolysis in New Zealand obese mice and American obese hyperglycemic mice. *Diabetologia* 6:300–305.

Hofmann C, Lorenz K, Braithwaite SS, et al. (1994) Altered gene expression for tumor necrosis factor-α and its receptors during drug and dietary modulation of insulin resistance. *Endocrinlogy* 134:264–270.

Hofmann C, Lorenz K, Colca JR. (1991) Glucose transport deficiency in diabetic animals is corrected by treatment with oral antihyperglycemic agent pioglitazone. *Endocrinology* 129:1915–1925.

Hotamisligil GS, Budavari A, Murray D, et al. (1994) Reduced tyrosine kinase activity of the insulin receptor in obesity-diabetes. Central role of tumor necrosis factor-α. *J Clin Invest* 94:1543–1549.

Hotamisligil GS, Spiegelman BM. (1994) Tumor necrosis factor: A key component of the obesity-diabetes link. *Diabetes* 43:1271–1278.

Iida M, Murakami T, Ishida K, et al. (1996) Substitution at codon 269 (glutamine→proline) of the leptin receptor (OB-R) cDNA is the only mutation found in the Zucker fatty (*fa/fa*) rat. *Biochem Biophys Res Commun* 224:597–604.

Ikeda H, Taketomi S, Sugiyama Y, et al. (1990) Effects of pioglitazone on glucose and lipid metabolism in normal and insulin resistant animals. *Arzneim Forsch/Drug Res* 40:156–162.

Iwatsuka H, Ishikawa E, Shimakawa K, et al. (1974) The role of genetic background of KK strain in pathogenesis of diabetes: Polygenic inheritance of a sensitivity to a diabetogenic action of obesity. *J Takeda Res Lab* 33:248–251.

Iwatsuka H, Ishikawa E, Shino A. (1974) Remission of diabetic syndromes in advanced age of genetically obese and diabetic mice, yellow KK. *J Takeda Res Lab* 33:203–212.

Iwatsuka H, Matsuo T, Shino A, et al. (1970) Metabolic disturbance of KK mice in chemical diabetes. *J Takeda Res Lab* 29:685–692.

Iwatsuka H, Shino A. (1970) Studies on diabetogenic action of obesity in mice: Congenital insulin resistance of KK mice. *Endocr Jpn* 17:535–540.

Iwatsuka H, Shino A, Suzuoki Z. (1970) General survey of diabetic features of yellow KK mice. *Endocr Jpn* 17:23–35.

Iwatsuka H, Taketomi S, Matsuo T, et al. (1974) Congenitally impaired hormone sensitivity of the adipose tissue of spontaneously diabetic mice, KK. Validity of thrifty genotype in KK mice. *Diabetologia* 10:611–616.

Kondo K, Nozawa K, Tomita T, et al. (1957) Inbred strains resulting from Japanese mice. *Bull Exp Animals* 6:107–112.

Lee GH, Proenca R, Montez JM, et al. (1996) Abnormal splicing of the leptin receptor in diabetic mice. *Nature* 379:632–635.

Masuzaki H, Ogawa Y, Aizawa-Abe M, et al. (1999) Glucose metabolism and insulin sensitivity in transgenic mice overexpressing leptin with lethal yellow agouti mutation. Usefulness of leptin for the treatment of obesity-associated diabetes. *Diabetes* 48:1615–1622.

Matsuo T, Furuno K, Shimakawa K. (1971) Factors affecting development of hyperglycemia in KK mice. *J Takeda Res Lab* 30:307–313.

Matsuo T, Iwatsuka H, Suzuoki Z. (1971) Metabolic disturbance of KK mice in overt diabetes. *Endocr J* 18:501–506.

Matsuo T, Shino A. (1972) Induction of diabetic alterations by goldthioglucose obesity in KK, ICR and C57BL mice. *Diabetologia* 8:391–397.

Matsuo T, Shino A, Iwatsuka H, et al. (1970) Induction of overt diabetes in KK mice by dietary means. *Endocr J* 17:477–488.

Michaud EJ, Bultman SJ, Klebig ML, et al. (1994) A molecular mode for the genetic and phenotypic characteristics of the mouse lethal yellow (A^y) mutation. *Proc Natl Acad Sci USA* 91:2562–2566.

Nakamura M. (1962) A diabetic strain of the mouse. *Proc Jpn Acad* 38:348–352.

Nakamura M, Yamada K. (1963) A further study of the diabetic (KK) strain of the mouse: F_1 and F_2 offspring of the cross between KK and C57BL/6J mice. *Proc Jpn Acad* 38:489–493.

Nakamura M, Yamada K. (1967) Studies on diabetic (KK) strain of the mouse. *Diabetologia* 3:212–221.

Neel JV. (1962) Diabetes mellitus: A "thrifty" genotype rendered detrimental by "progress"? *Am J Hum Genet* 14:353–362.

Nishimura M. (1969) Breeding of mice strains for diabetes mellitus. *Exp Anim (Tokyo)* 18:147–157.

Ogawa Y, Masuzaki H, Hosoda K, et al. (1999) Increased glucose metabolism and insulin sensitivity in transgenic skinny mice overexpressing leptin. *Diabetes* 48:1822–1829.

Olefsky JM. (1977) Mechanism of decreased insulin responsiveness of large adipocytes. *Endocrinology* 100:1169–1177.

Phillips MS, Liu Q, Hammond HA, et al. (1996) Leptin receptor missense mutation in the fatty Zucker rat. *Nat Genet* 13:18–19.

Rask E, Olsson T, Soderberg S, et al. (2001) Tissue-specific dysregulation of cortisol metabolism in human obesity. *J Clin Endocrinol Metab* 86:1418–1421.

Reddi AS, Oppermann W, Patel DG, et al. (1978) Diabetic microangiopathy in KK mice. III. Effect of prolonged glyburide treatment on glomerulosclerosis. *Exp Mol Pathol* 29:92–101.

Richardson DK, Czech MP. (1978) Primary role of decreased fatty acid synthesis in insulin resistance of large rat adipocytes. *Am J Physiol* 234:E182–E189.

Sasaki M, Uehara S, Ohta H, et al. (2004) Losartan ameliorates progression of glomerular structural changes in diabetic KKA^y mice. *Life Sci* 75:869–880.

Seckl JR, Chapman KE. (1997) The 11β-hydroxysteroid dehydrogenase system, a determinant of glucocorticoid and mineralocorticoid action. Medical and physiological aspects of the 11β-hydroxysteroid dehydrogenase system. *Eur J Biochem* 249:361–364.

Shike T, Gohda T, Tanimoto M, et al. (2005) Chromosomal mapping of a quantitative trait locus for the development of albuminuria in diabetic KK/Ta mice. *Nephrol Dial Transplant* 20:879–885.

Shike T, Hirose S, Kobayashi M, et al. (2001) Susceptibility and negative epistatic loci contributing to type 2 diabetes and related phenotypes in a KK/Ta mouse model. *Diabetes* 50:1943–1948.

Shino A, Iwatsuka H. (1970) Morphological observations on pancreatic islets of spontaneous diabetic mice, yellow KK. *Endocr Jpn* 17:459–476.

Soret MG, Peterson T, Wyse BM, et al. (1977) Renal amyloidosis in KK mice that may be misinterpreted as diabetic glomerulosclerosis. *Arch Pathol Lab Med* 101:464–468.

Stewart PM, Krozowski ZS. (1999) 11β-Hydroxysteroid dehydrogenase. *Vitam Horm* 57:249–324.

Suto J, Matsuura S, Imamura K, et al. (1998a) Genetic analysis of non-insulin-dependent diabetes mellitus in KK and KK-A^y mice. *J Eur Endocr* 139:654–661.

Suto J, Matsuura S, Imamura K, et al. (1998b) Genetics of obesity in KK mouse and effects of A^y allele on quantitative regulation. *Mammalian Genome* 9:506–510.

Suto J, Sekikawa K. (2002) A quantitative trait locus that accounts for glucose intolerance maps to chromosome 8 in hereditary obese KK-A^y mice. *Int J Obesity* 26:1517–1519.

Taketomi S, Fujita T, Yokono K. (1988) Insulin receptor and postbinding defects in KK mouse adipocytes and improvement by ciglitazone. *Diabetes Res Clin Pract* 5:125–134.

Taketomi S, Ikeda H, Ishikawa E, et al. (1982) Determination of overall insulin sensitivity in diabetic mice, KK. *Horm Metab Res* 14:14–18.

Taketomi S, Tsuda M, Matsuo T, et al. (1973) Alterations of hepatic enzyme activities in KK and yellow KK mice with various diabetic states. *Horm Metab Res* 5:333–339.

Taylor BA, Tarantino LM, Phillips SJ. (1999) Gender-influenced obesity QTLs identified in a cross involving the KK type II diabetes-prone mouse strain. *Mammalian Genome* 10:963–968.

Tontonoz P, Hu E, Spiegelman BM. (1994) Stimulation of adipogenesis in fibroblasts by PPAR$_2$, a lipid activated transcription factor. *Cell* 79:1147–1156.

Treser G, Oppermann W, Ehrenreich T, et al. (1968) Glomerular lesions in a strain of genetically diabetic mice. *Proc Soc Exp Biol Med* 129:820–823.

Wehner H, Hohn O, Faix-Shade U, et al. (1972) Glomerular changes in mice with spontaneous hereditary diabetes. *Lab Invest* 27:331–340.

Wilson TM, Cobb JE, Cowan DJ, et al. (1966) The structure–activity relationship between peroxisome proliferation-activated receptor γ agonism and antihyperglycemic activity of thiazolidinediones. *J Med Chem* 39:665–668.

Wyse BM, Dulin WE. (1974) Further characterization of diabetes-like abnormalities in the T-KK mouse. *Diabetologia* 10:617–623.

Yamauchi T, Kamon J, Waki H, et al. (2001) The fat-derived hormone adiponectin reverses insulin resistance associated with both lipoatrophy and obesity. *Nat Med* 7:941–946.

Zhang YR, Proenca R, Maffei M, et al. (1994) Positional cloning of the mouse obese gene and its human homologue. *Nature* 372:425–432.

17 Animal Models to Study Obesity and Type 2 Diabetes Induced by Diet

Tamer Coskun, Yanyun Chen, Dana Sindelar, and Mark Heiman

CONTENTS

INTRODUCTION

Thermodynamic laws state that energy must be supplied for work to be performed by living cells. That energy is supplied as food and the processes of converting consumed food to usable fuel also require energy. Some of the acquired external energy is used for work and some is converted to heat; a smaller portion is lost as waste products and the remainder is stored as fat in adipose tissue. In wealthy societies, food is abundant and this is most evident by levels of stored fat carried by individuals in those civilizations. The linkage of external energy supply to obese societies makes it difficult to think that most types of obesity are a consequence of genetic mutations.

Homeostatic mechanisms exist in mammals to regulate energy balance.[1] Long-term and short-term feedback mechanisms are present that sense the energy level and signal behaviors to consume more energy when fuel levels are low or to cease eating when the fuel level is sufficient. However, systems evolved to prevent starvation rather than to protect against obesity. Nonhomeostatic mechanisms also exist.[1] Most involve some central reward mechanisms associated with pleasure obtained by consumption of food, especially caloric-dense food. Such tasty food, loaded with energy, can trigger similar brain centers as do some addicting drugs that are abused. These nonhomeostatic mechanisms can also be thought of as a means to protect against starvation.

Thus, when very palatable food is abundant, the animal or human will tend to consume more energy than needed and will store the excess as fat. This can be studied in experimental animals and is known as diet-induced obesity (DIO). Moreover, like in modern human societies, DIO occurs in most species without known genetic mutations. Further, the newly acquired lipid is best stored in adipose tissue, but until fat cells proliferate, the lipid is often first stored in liver and then muscle or even pancreas, causing insulin resistance and even diabetes. This chapter describes some of the DIO animal models.

RAT

Rat strains differ in their susceptibility to obesity when fed experimental diets. In a 1969 study, Schemmel and colleagues tested seven strains of rats, including Osborne–Mendel, Sprague–Dawley, Hoppert, Wistar–Lewis, hooded, gray, and S5B/P1, with a diet extremely high in fat. At weaning, body weight for most strains was similar. They found that, except for the S5B/P1 strain, rats within the same age, sex, and strain that were fed a high-fat diet were heavier and had a higher percentage of body fat than rats fed the high-grain chow. Fat accounted for much of the additional body weight. S5B/P1 rats were resistant to DIO. Osborne–Mendel, Sprague–Dawley, and Wistar–Lewis rats were the three most susceptible strains, with 50 to 25% body weight gain on the diet. All animals of these strains were unable to adjust their caloric intake to their energy requirement when the only food available was the high-fat diet. Osborne–Mendel, Sprague–Dawley, and Wistar–Lewis rats increased caloric intake by 25, 20, and 12%, respectively.[2,3]

Another commonly used obesity-prone strain is the Long–Evans rat. Long–Evans rats fed a high-fat diet weigh more than rats fed a low-fat diet and have more body fat, as well as significant hyperleptinemia and insulin resistance.[4,5] Only 50–60% of male Charles River Sprague–Dawley rats develop obesity, while all male Fischer F-344 become obese with half of the severity seen in Sprague–Dawley rats on a high-fat (high-energy) diet.[6] Thus, genetic background strongly influences susceptibility to dietary obesity.

Certain macronutrients, such as carbohydrate and fat, have long been implicated as weight-promoting agents in rats. Some strains develop obesity when fed low-fat, high-carbohydrate diets. Others develop obesity on high-fat diets. The most effective fat-inducing diet is a caloric-dense diet rich in fat and in carbohydrate, and it must be considered palatable by the species.

Studies demonstrate that rats are very attracted to the taste of sucrose. Consumption of the sugar alone may not be sufficient to cause obesity. Some studies report that high-sucrose diets produce overeating, overweight, and obesity, while more studies find such diet does not alter caloric intake.[7] In an earlier study, Oscai suggested that small increases in body adiposity seen with high-carbohydrate diets might be due to the short-term nature of the experiments. His long-term study found that habitual consumption of a sugar-rich diet caused severe obesity.[8] These results are more consistent when sucrose is provided as a solution in addition to water and a nutritionally complete diet.

In general, rats offered a sugar solution consume about half of their calories as sugar and increase their total caloric intake by 8–20% compared to control rats given only chow and water. The rats usually gain more weight and more body fat.[9–11] In general, rats increase caloric intake, body weight, and body fat more when given a sugar solution than when fed a high-sugar composite diet.[12] However, this may be strain dependent. Fischer rats fed a high-sucrose diet consume more energy for only one week. In contrast, adding water to a high-starch diet stimulates energy intake for at least 10 weeks; once water is removed from the diet, hyperphagia subsides.[13] In addition, rats prefer sugar in solution or solid-gel forms rather than in powder form.[7] These results indicate that dietary texture and/or palatability are important factors in promoting obesity in rats.

Dietary fat is also considered appetizing to rats and, because caloric density of fat is twice that of carbohydrate or protein, it is a significant contributor to hyperphagia, weight gain, and adiposity in many strains of rats, including Osborne–Mendel, Wistar–Lewis, and Long–Evans.[2,4] An overview of rodent studies indicates that high-fat diets induce greater food intake and weight gain than high-carbohydrate diets. The early studies of Schemmel and Mickelsen using extreme high-fat diets (60–85% fat) produced morbid obesity.[2,3] Later experiments showed that moderate high-fat diets (30–45% fat) can produce severe obesity. Furthermore, body fat increases in direct proportion to the fat content of the diet.[14] More careful analysis of high-fat diets that induce hyperphagia and adiposity revealed that these diets were usually rich in carbohydrate content as well. In experiments performed by Ramirez and Friedman,[11] energy intake and weight gain were lower in rats given an isoenergetic high-fat diet than in those given a high-carbohydrate diet. They concluded that rats fed carbohydrate overindulged and gained the most weight and fat. Rats fed high-carbohydrate diets differing in fat content consumed more calories and gained body weight as a function of caloric density, regardless of the fat content. In summary, hyperphagia of high fat diets does not facilitate excess adipocity as a result of the high dietary fat content alone; rather, it results from an interaction of caloric density of fat and carbohydrate.

Levin and colleagues selectively bred Sprague–Dawley rats to be resistant or prone to obesity when fed a palatable liquid diet (Ensure®).[15] Those critical studies demonstrated the central interaction of diet and genetic framework. However, this study as well as most other DIO investigations measured rate of adipose accumulation. Indeed, all rats will gain body fat with age regardless of the diet. Spontaneous obesity even occurs in rats maintained *ad libitum* on low-fat diets. Nonetheless, rate of weight and fat accretion with age is much faster for rats fed a high-fat diet than for rats fed regular chow.[2]

Keenan et al. showed that *ad libitum* access to a relatively low-fat rodent chow eventually produced an obese phenotype in aged outbred Charles River Sprague–Dawley rats. That phenotype is characterized by central obesity, hypertriglyceridemia, hyperinsulimia, and hypercholesterolemia. As animals age, they appear to become more efficient at storing calories since they continue to gain body weight even though mean relative food consumption per gram of body weight is decreased. Further, the gain is mostly due to increased fat content, which is located in the abdominal cavity, the subcutaneous tissue of the abdomen, and the lower thoracic area.[16]

Management of housing conditions can also accelerate rate of DIO. Manipulations such as raising small litter sizes can provide the sucking offspring more calories at an early age. Single-housing rats prevent competition for food and establishment of a feeding hierarchy and the stress associated with these living conditions. Rodents have a large surface area and, consequentially, high heat loss. This is exacerbated by modern animal facilities that present high air turnover rooms and extremely vented caging systems. Thus, much energy can be spent to maintain core body temperature. Increasing room temperature just a few degrees can preserve such energy expenditure and contribute to DIO.

MICE

Mice are becoming widely used experimental animals because of available tools for genetic manipulation and accessibility to numerous inbred strains. It is easy to access several reviews on DIO animal models.[17–19] In this review, we will focus on the effect of dietary manipulations (high-fat and high-carbohydrate diets) in several inbred mouse strains.

Diet manipulation in mice was performed by Fenton et al. using diets with extremely high fat content (70–80% of total energy derived from fat), which advance onset of obesity.[20–22] However, as pointed out in a related review by West and York,[23] the level of obesity attained depends on the mouse strain. Several investigators[24–26] have identified which mouse strains are prone or resistant to DIO. West et al.[26] fed mice chow (11.6, 26.4, and 62.0 kcal as fat, protein, and carbohydrate, respectively) or sweetened condensed milk (CM) (32.6, 15.0, and 52.4 kcal as fat, protein, and carbohydrate, respectively) and found that six out of nine mice strains (AKR/J, C57L/J, A/J, C3H/HeJ, DBA/2J, and C57BL/6J) presented with significantly increased adiposity with CM diet. Despite consuming more energy than their chow-diet controls, three strains (I/STN, SWR/J, and SJL/J) did not present with increased adiposity. This study clearly demonstrates the importance of genetic background when studying DIO with inbred strains.

While it is clear in some strains of mice (AKR/J and C57BL/6J) that a high-fat diet determines the onset of obesity without changing physical activity or total energy intake,[27–31] more studies are needed to understand the contribution of different fats or carbohydrates fully. Recent advances indicate that not all fats are contributors.[32] Ikemoto et al. demonstrated that C57BL/6J mice become more resistant to obesity induced by high fat when the fat is substituted with fish oil,[33] and Jurgens et al. showed that high-fructose consumption in NMRI mice also increases adiposity.[34]

COMPANION ANIMALS

Veterinary practices document that development of obesity in companion animals is a problem, with ~30–40% of canines being obese.[35] Canine obesity is attributed to a decrease in physical activity and the use of highly palatable diets. Certain breeds are more prone to obesity.[36] Therefore, as in rodents and humans, genetic predisposition and environmental factors contribute to obesity in canines. Various diets promote weight gain in canines; almost all added fat calories. Only a few studies have focused on use of isocaloric diets with an increase in the fat percentage. Moreover, obese canine models are primarily studied for metabolic disturbances (hypertension, insulin resistance, etc.) associated with an increase in body weight and fat mass. However, measurement of fat mass in canines is limited to the lab setting, mostly due to limited techniques. For those investigators that do measure fat mass, isotope dilution techniques have been applied in some studies, while newer methodologies use dual-emission x-ray absorption (DEXA) or magnetic resonance imaging (MRI).

Several investigators added fat to the diet to induce a weight gain canine model in order to investigate hypertension or other metabolic pathology. Rocchini's investigations into the cardiovascular and renal dysfunction caused by obesity used a canine model fed 0.8–0.9 kg of cooked beef fat added to a typical chow. Those studies showed that body weight can increase anywhere from 3 to 8 kg over a five- to six-week period in mongrel dogs with addition of less than 1 kg of fat in the diet. Most of the increase in body weight was presumed to be due to an increase in fat mass; however, this was not confirmed with an independent determination of fat mass.[37–40] A very similar model[41] used addition of 0.5–0.9 kg of cooked beef fat for six weeks that induced an incredible increase of 16.9 kg in mongrel dogs. However, no measurement of body fat was made. Fong et al.[42] added 30% cholesterol-enriched lard to a normal canine chow diet and fed dogs enough of the diet so that caloric intake matched that of control animals (5.56 kcal/day). This paradigm produced only a modest increase in body weight to mongrel dogs (0.6 kg), but significantly stimulated abdominal fat cell weight (no measurement of total fat mass was made).

Increasing carbohydrate consumption can also be used to induce obesity. Supplementation of a standard canine diet with 60% fructose[43] does not increase body weight (19.9 to 19.2 kg) over 25 days in mongrel dogs; however, the metabolic parameters associated with high-fructose feeding, such as circulating triglycerides and insulin, are elevated equivalent to threefold along with the development of hypertension. Moreover, a slight increase (18.3 to 19.6 kg) in weight is realized when chow is supplemented with 60% dextrose. However, interpretation of whether elevated carbohydrate (dextrose or fructose) induces obesity in dogs was limited since no measurement of body fat was made.

To study changes in body composition, other investigators use isotope dilution techniques or DEXA. One of the more complete studies that examine dietary fat content is reported by Romsos et al.[44] They customized diets fed to beagles with fat content ranging from 13 to 76% (no carbohydrate). After eight months, dogs consuming 38 and 55% fat in the diet accrued significantly greater fat mass (determined by isotope dilution techniques) when compared to animals fed low-fat diets. Interestingly, those on the 76% fat diet gained slightly less fat mass, perhaps due to the lack of carbohydrate in the diet. There was no major difference in fat-free mass. In a similar 25-week study,[45] the same investigators again showed that a diet containing 51% fat (1.43 kcal/g) induced a greater body weight gain (3.5 kg) and fat mass gain (2.8 kg) than a diet containing only 23% fat (1.16 kcal/g). The latter group gained only 1.6 and 1.26 kg body fat.

Truett and West[46] compared mongrel dogs fed dry dog food with 12% fat in one group to that diet supplemented with lard or corn oil to raise caloric content to 65% fat (percent kilocalories). This increase in fat intake boosted body weight to 29.5 kg while the low-fat diet group weighed only 19.8 kg. Further, this difference in weight was due to a specific increase in fat mass (14.7 kg) while the low-fat diet group maintained a low-fat mass of 3.67 kg (determined by DEXA). Again, lean mass was not different between groups (15.1 kg high fat and 15.2 kg low fat). A limitation of these investigations is that the comparisons were made between groups and measurement of body composition was not made from baseline.

More current studies[47] have examined change from baseline and illustrated that gain in body weight induced by a high-energy diet (4420 kcal metabolizable energy per kilogram) will increase fat mass (determined with isotope dilution methods) from 17 to 32% in beagles while not affecting lean mass. Likewise, Kaiyala et al.[48] demonstrated that increasing calories from 3.47 kcal/g (standard diet) to 4.13 kcal/g, using the addition of lard to the diet and with the food provided *ad libitum* for seven weeks, increased body weight from 35.9 to 42.2 kg (18% increase). Moreover, this increase was entirely a consequence of fat mass (increase from 5.7 to 12.5 kg). Thus, DIO can be observed in dogs.

Use of the isotope dilution technique or DEXA does not allow the examination of specific anatomical adipose depots such as visceral or subcutaneous fat mass. One of the biggest advancements is use of MRI to obtain such fat mass measurements. Bergman and colleagues supplemented a normal maintenance chow diet with cooked bacon grease and fed it to dogs for 12 weeks. This resulted in a dietary increase of ~500–600 kcal/day more than the maintenance diet.[49–51] MRI imaging was performed prior to the experiment and at eight weeks to examine changes in individual dogs, adding more power to the investigation. Only a moderate increase in body weight (27.8 to 28.9 kg) after 12 weeks was observed; however, trunk body fat increased from 12.3% (baseline) to 18.9% (four weeks) and to 17.8% by the eighth week (final measurement). This was a result of an increase in the abdominal and subcutaneous fat depots.[49] In a second study,[50] the same group found that a body weight increase from 26.7 kg (control diet) to 30.5 kg (fat-supplemented diet) was accompanied by an increase in abdominal fat (22.9 to 26.8%) and subcutaneous fat (11.2 to 14.5%).

Finally, Kim and Bergman[51] explained that increasing fat calories by only 8% without a large (~100 kcal/day) increase in total calories over 12 weeks increased body weight slightly (27.5 to 29 kg), but increased abdominal fat by 76% and subcutaneous fat by 182%. Such studies clearly demonstrate that with even a slight increase in dietary fat, canines increase their abdominal and subcutaneous fat mass without a major increase in their body weight.

This review of the literature for DIO in a canine model indicates that a majority, if not all, of the increase in weight is due to an increase in fat mass with very little change in lean mass. Addition of fat to the diet is the consistent method to develop body weight gain leading to obesity and the metabolic disturbances associated with increased weight. With further development and refinement of imaging techniques for fat distribution, more investigations to examine the sequence of lipid distribution are needed to best define the progression from lean to obese phenotype.

REFERENCES

1. Berthoud, H. R., Mind versus metabolism in the control of food intake and energy balance, *Physiol Behav* 81, 781–793, 2004.
2. Schemmel, R., Mickelsen O., Gill, J. L., Dietary obesity in rats: Body weight and body fat accretion in seven strains of rats. *J Nutr* 100, 1041–1048, 1970.

3. Schemmel, R., Mickelsen, O., Tolgay, Z., Dietary obesity in rats: Influence of diet, weight, age, and sex on body composition, *Am J Physiol* 216 (2), 373–379, 1969.
4. Woods, S. C., Seeley, R. J., Rushing, P. A., et al., A controlled high-fat diet induces an obese syndrome in rats, *J Nutr* 133 (4), 1081–1087, 2003.
5. Woods, S. C., D'Alessio, D. A., Tso, P., et al., Consumption of a high-fat diet alters the homeostatic regulation of energy balance, *Physiol Behav* 83 (4), 573–578, 2004.
6. Levin, B. E., Triscari, J., Sullivan, A. C., Relationship between sympathetic activity and diet-induced obesity in two rat strains, *Am J Physiol* 245 (3), R364–371, 1983.
7. Sclafani, A., Carbohydrate taste, appetite, and obesity: An overview, *Neurosci Biobehav Rev* 11 (2), 131–153, 1987.
8. Oscai, L. B., Miller, W. C., Arnall, D. A., Effects of dietary sugar and of dietary fat on food intake and body fat content in rats, *Growth* 51 (1), 64–73, 1987.
9. Kanarek, R. B., Orthen-Gambill, N., Differential effects of sucrose, fructose and glucose on carbohydrate-induced obesity in rats, *J Nutr* 112 (8), 1546–1554, 1982.
10. Kanarek, R. B., Marks-Kaufman, R., Developmental aspects of sucrose-induced obesity in rats, *Physiol Behav* 23 (5), 881–885, 1979.
11. Ramirez, I., Friedman, M. I., Dietary hyperphagia in rats: Role of fat, carbohydrate, and energy content, *Physiol Behav* 47 (6), 1157–1163, 1990.
12. Ramirez, I., Feeding a liquid diet increases energy intake, weight gain and body fat in rats, *J Nutr* 117 (12), 2127–2134, 1987.
13. Ramirez, I., Strain differences in dietary hyperphagia: Interactions with age and experience, *Physiol Behav* 49 (1), 89–92, 1991.
14. Boozer, C. N., Schoenbach, G., Atkinson, R. L., Dietary fat and adiposity: A dose–response relationship in adult male rats fed isocalorically, *Am J Physiol* 268 (4 Pt 1), E546–550, 1995.
15. Levin, B. E., Dunn-Meynell, A. A., Reduced central leptin sensitivity in rats with diet-induced obesity, *Am J Physiol Regul Integr Comp Physiol* 283 (4), R941–948, 2002.
16. Keenan, K. P., Hoe, C. M., Mixson, L., et al., Diabesity: A polygenic model of dietary-induced obesity from *ad libitum* overfeeding of Sprague–Dawley rats and its modulation by moderate and marked dietary restriction, *Toxicol Pathol* 33 (6), 650–674, 2005.
17. Tschop, M., Heiman, M. L., Rodent obesity models: an overview, *Exp Clin Endocrinol Diabetes* 109 (6), 307–319, 2001.
18. Inui, A., Transgenic approach to the study of body weight regulation, *Pharmacol Rev* 52 (1), 35–61, 2000.
19. Bray, G. A., York, D. A., Hypothalamic and genetic obesity in experimental animals: An autonomic and endocrine hypothesis, *Physiol Rev* 59 (3), 719–809, 1979.
20. Fenton, P. F., Chase, H. B., Effect of diet on obesity of yellow mice in inbred lines, *Proc Soc Exp Biol Med* 77 (3), 420–422, 1951.
21. Fenton, P. F., Dowling, M. T., Studies on obesity. I. Nutritional obesity in mice, *J Nutr* 49 (2), 319–331, 1953.
22. Fenton, P., Carr, C., The nutrition of the mouse. XI. Response of four strains to diets differing in fat content, *J Nutr* 45, 225–233, 1951.
23. West, D. B., York, B., Dietary fat, genetic predisposition, and obesity: Lessons from animal models, *Am J Clin Nutr* 67 (3 Suppl), 505S–512S, 1998.
24. Lemonnier, D., Suquet, J. P., Aubert, R., et al., Metabolism of the mouse made obese by a high-fat diet, *Diabete Metab* 1 (2), 77–85, 1975.
25. Salmon, D. M., Flatt, J. P., Effect of dietary fat content on the incidence of obesity among *ad libitum* fed mice, *Int J Obesity* 9 (6), 443–449, 1985.

26. West, D. B., Boozer, C. N., Moody, D. L., Atkinson, R. L., Dietary obesity in nine inbred mouse strains, *Am J Physiol* 262 (6 Pt 2), R1025–1032, 1992.
27. Smith, B. K., West, D. B., York, D. A., Carbohydrate versus fat intake: Differing patterns of macronutrient selection in two inbred mouse strains, *Am J Physiol* 272 (1 Pt 2), R357–362, 1997.
28. Surwit, R. S., Feinglos, M. N., Rodin, J., et al., Differential effects of fat and sucrose on the development of obesity and diabetes in C57BL/6J and A/J mice, *Metabolism* 44 (5), 645–651, 1995.
29. Brownlow, B. S., Petro, A., Feinglos, M. N., Surwit, R. S., The role of motor activity in diet-induced obesity in C57BL/6J mice, *Physiol Behav* 60 (1), 37–41, 1996.
30. Petro, A. E., Cotter, J., Cooper, D. A., et al., Fat, carbohydrate, and calories in the development of diabetes and obesity in the C57BL/6J mouse, *Metabolism* 53 (4), 454–457, 2004.
31. Van Heek, M., Compton, D. S., France, C. F., et al., Diet-induced obese mice develop peripheral, but not central, resistance to leptin, *J Clin Invest* 99 (3), 385–390, 1997.
32. Huang, X. F., Xin, X., McLennan, P., Storlien, L., Role of fat amount and type in ameliorating diet-induced obesity: Insights at the level of hypothalamic arcuate nucleus leptin receptor, neuropeptide Y and pro-opiomelanocortin mRNA expression, *Diabetes Obesity Metab* 6 (1), 35–44, 2004.
33. Ikemoto, S., Takahashi, M., Tsunoda, N., et al., High-fat diet-induced hyperglycemia and obesity in mice: Differential effects of dietary oils, *Metabolism* 45 (12), 1539–1546, 1996.
34. Jurgens, H., Haass, W., Castaneda, T. R., et al., Consuming fructose-sweetened beverages increases body adiposity in mice, *Obesity Res* 13 (7), 1146–1156, 2005.
35. Hand, M., Armstrong, P., Allen, T., Obesity: occurrence, treatment, and prevention, *Vet Clin North Am* 19, 447–474, 1989.
36. Edney, A. T., Smith, P. M., Study of obesity in dogs visiting veterinary practices in the United Kingdom, *Vet Rec* 118 (14), 391–396, 1986.
37. Rocchini, A. P., Moorehead, C., Wentz, E., Deremer, S., Obesity-induced hypertension in the dog, *Hypertension* 9 (6 Pt 2), III64–68, 1987.
38. Rocchini, A. P., Marker, P., Cervenka, T., Time course of insulin resistance associated with feeding dogs a high-fat diet, *Am J Physiol* 272 (1 Pt 1), E147–154, 1997.
39. Rocchini, A. P., Mao, H. Z., Babu, K., et al., Clonidine prevents insulin resistance and hypertension in obese dogs, *Hypertension* 33 (1 Pt 2), 548–553, 1999.
40. Rocchini, A. P., Yang, J. Q., Gokee, A., Hypertension and insulin resistance are not directly related in obese dogs, *Hypertension* 43 (5), 1011–1016, 2004.
41. Hall, J. E., Brands, M. W., Zappe, D. H., et al., Hemodynamic and renal responses to chronic hyperinsulinemia in obese, insulin-resistant dogs, *Hypertension* 25 (5), 994–1002, 1995.
42. Fong, B. S., Despres, J. P., Julien, P., Angel, A., Interactions of high-density lipoprotein subclasses (HDL2 and HDLc) with dog adipocytes: Selective effects of cholesterol and saturated fat feeding, *J Lipid Res* 29 (5), 553–561, 1988.
43. Martinez, F. J., Rizza, R. A., Romero, J. C., High-fructose feeding elicits insulin resistance, hyperinsulinism, and hypertension in normal mongrel dogs, *Hypertension* 23 (4), 456–463, 1994.
44. Romsos, D. R., Belo, P. S., Bennink, M. R., et al., Effects of dietary carbohydrate, fat and protein on growth, body composition and blood metabolite levels in the dog, *J Nutr* 106 (10), 1452–1464, 1976.

45. Romsos, D. R., Hornshuh, M. J., Leveille, G. A., Influence of dietary fat and carbohydrate on food intake, body weight and body fat of adult dogs, *Proc Soc Exp Biol Med* 157 (2), 278–281, 1978.
46. Truett, A. A., West, D. B., Validation of a radiotelemetry system for continuous blood pressure and heart rate monitoring in dogs, *Lab Anim Sci* 45 (3), 299–302, 1995.
47. Blanchard, G., Nguyen, P., Gayet, C., et al., Rapid weight loss with a high-protein low-energy diet allows the recovery of ideal body composition and insulin sensitivity in obese dogs, *J Nutr* 134 (8 Suppl), 2148S–2150S, 2004.
48. Kaiyala, K. J., Prigeon, R. L., Kahn, S. E., et al., Reduced beta-cell function contributes to impaired glucose tolerance in dogs made obese by high-fat feeding, *Am J Physiol* 277 (4 Pt 1), E659–667, 1999.
49. Mittelman, S. D., Van Citters, G. W., Kim, S. P., et al., Longitudinal compensation for fat-induced insulin resistance includes reduced insulin clearance and enhanced beta-cell response, *Diabetes* 49 (12), 2116–2125, 2000.
50. van Citters, G. W., Kabir, M., Kim, S. P., et al., Elevated glucagon-like peptide-1-(7-36)-amide, but not glucose, associated with hyperinsulinemic compensation for fat feeding, *J Clin Endocrinol Metab* 87 (11), 5191–5198, 2002.
51. Kim, S. P., Ellmerer, M., Van Citters, G. W., Bergman, R. N., Primacy of hepatic insulin resistance in the development of the metabolic syndrome induced by an isocaloric moderate-fat diet in the dog, *Diabetes* 52 (10), 2453–2460, 2003.

Index

C

D

E

F

G

H

I

J

K

L

M

N

O

P

Q

R

S

T

V

W

Z